Science Education in East Asia

Myint Swe Khine
Editor

Science Education in East Asia

Pedagogical Innovations and Research-informed Practices

Springer

Editor
Myint Swe Khine
Emirates College for Advanced Education
Abu Dhabi, United Arab Emirates

ISBN 978-3-319-16389-5 ISBN 978-3-319-16390-1 (eBook)
DOI 10.1007/978-3-319-16390-1

Library of Congress Control Number: 2015949925

Springer Cham Heidelberg New York Dordrecht London

Printed on acid-free paper

Springer International Publishing AG Switzerland is part of Springer Science+Business Media (www.springer.com)

Contents

Part I
Introduction

Chapter 1
Research and Development in Science Education: East Asian Perspectives

Myint Swe Khine

Introduction

The results from the international benchmarking tests on mathematics and science achievements such as the Trends in International Mathematics and Science Study (TIMSS) and Programme for International Student Assessment (PISA) revealed that students from East Asian countries consistently outperformed their counterparts in each and every cycle of the study. Among the high-achieving countries, the performances of the students in Singapore, Taiwan, Korea, Japan, and Hong Kong are recognized by the international education community. Some educators believe that the success of these countries is deeply rooted in eastern philosophy and cultural notion of education and influence of parental encouragements. Some think that the reasons are multifaceted including instructional practices, classroom climates, and the curriculum. Scientific explanations will provide insights into how these educational systems are functioning in improving the standard and achievement in science subjects in schools. This book attempts to bring together prominent science educators and researchers from the East Asia to share their experience and findings, reflection and vision on emerging trends, research-informed pedagogical innovations, and best practices in science education in the region.

Topics covered in the book are innovations in teaching and learning science; novel approaches to science curriculum; preservice and in-service teacher education; integration of technology in science education, social, cultural, and contextual factors in promoting science education; and student characteristics and improving science education through research.

The book is organized in eight parts. Part I covers introduction to the book, and Part II contains four chapters in Learning: The East Asian way. Part III deals with

M.S. Khine (✉)
Emirates College for Advanced Education, PO Box 126662, Abu Dhabi, United Arab Emirates
e-mail: dr.mkhine@gmail.com; mskhine@ecae.ac.ae

M.S. Khine (ed.), *Science Education in East Asia*,
DOI 10.1007/978-3-319-16390-1_1

science curriculum changes and reforms in China and Korea. Pedagogical innovations in science education is the topic for Part IV, and this part presents innovative programs in Taiwan, China, Japan, Korea, and Hong Kong in advancing science-related knowledge among students. Part V focuses on inquiry-based science instruction, and researchers from Singapore, Japan, and Taiwan share their experience in using this approach. One of the key elements in education is teachers' professional development. Chapters in Part VI touch on the aspects of professional development initiatives for science teachers in Singapore, Japan, China, and Korea. The international benchmarking tests are held at regular intervals to gauge the standards of science and mathematics in participating countries. The chapters in Part VII explore PISA and TIMSS results in high-performing countries using multilevel analysis. Finally, the emerging research studies in science education in East Asian countries are described in Part VIII. In general chapters in this book represent science education in East Asian countries encompassing China, Hong Kong, Japan, Korea, Singapore, and Taiwan.

The eastern philosophies of education such as Confucianism (Lee & Sriraman, 2013) and myths about Chinese learners (Örtenblad, Babur, & Kumari, 2012; Rao & Chan, 2010) have been extensively discussed in the literature. It is a common belief that Chinese students are rote learners and choose passive approach to learning. In Chap. 2, Lau, Ho, and Lam discussed about the effective classroom pedagogy for scientific literacy and explored the "Chinese paradoxes" using the secondary data from PISA. The results pointed to the fact that the western students are relatively better in understating the process and the nature of science, while the East Asian students are relatively better in science contents than science process.

Another dimension in educational success in children is closely related to the parenting style. The studies consistently found strong associations between parenting style and a wide range of factors such as self-esteem and school results (Chan & Koo, 2011). Chapter 3 by Huang and Gove explored how Confucianism and Asian descendent parenting practices influence children's academic achievement. The study involved fifteen parents, and qualitative research method (individual interviews and focus-group interviews) was used to collect data. The authors noted that Confucianism was a driving force for parents in setting high expectations of their children. The Chinese classic philosophy of learning and teaching is the topic in Chap. 4 by Zhang and Bai. "Essay on Learning" is the earliest scholarly record in the Chinese history, and the authors discussed how certain aspects of eastern philosophy and culture influence learning and teaching, and some of the principles can be applied to science education. Lee and Nie in Chap. 5 highlighted that parents can play pivotal role in facilitating their children's science learning in informal way. The authors presented both theoretical aspects on engaging parents in their children's science learning and practical examples on how parents can support their children at home. They concluded that willingness, interest, and commitment of time and effort from the parents are crucial in helping the children. Above all stronger bonding between parent and child is essential. In Chap. 6, Yeung reports findings from the study on the students' affective domain and choices of science learning in China under the projects Relevance of Science Education (ROSE) and

Interests and Recruitment in Science (IRIS). The study provided more accurate and objective description and characteristics of the Chinese learners.

Part III covers chapters concerned with science curriculum changes and reforms in some of the East Asian countries.

This part begins with Chap. 7 in which Kim, Chu, and Lim examined the science curriculum changes and STEM education in Korea, Hong Kong, Japan, Singapore, and Taiwan. The authors observed that science curriculum reforms took place in different countries with different rates, notably average about seven years in Korea, ten years in Japan, and more than ten years in Singapore and Hong Kong. This comprehensive analysis concluded that when improving the quality of science teaching and learning in schools, mathematics, technology, and engineering should be considered since these subjects are interrelated.

Yin, Guo, and Wang presented the science education reforms in China with the focus on Shanghai in Chap. 8. The authors explored the educational context, science textbooks adopted, assessment of science education, promotion of teaching innovations, science teachers' practices, and students' science learning in Shanghai schools. The chapter concluded with the implications that can be drawn from the Shanghai experience to the larger science education community.

Chapters in Part IV describe pedagogical innovations in science education. International attempts to research in educational neuroscience (or mind, brain, and education) are growing rapidly in recent years. Educators are hoping that outcomes from these studies produce useable results to inform the educational policy makers and practitioners (Hruby, 2012; Patten & Campbell, 2011). Liu and her team presented (Chap. 9) how they apply developments in neuroscience to reveal the underlying physiological mechanisms involved in science learning process. The authors first examined how the brain functions and reacts to daily learning routines, followed by recent studies related to educational neuroscience. The authors also reviewed the neuroscience research conducted in the past few decades with regard to the whole brain, brain-based curricula, building neural pathways, and how these elements interacts each other for optimum learning. It was noted that in order to capitalize findings from neuroscience research, educators and scientists must collaborate to further understand the complex phenomena. Sumida in Chap. 10 described the talent developments in STEM from early childhood years in Japan. To overcome the "science phobia" among junior and senior high school students in Japan, Kids Science Academy project was initiated for 5–8-year-old children. The project is aimed to develop the intellectual curiosity of young children into scientific inquiry using technological tools. The author presented the outline of the project, activities, and examples. The efficacy of the program was surveyed involving parents. The chapter concluded by highlighting the importance of science learning at young age and engaging them in collaborative activities.

In recent years, smart devices that combine telephone, computers, and advanced sensing technologies have been emerging as consumer products. Ryu in Chap. 11 identified smart technologies as devices that have higher computation powers, touch screen, and networking capabilities and support ubiquitous computing. Some devices have capabilities of handwriting recognition and multimodal interfaces. In her chapter she presented case studies involving the use of smart technologies in science lessons that afford collaboration among students and foster argumentation in the learning process. The first case study described how three science classrooms in a city, farm village, and fishing village are connected through these devices, how students in each class participated in collaborative lesson, and how students in city and village share their life experiences. In the second case study, the author presented how mobile Internet and smartphone were used in afterschool science and mathematics class. Similar to the first case study, the smart devices in this experiment allow students to interact each other and facilitate collaborative argumentation. In the last chapter in Part VI, So and Ching shared their study on the use of Internet resource-based e-learning platform in teaching specific science topic in a primary school. The quasi-experimental study involved two classes of sixth graders in resourced-based e-learning environments. The results showed that the students enjoyed the learning when they were engaged with Internet resources and gained better understanding of the topic.

The effectiveness and the nature of inquiry-based science lessons have been explored by the science educators in the past. For example, Minner, Levy, and Century (2010) provided a research synthesis from the studies conducted from 1984 to 2002. A meta-analysis by Furtak, Seidel, Iverson, and Briggs (2012) included studies conducted from 1996 to 2006. The authors concluded that there is a positive effect on student learning favoring inquiry-based instructional practices, particularly instruction that emphasizes student active thinking and drawing conclusions from data. In this regard Forbes (2011) stated the essential features of inquiry-based lessons and reminded the importance of curriculum adaptation. Chapters in Part V discuss inquiry-based science instruction, and the part begins with the chapter by Chue and Lee. In Chap. 13, Chue and Lee examined one high school chemistry teacher's attempt in teaching inquiry-based lesson. The study reports a discursive analysis of teacher's instructional talk. In Chap. 14, Kamon and Fujii presented the practice and effect of inquiry-based chemistry lesson in a high school in Japan. An 18-item questionnaire was used to measure the attitudes toward chemistry. The study reports the differences in attitudinal outcomes between control and experimental groups after the students were exposed to an inquiry-based lesson. Hsu and her team presented their experience in developing technology-infused inquiry-based science modules in Taiwan. Chapter 15 covers the rationale for the instructional design and the development process of the inquiry learning modules. The topics on plate tectonics and buoyancy were described as specific examples. As pointed out by Dorier and García (2013), there are challenges as well as opportunities for the implementation of inquiry-based learning in day-to-day teaching.

Part VI addresses issues in teachers' professional development for science educators in Singapore, China, Japan, and Korea. This part begins with the chapter by Salleh and Tan (Chap. 16). Their chapter describes how the professional development is organized through Professional Learning Communities (PLCs) in Singapore and Shanghai. The authors examined PLCs as key platform and how these nurture a community of educators to engage in reflective inquiry practices to continually develop content knowledge and pedagogical knowledge. The chapter also describes the key approaches and practices in teachers' professional development in these contexts. Yamaguchi in Chap. 17 examined Japanese elementary teachers' ability to learn how to teach science from curriculum materials. The case study involved 84 Japanese preservice elementary teachers. The results showed that while preservice teachers can learn pedagogical content knowledge (PCK) for elementary science teaching from curriculum materials, they have difficulty in learning topic-specific PCK. The author argued that preservice teachers need support in developing topic-specific PCK for science teaching.

In Chap. 18, Ma described Chinese secondary school science teachers' perceptions of the nature of science and Chinese native knowledge. The case study involved three teachers, and it was found that teachers hold different perceptions of school science and native knowledge. These views were classified as universalist scientism, realist multiculturalism, and compromised constructivism. The chapter discussed implications for science teaching and science teacher education. Another case study on teachers' professional development is on science teachers' pedagogical reasoning through reflective practices, presented by Park and Kim in Chap. 19. The authors traced the background of overseas science teacher professional development program and the model in organizing the program for biology teachers in Korea. The professional development program took place in the USA and exposed teachers from Korea on aspects of biology education, biological science, and cultural experiences through school and museum visits. The study concluded that reflective journal writing is an effective way in professional development.

Part VII covers emerging research in science education. This part includes a collection of studies aimed to improve science education in selected East Asian countries. Lam and Chan in Chap. 20 presented their experience in designing computer-supported knowledge building to promote conceptual change among high school students in Hong Kong. The study involved two classes of grade ten students in a girl's high school who do not have prior experience of using Knowledge Forum, an online learning management system. One of the classes was engaged in knowledge building inquiry mediated by the computer-supported collaborative learning platform, and another class was taught using conventional teaching approach. The study found that the knowledge building class experienced greater conceptual change and students made deeper inquiry overtime. The authors also noted that students' inquiry on Knowledge Forum was correlated with conceptual changes and achievement.

In Chap. 21 Bo Chen and Bing Wei presented their experience in the implementation of standard-based curriculum materials in the classroom in China. The study explored the interaction of the teacher with the curriculum materials and how these are delivered in the classroom. The study employed case study method with classroom observations and interviews with five secondary chemistry teachers. The authors highlighted the curriculum representations and discrepancies between operational and intended curricula. In another study, Chien and Chang (Chap. 22) discussed the development of Cloud Classroom (CCR), a Web-based instant response system that facilitates socio-scientific argumentation in the class. The system allows students to respond to teacher's questions instantly using handheld devices. The study found that the heterogeneous grouping method was more effective in bringing out students' rebuttals to refute counter argument.

The international benchmarking tests provide an opportunity for the stakeholders to gauge whether the school systems in the countries are offering quality education to their learners and the levels of achievement are comparable to international standards (Fensham, 2013). Among these, PISA and TIMSS are common benchmarking tests in science and mathematics achievements in schools. The last part of the book, Part VIII, the studies on performance of East Asian countries in international benchmarking tests, are reported.

Chapter 23 by Yu and his colleagues identified crucial and malleable factors of successful learning by using secondary data from PISA 2012. The authors analyzed student variables, home-environment variables, and school environment variables of top-performing countries, Hong Kong, Singapore, Japan, Korea, and Vietnam. The study found that attitudinal variables of the students, home-environment variables, and technological variables outside of schools are determinants of PISA science test performance in those countries.

Another analysis by Fah and his colleagues (Chap. 24) explored the predictive effects of the eighth-grade students' dispositions toward science on science achievement in five of the highest-performing education systems – Hong Kong, Taiwan, Japan, Korea, and Singapore. The results indicated that eighth-grade students' science-related attitudes such as liking and valuing of learning science and confidence in learning science were positively related with their science achievement in five countries. In Chap. 25, Areepattamannil and his team present a multilevel exploration of correlates of science achievement in Singapore. When considering the associations between student-level and school-level factors with achievement scores, it was found that school-level factors such as school size and school socioeconomic status are linked to science achievements and accounted for 78 % of the variance. The student-level factors such as student background characteristics and students' perceptions of school climate accounted for only 12 % of the variance. The chapter concluded with implications for policy makers and other stakeholders in the education system.

The chapters in this book introduce the educational reforms and promotion of science education in East Asian countries, discuss the pedagogical innovations in science education, and present the research-informed practices that lead to the successful teaching and learning of science in East Asian context. It is hoped that this comprehensive volume will help readers develop their understanding of the status, curriculum reform, and teaching and learning approaches in science education in East Asia.

References

Chan, T. W., & Koo, A. (2011). Parenting style and youth outcomes in the UK. *European Sociological Review, 27*(3), 385–399.

Dorier, J. L., & García, F. J. (2013). Challenges and opportunities for the implementation of inquiry-based learning in day-to-day teaching. *ZDM, 45*(6), 837–849.

Fensham, P. J. (2013). International assessments of science learning: Their positive and negative contributions to science education. In D. Corrigan, R. Gunstone, & A. Jones (Eds.), *Valuing assessment in science education: Pedagogy, curriculum, policy* (pp. 11–31). Dordrecht, The Netherlands: Springer.

Forbes, C. T. (2011). Preservice elementary teachers' adaptation of science curriculum materials for inquiry-based elementary science. *Science Education, 95*(5), 927–955.

Furtak, E. M., Seidel, T., Iverson, H., & Briggs, D. C. (2012). Experimental and quasi-experimental studies of inquiry-based science teaching: A meta-analysis. *Review of Educational Research, 82*(3), 300–329.

Hruby, G. G. (2012). Three requirements for justifying an educational neuroscience. *British Journal of Educational Psychology, 82*(1), 1–23.

Lee, K., & Sriraman, B. (2013). An Eastern learning paradox: Paradoxes in two Korean mathematics teachers' pedagogy of silence in the classroom. *Interchange, 43*(2), 147–166.

Minner, D. D., Levy, A. J., & Century, J. (2010). Inquiry-based science instruction—What is it and does it matter? Results from a research synthesis years 1984 to 2002. *Journal of Research in Science Teaching, 47*(4), 474–496.

Örtenblad, A., Babur, M., & Kumari, R. (2012). Learning in Asia. *Asia Pacific Journal of Education, 32*(2), 131–136.

Patten, K. E., & Campbell, S. R. (2011). Introduction: Educational neuroscience. *Educational Philosophy and Theory, 43*(1), 1–6.

Rao, N., & Chan, C. K. (2010). Moving beyond paradoxes: Understanding Chinese learners and their teachers. In N. Rao & C. Chan (Eds.), *Revisiting the Chinese learner* (pp. 3–32). Comparative Education Research Centre, University of Hong Kong, China.

Part II
Learning: The East Asian Way

Chapter 2
Effective Classroom Pedagogy and Beyond for Promoting Scientific Literacy: Is There an East Asian Model?

Kwok-chi Lau, Esther Sui-chu Ho, and Terence Yuk-ping Lam

Introduction

East Asian countries/regions including Hong Kong, Shanghai, Singapore, Taiwan, Japan, and Korea have consistently ranked top in international assessments of science and mathematics, such as the Programme for International Student Assessment (PISA) and the Third in International Mathematics and Science Study (TIMSS). The outstanding performances of the East Asian regions have created the "paradox of the Chinese learner" and the "paradox of the Chinese teacher" (Chan & Rao, 2009; Watkins & Biggs, 2001). The "Chinese" here does not denote the ethnic Chinese, but those who are under the influence of the Confucian Heritage Culture (CHC) (Watkins & Biggs, 1996), including both Chinese (China, Hong Kong, Singapore) and non-Chinese (Japan, Korea, Vietnam) regions (Ho, 1994). The paradoxes, simply put, are concerned with why the Chinese learners, who are commonly believed to be learning mainly by rote and be taught by the authoritarian Chinese teachers using expository and didactic pedagogy, can outperform their western counterparts who learn in the more progressive, constructivist classroom. More puzzling is that these international assessments call for not only memorization of facts but also application of concepts in novo contexts, problem-solving abilities, and understandings of scientific processes and the nature of scientific knowledge.

The international assessments, notably PISA, and the accompanying paradoxes revealed, have gradually become the game changer in what counts as "good" education in the international arena. In 2014, the Education Minister of the UK

K.C. Lau (✉) • E.S.C. Ho
Faculty of Education, The Chinese University of Hong Kong, Hong Kong, China
e-mail: laukwokchi@cuhk.edu.hk; estherho@cuhk.edu.hk

T.Y.P. Lam
Hong Kong Centre for International Student Assessment, The Chinese University of Hong Kong, Hong Kong, China
e-mail: yplam@cuhk.edu.hk

M.S. Khine (ed.), *Science Education in East Asia*,
DOI 10.1007/978-3-319-16390-1_2

led a team of educational experts to visit Shanghai hoping to learn how to boost their students' math performance. This would have never happened should there were no such international assessment results undeniably showing that the Shanghai students were "streets ahead" of the UK children in math performance. The notion that the western, progressive education is necessarily superior to the Asian, traditional one is therefore called into question.

This study aims to explore further the "Chinese paradoxes" and extend to East Asian societies using mainly the data from the assessment of scientific literacy in PISA 2006. The study will seek to identify if there exists an *East Asian model* of teaching and learning in science through comparing the top five East Asian regions in PISA 2006 – Hong Kong, Taipei, Japan, Korea, and Macau – with the top five western countries – Finland, Canada, New Zealand, Australia, and the Netherlands. It is not the purpose of the study to argue whether the East Asian or the western ways of science education are superior; instead, through comparing the top East Asian and outstanding western countries, we hope to shed light on the different ways that high quality of science education can be brought about in different cultures.

Data of PISA 2006 is used since it is so far the only cycle where science is the major domain and many major science-related factors were measured. Shanghai and Singapore, the two strongly performing East Asian regions, are not included in the main analysis since they had not yet joined the PISA assessment in 2006. But their characteristics found in later PISA assessments would nevertheless be taken into account in discussing the East Asian model. Although the 2006 data may not be able to reflect the most updated situations of these countries/regions, data in 2009 and 2012 will also be used wherever appropriate. Moreover, the analysis would shed important light on the analysis of the upcoming PISA 2015 data where science becomes the major domain of assessment again.

This study will use *hierarchical linear modeling* (HLM) to analyze the associations of the four pedagogies (application, hands-on activities, investigation, and interaction) with performance and attitudes at the personal and school levels within a country/region. This method is thought to be appropriate since many student factors are nested in classes and schools. For each country/region, the gender and socioeconomic background of the students, the four kinds of pedagogy at both individual and school levels, and the after-school science activities will be analyzed for their associations with the overall scores, the three component competencies, and two attitudes: self-concept in science, enjoyment of science, and interest in science. It aims to explore how the East Asian regions differ from the western countries, if any, in their associations between the pedagogy and the cognitive and noncognitive outcomes.

Under the research aim of identifying the East Asian model of science teaching and learning, the specific research questions that the study seeks to answer are:

1. What are the differences and similarities in performance and other characteristics between the top five East Asian regions and the top five western countries in scientific literacy of PISA 2006?
2. How do the various forms of pedagogies inside school associate with performances and attitudes in these East Asian regions and western countries, and how are these associations different between the two groups of countries/regions?

3. How do the informal learning outside school through science activities associate with performances and attitudes in these East Asian regions and western countries, and how are these associations different between the two groups of countries/regions?
4. What accounts for the above similarities and differences between the East Asian and the western countries/regions?

Literature Review

The review will be mainly done on two lines of research: the studies on PISA science, particularly those pertaining to the differences in pedagogy and performance among countries, and the studies on the "Chinese" learners.

Ho (2009) analyzed the four East Asian regions, Hong Kong, Macau, Japan, and Korea, using the data of PISA+ and PISA 2003. These East Asian regions were found having some common characteristics: high achievement, high aspiration, orderly disciplinary climate in school, low self-concept, and disengaged learning climate. However, Japan and Korea are distinct in their students reporting less use of learning strategies like elaboration and memorization, having high anxiety, and less intrinsically motivated. A study grouped the OECD countries according to their performances on different domains of knowledge in PISA science (Olsen & Lie, 2009). Japan and Korea belong to the group relatively strong in knowledge about science and physical systems, but weaker in living systems. Most of the English-speaking countries were grouped together for their better performance in Earth and space systems but weaker in physical systems. Finland belongs to another group that is good at living systems but poor in Earth and space systems. Disappointingly, the study did not include any Chinese regions. Another study looked into how the pedagogies differ between two strong performers (Canada and Czech) and two low performers (USA and Slovak) (Korsnakova, McCrae, & Bybee, 2009). One interesting finding is that the US lessons were filled with all four kinds of pedagogy in much higher percentages, particularly hands-on activities and investigations. This, together with the findings from the TIMSS 1999 Video Study that most of the activities in the US classrooms were not connected with content learning, suggested the likely cause of the low performance of the USA. A similar study sought to identify the classroom pedagogies that are associated with good performance (Taylor, Stuhlsatz, & Bybee, 2009). Instead of using the four constructs of pedagogy developed in PISA 2006, the study worked on individual questions and found three significantly correlated with high performances: students explaining their ideas, students drawing conclusions from experiment, and teachers explaining the applications of science. The top five performing countries including Japan had higher percentages of these three pedagogies than the bottom five countries. But the non-OECD Chinese regions were not included in the study. Another study compared the findings of the TIMSS 1999 Video Study with that of PISA 2006 in five countries (Thomson, 2009). The PISA surveys on pedagogy were found basically correspondent with the class observations in TIMSS. Class observations showed that the high-achieving countries were highly content focused in their teaching.

The paradoxes of the Chinese learners and the Chinese teachers were raised and thoroughly discussed in the three books: *The Chinese Learner* (Watkins & Biggs, 1996), *Teaching the Chinese Learner* (Watkins & Biggs, 2001), and *Revisiting the Chinese Learner* (Chan & Rao, 2009), in which some of the myths of the stereotyped Chinese learners and teachers were debunked. Chinese students do not learn by rote but seek understanding through repetition and memorization. They are motivated both extrinsically and intrinsically by not only individual interest and ambition but also by social, family, and peer expectations. Despite more extrinsically motivated, Chinese students do not necessarily use surface approach to learning; instead, many were found using deep learning approach.

Chinese teachers are indeed more authoritarian in the classroom, but it was found that good teacher-student relationships are built up through informal interactions outside the classroom (Ho, 2001). Given this and the Confucian culture of obedience and respect for teachers, a more teacher-controlled classroom may not necessarily be negative for the Chinese students as construed from the western perspectives; rather, it can make learning more efficient. Besides, a more teacher-controlled lesson is not necessarily passive and didactic; instead, the Chinese teachers carefully orchestrate direct teaching, whole class discussion, and group activities in the lesson, which, albeit teacher centered, can engage students actively in learning (Mok et al., 2001). It is also a wrong belief that Chinese teachers only aim for memorization without understanding. Mok et al. (2001), after studying numerous Chinese lessons, argued that the key of the "success" of the Chinese teaching may lie in the variation of the ways that the teacher deal with the content to facilitate student understanding. At last, the intense examination pressure in most Asian education systems might not be negative since it is to some extent in alignment with the Confucian culture that stresses efforts, diligence, and personal success through education.

Constructs of Pedagogy, Attitudes, and Science Activity in PISA

In this study, the four pedagogies are represented by the international indexes calculated from four or three student questionnaire items in PISA 2006 (Table 2.1). These four constructs of pedagogy have reliabilities ranging from 0.78 to 0.82 and show model fit for most of the countries/regions. Some studies on the pedagogies of PISA science, however, do not use the scaled scores but the percentages of the students reporting that a particular pedagogy occurs in "all or most" or "most or some" lessons (Korsnakova et al., 2009; Taylor et al., 2009) and individual

Table 2.1 Questionnaire items for the four constructs of science pedagogy in PISA 2006

Pedagogy	Questionnaire items
	When learning school science topics at school, how often do the following activities occur?
	(*In all lessons/In most lessons/In some lessons/Never or hardly ever*)
Application	1. The teacher explains how a < school science > idea can be applied to a number of different phenomena (*e.g.*, the movement of objects, substances with similar properties)
	2. The teacher uses science to help students understand the world outside school
	3. The teacher clearly explains the relevance of science concepts to our lives
	4. The teacher uses examples of technological application to show how science is relevant to society
Interaction	1. Students are given opportunities to explain their ideas
	2. The lessons involve students' opinions about the topics
	3. There is a class debate or discussion
	4. The students have discussions about the topics
Hands-on activities	1. Students spend time in the laboratory doing practical experiments
	2. Students are required to design how a science question could be investigated in the laboratory
	3. Students are asked to draw conclusions from an experiment they have conducted
	4. Students do experiments by following the instructions of the teacher
Investigation	1. Students are allowed to design their own experiments
	2. Students are given the chance to choose their own investigations
	3. Students are asked to do an investigation to test out their own ideas

questionnaire items rather than the four pedagogical constructs (Taylor et al., 2009). We choose to use the international indexes because we think they are reliable measures for comparison across countries.

PISA 2006 had put a great emphasis on the assessment of students' attitudes toward science because it considers attitudes the crucial component of scientific literacy: *willingness to engage with science-related issues and with the ideas of science, as a reflective citizen* (OECD, 2006). Among the various sub-constructs of attitudes toward science in PISA, enjoyment of science was found central in connecting achievement (Ainley & Ainley, 2011; Lam & Lau, 2014). Self-concept is also included in the study since it is the attitudinal construct that East Asian countries/regions tend to be lower. The items for measuring the constructs of enjoyment of science learning and self-concepts in PISA 2006 are shown in Table 2.2. Apart from pedagogy, after-school science activities are thought to be an important factor in science learning as both a cause and consequence. In PISA 2006, students were asked to report the science activities they do after school (Table 2.3).

Table 2.2 Questionnaire items for the self-concept and enjoyment of science learning constructs in PISA 2006

Attitudes	Questionnaire items
Self-concept in science	*How much do you agree with the statements below?*
	(Strongly agree/Agree/Disagree/Strongly disagree)
	(a) Learning advanced science topics would be easy for me
	(b) I can usually give good answers to test questions on science topics
	(c) I learn science topics quickly
	(d) Science topics are easy for me
	(e) When I am being taught science, I can understand the concepts very well
	(f) I can easily understand new ideas in school science
Enjoyment of science learning	*How much do you agree with the statements below?*
	(Strongly agree, Agree, Disagree, Strongly disagree)
	(a) I generally have fun when I am learning science topics
	(b) I like reading about science
	(c) I am happy doing science problems
	(d) I enjoy acquiring new knowledge in science
	(e) I am interested in learning about science

Table 2.3 Questionnaire items for the construct of science activities in PISA 2006

	Questionnaire items
Science activities	*How often do you do these things?*
	(Very often/Regularly/Sometimes/Never or hardly ever)
	(a) Watch TV programs about science
	(b) Borrow or buy books on science topics
	(c) Visit web sites about science topics
	(d) Listen to radio programs about advances in science
	(e) Read science magazines or science articles in newspapers
	(f) Attend a science club

Methods

Using hierarchical liner modeling (HLM), the top five East Asian and the top five western countries/regions in PISA 2006 science were analyzed for the associations of the four kinds of pedagogy and science activity with the students' cognitive and affective performances within each country/region.

The analysis was divided into four parts: (1) analysis of the variation of students' literacy performance among schools using a null model; (2) analysis of the effects of student background characteristics and school contextual factors on students' literacy performance in Model 1; (3) analysis of the effects of parental investment after controlling for student's background characteristics and school contextual factors in Model 2; and (4) analysis of the effects of parental involvement, parental investment, and parental satisfaction after controlling for student's background characteristics and school contextual factors in Model 3.

First, for each country, a null model was used to partition the variance of literacy performance into within- and between-school portions. The model is represented by Eqs. (2.1) and (2.2):

$$\text{Scie}_{ij} = B_{0j} + R_{ij} \tag{2.1}$$

$$B_{0j} = G_{00} + U_{0j} \tag{2.2}$$

where Scie_{ij} is the overall score of science of student i in school j. (Similar analyses are conducted for three subscales of science performance and science self-concept and interest in science.) B_{0j} is the average science score of school j without any adjustment, and G_{00} is the grand mean of science score. The variance (R_{ij}) is the within-school variance of science score, and the variance (U_{0j}) is the between-school variance.

Model 1 builds on the null model by adding the student background factors and the school contextual factors. This model examines the effect of the student background and the school background on students' science performance. It is represented by Eqs. (2.3) and (2.4):

$$\text{Scie}_{ij} = B_{0j} + B_{1j}(\text{girl}) + B_{2j}(\text{ESCS}) + R_{ij} \tag{2.3}$$

$$B_{0j} = G_{00} + G_{01}(\text{school mean ESCS}) + U_{0j} \tag{2.4}$$

Then, the four kinds of pedagogy constructs, namely, "application," "hands-on," "interaction," and "investigation," both at student level and school level included to construct Model 2, which is represented by Eqs. (2.5) and (2.6):

$$\begin{aligned}\text{Scie}_{ij} = {} & B_{0j} + B_{1j}(\text{girl}) + B_{2j}(\text{ESCS}) + B_{3j}(\text{"application"}) + \\ & B_{4j}(\text{"hands-on"}) + B_{5j}(\text{"interaction"}) + B_{6j}(\text{"investigation"}) + R_{ij}\end{aligned} \tag{2.5}$$

$$\begin{aligned}B_{0j} = {} & G_{00} + G_{01}(\text{school mean ESCS}) + G_{02}(\text{school mean "application"}) \\ & + G_{03}(\text{school mean "hands-on"}) + G_{04}(\text{school mean "interaction "}) \\ & + G_{05}(\text{school mean "investigation"}) + U_{0j}\end{aligned} \tag{2.6}$$

Finally, variable of outside school learning through science activities is incorporated into the Model 3. This full model is represented by Eqs. (2.7) and (2.8):

$$\begin{aligned}\text{Scie}_{ij} = {} & B_{0j} + B_{1j}(\text{girl}) + B_{2j}(\text{ESCS}) + B_{3j}(\text{"application"}) \\ & + B_{4j}(\text{"hands-on"}) + B_{5j}(\text{"interaction"}) + B_{6j}(\text{"investigation"}) \\ & + B_{7j}(\text{science activity}) + R_{ij}\end{aligned} \tag{2.7}$$

$$\begin{aligned}B_{0j} = {} & G_{00} + G_{01}(\text{school mean ESCS}) + G_{02}(\text{school mean "application"}) \\ & + G_{03}(\text{school mean "hands-on"}) + G_{04}(\text{school mean "interaction "}) \\ & + G_{05}(\text{school mean "investigation"}) + U_{0j}\end{aligned} \tag{2.8}$$

Therefore, the full model will include both student and school-level pedagogy after controlling for the gender and socioeconomic factors. The full model of the HLM analysis of Hong Kong is shown to illustrate how the analysis is done (Table 2.6). For clarity, only the results of the full model are shown for the rest of the countries in Tables 2.7 and 2.8.

Results and Discussions

Descriptive Analysis of Major Constructs

Performance

The five East Asian regions and the five western countries, despite all top performers in overall PISA science, differ in their performances on different knowledge domains. All East Asian regions, especially the three Chinese regions (Hong Kong, Taipei, Macau), show relatively weaker performance in *identifying scientific issue*, whereas most western countries (except Finland) are relatively weaker in *explaining phenomena scientifically* (see column four of Table 2.4). However, when looking at the *actual scores* of the competencies, the East Asian regions did perform not as good as their western counterparts in identifying scientific issues (see column five of Table 2.4), but their performance in explaining phenomena scientifically is no better than the western countries. Since identifying scientific issues draws on students' *knowledge about science* (the nature and process of science) while explaining scientific evidence is mainly about *knowledge of science* (science concepts and facts), the western countries seem to have struck a better balance in fostering students' understanding of the two domains of science knowledge than the East Asian countries/regions. In Hong Kong, the practical activities are predominately recipe type (Yip, 2008; Yip & Cheung, 2004), in which students are not required to design experiments and control variables. This is likely contributing to their weaker performance in identifying scientific issues that call for these process skills.

Attitudes

When Korea and Japan are set aside, the three Chinese regions show an attitudinal pattern that is rather different from that of the western countries. As compared to the western students, the Chinese students tend to have greater enjoyment of science learning, but lower self-concept in science (Table 2.5). Japan and Korea, on the contrary, are below the OECD averages in enjoyment of science learning as well as having the lowest self-concept among all the countries/regions. Only two western countries, Finland and Canada, are positive in both enjoyment of science learning

Table 2.4 Performance of scientific literacy of the countries/regions in PISA 2006

Countries/regions	Rank[a]	Mean overall score[b]	Identify scientific issue		Use scientific evidence		Explain phenomena scientifically	
			Relative performance[c]	Actual scores[d]	Relative performance	Actual scores	Relative performance	Actual scores
Hong Kong	2	542	−14.4	528	0.2	542	7.0	549
Taipei	4	532	−23.8	509	−0.6	532	12.7	545
Macau	17	511	−20.8	490	0.7	512	9.2	520
Japan	6	531	−9.3	522	1.0	544	−4.1	527
Korea	11	522	−3.1	519	16.3	538	−10.5	512
Finland	1	563	−8.4	555	4.1	567	2.8	566
Canada	3	534	−2.6	532	7.1	542	−3.6	531
New Zealand	7	530	5.8	536	6.4	537	−8.2	522
Australia	8	527	8.4	535	4.4	531	−6.6	520
Netherlands	9	525	7.7	533	0.7	526	−3.1	522

Note:
[a]Ranks are based on the mean scores in scientific literacy of a country/region
[b]Mean overall score of a country/region is a scaled score with the OECD average set as 500
[c]Relative performance of a competency is the difference in score points (with the OECD average scaled as 500) between this competency and the mean overall scores of the country/region
[d]The actual score of a competency is a scaled score with the OECD average set as 500

Table 2.5 Attitudes and pedagogy of the countries/regions

	Attitudes		Pedagogy			
Countries/regions	Enjoyment of science learning	Self-concept	Application	Interaction	Hands-on	Investigation
Hong Kong	0.38	−0.25	−0.003	−0.291	0.288	0.215
Taipei	0.17	−0.40	0.130	−0.061	−0.101	0.090
Macau	0.41	−0.11	−0.185	−0.410	−0.191	0.023
Japan	−0.26	−0.87	−0.931	−1.140	−0.519	−0.241
Korea	−0.17	−0.71	−0.335	−1.039	−0.419	−0.205
Finland	0.11	0.06	−0.135	−0.135	0.020	−0.268
Canada	0.17	0.27	0.381	0.209	0.371	0.154
New Zealand	−0.01	−0.06	0.175	0.127	0.314	0.000
Australia	−0.06	−0.03	0.211	0.177	0.385	0.183
Netherlands	−0.32	−0.33	−0.259	−0.260	0.089	−0.164

Note: Figures in the table are the indexes of each construct with the OECD average set as zero and self-concept in science, whereas the rest are all negative in these two attitudinal constructs.

Since enjoyment of science learning pertains to intrinsic motivation, the data do not support the common belief that the East Asian students are less intrinsically motivated than their western counterparts. But the finding that the East Asian students have low self-concept is consistent with most studies on the Chinese learners. One of the likely causes is that the Chinese teachers are "extraordinarily mean in their marking, a deliberate strategy for extracting ever more efforts from students" (Watkins & Biggs, 2001, p. 16). This belief of emphasizing effort over ability has its root in the Confucianism so that the Chinese learners could still perform well despite low self-concept. That also explains why the association of self-concept with performance is not so noticeable among the East Asian students than their western counterparts (OECD, 2009, p. 65).

Pedagogy

The PISA 2006 surveys indicate that science lessons in the East Asian regions, particularly Japan and Korea, are less interactive, less hands-on, and less application focused as compared to the western lessons (Table 2.5). The difference is salient for hands-on activity. In other words, science teaching in East Asia lesson tends to be more traditional – expository, authoritarian, teacher centered, and content based. This picture is largely in line with the previous findings about CHC classroom (Watkins & Biggs, 2001).

HLM Analysis

Pedagogy, Performance, and Attitudes

Application

Using HLM, the associations of the four kinds of pedagogy with student performance and attitudes at both student and school levels are explored. Table 2.6 shows the full model analysis of Hong Kong, and the analyses of other countries are summarized in Tables 2.7 and 2.8.

At student level (Table 2.7), application is the only pedagogy that is positively associated with performance in all East Asian and western countries/regions. Hong Kong students increase by an average of 23 points in total scores for each additional unit of application in class, but it is only about 7 points for Japanese students. Apart from cognitive performance, application is also positively correlated with enjoyment of science and self-concept in science in all East Asian and western countries/regions. It suggests that teaching with greater focus on the application of science is conducive to science learning not only through the application and elaboration of

Table 2.6 The full model of HLM analysis of Hong Kong in PISA 2006

	Total scores in scientific literacy		Competency – explaining phenomena scientifically		Competency – identifying scientific issues		Competency – using scientific evidence		Self-concept in science		Enjoyment of science	
	Coeff.	S.E.	Coeff.	S.E.	Coeff.	S.E.	Coeff.	S.E.	Coeff.	S.E.	Coeff.	S.E.
Intercept	555.51***	2.82	562.52***	2.76	540.77***	3.26	555.44***	3.05	−.267***	.016	.421***	.011
Student level												
Girl (relative to boy)	−19.77***	2.33	−32.72***	2.44	−3.85	2.52	−17.25***	2.40	−.355***	.026	−.211***	.024
SES	3.62*	1.51	5.23**	1.48	5.26**	1.55	4.85**	1.62	.018	.016	−.014	.015
Pedagogy												
Application	17.39***	1.96	16.26***	2.00	18.30***	2.13	18.19***	2.11	.170***	.026	.172***	.020
Hands-on	4.76*	2.17	4.69*	2.35	5.51*	2.23	5.53*	2.26	.053*	.026	.015	.026
Interaction	−5.47*	2.10	−5.63*	2.38	−5.90*	2.49	−4.94*	2.37	.097**	.030	.042	.024
Investigation	−25.25***	2.25	−25.38***	2.29	−26.98***	2.24	−25.73***	2.31	−.032	.024	−.145***	.022
School level												
School mean SES	44.60***	6.07	33.38***	5.80	50.02***	7.43	51.86***	7.04	.074	.043	.078*	.032
Pedagogy												
School mean application	48.95*	22.08	55.50**	20.94	62.54*	24.83	46.01*	22.50	−.035	.141	.042	.112

School mean hands-on	44.82*	17.28	39.32*	15.81	52.02*	20.53	45.20*	17.67	−.052	.092	.148	.088
School mean interaction	−10.02	23.06	−19.90	22.15	−15.61	27.35	−.69	23.99	−.035	.150	−.268*	.103
School mean investigation	−63.61***	17.99	−60.83**	17.28	−71.63**	21.57	−70.73***	18.95	.248*	.099	.121	.086
Within-school variance explained (%)	14.01		17.27		9.85		11.26		27.80		39.72	
Between-school variance explained (%)	63.13		59.71		63.05		63.59		34.53		90.48	
Total variance explained (%)	31.87		29.91		31.73		30.87		27.96		40.65	

Table 2.7 Summary of coefficients of the *student-level* pedagogies with overall scores, self-concept,

	Application			Interaction		
	Overall score	Self-concept in science	Enjoyment of science learning	Overall score	Self-concept in science	Enjoyment of science learning
HKG	17.39***	.170***	.172***	−5.47*	.097**	0.042
Taipei	13.05***	.054***	.063***	−2.86**	.028*	0.013
Macau	23.18***	.123***	.107***	−9.43***	−0.015	−0.032
Korea	16.61***	.100***	.124***	−14.45***	−0.015	−.049**
Japan	7.13***	.102***	.087***	−15.31***	−0.026	−0.025
Finland	16.34***	.179***	.129***	−13.39***	0.016	0.012
Canada	16.77***	.192***	.161***	−0.3	.078***	.067***
Australia	11.41***	.150***	.158***	6.53***	.071***	.059***
New Zealand	12.47***	.136***	.161***	3.43	.081***	.053**
Netherlands	8.97***	.149***	.101***	−8.31***	−0.002	−0.007

$*p < 0.05$, $**p < 0.01$, $***p < 0.001$

and enjoyment of science learning of the ten countries/regions based on HLM analysis

Hands on activity			Investigation			Science activity		
Overall score	Self-concept in science	Enjoyment of science learning	Overall score	Self-concept in science	Enjoyment of science learning	Overall score	Self-concept in science	Enjoyment of science learning
4.76*	.053*	0.015	−25.25***	−0.032	−.145***	15.96***	.323***	.501***
−5.28***	.040*	−0.014	−17.96***	0.02	−.053***	10.14***	.417***	.561***
0.2	.041*	−.047*	−25.10***	.070**	−.069***	13.09***	.291***	.528***
−2.06	−0.014	−0.005	−18.27***	0.013	−.058**	17.79***	.394***	.536***
1.92	0.034	0.029	−10.47***	0.037	−.056*	18.52***	.351***	.601***
14.32***	.073***	.037*	−35.19***	−.087***	−.094***	22.82***	.344***	.567***
−0.77	.044***	0.018	−33.00***	−.111***	−.131***	20.32***	.374***	.560***
−2.04	.035*	0.023	−25.43***	−.044***	−.098***	22.03***	.395***	.557***
2.84	.070**	.079***	−32.74***	−.035*	−.101***	21.05***	.382***	.523***
−2.92*	.074**	.070***	−12.63***	−0.008	−.070***	10.37***	.294***	.504***

Table 2.8 Summary of coefficients of *school-level* pedagogies with overall scores, self-concept, and enjoyment of science learning of the ten countries/regions based on HLM analysis

	Application			Interaction			Hands on activity			Investigation		
	Overall score	Self-concept in science	Enjoyment of science learning	Overall score	Self-concept in science	Enjoyment of science learning	Overall score	Self-concept in science	Enjoyment of science learning	Overall score	Self-concept in science	Enjoyment of science learning
HKG	48.95*(+)	***-.035(+)***	***.042***(+)	***-10.02 (-)***	***-.035 (+)***	***-.268****	4.82*(+)	-***.052***(+)	.148	-63.61***(-)	***.248****	***.121***(-)
Taipei	41.38**(+)	***.004(+)***	.195**(+)	***1.18 (-)***	***-.043 (+)***	-.105	***7.46*(-)***	***.040***(+)	***.095****	-87.23***(-)	.067	-.194**(-)
Macau	71.22*(+)	***-.104(+)***	**-.089**(+)	-100.53** (-)	.107	-.147	***3.66****	***.095***(+)	***.326***(-)***	***-47.41***(-)	***-.096***(+)	-.239*(-)
Korea	46.33***(+)	***.094(+)***	.179**(+)	***25.70 (-)***	.097	***.100 (-)***	***1.43****	.011	.037	-71.36**(-)	-.066	***-.178***(-)
Japan	48.03***(+)	***-.011(+)***	.218**(+)	-53.10***(-)	-.042	***-.168****	***6.94*****	-.021	-.001	-43.54*(-)	-.038	***-.102***(-)
Finland	***11.31(+)***	***.039(+)***	***.069(+)***	***7.15 (-)***	-.041	.067	***11.29 (+)***	***-.044***(+)	***-.112***(+)	***-1.88***(-)	***.073***(-)	***-.125***(-)
Canada	***10.14(+)***	***.054(+)***	***.044(+)***	***10.98****	***-.055 (+)***	***.029 (+)***	***7.44******	***.006***(+)	-.019	-30.98***(-)	***.062****(-)	***-.041***(-)
Australia	22.87**(+)	***-.012(+)***	***.086(+)***	***-24.80**(+)***	***-.024 (+)***	***.024 (+)***	***8.78*****	***.102***(+)	-.001	***.35***(-)	***-.016***(-)	***.028***(-)
New Zealand	***-42.39***(+)***	***-.099(+)***	***.093(+)***	***23.68****	.197** (+)	.197** (+)	12.59	***.089***(+)	***-.035***(+)	***-1.89***(-)	***.004***(-)	***-.053***(-)
Netherlands	67.36***(+)	.237**(+)	***.163(+)***	-78.07*** (-)	***-.156****	***-.163****	***.83 (-)***	***-.069***(+)	***.075***(+)	***-19.13***(-)	.094	***.021***(-)

$*p < 0.05$, $**p < 0.01$, $***p < 0.001$

Bold and *italic* figures: those that change from insignificant to significant associations or vice versa, or, from positive to negative association or vice versa, from the student level to school level

(+)/(−): significant positive/significant negative associations at student level. The absence of it shows that the student-level association is insignificant concepts in everyday life contexts but also through enhancing the intrinsic motivation of students. Given that East Asian lessons are relatively less application focused as compared to the western ones, the science performance of the East Asian regions would likely improve further should a greater emphasis on the applications of science be given in the classroom.

Interaction

Interaction in classroom shows negative relationships with performance in all East Asian regions. Japan and Korea not only have the least interactive lessons but also see the largest negative associations: for every one additional unit of interaction in class, there is a drop of about 15 points in overall scores (Table 2.7). Interaction is also not positively related to self-concept and enjoyment of science learning in most of the East Asian countries/regions. However, interaction seems not so negative for the western countries. For Canada, Australia and New Zealand, interaction is positively associated with both enjoyment of science learning and self-concept, and it also does not have negative associations with performance as in the East Asian countries. Finland and the Netherlands, on the other hand, are closer to the East Asian countries. In general, whenever interaction has positive relations with attitudes (as in Canada, Australia, and New Zealand), its relations with performance are not negative. Seemingly, interaction is related to performance partly through making student enjoy more science learning and having good self-concept in it.

It would be difficult to interpret why interaction in class, a widely valued pedagogy, would relate to learning negatively in all of the East Asian countries/regions and even some western countries. One possible answer is that interaction, measured as student expression of ideas in class by PISA, may not necessarily lead to effective learning when it is not connected with the pedagogies and goals of the conceptual change model. Many studies show that there exists a huge gap between actual classroom practice and practice informed by the conceptual change theory (Duit & Treagust, 2003). Research on classroom talk also reveals that interaction in class is often authoritative rather than dialogic, and this kind of interaction is limited in probing into and working on students' ideas for concept construction (Scott, 1998; Mortimer & Scott, 2003). Even worst, interaction in class is sometimes not connected with any content learning but as a means of engaging and disciplining students. Interaction in these manners would at best enhance motivation, but may render learning nonsystematic and inefficient as a consequence. In the East Asian classrooms that emphasize obedience, teacher authority, and diligence, too much interaction is therefore construed as unnecessary impediment to learning by both the teacher and students. But some western teachers and students seem enjoying more interaction in class, though interaction may not necessarily improve the cognitive learning outcomes. Nonetheless, the interpretation can be the other way around: teachers tend to teach more interactively when faced with the less-disciplined, lower ability students in order to get them motivated in class. A case in point is Hong Kong where schools with low academic intake of students tend to

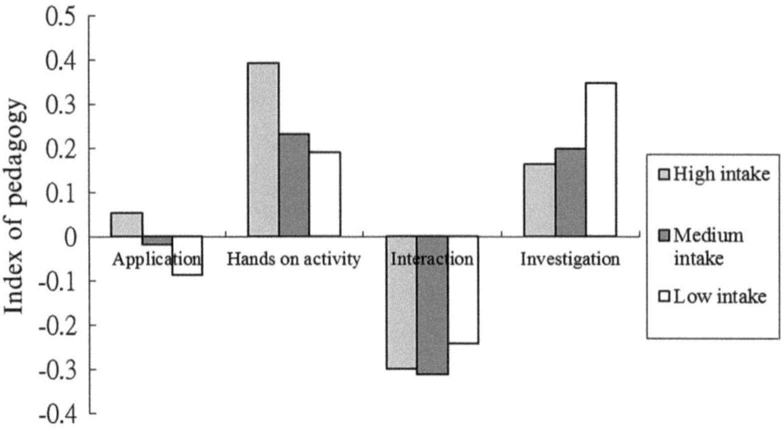

Fig. 2.1 A comparison of the reported pedagogies in Hong Kong schools with high, medium, and low academic intake of students as surveyed in PISA 2006

have lessons with significantly less hands-on activities, more investigation, and more interaction in class (Fig. 2.1).

Investigation

The construct of *investigation* used in PISA 2006 refers to those involving student decisions on the choice of experimental problems and design, while ordinary experiments are referred to as *hands-on activity* (Table 2.1). Investigation, a widely valued pedagogy for science teaching and learning, however, was found having the greatest negative associations with performance for all East Asian and western countries/regions (Table 2.7). For every additional unit of investigation in lessons, there is a drop of as high as 35 points in total scores. In addition, investigation also shows negative associations with enjoyment of science learning in all East Asian and western countries/regions. These findings are consistent with many other secondary analyses on investigation in PISA (Seidel et al., 2007; Taylor et al., 2009). Surprisingly, the western countries see even greater negative relations of investigation with performance, self-concept, and enjoyment of science learning than the East Asian countries/regions.

Hands-On Activity

Hands-on activity does not associate with performance in many of the East Asian and western countries/regions (Table 2.7). Only two countries/regions, Finland and Hong Kong, have positive associations with performance, whereas Taipei and the Netherlands have negative association. But for the attitudes, clearly different patterns were found between the East Asian and western countries/regions: it is positively associated with self-concept and enjoyment in science in almost all western countries, but the associations are mostly insignificant in East Asian regions. It seems that, similar to interaction, western students benefit more from hands-on activities than East Asian students affectively, albeit not cognitively.

Hands-on activity and investigation are widely seen as the integral part of science instruction, but the findings of PISA cast doubt on its role in the science learning of both East Asian and western students. Investigation, particularly, should be a crucial means to learn about the process and nature of science that constitutes an essential component of scientific literacy in the PISA framework. It is thus puzzling why it has obvious negative relationships with performance and attitudes. However, a review of the effects of practical work concludes that "most of the evidence does not support the argument that typical laboratory experiences lead to improved learning of science content" (Singer et al., 2006, p. 88). One main reason for that conclusion is that hands-on activities and investigation actually done in science class are far from desirable to produce the effects they purport to have on understanding about the content, process, and nature of science. Much of the practical work done in science class only involves manipulation of materials rather than ideas and does not interweave with other learning activities such as lecture and class discussion to form an *integrated learning unit* (Singer et al., 2006, p. 82). Integration was found an important factor positively related to the cognitive outcomes of practical activity (Henderson et al., 2000, in Singer et al., 2006, p. 97). The TIMSS 1999 Video Study also shed light on the problems of practical activity in science class: the US classrooms were found filled with a variety of activities that were not much connected with the learning of science ideas (Roth et al., 2006), which partly accounted for the unsatisfactory performance of the USA in the international assessments on science. Moreover, typical laboratory work often stresses procedure and manipulation skills more than the complex scientific inquiry processes such as formulating investigative questions and hypothesis, designing experiment, and critically evaluating the conclusion (Klopfer, 1990, in Singer et al., 2006). This kind of recipe-type experiments is limited in fostering the scientific reasoning skills of students. In addition, regarding the learning about the nature of science, the *implicit* approach through merely doing investigation has been shown to be largely ineffective in enhancing one's understandings about the nature of science (Lederman, 2007). Finally, even when scientific investigation is done properly with higher-order learning goals, it seems that it is not the more the better, but has to be in an "optimum" amount (Taylor et al., 2009). A case in point is France, which was found relatively strong in *knowledge about science* in PISA 2006 due to its curricular and instructional emphasis on investigation, but France was not ranked high in overall scientific literacy as a result of its low performance in *knowledge of science* (Olsen & Lei, 2009). It is likely because the student-directed, open-ended scientific investigation, when heavily conducted in class, may come at the expense of the time for content learning.

Investigation and practical activities in the class are also not necessarily enjoyed by students. Studies using the Science Laboratory Environment Inventory have shown that students' positive perceptions of laboratory experiences are associated with five dimensions of the laboratory environment: *student cohesiveness, open-endedness, integration, rule clarity, and material environment*, with particularly strong associations with cohesiveness (students are supportive of each another) and integration (laboratory work is integrated with theory learning) (Singer et al., 2006, pp. 95–98). Therefore, how the laboratory work is done is more important than its quantity in impacting students' attitudes. Since the attitudes of students are strongly associated with performance as a cause and consequence (OECD, 2009, p. 66), it is pivotal to identify ways that can make science teaching more engaging and interesting to students, particularly for the East Asian students.

Apart from the overall scores, the study also looks at the associations of the pedagogy with the three competencies: identifying scientific issues, using scientific evidence, and explaining phenomena scientifically. In both Asian and western countries/regions, the associations of the four pedagogies with the three competencies show no differences from their associations with the overall scores. Therefore, there is no evidence that a kind of pedagogy is particularly more powerful than others in fostering students' particular competency. The assumption that doing more investigation would foster students' abilities in identifying scientific issues and using scientific evidence seems not supported by the data, nor is the assumption that more interaction and application in class would make students more able to explain phenomena scientifically. Even though the western countries are relatively better in identifying scientific issues and the Asian regions better in explain phenomena scientifically, there is no evidence that the differences are the result of particular pedagogy. Though France's *relative* strength in knowledge about science has often been attributed to its emphasis on investigation and practical work in science class, its actual scores of *identifying scientific issues* and *using scientific evidence* are well below that of Japan and Korea, which reported below average use of investigation and hands-on activity in lessons (OECD, 2007, pp. 66–68).

Pedagogy at School Level

The school-level associations of pedagogy with performance and attitudes of the countries/regions are shown in Table 2.8. The data are highlighted to show the discrepancies with that at student level, by which the differences between the East Asian and western groups are discerned. Regarding the pedagogy of application, its positive associations with performance and enjoyment at student level are largely kept at school level for the East Asian regions, whereas for the western countries, many of these positive associations disappear at school level. The positive associations with self-concept become largely insignificant at school level in both the East Asian and western groups. Investigation sees its student-level negative associations with scores and attitudes largely gone at school level, which is more salient for the western countries. Hands-on activities, from student to school level, become more positive in its relations with scores, but less positive with attitudes, which is similar in both the East Asian and western countries. Interaction sees a more diverse changes in the direction of the coefficients from student to school level, and no clear differences are discerned between the East Asian and western countries.

There are many possible reasons to account for the discrepancies between the student-level and school-level associations of pedagogy with performance and attitudes. One is whether there exists a significant variation of pedagogy among schools. Further analysis is needed to examine if there is substantial variation in terms of pedagogy between schools, and this might be related to the highly academically segregated school system in East Asian societies. That is, the more academically segregated schools in East Asian countries may make teachers in a school employ similar teaching approach in response to the more uniform student characteristics. Figure 2.1 shows how different forms of school-level pedagogy of Hong Kong schools are associated with their academic intake of students. The western countries, on the contrary, tend to have more individualistic school culture: "Finnish teachers have been entrusted with considerable pedagogical independence in the classroom and schools have likewise enjoyed substantial autonomy in organizing their work within the constraints of the national core curriculum" (Välijärvi, 2003).

In summary, among the four pedagogies that are deemed desirable for science learning, only application is found positively associated with science performance and most of the attitudes for both East Asian and western countries/regions, whereas investigation is quite the opposite. Interaction has negative relations with performance and attitudes for the East Asian students, particularly in Japan and Korea, but is positive for some western students when it can enhance students' enjoyment of science learning. Hands-on activity is found not associated with performance for many countries/regions, but it does contribute to better self-concept and enjoyment of science learning for some western students. In the East Asian regions, the associations of pedagogy with performance and attitudes at student level tend to align more with that at school level than in the western countries. Nonetheless, it has to remain cautious to make conclusion about the impacts of these pedagogies on performance based on the correlational data. There may be many other extraneous factors at play or a reversed causation that student abilities influence the pedagogical approaches a teacher or school takes. In addition, the survey results could not reveal the cumulative classroom experiences of the students beyond the year the survey was done, but that may be significant for the student performance.

Science Activity

PISA surveyed students on the science-related activities done at their free time as a measure of their interest in science. However, this can also be construed as a kind of learning outside classroom that complements formal schooling. The science-related activities are found positively associated with performance and attitudes in all the countries/regions, which persist even after taking account of students' gender and socioeconomic background (Table 2.7). If these science activities are considered as a kind of out-of-classroom learning, they are as powerful as application in schools as a pedagogy. Moreover, science-related activities are moderately associated with enjoyment of science learning, which makes it a desirable means to promote lifelong learning. However, it is difficult to tell if science activity is a cause or consequence of better performance and enjoyment, and it is likely that the relation is reciprocal. Even so, science activity, particularly at the early ages, would be important. Ho (2003) found that in Hong Kong the parental arrangement in arranging science activities for children at the age of 10 was strongly correlated with science performance.

Conclusion

This study does not aim to find out whether the Asian or western ways of science teaching are superior since all the countries/regions were top performers in PISA. Instead, we aim to explore what makes these East Asian and western countries/regions top in PISA science and, more importantly, how and why the different ways East Asian and western teachers and students teach and learn science could produce similarly high-quality outcomes. Combining the descriptive data of the PISA 2006 surveys and other evidence with the HLM analysis done in this study, we can see some convergent and divergent characteristics of science teaching and learning in these top East Asian and western regions/countries of PISA (Table 2.9).

The western students are relatively better in understanding the process and nature of science, feel more confident in their ability to tackle science tasks, but do not enjoy learning science any greater than their East Asian counterparts. The science teaching of the western classrooms generally includes more hands-on activities and teacher-student interaction as well as greater emphasis on the application of science. Western students benefit more from interaction, both cognitively and affectively, as compared to the East Asian students.

The East Asian students are relatively better in science contents than science process. They have invariably low confidence in their ability to deal with science tasks. Despite that, the Chinese students enjoy learning science much more than those in Japan and Korea, and even the western students. In the East Asian classroom, science teachers tend to teach more traditionally: fewer hands-on activities, less interaction, and less emphasis on the application of science. Interaction seems not an effective pedagogy in the East Asian classroom – the more the lesson is interactive, the poorer the student performance, particularly for Japan and Korea. Hands-on activity, though not clearly conducive to science achievements, could make both Chinese and western students enjoy learning science (except for the Japanese and Korean students). Common to the East Asian and western are that student performance and enjoyment get enhanced when there is more application of science in class and more outside classroom science activities. Scientific

Table 2.9 A comparison of the East Asian and western groups of countries/regions using the descriptive and HLM data

	East Asian		
	[a]Chinese regions	[a]Japan and Korea	Western
[#]Performance	EPS		ISI
Attitudes			
Enjoyment	+	–	+/–
Self-concept	–		+/–
Pedagogy			
Application	–		+/–
Interaction	–		+/–
Hands-on	–		+
Investigation	+	–	+/–
HLM analysis			
Application			
Performance	+		+
Enjoyment	+		+
Self-concept	+		+
Interaction			
Performance	–		+/–/is
Enjoyment	is		+/is
Self-concept	+/is		+/is
Hands-on			
Performance	+/–/is		+/–/is
Enjoyment	is		+/is
Self-concept	+	is	+
Investigation			
Performance	–		–
Enjoyment	–		–
Self-concept	is		–
Science activities	+		+

[a]When the Chinese regions and Japan and Korea are shown separately, the three Chinese regions show a pattern different from Japan and Korea

[#]EPS or ISI: Performance in *explaining phenomena scientifically* or *identifying scientific issue* is relatively better

+ or –: 4/5 countries in the East Asian/western group are positive or negative in value (zero is the OECD average), or show significantly positive or significantly negative or insignificant correlation in HLM analysis

is: 4/5 countries in the East Asian/western group are insignificant in correlation

+/–/is: the East Asian/western group shows a mix of positive and negative values, or a mix of significantly positive, negative or insignificant correlation in HLM analysis

investigation, to the contrary, is negative to both performance and attitudes for all countries/regions.

The above pictures have depicted both the disparity and commonality between the East Asian and western countries/regions, but seemingly there are more disparities than commonalities. The findings seem to lend support to the East Asian model that education systems under the Confucian Heritage Culture have shared some distinctive features rooted in Confucianism emphasizing obedience, authority, and diligence (Watkins & Biggs, 2001). The didactic, teacher-centered instruction would not work without obedient students who respect teacher authority. The high academic achievement but low self-concept and enjoyment of learning (for Japan and Korea) has reflected that East Asian students are not motivated to learn mainly by enjoyment but the culture that emphasizes diligence, hard work, and success through education. Nonetheless, there are variations within the East Asian regions as a result of each region's unique sociocultural-political context and the "vernacular Confucianism" (Chang, 2000) developed under it. Japanese and Korean students are different from their Chinese counterparts in many ways: much lower enjoyment and self-concept in science, much less common of the four pedagogies in class, and relatively better in identifying scientific issues and weaker in explaining phenomena scientifically, which somehow place Japan and Korea in between the East Asian and the western countries/regions.

Among the western countries, Finland seems more "traditional" and "Asian" in view of its below average use of application, interaction, and investigation in science class, relative strength in explaining phenomena scientifically but weaker in identifying scientific issues, and negative associations of interaction with performance. In general, the countries in the "western" group of this study show greater diversity in their pedagogical and attitudinal characteristics than the countries/regions in the "East Asian" group. It is probably because these "western" countries are not sharing a common cultural heritage as the East Asian countries/regions do. Therefore, it may not be appropriate to give them a unitary label as "western."

The disparities between and within the East Asian and western countries/regions suggest that there is no single "formula" for high science achievement, but culturally appropriate ways to make the best out of a country's characteristic profiles of students, teachers, parents, schools, and educational systems. For instance, the East Asian students could benefit to some extent from the teacher-dominated, expository instruction in a disciplinary classroom, but these may be detrimental for the western students, who would otherwise learn better when the class is interactive and hands-on. The East Asian students, despite low in self-concept, could still enjoy learning and perform well. It seems that the pressure and challenges have driven them to put in greater efforts rather than to withdraw, but this may come at the costs of other valuable educational outcomes that will be discussed in the next paragraph. The western students, on the other hand, may benefit more from the relaxing classroom culture with less rigorous content standards so that their self-concept can be protected. An East Asian classroom can seat in more than 40 students, yet teaching and learning can proceed effectively, but it does not mean that such a large class size works for the western students. East Asian regions, notably Shanghai, are more top-down and collectivist in teaching and learning at school, having "teaching study groups," open lesson, lesson demonstration, and even competitions to promote the quality of teaching. These measures, however, may be met with high resistance and not workable in the more individualistic, liberal school culture of the western countries. Therefore, it is problematic for countries to seek quick remedies in education from the top-performing countries in PISA regardless of their sociocultural uniqueness, which amounts to introducing an exotic species into a local ecosystem.

The East Asian model, however, is not without problems. The low self-concept, overemphasis on memorization, highly examination-centered teaching, and submissive classroom climate may have a cost on students' creativity, problem-solving capability, autonomy in learning, and lifelong learning skills. Similarly, the "western" model emphasizing students' autonomy, freedom, and enjoyment of learning may end up with students putting inadequate effort into learning and having a weaker knowledge foundation. Therefore, to improve science education, we should not seek a "transplant" of the East Asian or western model, but a mix of both models in a way appropriate to a country's own sociocultural context. The findings of this study suggest that the widely advocated student-centered pedagogies for science, interaction, investigation, and hands-on activities, are not having the desirable outcomes they are expected to produce in many top-performing countries, at least in terms of the cognitive and affective outcomes measured by PISA. On the contrary, many top performers like Finland and many East Asian countries/regions tend to employ more teacher-directed instruction (Kesler & Lavonen, 2009). Seemingly, the student-centered, interactive, and activity-based lessons are not necessarily good, while the teacher-directed, exposition-based lessons are not necessarily bad. It is apparent that teaching at both extremes is not desirable regardless of the characteristics of the teacher and learner. More important is whether the pedagogy and activities, be it teacher or student centered, are closely connected with meaningful and effective learning. In the TIMSS Video Study, the US classrooms were found spending a lot of time on activities that had minimum learning in science content (or even the process of science!), while the high-achieving countries all put science content and high content standards at the core of instruction (Roth et al., 2006). That explains the below average performance of the USA in PISA. France put inquiry at the core of its curriculum and instruction, which did make France relatively good in knowledge about science in PISA, but its below average performance in knowledge of science made France mediocre in its overall performance. Therefore, what matters is "learning centered" – that is, work beyond student and teacher centeredness – and a balance between the process and content of science. What is a right balance is in turn a judgment made with reference to the sociocultural contexts of the country and through the lens of international assessments.

Apart from the pedagogy, the East Asian education systems are different from the western countries in many ways. Most noticeable is the high examination pressure and competitive school culture. These competitive climates, together with the CHC culture of the societies, drive students to work hard both in school and after school. For instance, the East Asian students virtually did not skip any classes as compared to the 18 % of the OECD average (OECD, 2014), and after-school tutorial class is highly popular (Mark & Chad, 2012). However, as mentioned previously, this kind of examination and competition-driven learning may come at a cost of self-concept, enjoyment of learning, creativity, and other lifelong learning skills (Ho, 2009).

All the top East Asian countries/regions have also put in great efforts in curriculum reforms in recent years: Japan's *Zest for Living* curriculum emphasizing the independent thinking of students; Korea's *Smart Education* with strong focus on ICT; Singapore's *Thinking School, Learning Nation* reform with particular focus on math, science, and technical skills; Shanghai's centrally administered *Teaching Study* to enhance classroom teaching; and Hong Kong's 2001 *Learning for Life, Learning through Live* education reform with drastic changes in curriculum and school structures (Strong Performers, 2013). Most of these reforms have their goals aligned, at least rhetorically, with some of the western progressive education ideals such as thinking skills, creativity, independent learning, joyful learning, and ICT skills. These reform efforts, together with the high-quality, professional teachers, have made some East Asian countries/regions improve significantly in the recent years, which is one of the important lessons learned from the PISA studies over the past decade.

References

Ainley, M., & Ainley, J. (2011). Student engagement with science in early adolescence: The contribution of enjoyment to students' continuing interest in learning about science. *Contemporary Educational Psychology, 36*(2011), 4–12.

Chan, C. K. K., & Rao, N. (2009). Moving beyond paradoxes: Understanding Chinese learners and their teachers. In K. K. C. Chan & N. Rao (Eds.), *Revisiting the Chinese learner: Changing contexts, changing education* (pp. 3–34). Dordrecht, The Netherlands: Springer.

Chang, W. C. (2000). In search of the Chinese in all the wrong places! *Journal of Psychology in Chinese Societies, 1*(1), 125–142.

Duit, R., & Treagust, D. (2003). Conceptual change: A powerful framework for improving science teaching and learning. *International Journal of Science Education, 25*, 671–688.

Henderson, D., Fisher, D., & Fraser, B. (2000). Interpersonal behavior, laboratory learning environments, and student outcomes in senior biology classes. *Journal of Research in Science Teaching, 37*, 26–43.

Ho, A. S. P. (2001). A conceptual change approach to university staff development. In D. A. Watkins & J. B. Biggs (Eds.), *Teaching the Chinese learner: Psychological and pedagogical perspectives* (pp. 239–254). Hong Kong, China/Melbourne, VIC: Comparative Education Research Centre, The University of Hong Kong/Australian Council for Educational Research.

Ho, D. Y. F. (1994). Cognitive socialization in Confucian-heritage cultures. In P. Greenfield & R. Cocking (Eds.), *Cross cultural roots of minority child development* (pp. 285–314). Hillsdale, NJ: Lawrence Erlbaum Associates.

Ho, E. S. C. (2003). Effects of parent factors on student's literacy performance. In *HKPISA Centre, The third HKPISA report: PISA 2003. Monitoring the quality of education in Hong Kong from an international perspective*. Hong Kong, China: HKPISA Centre.

Ho, E. S. C. (2009). Characteristics of East Asian learners: What we learned from PISA. *Educational Research Journal, 24*, 2.

Kesler, M., & Lavonen, J. (2009). What lies behind finnish students' success in PISA science? In R. W. Bybee & B. J. McCrae (Eds.), *PISA science 2006: Implications for science teachers and teaching* (pp. 59–68). Arlington, VA: NSTA Press.

Klopfer, L. E. (1990). Learning scientific enquiry in the student laboratory. In E. Hegarty-Hazel (Ed.), *The student laboratory and the science curriculum* (pp. 95–118). London: Routledge.
Korsnakova, P., McCrae, B. J., & Bybee, R. W. (2009). Improving science teaching and learning. In R. W. Bybee & B. J. McCrae (Eds.), *PISA science 2006: Implications for science teachers and teaching* (pp. 59–68). Arlington, VA: NSTA Press.
Lam, T. Y. P., & Lau, K. C. (2014). Examining factors affecting science achievement of Hong Kong in PISA 2006 using hierarchical linear modeling. *International Journal of Science Education, 36*(15), 2463–2480. doi:10.1080/09500693.2013.879223.
Lederman, N. G. (2007). Nature of science: Past, present, and future. In S. K. Abell & N. G. Lederman (Eds.), *Handbook of research on science education* (pp. 831–880). Mahwah, NJ: Lawrence Erlbaum Associates Publishers.
Mark, B., & Chad, L. (2012). *Shadow education: Private supplementary tutoring and its implications for policy makers in Asia*. Hong Kong, China: Comparative Education Research Centre (CERC) in collaboration with Asian Development Bank (ADB).
Mok, I., Chik, P. M., Ko, P. Y., Kwan, T., Lo, M. L., Marton, F., et al. (2001). Solving the paradox of the Chinese teacher? In D. A. Watkins & J. B. Biggs (Eds.), *Teaching the Chinese learner: Psychological and pedagogical perspectives* (pp. 161–179). Hong Kong, China/Melbourne, VIC: Comparative Education Research Centre, The University of Hong Kong/Australian Council for Educational Research.
Mortimer, E., & Scott, P. (2003). *Meaning making in secondary science classrooms*. Maidenhead, UK/Philadelphia, PA: Open University Press.
OECD. (2006). *Assessing scientific, reading and mathematical literacy: A framework for PISA 2006*. Paris: OECD Publications.
OECD. (2007). *PISA 2006: Science competencies for tomorrow's world* (Vol. 1). Paris: OECD Publications.
OECD. (2009). Experiences, attitudes and motivations for excellence. In *Top of the class: high performers in science in PISA 2006*. Paris: OECD Publishing. http://dx.doi.org/10.1787/9789264060777-6-en
OECD. (2013). *Strong performers and successful reformers in education*. Retrieved April 9, 2014, from http://www.pearsonfoundation.org/oecd/index.html
OECD. (2014). Who are the school truants? (PISA in focus, no. 35). Paris: OECD Publishing. doi:10.1787/5jzb019jwmd5-en
Olsen, R. V., & Lie, S. (2009). Knowledge of and about science. In R. W. Bybee & B. J. McCrae (Eds.), *PISA science 2006: Implications for science teachers and teaching* (pp. 59–68). Arlington, VA: NSTA Press.
Roth, K. J., Druker, S. L., Garnier, H. E., Lemmens, M., Chen, C., Kawanaka, T., Rasmussen, D., Trubacova, S., Okamoto, Y., Gonzales, P., Stigler, J., & Gallimore, R. (2006). *Highlights from the TIMSS 1999 video study of eighth-grade science teaching* (NCES 2006-17). U.S. Department of Education, National Center for Education Statistics. Washington, DC: U.S. Government Printing Office.
Seidel, T., Prenzel, M., Wittwer, J., & Schwindt, K. (2007). Unterricht in den naturwissenschaften [Instruction in the natural sciences]. In M. Prenzel, C. Artelt, J. Baumert, W. Blum, M. Hammann, K. Klieme, & R. Pekrun (Eds.), *PISA 2006. Die ergebniss der dritten internationalen vergleichsstudie [PISA 2006: The results of the third international comparative study]* (pp. 147–179). Munster, Germany: Waxmann.
Scott, P. (1998). Teacher talk and meaning making in science classrooms: A Vygotskian analysis and review. *Studies in Science Education, 32*, 45–80.
Singer, S. R., Hilton, M. L., & Schweingruber, H. A. (Eds.). (2006). *America's lab report investigations in high school science*. Washington, DC: National Academies Press.
Taylor, J. A., Stuhlsatz, M. A. M., & Bybee, R. W. (2009). Windows into high-achieving science classrooms. In R. W. Bybee & B. J. McCrae (Eds.), *PISA science 2006: Implications for science teachers and teaching* (pp. 59–68). Arlington, VA: NSTA Press.

Thomson, S. (2009). Teaching and learning science: PISA and the TIMSS video study. In R. W. Bybee & B. J. McCrae (Eds.), *PISA science 2006: Implications for science teachers and teaching* (pp. 59–68). Arlington, VA: NSTA Press.

Watkins, D. A., & Biggs, J. B. (Eds.). (1996). *The Chinese learner: Cultural, psychological and contextual influences*. Hong Kong, China/Melbourne, VIC: Comparative Education Research Centre, The University of Hong Kong/Australian Council for Educational Research.

Watkins, D. A., & Biggs, J. B. (2001). The paradox of the Chinese learner and beyond. In D. A. Watkins & J. B. Biggs (Eds.), *Teaching the Chinese learner: Psychological and pedagogical perspectives*. Hong Kong, China/Melbourne, VIC: Comparative Education Research Centre, The University of Hong Kong/Australian Council for Educational Research.

Välijärvi, J. (2003). The System and How Does It Work: Some Curricular and Pedagogical Characteristics of the Finnish Comprehensive School. *Education Journal*, *31*(2) and *32*(1), 31–55.

Yip, D. Y. (2008). Hong Kong students' performance in scientific literacy. HKPISA centre. Retrieved June 26, 2008, from http://www.fed.cuhk.edu.hk/~hkpisa/events/2006/events2006_20071210.htm

Yip, D. Y., & Cheung, S. P. (2004). Scientific literacy of Hong Kong students and instructional activities in science classrooms. *Education Journal, 32*, 2.

Chapter 3
Confucianism, Chinese Families, and Academic Achievement: Exploring How Confucianism and Asian Descendant Parenting Practices Influence Children's Academic Achievement

Grace Hui-Chen Huang and Mary Gove

Introduction

Our interest in Asian parenting style drew from the media attention given to Amy Chua's (2011) book, *Battle Hymn of the Tiger Mother*. Chua's book is a memoir of her own philosophy and practices of parenting raising two high achieving daughters. She placed great emphasis on her children's academic achievement. Chua's views and practices raised much controversy regarding parent–child interaction. Public media began debating and discussing different views and practices about parenting (Chang, 2011; Chua, 2011; Flanagan, 2011; Husbands, 2012; Kohler, Aldridge, Christensen, & Kilgo, 2012; Kohler, Kilgo & Christensen, 2012; Ninh, 2011; Poon, 2011; Wang, 2011). Chua connects her parenting experiences with her children's academic achievement.

Secondly, we considered the achievement gap between Asian Americans and Caucasian Americans (Grissmer, 1994; Hsin & Xie, 2014; Kao, 1995; Konstantopoulos, 2009; Pang, Han, & Pang, 2011; Yong & Wei, 2009). The US media touts an achievement gap between white and African Americans and between white and Hispanic Americans (California Department of Education, 2010; Hemphill & Vanneman, 2011; Noguera, 2012; Noguera, 2008; O'Donnell & Kirkner, 2014; Schott Foundation, 2010). Because of this, the general public tends to be more aware of these achievement gaps. Less media attention is given to the achievement gap we focus on, the gap between Asian Americans and Caucasian Americans. Asians outscore whites, especially in mathematics, science, and other technical areas (Pang et al., 2011; Peng & Wright, 1994). In the USA, Asian Americans score higher on the SAT and ACT (Hsia, 1988; Hsin & Xie, 2014) and are overrepresented among winners of different prestigious scholarships, such as National Merit, US Presidential, and Westinghouse Science talent Search

G.H.-C. Huang (✉) • M. Gove
Department of Teacher Education, Cleveland State University, Cleveland, OH, USA
e-mail: g.h.huang@csuohio.edu; m.gove@csuohio.edu

M.S. Khine (ed.), *Science Education in East Asia*,
DOI 10.1007/978-3-319-16390-1_3

Scholars (Yong & Wei, 2009). Hildebrand, Phenice, Gray, and Hines (2008) indicated that Asian Americans have become the largest ethnic minority group in many career fields and at many elite colleges. In 2003, 51 % of Asian American males and 44 % of females aged 25 and older had a bachelor's degree or higher compared with 32 % of non-Hispanic white males and 27 % of non-Hispanic white females. In 1980, Asian Americans comprised only 1.5 % of the population, but Asian Americans consisted of 5 % of all engineers and 8 % of all doctors in the United States (Hildebrand et al., 2008). Similar trends occur in international data. The latest PISA (Program for International Student Assessment) survey, an international educational survey conducted every three years by the Organization for Economic Cooperation and Development (OECD) released in 2012, indicated the top five overall ranking were Asian students from Shanghai (China), Singapore, Hong Kong, Taiwan, and South Korea (Brown, 2013).

Thirdly, coming from our own personal experiences, as Asian American and European American researchers, we each raised successful children who demonstrate strong academic performance. Yet we have similarities and differences in our parenting practices. We wondered how parenting associates with children's performance.

These three phenomena motivated our conceptual inquiry and qualitative interview study about parenting styles and academic performances. In this chapter, we explore how parenting styles associate with academic performance through the lens of Eastern and Western literature and a qualitative research conducted by our research team through Asian descendants' experiences and views.

The two overarching questions we developed for this conceptual framework and the follow-up qualitative research study are as follows:

1. How do Asian descendant parenting styles associate with children's attaining high academic performance?
2. Are Asian parenting styles related to the age-old Confucian philosophy that is part of the Asian culture?

In our conceptual framework, we examined concepts involved in Confucianism as well as western parenting style literature to shed light on Asian descendants' high academic performance. First we turn to how Confucianism, embedded in Asian's thinking, is related to Asian descendants' ideas about parenting.

Conceptual Framework: Eastern and Western Lenses

Confucianism and Asian Descendant Families

Though Confucianism was developed in China, the philosophy has influenced many Asian descendants in other countries, especially in East Asia (Slote & De Vos, 1998). Confucianism, developed by the greatest Chinese philosopher,

Confucius, is a philosophy which focuses on the conduct and practices of people in daily life. For nearly 2,000 years, Confucianism has shaped the social, ethical, and political aspect of Chinese cultures. It is a complex set of ethical and moral rules that dictate how a person relates to others and the world. The Confucianism philosophy and principles play a key role in forming the norms of social morality which influence the culture in personal, familial, and social relationships.

Additionally, Confucianism has a significant influence on education and families' educational practices (Huang & Gove, 2012). In the following, we will further explore the relationship between Confucianism and education at the societal and family levels, including education and social class, education as a family business, and the role of education in family hierarchy and family harmony.

Social Morality

The core value of Confucianism is benevolence "仁" (Slote & De Vos, 1998). This central philosophy includes many virtues, such as compassion, kindness, and generosity toward important matters and people surrounding them. Regarding the relationship between benevolence and learning, Confucianism places high value on benevolence. It is considered more important than learning and education. Several verses articulate Confucius' philosophy well. For example, "当仁, 不让于师" (Confucius & Waley, 2005, 15.35). When one carries out the act of benevolence, respecting teachers is valued; however, it is not the top priority. In other words, benevolence is highly valued, even more than the respect for teachers. Another example is "弟子入则孝, 出则弟, 谨而信, 汎爱众, 而亲仁, 行有余力, 则以学文" (Confucius & Waley, 2005, 1.6). This verse states that a person is expected to have filial piety toward the parents (Huang & Gove, 2012), i.e., respect others, be honest, be trustworthy, behave properly, show compassion toward people, and be close to people with benevolence. Most important of all, if one performs all the qualities above and has energy to spare, he can enrich himself through studying. This illustrates the priority of benevolence and learning.

Education and Social Class

Asian education is associated with a person's social class. In ancient Chinese society (770 BC to 1910s), success is defined by one's social class, not necessary by a person's wealth (Huang & Gove, 2012; Lien, 2006). Social class is associated with occupations as well as the moral character related to the occupation. According to Confucius, out of four social strata which includes scholars *(Shi)*, farmers *(nong)*, workers *(gong)*, and businessmen *(Shang)* based on occupation (Park & Chesla, 2007), the scholars strata is considered the highest class. Scholars did "mental labor" and usually made decisions that influenced the whole society. Based on Confucianism, leaders needed to be intelligent, have high standards of ethics, and learn scholarly work (Tu, 1998). In terms of the importance of scholarly work and key virtues, a verse from Confucius described the relationship well. "士不可以不弘毅，任重而道远。仁以为己任，不亦重乎？死而后已，不亦远乎?" (Confucius & Waley, 2005, 8.7). The scholars are expected to not only practice key virtues, benevolence, but also to take on this responsibility to impact the society with this core philosophy, until death. Therefore, scholarship was associated with high social class, leadership, and high moral character.

Today, Asian society and families still retain the idea that scholars belong to higher social class and education provides a route to reach the "scholarly" status (Huang & Gove, 2012). Highly educated people will become leaders and contribute to society. This value of education and social class influences the everyday life of Asian families. Many Asian descendants believe that educational success leads to a better life, including higher social status, getting a good job, or a better marriage and relationships (Cheon, 2006; Hildebrand et al., 2008; Louie, 2004; Lien, 2006). Therefore, education is central to most Asian families' daily life. In fact, the various statistics indicated in the beginning of this paper is the evidence of educational emphasis within the family. Asian parents, such as Amy Chua (2011), place a high priority on educational success and that success is often measured by the test scores (Huang & Gove, 2015).

Education as a Family Business

At the macro-system level (Bronfenbrenner, 1979), Asian cultures value the collectivist ideology (Lu & Shih, 1997) which affects family functions and parent–child interaction. The family unit functions as a collective and de-emphasizes individualism which impacts on the daily practices, including education. Based on Confucianism, there are five basic human relationships: father–son, emperor–subject, husband–wife, elder–younger, and friend–friend. Three out of five bases of these relations occur in the family which indicates the importance of family interactions and relationships (Chang & Holt, 1991). Chinese cultures advocate that each individual should strive to expand the prosperity and vitality of one's family (Lu & Shih, 1997; Miller & Yang, 1997). In other words, individual's development and performance are to achieve the success of the family. Family, then, is in the center of an individual's life and everyday existence, including educational practice.

Education is considered a family business, an interdependent process for many Asian families. Though children are responsible for their own educational success, Asian parents believe that their children's educational achievement is greatly influenced by their parenting practices as well. They believe that they have the obligations and responsibilities to help build their children's success in education (Huang & Gove, 2012).

Parents possess passion and zeal for their children's education (Lien, 2006). A famous real-life story known by Chinese descendants concerns a mother who took much effort to find a proper learning environment for her son. This son later became a well-known scholar, Mencius, who continued the spread of the philosophy of

Confucianism. The story is called 孟母三遷 (mèng mǔ sān qiān; literal translation: "Mencius' mother, three moves"). Mencius' mother literally moved her family three times before finding a location where she felt suitable for her child's upbringing. During this ancient time, moving from one place to another was a difficult task. They moved from the neighborhood of a cemetery, to a local market, and finally to a school. This story illustrates the importance of education to the Chinese families.

Education is an intensive process, almost like a religious practice which affects the whole family's daily interactions and routines. In Asian descendant families, a high priority is placed in any education-related topics during daily activities and conversations. For example, parents may converse with their children concerning homework and test results as soon as children come home from school. In Chua's case, her daughters were not allowed to have playdates as they would interfere with their academic endeavors.

Asian parents are highly demanding and exert high control over children's academic practice which relates to the dimension of the ethnic minority parenting style: high demandingness (Baumrind, 1971; Maccoby & Martin, 1983). During the daily parent–child interaction, parental involvement on academic practice is a focused and intense practice. Asian family takes academic success seriously. If parents have a difficult time exerting influence over their children's academic success, they may regard themselves as a failure as a parent and feel disappointed, anxious, and embarrassed (Kim, 2006). Asian parents hold themselves accountable for their children's academic performance. They take their children's academic failure as their own responsibility because they feel that they did not do their best to work with their children. Pressures can be high on both parents and children. Low achievement brings shame and embarrassment to the family (Chen, Miller, Wang, & Mark, 1996). On the other hand, if the child succeeds in academics, it represents the triumph of the entire family and is considered a family achievement, collective effort of the entire family. In Chua's narrative, she drilled her daughters on homework from school and on their music lessons, which she also deemed quite important to success (Huang & Gove, 2015).

The Role of Education in Family Hierarchy and Family Harmony

Asian cultures value family hierarchy and harmony which are central concepts in Confucianism (Huang & Gove, 2012; Leung, Wong, Wong, & McBride-Chang, 2010). The dimension of high demandingness within the ethnic minority parenting style is a reflection of these values (Leung et al., 2010). Based on Confucianism, the family system has a clear and highly structured hierarchy. Confucius' doctrine distinctly stated the order: *Let the prince be a prince, the minister a minister, the father a father, and the son a son* (Kao, 2006). Each individual's role and responsibilities are clearly defined (Huang & Gove). For example, one of the virtues of family hierarchy, filial piety, portrays the relationship between parent and child. Filial piety is an expression of respect from the child toward parent and elderly. In Confucianism, this virtue is associated with a person's citizenship and leadership in the society. Asians believe that children who respect their parents will become good citizens and leaders (Huang & Gove). Therefore, filial piety is highly valued. In daily life, filial piety is reflected in children's respect and obedience on parental demands and authority (Leung et al.). In most cases, children do not have many choices but to comply with parents' expectations. Rebellion against parents' authority is unacceptable and will result in harsh consequences. Being "unfilial" was considered a crime in the traditional Chinese society. As Huang and Gove indicated, it could result in death as a punishment. The virtue, filial piety, is interwoven with family's educational practice on a daily basis.

Additionally, Confucianism promotes social harmony in human relationships, including family relationships (Chang & Holt, 1991; Leung et al., 2010). To reach harmony within the context of family, it often relies on considering parent's authority, conformity, and role and responsibilities of each individual (Lu & Shih, 1997; Miller & Yang, 1997). Thus, the decisions children make regarding education are associated with family harmony as well. Asian children learn more behavioral rules than Westerner children do. Asian children are comfortable following instructions (Moneta, 2004). Today, Asian parent's authority can be observed in educational decision-making and behaviors in homework, exams, school selections, career choices, and so on (Huang & Gove, 2012).

Because of this emphasis on filial piety and harmony of the family, Asian families are able to demand academic excellence. However, Asian families tend to be less supportive of children's social needs or other interests the children may want to pursue. As stated earlier,

Lulu, Chua's younger daughter, wanted to learn tennis, and she had to battle her mother in order to pursue this interest (Chua, 2011). Yet, once Lulu pursued her desire to play tennis, she drew from the discipline she had acquired by working with her mother on academics and playing classical musical instruments. Eventually, she was successful at tennis. We speculate that Chua's parenting facilitated Lulu's self-motivated disposition (Huang & Gove, 2015).

Next we turn to an analysis of how the Western parenting style literature explains the relationship between Asian parenting style and the high achievement of their children.

Parenting Styles: Theories and Literature

Based on Maccoby and Martin's (1983) work, the two primary factors or dimensions to determine parenting style are responsiveness and demandingness. Parental responsiveness refers to the degree parents respond to the child's needs. Parental demandingness or parental control is the degree of demands, control, or expectations parents have toward children (Maccoby & Martin). Analyzed by these two factors, 6 different types of parenting styles emerged. They are authoritative, authoritarian, permissive, ethnic minority, indifferent, and inconsistent parenting styles (Baumrind, 1971; Maccoby & Martin, 1983). Since our purpose is looking at

Asian parenting styles and achievement, we considered three of these: authoritative because the literature indicates it leads to high achievement in children and ethnic minority as a subset of authoritarian style because that is the style most used by Asian families.

In this conceptual framework, we focus on authoritative, authoritarian, and ethnic minority parenting styles and how they relate to children's academic achievement. Based on the findings, the authoritative style is recognized as related to high achievement, while authoritarian style relates to low achievement (Sung Seek & Joohi, 2009). Ethnic minority style is less often considered in the literature and is considered a subset of authoritarian parenting, but reached the public arena through Chua's memoir (Baumrind, 1971; Baumrind, 1987; Chao, 1994; Chao & Tseng, 2002; Chau, 2001; Chua, 2011; Fuligni & Tseng, 1999; Giarrusso, Du, Silverstein, & Bengtson, 2001; Heath, 2012; Hsai & Scanzoni, 1996; McLoyd & Smith, 2002; Shek & Chen, 1999).

In this analysis, we look primarily at the demandingness/control dimension as well as the extent parents support the interests of children for these two dimensions seem to be distinguishing dimensions between ethnic minority and authoritative parenting style.

With the authoritative parenting style, the parent has reasonable control, uses reasoning with the children, is responsive to the children's needs and interests, and is strongly related to high achieving children (Baumrind, 1971; Chau, 2001; Garg, Levin, Urajnik, & Kauppi, 2005; Heath, 2012; McLoyd & Smith, 2002; Steinberg & Levine, 1997). The authoritarian parenting style emphasizes the hierarchical relationships between adults and children and uses this to maintain control, sometimes with forceful means, may restrict children's autonomy, and is related to low achieving children. In terms of the responsiveness dimension, authoritarian parenting style is low in supporting children's needs and interest.

The ethnic minority parenting style as articulated by Heath (2012) is considered a subset of the authoritarian style and is less prominent in the literature of parenting styles. However, we, the authors, realize that authoritarian style has a negative connotation in mainstream American culture. Additionally, we see the ethnic minority parenting style as a category of parenting style by itself because of the significant evidence of the exceptional high academic achievement especially among Asian children. Therefore, we would like to consider this parenting style as a separate entity.

Ethnic minority parenting style, at least in Asian American families, is related to high achievement as noted earlier as is authoritative parenting style, which is clearly supported in the literature (Heath, 2012). This style differs from authoritative style in that parents with ethnic minority style have high demandingness or control and do not place children's interests as a priority.

Comparing the ethnic minority style to strict authoritarian parenting style, both styles are high in the dimension of demandingness or control. On the other hand, ethnic minority parenting style is differentiated from the authoritarian parenting style in the high parents' responsiveness toward children's needs (Leung et al., 2010). Even though Asian parents tend to be highly responsive to their children's needs in relation to academic achievement issues, we speculate that Asian

Table 3.1 Parenting style analysis

Parenting styles	Dimensions			Children's academic performance
	Demandingness/ control	Responsiveness to child's needs	Responsiveness to child's interests	
Authoritative	Medium	High	High	High
Authoritarian	High	Low	Low	Low
Ethnic minority	High	High	Low	High

descendants tend to have a lower priority in supporting the child's interest. If an Asian child wants to pursue an interest not perceived by the parent as viable, will the child be supported in participating in this interest? For example, Lulu, Chua's younger daughter, wanted to learn tennis and she had to battle her mother in order to pursue this interest (Chua, 2011; Huang & Gove, 2015).

From this analysis, we come to the conclusion that parenting that is medium (authoritative parenting style) to high in demandingness (ethnic parenting style) and high in responsiveness (both authoritative and ethnic parenting style) tends to produce high academic performance in children. Table 3.1 summarizes the above analysis of the literature, showing how parenting styles relate to the factors of demandingness, responsiveness, interests, and academic performance (Huang & Gove, 2015).

Conclusions from Our Conceptual Framework

We believe that Confucianism plays a role in parenting style of Asian descendants which associate with children's high academic performance. Additionally, we see the ethnic minority parenting style as related to children's strong academic achievement. We speculate that the Asian parenting style is influenced by values and philosophies of Confucianism embedded in the Asian culture. Confucianism places high value on education, filial piety, family hierarchy, and family harmony. These important family parenting traits may lead to the high academic achievement Asian descendant youth exhibit. These traits are reflected in the parenting style literature (ethnic minority parenting style) as high demandingness and high responsiveness especially toward concerns related to school and education. We further speculate, as Amy Chua's parenting illustrates (Chua, 2011), this kind of parenting tends to foster self-motivated disposition in the children over time. We will explore this further through Asian parents' lived parenting experiences in our qualitative research below.

Further, Western parenting style literature emphasizes that the authoritative style with its medium demandingness emphasizing reasoning and high responsiveness to both needs and interests leads to high achievement in children. However, Asian descendant families foster high academic achievement in children and their parenting style does not fit into the authoritative style. It is considered in the category of ethnic minority parenting style in the literature. Thus, in our analysis, we looked at the demandingness/control dimension as well as the responsiveness dimension, the extent parents support the needs and interests of children. These two dimensions differentiate ethnic minority and authoritative parenting style. Concerning issues relating to academic endeavors, we concluded that parenting with medium to high demandingness and high responsiveness is related to high academic achievement in children.

We have reached three conclusions in this conceptual framework. First, Confucianism,

as a belief system which is part of the heritage of Asian descendants, seemed to foster the high achievement of children especially its emphasis on education and family dynamics. The impact of Confucianism and its association with parenting and academic performance was further explored in our qualitative study below. Second, the ethnic minority parenting style needs to be more prominent in the parenting style literature, and more research needs to be conducted concerning this style, particularly with Asian descendants. In this qualitative study, we began to tease out how ethnic minority parenting style differs from the authoritarian parenting style in the literature. Third, in light of the fact that Asian descendants tend to excel in academic arenas, we believe that Asian parenting style needs to be singled out and examined with an in-depth look at its relationship with children's academic performance. This, too, was one of the goals of our qualitative research study below.

Our Qualitative Research Study

In order to move beyond our literature review and personal experiences, we conducted a research study to closely examine the Asian parents' perception of how Confucianism affected their parenting and how their perception of their parenting fit into the ethnic minority parenting style which does not have extensive literature describing it. In our research team, two researchers are European American women, including a university professor and an undergraduate research assistant. The other two researchers are Chinese descendant women, including a university professor and an undergraduate research assistant. In this study, we each took the role of researchers, while reflecting on our own experiences, backgrounds, and views relating to the topic as we learned from Asian descendant parents as we interviewed them. Our intention of this research was to build on our thoughts, reflect on ourselves, as well as contribute to the research community in the area of parenting and academic performance.

Research Purpose and Question Statements

The purpose of our qualitative research was to learn about the Asian descendants' perceptions concerning their parenting experiences and its association with academic performance, as well as the impact of Confucianism on their parenting practices.

How do Asian descendant parents' perceptions and experiences inform the following central questions?

1. How do Asian descendant parents' parenting style relate to children's academic performance?
2. How does Confucianism impact Asian descendant parents' parenting experiences?

Methodology and Research Context

Descriptions of the Participants

This study, conducted in a midsize city in the Midwest, has 2.57 % Asian descendant population as of 2010 (USA.com, 2010) and 2.9 % in 2013 (USA.com, 2013). Among the Asian population, nearly 28 % are Chinese (USA.com, 2013). Chinese American is the second largest Asian population in this city, next to Indian population.

Fifteen parents participated in this study. Among them, ten female participants who participated in individual interviews were Chinese descendant parents. Ten participants, including five fathers and five mothers, participated in a follow-up focus group interview. One of them was born in the USA, while the others were born in either Taiwan or China. All female parents were Chinese language teachers who taught in k-12 public or private schools, majority in suburban areas. Their children's ages ranged from two to twenty-five. The basic information of the participants was summarized in Table 3.2.

Table 3.2 Participant demographics

Demographics	Number or % of the participants
Years in the USA	1–10 years: 5 p.p.
	11–20 years: 6 p.p.
	21 years and above: 4 p.p.
Family income	Under $60,000: 3 p.p.
	$60,000–$100,000: 5 p.p.
	$100,000–$150,000: 4 p.p.
	$150,000 and more: 3 p.p.
Education	High school: 1 p.p.
	Bachelor: 2 p.p.
	Master: 9 p.p.
	Ph.D.: 5 p.p.
Children's age level	Age 2–4: 5 %
	Age 5–18: 65 %
	Age 19–25: 30 %

Methods

This qualitative research endeavored to make sense of and interpret parenting phenomena in terms of the meanings people bring to them (Creswell, 2007; Denzin & Lincoln, 2005). Specifically, we used a phenomenological approach, the study of everyday parenting experiences, and the meanings we co-constructed with the interviewed participants (Hein & Austin, 2001; Van Manen, 1990). We intended to explore participants' parenting experiences, their children's academic performance, and the impact of Confucianism on their parenting experiences.

As to the method, both the interview and focus group were used to explore participants' experiences. We conducted 10 individual interviews with the average time of 1.5 h. We also conducted a two-hour follow-up focus group session. The semi-structured interview allowed participants to talk in depth about their experiences. All interviews were audio-recorded. The recordings were transcribed, and the Mandarin-to-English transcript translations were done by the research team. All names listed are pseudonyms.

The focus group was used to identify participants' perceptions and thoughts regarding emerging preliminary themes from the individual interview data (Milena, Dainora, & Alin, 2008; Morgan, 1993). The focus group moved beyond participants' own personal experiences and through the interaction among the group participants, the researchers had the opportunity to study the collective sense making and to learn the language and vocabulary used by the participants (Överlien, Aronsson, & Hydén, 2005). The focus group questions were structured around themes that emerged from the preliminary individual interview data analysis. The overall questions that evolved are the following:

- How does Confucianism impact on Chinese families, children, and education?
- What is your view regarding the relationship between parenting practices and academic performance, if any?

We addressed two types of research validity, including inner-subjective and intersubjective validities. Inner-subjective validity primarily relies on the researcher's self-disciplined judgment and interpretations (Lavery, 2003; Moustakas, 1994). The immediate intersubjective validity was checked during the individual interviews. Also, a considerable amount of verbatim materials was quoted to allow readers to evaluate interpretations. Furthermore, through the focus group, the researchers returned to the participants, seeking their assessment for accuracy on the meanings and essences analyzed from the individual interviews, which is intersubjective validity (Laverty, 2003; Moustakas, 1994; Yardley, 2000).

Data Analysis

We used inductive data analysis, coding notes, classifying the coding, and building patterns, and themes from the "bottom-up," considering in terms of frequency, quality, and significance of data evidence. When the patterns were named, each researcher further examined data sets related to specific categories. Interpretations were shared across the research team as a form of triangulation through which individual interpretations were confirmed or unconfirmed. Also, this inductive process included the researchers working back and forth between the themes until we established a comprehensive set of themes. We also collaborated with the interviewees interactively in the focus group session, so that they had a chance to shape the themes that emerged from the process (Creswell, 2007). A word processing program and NVivo analysis program were used to organize the analyzed texts relating to each code.

Through thematic analysis four themes emerged from the interviews. They are *Confucianism Impact*, *Demandingness as Expectations*, *Responsiveness to Academics Needs and Interests*, and *Balance Between Expectations and Responsiveness*.

Themes: Lessons We Learned

Confucianism Impact

All participants believed that Confucianism had great impact on Chinese culture and their own families. In terms of their own parenting, the influence could be seen in two areas, education and proper social behaviors. Two subthemes under Confucianism impact include *Importance of Education and The Emphasis on Proper Social Behaviors.*

Importance of Education

All participants emphasized the importance of education and the educational impact of Confucianism. Two mothers below described its importance.

> No matter what you do in the future, first thing is to do a good job in learning. If you can't even do well in this, you don't deserve to be a student. (Yuki)
>
> I think that no matter (if) one admits it or not, Chinese families or Asian families are deeply impacted by Confucius philosophies and principles.... no matter what Chinese parents' professions are, most parents, 90 % of the parents [refer to parents in China], especially concern and pay attention to the schooling and learning of their own children... Everyone [Chinese parents] would think that nothing is more important than learning [education] for a student. (Yuki)

The emphasis of education is also shown by parents' financial expenses on children's education and learning.

> There are so many parents who are willing to spend the time and money to prepare their children [in education]... No matter how poor the family is, how difficult it is, they would try all methods to give their children the best education they can provide... In addition to

school, parents are willing to spend money for their children to attend all kinds of learning enrichment programs. (Yuki)

Some parents brought up the concept of parental sacrifice to support children's education and learning. This concerns the parents' willingness to give up other things that are important to them. In the focus group, participants agreed that the phenomenon of sacrifice was common in Chinese culture. For example, one participant said,

We have a lot of Asian people (that) don't have very high education. They work in kitchen [restaurant]. They are not well educated, but for some reason, Chinese culture is...we as parents, we sacrifice our own business or our own financial situation to save money for the children even though those are lower social class. They are still willing to do that. They don't say that, oh, we don't have money and I'm not going to invest my money on your [my child's] education. I'm not goanna spend too much money on the car. I'm not goanna spend too much money on vacation. But I will spend money for my children's education. And that is first, so I would say that is sacrifice. (Celina)

The education focus is across Chinese descendant families of all backgrounds. This mother gave a concrete example of Chinese parents' wealth of knowledge about schooling and children's learning regardless of ones' professions in the context of China.

Chinese parents pay a lot of attention on school work. For example, when I was in Beijing, even the shoemaker who fix my shoes or the cleaning people, no matter what type of work parents do, when you chat with them about their children, they could talk about very specific and concrete things [about their children's education]. He could talk about his child's midterm; he would remember everything. He knows his child's learning schedule ... Everyone knows which school is good. (Yuki)

Education as Family Business

As mentioned in the conceptual framework, education is considered a family business. A child's academic success is the triumph of the family. A mother mentioned about her strategy of using family-centered incentive to encourage her children's academic success. She elaborated on providing incentives that are family occasions rather than an individual reward for the child.

When they [children] bring their report card back, we [parents] will give them something like, "Oh you did very well! So we will all eat outside or you can choose a place you would like to go; we can go together!" So that's the ... way we show them, we emphasize and really see that academic work is very important to them. So that's the way we do it. We never give them money or special prize because they have good grades. We usually just [have] family events. (Nova)

In this case, the child achieved and met the expectation. Then the whole family celebrated together.

The Emphasis on Proper Social Behaviors

As stated in the conceptual framework, Confucius emphasized that there were some key virtues that are fundamental to the greater good. While all participants emphasized the importance of education, many parents also pointed out that academics were not the only essential aspect of a child's life. Social development, specifically, proper social behaviors, was critical or even more important than academic learning for their children's development. These parents indicated that their emphases on children's social development involved principles and virtues associated with being a good human being who benefits and contributes to the larger society. Those qualities consist of both intrapersonal and interpersonal components. Here is an example from one participant concerning the importance of proper social behaviors.

> I think people talk a lot about success; then I question the word, what is success? Does that mean success in academic, if your child go to a good college or have good job? Does that mean success? I think that's not enough. I think if your child learn to be a good citizen, how to be citizen of the world, that's good success. How to deal with frustration, how to love people, how to have empathy and sympathy to people, to those who are poor, who are underprivileged. So those are another aspect of success. So that's very important. (Quincy)

Specifically, the term *ZuoRen De Dao Li* (the philosophy of *ZuoRen*) was brought up often which seemed to be crucial to these parents. To these parents, *ZuoRen De Dao Li* contains a set of virtues that a good person should have. Among the virtues mentioned, there were shared qualities, such as honesty, modesty, respect, humanity, kindness, filial piety, and positive social interaction with others. They were important attributes that these parents wanted to instill in their children and to emulate with people around them. As they examined these virtues, some participants indicated that they related to what Confucius advocated. Here are some examples from the participants.

> For my perspective, I think ZuoRen is the attitude toward different people. As [to] Confucianism, family is important; and you have to trust your friends, and those are ZuoRen ... you have to be honest, trust [trustworthy], be a valuable person. (Celina)
>
> I remember, when I first experienced the sentence about "吾日三省吾身" [Every day, I examine myself three times], I remember, [in the] beginning, I almost, you know ... Before I went to bed, I will think about what I did today. Did I fulfill? The kind of things that Confucianism (contains) ... I think .... [that we are] deeply influenced by Confucianism.... Because I deeply believe it, so I try to educate my kids; when there is an opportunity, I told them this is what you have to do... I hope my kids can have that attitude, you know, when they grow up to be useful for the society. If they have extra resources, they can help people. (Henry)

The participants seemed to think that *ZuoRen* has to be nurtured starting when children are young. A mother contrasted academic learning with learning of *ZuoRen*.

> To a child, *ZuoRen* is especially important because studying is a gradual and slow process. When children are young, they learn minimal [at school]. If one wants to study well, the learning process can continue in the future. However, regarding *ZuoRen*, if one doesn't develop appropriate [proper] behaviors, it's hard to change when one grows up. (Yuki)

The Goal of ZuoRen

During the focus group, these participants reached a consensus that the ultimate goal of *ZuoRen* is to nurture their children to contribute to the larger society and academic success is part of this. Here are some examples:

> If you behave correctly, everything should fall into place. So they [philosophers] keep saying this, "仁者不忧" [a person with benevolence, there is no worries]. . .. [based on] Confucius' idea, you are not on the island. You are always in the society, and then that is your job to behave correctly. And then you can expand your influence to society and be a contributing member to the society. (Sam)
>
> [We are] deeply influenced by Confucianism . . . if everybody does that, it would be peaceful, should be very happy society. Because I deeply believe it, so I try to educate my kids; when there is an opportunity, I told them this is what you have to do.... I hope my kids can have that attitude, you know, when they grow up to be useful for the society. (Henry)

These participants made statements to show that they were deeply impacted by Confucianism on their parenting practices. These influences included the significant emphasis on the importance of education and treating education as a family matter. Additionally, these participants placed great importance on developing virtues in their children, such as *ZuoRen De Dao Li,* the principles of proper social behaviors. They put effort into enhancing children's proper social behaviors as well as academic excellence with the intention of benefiting the greater society as a whole.

Next, we explored one of the key dimensions of parenting styles: demandingness.

Demandingness as Expectations

Academic Expectation in Tune with Child's Ability

> One way that demandingness is played out in the participants' description of their parenting was through high expectations concerning academics. All Chinese descendant participants spoke about high academic expectations for their children, and this high expectation was *always* tempered by an assessment of the child's ability level. A typical response was "I expect an A because he can do it. . . I'm not expecting too much from him. That is [his] ability." (Quincy)

Another example of this pairing concerns a parent's differentiating academic expectations of her two sons due to their ability.

> I think every child would be different. What I expect of Jack. I always expect all the A's. Um, for Zac, I think he is more [capable in] sports than academic. But I will not accept below B-. Will be B and up! This is my lowest one. (Eva)
>
> We found another occurrence in the interview data that supports the fact that the Chinese parents have higher academic expectation for their children than other ethnic groups. The two mothers who were a part of interracial couples (Chinese mother, Caucasian father) held higher expectations for their children's academics than their husbands. None of the other participants brought up this idea in the interview. For example, a mother of interracial couple said, "There's no B-. If they have B-, they get [into] trouble [from me]. (laughter) My husband, that's very different. He thinks it is okay to have B-." (Eva)

How Academic Expectation Was Conveyed

These parents communicated their academic expectations to their children in the early years and they perceived that the children internalized their expectations. One of the parenting practices that over half of the participants mentioned was that as the children grew older, not a great deal needed to be said. For example, this mother said,

> Growing up, all the way to high school, it's very much likely they followed how I wanted them ... During High School, they kind of did everything they could to achieve academically to pass SAT, ACT tests [by] themselves. But we never said anything verbally. (Hannah)

Another mother said she needed at times to remind her children to practice, but they had internalized their parents' expectations through experiences in the early years.

> I need to remind them sometimes. Sometimes they will do it by themselves (laughter), but usually I need to remind them. So I'll say, "Okay, time to practice"; they will do that.... I think maybe part from me and my husband; we have some expectations for them, but most part is from themselves. I feel like when they were young, we would just spend a lot of time with them; we do reading; we do math together. They know what we expect. (Nova)

These parents explicitly set academic expectations for their children early on through working with them on learning experiences. Through this process, the children internalized parents' expectations.

Views and Practices About Incentives

Some participants also mentioned about incentives in their parenting practices to encourage children to perform well academically. They tended to de-emphasize personal rewards because quality academic work was expected. For example, Celina said, "Yeah, I do give them [rewards] . . . Occasionally, but not a lot. . .unless they do quite good, doing well in academics is the basic. That's what you're supposed to do. [It's] not worth to celebrate."

In the previous section, *Education as Family Business*, a mother described how she used family occasions as an incentive for specific academic achievement. The parents interviewed de-emphasized extrinsic personal rewards for academic achievement. Instead, children were rewarded with family events.

Children Not Measuring Up to the Academic Expectation

> With these high academic expectations expressed by all participants, a question we had for the interviewees was the following: what if their children did not meet their high academic expectations? One parent reported "When I see the B- with Jack, I will yell at him. He doesn't have B-. [If he gets] B or B +, I'll still yell at him. I'll say, "See, you have 92 %, now you have a B+, but you only need 1 % to get you to the A-." (Eva)

Another interviewee had a different tack of what to do when a high academic achievement was not met. She said,

> I think both my husband and I knew that they [children] are at a level we think they could be the best they could and be the top of the class. We believe that they could do that. So that will be our expectation the majority of time that they should get A's.. . . Once she got a C . . . and that was okay for me because I knew she is self-motivated so she will rebound. . .. One time she said, "Mommy, you don't seem to care that much if I got a bad grade." I said, "I never worry about you, I only worry that you work too hard." So in my mind, even though I expect her to get an A and for her, she expects herself to get an A also. The fundamental expectation is if she knew where the mistakes are, why she made that mistake, then for me, it is fine. (Glenda)

This mother was more interested in the learning process than the grade for she realized that her daughter was self-motivated and had internalized the high expectations for her own academics achievement.

From these two parents, we realized that there were at least two different strategies that these parents used when their children didn't reach the high expectation. One showed displeasure through explicit one-way communication, telling the child about not measuring up; the other had two-way communication and emphasized that the child had already internalized high expectations.

Responsiveness to Academics Needs and Interests

High Responsiveness in Learning Needs

In the conceptual framework, we defined parental responsiveness as the degree parents respond to the child's needs. In the aspect of parents' responsiveness toward children's learning needs, overall, we found that these parents were responsive to their children. They supported children's academic and other achievement needs. We see one way that these parents were responsive to their children's learning was that they tempered their expectations with their assessment of the child's capability which was emphasized in the earlier section on *Academic Expectation in Tune with Child's Ability* in the theme on *Demandingness as Expectations.*

These parents described four approaches that they responded to children's needs concerning learning. They include (1) parents responding to specific knowledge about children's learning needs, (2) providing learning enrichment, (3) addressing children's special learning needs, and (4) being present in the activities that are important to the child.

Parents Responding to Specific Knowledge About Children's Learning Needs

A clear example of how these parents are responsive to the children's academic needs is the following:

> So for me, I also use the test for, to compare their abilities, because I usually work with them and their homework. One time their science score was like a B- haha . . . I basically just have to ask her to basically bring all the subjects home and basically teach her, to learn it . . . haha. Immediately, after I intervene, the scores go up. (Sam)

The other example is parents' effort to seek help through her social network. A mother felt that she did not always have the knowledge and skills to help her children. She expressed that she would respond to their children's learning needs by reaching out to find support.

> I often told my children "I can't help you!" My English is not good . . . math and reading are difficult, so I can't help. However, if you really don't understand, you need to tell me. I will find someone to help you. I would ask my friends regarding the problem they [children] don't understand and see if they [my friends] can help my children. (Yolanda)

Both parents zero in on ways to support their children when they perceived challenges in children's learning.

Providing Learning Enrichment

Also, these parents commonly provided enrichment experiences for their children to respond to their learning needs and interests, both academic and extracurricular. There are various hands-on approaches, such as paying for lessons, discussing parents' work, and developing structured academic activities for their children. The participants related the following:

> In addition to school, Chinese parents are willing to spend money for their children to attend all kinds of learning enrichment [outside the school]. (Yuki)
>
> But for math and science, he [child] just loves it naturally . . . he would ask my husband, "What are you studying? Why you came home so late tonight? Are you doing some experiments?" My husband would explain to him, and he just listen and listen, So attentive. Also, about math, also my husband just gives him a problem to solve, and he could just sit still for hours to solve the problem. (Quincy)

Five participants mentioned that they created additional homework for their children to strengthen their knowledge and skills. The following example described a mother's efforts of designing summer enrichment activities.

> Every summer, I never sent them to the camp. I don't believe in the camp because one thing, I don't want to spend the money to have somebody to do something during the summer time. The other thing is like, I plan every morning activities because every summer, I always make sure that they do academic work: to review, to read from kindergarten, so they don't have idling summer. One of the research that I know, I kind of read somewhere that students forget like 85 % [of what they had learned] during the summer. So actually, without knowing that, I have been doing these for 18 years. Every

summer, they three got up; I let them sleep a little bit later. But they get up at 8 or 9, and they just read or do worksheets. I prepare worksheets for them. They all have to sit down and do the worksheets, social studies, math, and I just lined up Monday through Friday, just like they are going to school. In the afternoon, I let them go out to swim, play basketball, play tennis ... Even if we travel to Taiwan, they still do the homework. (Hannah)

Addressing Children's Special Learning Needs

Two of the parents interviewed expressed that they were responsive to children's learning disability. One recognized the learning needs, while the second parent explained how she used different strategies to support the child's academic growth when she noticed that his child didn't like to read.

I have sympathy with my younger one because he is struggling with school; he doesn't like school; he doesn't like reading. It's not that he doesn't like it. It's just that he has a problem with it physically. He can't focus; he reads letters left and right, all sorts of things. (Hebe)

I try to find some books that he is interested in. I saw a lot of topics that he might be interested. Some of them are adventurous and some funny books. So I begin with audio book. First he just listened. It is so funny. Then I give him the book; then he can read. (Quincy)

Being Present for the Activities that Are Important to the Child

Some parents indicated that they would often attend children's activities. A mother said,

My husband devoted most of his time at his work. That's his baby. In average, he works 10–12 h a day and 6 days a week. It was always a challenge to ask him to do something outside of his work, like vacation. However, he would attend her daughter's activities, 90–100 % of the time ... Yesterday, I called him at 4 pm, and he was at the school tennis court watching his daughter's tennis game, spent an hour just to watch her play. After the game, he would go back to his work and work for another 4 h. I have to say his commitment to his children's school activities is way up there ... He participates in his daughter's school activities and music performance, 100 % of the time if he is in town ... He would drive her daughter to Barnes and Nobles to buy books, but not Traders Joes to get groceries [laugh]. (Glenda)

These parents showed their support and responsiveness toward children's activities by their presence.

In the conceptual framework, we speculated that even though Asian descendant parents tended to be highly responsive to their children's needs in relation to academic achievement. We thought that they tended to have a lower priority in supporting children's interests. If a child wants to pursue an interest not perceived by the parent as viable, will the child be supported in participating in this interest? For example, Lulu, Chua's younger daughter, wanted to learn tennis and she had to battle her mother in order to pursue this interest (Chua, 2011; Huang & Gove, 2015).

Interestingly, we found that the parents we interviewed were responsive and supportive of their children's interests. However, there often was compromise or negotiation involved.

An example is the following:

> I would try to persuade him to consider joining the orchestra ... If he decides not to, I will be sad [disappointed]. I will try to persuade [him]. Like soccer, I couldn't persuade him [child], so I provided two choices for him to choose one. He chose swimming and got rid of cello lessons which he had played for a long time. (Yolanda)

Most of our participants tended to make decisions at first by providing specific resources, for example, piano lessons. As children began to voice their preferences, the parents then negotiated with the children concerning the choices of learning activities the children would participate in. Through the process, parents allowed children to select the ones that they were interested in. Parents then supported children's decision.

Another way these parents negotiated with the children and were responsive to children's interests was by setting a priority of finishing homework before supporting the child's request.

> We will try to talk about it [children's interests] and usually fulfill their needs. Like if they would like to go to search the musical instrument, or they would like to go to a friend's party or something, we say, "That's fine." But make sure you finish all your homework first, and everything else [music instruments daily practices]." (Henry)

Another mother responded to the child's request to reinforce his already strong ability in math, even though his reading and writing were not as strong.

> An example is that he is very interested in math. This year, he requested that he want a private tutor. So I found a private tutor for him, to work on enrichment. He felt that if he didn't work on it, he would move back. He would not stay at straight A or A+ ... In the reading area, he doesn't want to do it [having a tutor]. His reading and writing are not so good. So I think, ok, as long as you have one [area] that you are interested, then go do what you want to do. (Eva)

Parents in our study were responsive to children's interests and preferences in social and academic pursuits. However, negotiation was often involved.

Balance Between Expectations and Responsiveness

The interviewed participants seemed to find a balance between the high expectations they set and responsiveness to their children's needs and interests. One mother provided a metaphor of flying a kite concerning how she had high expectations, yet nurtured her child around her needs. Here is what she said,

> You need to pull the kite, in order for the kite to fly higher. If you want them [children] to expand their abilities or strengths, you need to do a little pull, let go, then pull again. That's a metaphor for my parenting. (Hannah)

Another parent, discussing this kind of parenting practice referred to adjusting to the child's ability, said,

> I give my children space though I am strict. I try not to push them harder than they could sustain. I know my children's ability. If they could only receive a C, I will be happy, knowing that they do their best. If they could only do so much, I will reevaluate myself of what way I need to enforce and encourage them. (Celina)

Another mother related a specific example of pushing her daughter and then letting up:

> My daughter's grade went down when she was in 4th grade. I was stricter at that time. We had a huge conflict. After that incident, I didn't monitor her that much. I asked her to do few things, but she didn't work on them. She had Kumo (Japanese educational program) homework. She didn't write the homework. I was upset. I hit the chair with a [ruler]. The ruler broke "Why don't you just take few minutes to finish it? It wouldn't take you too much time." "I don't like to do it!" [the daughter said]. I was very upset and she was sad. After half an hour, I thought about it and went back and told her that "I am sorry. I shouldn't be so upset. You decide what you want to do. If you don't want to learn [Kumo], then don't take the lesson. However, you have to take care of your school work. I won't monitor you, but you need to maintain your grades. If your grades go down, we will go back to Kumo." Since then, I don't know why, she [the daughter's grade] got better. (Yolanda)

From this example, using the kite metaphor, we learned that the mother pulled hard for her daughter to achieve academically even in an enrichment activity while the daughter put up a fight. Because of this, the mother responded by letting go. Then the daughter responded in time and reached the mother's academic expectations. Several interviewed participants mentioned that learning to let go could be difficult, but they did so when great tension arouse between themselves and their children.

Below is another example where a father described the tension between the parents and their child; the response to this tension was to let go and allowed the child space to make decisions. In turn, the child continued to keep up communications about her life decisions with her parents:

> We started to feel like there is a little bit of tension [with his daughter] in senior year. That is the time she gets very busy; she has to write essays to apply for college. We had a conflict and she told us "this is my life!" I thought I need to give the kids some more space than they used to have. So when she go to college, from time to time, we feel like she is more willing to talk to us about what her life is going on and she will consult with me about her career choice, what do I think about it, what course to take. So, I think, at different age, we have to give them different space. For my older daughter, we pretty much give a hundred percent space, even though from time to time she will contact us about her career path, and we always give her suggestions, and we never say you just do this and you should do that. So I think it's more like mutual dialogue; we totally respect her choice. At the same time, I know she has her high expectations as well, but we don't put the pressure on her. Now she's totally enjoying herself, and at the same time she is maintaining some sort of academic excellence. (Henry)

We noticed that in both examples above, after the parents let go, that is, became less demanding, the children's academic performance continued to be high. It seems to us that the self-motivation instilled in the early years kicked in.

From these cases, we saw these Asian descendant parents want to help their children develop to their fullest potential and be successful. They danced around expectations and responsiveness, learned the proper timing to pull, and let go. In order to have a successful dance or kite flying, these parents continually set high expectations for their children, provided resources in tune with their child's ability, provided choices, and let go of the expectations at times.

This combination of high expectation and responsiveness to academic needs seems to be one of the factors that is associated with fostering self-motivation within the child. Over time, the children's self- motivation seems to benefit from this balance between their parents setting high expectations and pulling back from the demandingness when they saw this was needed.

Discussion and Implications

Learning through the thematic analysis, we believe that Confucianism set foundation and was a driving force for parents to foster the dimensions of high expectations and responsiveness to children's academic needs. However, the ultimate goal for these parents was to thrust their children to serve the larger society. The way these parents approached their parenting was to nurture and demand both academic and proper social behavior learnings. We found that these parents' perceptions and practices were deeply rooted in Confucianism philosophy.

We also learned that parents were highly proactive toward those two parenting dimensions, including setting academic expectations and being responsive toward children's learning needs. We found that every participant was preemptive in setting high expectations for their children starting in the early years. Overtime, their children internalized their parents' expectations. Also, we learned that parents used few extrinsic incentives and the ones they provided were family-oriented rather than individual incentives.

Concerning parents' responsiveness to children's academic needs, we found that parents were proactive in providing enrichment experiences as well as hands-on involvement in such experiences as homework and developing exercises to challenge their academic knowledge and skills. As children grew older, the parents monitored the homework less because they thought that their children were more self-motivated. In contrast to some ethnic minority parenting style literature, we concluded that parents in this study were responsive to children's interests in both social and academic pursuits. However, negotiation was often involved concerning how parents supported the children's interests. The pattern we learned was that these parents took the lead to introduce their children at the young age to many enrichment experiences, such as music, sports, and tutoring. As the children got older, parents responded to children's interests by offering choices which often involved negotiation. In one of the examples, the child decided to no longer play the cello but rather to play soccer. After discussing with the child, the mother respected his decision. The one kind of situation that the parents were not as proactive was when heightened tension occurred between parents and older children. In this case, parents tended to react by letting go and giving the child space to make their own decisions. Interestingly, the children responded by meeting their parents' expectations. This process seemed to foster self-motivation within the child. This link needs to be further investigated.

This research had limitations. The primary purpose of the qualitative research was to understand the interviewed Asian descendant parents' experiences and perceptions. It was not meant for generalization to the larger Asian descendant population, especially given the small number of participants and their demographics: these participants came primarily from the middle class, and a majority of them had graduate degrees from US universities. On the other hand, our findings illustrated the contributions in terms of rich text examples these parents offered to the research community. Moreover, although our findings provided us with a better understanding of how these Asian descendant parents invested in their children's education and learning, it is critical to keep the factors of social class and educational status in mind.

We believe that our conceptual framework can be used for future study of Asian descendant parenting practices. Our qualitative research examined how parenting practices foster children's high academic achievement. Further, we were able to describe the possible process through which Asian descendant parents foster highly motivated children, who reach high academic levels, to serve the greater society which was the main goal of the parents. This process needs to be further studied with both quantitative and qualitative approaches. In summary, our research and conceptual framework highlights the significant role that Asian descendant parents play and how their practices are deeply rooted in the philosophy of Confucianism.

References

Baumrind, D. (1971). Current patterns of parental authority. *Developmental Psychology Monographs, 4*(1, part 2), 1–103.

Baumrind, D. (1987). A developmental perspective on adolescent risk taking in contemporary America. In C. E. Irwin (Ed.), *Adolescent social behavior and health* (pp. 93–125). San Francisco, CA: Jossey-Bass.

Bronfenbrenner, U. (1979). *The ecology of human development: Experiments in nature and design*. Cambridge, MA: Harvard University Press.

Brown, E. (2013). PISA results show flat U.S. Student performance. *Education Daily, 46*(208), 1–4.

California Department of Education. (2010). *Cohort outcome summary by ethnicity for the class of 2009–2010: Statewide results*. Retrieved from http://dq.cde.ca.gov/dataquest/cohortrates/Grad-Rates.aspx?cds=00000000000000&TheYear=2009-10&Agg=T&Topic=Dropouts&%20RC=State&SubGroup=Ethnic/Racial

Chang, H. C., & Holt, G. R. (1991). More than relationship: Chinese interaction and the principle of Kuan-His. *Communication Quarterly, 39*(3), 251–271.

Chang, M. (2011). Battle hymn of the model minority myth. *Amerasia Journal, 37*(2), 137–143.

Chao, R. K. (1994). Beyond parental control and authoritarian parenting style: Understanding Chinese parenting through the cultural notion of training. *Child Development, 65*, 1111–1119.

Chao, R. K., & Tseng, V. (2002). Parenting in Asians. In M. H. Bornstein (Ed.), *Handbook of parenting* (Vol. 4, pp. 59–93). Mahwah, NJ: Lawrence Erlbaum Associates Publishers.
Chau, R. (2001). Extending research on the consequences of parenting style for Chinese Americans and European Americans. *Child Development, 72*, 1832–1843.
Chen, M., Miller, G., Wang, B. I., & Mark, M. (1996). Counseling Chinese in Republic of China. In W. Evraiff (Ed.), *Counseling in Pacific rim countries* (pp. 123–136). San Mateo, CA: Lake Press.
Cheon, H. (2006, November). *Korean Kirogi families: Separation so the children can study abroad.* Paper presented at the annual conference of the National Council on Family Relations, Minneapolis, MN.
Chua, A. (2011). *Battle hymn of the Tiger Mother.* New York, NY: Penguin Group.
Confucius, & Waley, A. (2005). *The analects of Confucius* (A. Waley, Trans.). New York, NY: Routledge.
Creswell, J. W. (2007). *Qualitative inquiry & research design: Choosing among five approaches.* Thousand Oaks, CA: Sage.
Denzin, N. K., & Lincoln, Y. S. (2005). *The SAGE handbook of qualitative research.* Thousand Oaks, CA: Sage.
Hein, S., & Austin, W. (2001). Empirical and hermeneutic approaches to phenomenological research in psychology: A comparison. *Psychological Methods, 6*, 3–17. doi:10.1037/1082-989X.6.1.3.
Flanagan, C. (2011). The Ivy delusion: The real reason the good mothers are so rattled by Amy Chua. *Atlantic Monthly, 307*(3), 92–99.
Fuligni, A. J., & Tseng, V. (1999). Attitudes toward family obligations among American adolescents with Asian. *Child Development, 70*(4), 1030.
Garg, R., Levin, E., Urajnik, D., & Kauppi, C. (2005). Parenting style and academic achievement for East Indian and Canadian adolescents. *Journal of Comparative Family Studies, 36*(4), 653–661.
Giarrusso, R., Du, F., Silverstein, M., & Bengtson, V. L. (2001). Grandparent-adult grandchild affection and consensus. *Journal of Family Issues, 22*(4), 456–477.
Grissmer, D. W. (1994). *Student achievement and the changing American family.* Santa Monica, CA: Rand Corp.
Heath, P. (2012). *Parent–child relations: Context, research, and applications* (3rd ed.). Columbus, OH: Pearson.
Hemphill, F. C., & Vanneman, A. (2011). *Achievement gaps: How Hispanic and White students in public schools perform in mathematics and reading on the National Assessment of Educational Progress (NCES 2011–459).* Washington, DC: National Center for Education Statistics, Institute of Education Sciences, U.S. Department of Education.
Hildebrand, V., Phenice, L. A., Gray, M. M., & Hines, R. P. (2008). *Knowing and serving diverse families* (2nd ed.). Upper Saddle River, NJ: Merrill.
Hsai, H. C., & Scanzoni, J. (1996). Rethinking the roles of Japanese women. *Journal of Aging Studies, 27*, 309–329.
Hsia, J. (1988). *Asian Americans in higher education and at work.* Hillsdale, NJ: Lawrence Erlbaum.
Hsin, A., & Xie, Y. (2014). Explaining Asian Americans' academic advantage over Whites. *Journal of PNAS, 111*(3), 8416–8421.
Huang, G. H. C., & Gove, M. (2012). Confucianism and Chinese families: Values and practices in education. *International Journal of Humanities and Social Science, 2*(3), 10–15.
Huang, G. H. C. & Gove, M. (2015). Asian parenting styles and academic achievement: Views from eastern and western perspectives, *Education, 135*(3), 389–397.
Husbands, C. (2012). What we need to learn from the Tiger Mothers. *Times Educational Supplement, 4977*, 48–49.
Kao, G. (1995). Asian American as model minorities? A look at their academic performance. *American Journal of Education, 103*, 121–159.

Kao, S. C. (2006, November). *Filial piety of the Confucian doctrine in relation to family relationship in Taiwan.* Paper presented at the annual conference of the National Council on Family Relations, Minneapolis, MN.

Kim, S. (2006, November). *The experience of Korean parents' educational zeal.* Paper presented at the annual conference of the National Council on Family Relations, Minneapolis, MN.

Kohler, M., Kilgo, J., & Christensen, L. M. (2012). The Tiger Mom phenomenon. *Childhood Education, 88*(1), 69–71.

Kohler, M., Aldridge, J., Christensen, L. M., & Kilgo, J. (2012). Tiger Moms: Five questions that need to be answered. *Childhood Education, 88*(1), 52–53.

Konstantopoulos, S. (2009). The mean is not enough: Using Quantile regression to examine trends in Asian-white differences across the entire achievement distribution. *Teachers College Record, 111*(5), 1274–1295.

Laverty, S. (2003). Hermeneutic phenomenology and phenomenology: A comparison of historical and methodological considerations. *International Journal of Qualitative Methods, 2*(3), 1–29.

Leung, N. A., Wong, S., Wong, I., & McBride-Chang, C. (2010). Filial piety and psychosocial adjustment in Hong Kong Chinese early adolescents. *Journal of Early Adolescence, 30*(5), 651–667. doi:10.1177/0272431609341046.

Lien, L. Y. (2006, November). *Creating a successful educational path for children: The experience of Taiwanese mothers.* Paper presented at the annual conference of the National Council on Family Relations, Minneapolis, MN.

Louie, V. S. (2004). *Compelled to excel: Immigration, education, and opportunity among Chinese Americans.* Stanford, CA: Stanford University Press.

Lu, L., & Shih, J. B. (1997). Sources of happiness: A qualitative approach. *The Journal of Social Psychology, 137*(2), 181–187.

Maccoby, E. E., & Martin, J. A. (1983). Socialization in the context of the family: Parent–child interaction. In E. M. Hetherington & P. H. Mussen (Eds.), *Handbook of child psychology* (Socialization, personality, and social development, Vol. 4, pp. 1–101). New York, NY: Wiley.

McLoyd, V. C., & Smith, J. (2002). Physical discipline and behavior problems in African American, European American, and Hispanic children: Emotional support as a moderator. *Journal of Marriage & Family, 64*(1), 40–53.

Milena, Z., Dainora, G., & Alin, S. (2008). Qualitative research methods: A comparison between focus-group and in-depth interview. *Annals of the University of Oradea, Economic Science Series, 17*(4), 1279–1283.

Miller, G., & Yang, J. (1997). Counseling Taiwan Chinese in America: Training issues for counselors. *Counselor Education& Supervision, 37*(1), 22–35.

Moneta, G. B. (2004). The Flow model of intrinsic motivation in Chinese cultural and personal moderators. *Journal of Happiness Studies, 5*, 181–217.

Morgan, D. L. (1993). *Successful focus groups: Advancing the state of the art.* Newbury Park, CA: Sage.

Moustakas, C. (1994). *Phenomenological research methods.* Thousand Oaks, CA: Sage.

Ninh, K. (2011). Advice on how not to misread the Tiger Mother. *Amerasia Journal, 37*(2), 123–129.

Noguera, P. (2008). *The trouble with black boys and other reflections on race, equity, and the future of public education.* San Francisco: Wiley.

Noguera, P. (2012). Saving black and Latino boys. *Phi Delta Kappan, 93*(5), 8–12.

O'Donnell, J., & Kirkner, S. L. (2014). The impact of a collaborative family involvement program on Latino families and children's educational performance. *School Community Journal, 24*(1), 211–234.

Överlien, C., Aronsson, K., & Hydén, M. (2005). The focus group interview as an in-depth method? Young women talking about sexuality. *International Journal Of Social Research Methodology, 8*(4), 331–344. doi:10.1080/1364557042000119607.

Pang, V., Han, P. P., & Pang, J. M. (2011). Asian American and Pacific Islander Students: Equity and the achievement gap. *Educational Researcher, 40*(8), 378–389. doi:10.3102/0013189X11424222.
Park, M., & Chesla, C. (2007). Revisiting Confucianism as a conceptual framework for Asian family study. *Journal of Family Nursing, 13*(3), 293–311.
Peng, S. S., & Wright, D. (1994). Explanation of academic achievement of Asian American students. *Journal of Educational Research, 87*(6), 346.
Poon, O. A. (2011). Ching Chongs and Tiger Moms: The "Asian Invasion" in U.S. Higher education. *Amerasia Journal, 37*(2), 144–150.
Schott Foundation. (2010). *Yes, we can: The 2010 Schott 50-state report on public education of black males*. Cambridge, MA: Author. http://blackboysreport.org
Shek, D. T. L., & Chen, L. K. (1999). Hong Kong Chinese parents' perceptions of the ideal child. *The Journal of Psychology, 87*, 787–793.
Slote, W. H., & De Vos, G. A. (1998). *Confucianism and the family*. Albany, NY: State University of New York Press.
Steinberg, L., & Levine, A. (1997). *You and you adolescent: A parent's guide for ages 10 to 20* (Rev ed.). New York, NY: Simon & Schuster.
Sung Seek, M., & Joohi, L. (2009). Multiple predictors of Asian American children's school achievement. *Early Education & Development, 20*(1), 129–147. doi:10.1080/10409280802206635.
Tu, W. M. (1998). Confucius and Confucianism. In W. H. Slote & G. A. DeVos (Eds.), *Confucianism and the family* (pp. 3–36). New York, NY: State University of New York Press.
USA.com demographic. (2010). *Cuyahoga County population and races*. http://www.usa.com/cuyahoga-county-oh-population-and-races.htm#PopulationbyRaces
USA.com demographic. (2013). *Cuyahoga County population and races*. http://www.usa.com/cuyahoga-county-oh-population-and-races.htm
Van Manen, M. (1990). *Researching lived experience: Human science for an action sensitive pedagogy*. Albany, NY: State University of New York Press.
Wang, G. (2011). On Tiger Mothers and music moms. *Amerasia Journal, 37*(2), 130–136.
Yang, K. S. (1981). The formation and change of Chinese personality: A cultural-ecological perspective. *Acta Psychologica Taiwanica, 23*, 39–55.
Yardley, L. (2000). Dilemmas in qualitative health research. *Psychology and Health, 15*, 215–228.
Yong, Z., & Wei, Q. (2009). How good are the Asians? *Phi Delta Kappan, 90*(5), 338–344.

Chapter 4
The Chinese Classic Philosophy of Learning and Teaching: Exploring Insights of the "Essay on Learning"

Yenming Zhang and Yongxiao Bai

Introduction

Backgrounds

"The Essay on Learning" (*Xue Ji* 学记) is one of the passages of *The Rites* (*Li Ji*《礼记》), the earliest scholarly record in the Chinese history on the philosophy of education. *The Rites* is the primal collection in the ancient times on Chinese institutions, including decrees and regulations, which comprises 49 passages in 20 volumes, and has been used as the authoritative textbook of rites and regulations for thousands of years in China. Compiled by Confucian scholars, the Essay on Learning records systematically Chinese classic philosophy with a predominant feature of embodying the Confucian philosophy on education and has been influential ever since the *Spring and Autumn Period* (c. 770–476 B.C.).

In this chapter, we examine the fundamental perspectives of educational philosophy and the key aspects of the Confucian model, i.e., the rationale of the state leader as educator, the purpose of education, the general principles of curriculum management, the guiding principles of teaching and learning, and the management philosophy of schools. In our exploration, we will emphasize the general guidance for educational management, elaborate the principle of building reciprocal and beneficial relationships, and highlight the value of respect for the teacher and the value of learning. Empirical examples are provided from our respective profes-

Y. Zhang (✉)
National Institute of Education, Nanyang Technological University, Singapore, Singapore
e-mail: yenming.zhang@nie.edu.sg

Y. Bai
Faculty of Teacher Education and Mathematics, Beijing Institute of Education, Beijing, China
e-mail: baiyxiao@163.com

M.S. Khine (ed.), *Science Education in East Asia*,
DOI 10.1007/978-3-319-16390-1_4

sional experiences, for the purpose of readers gaining a broader understanding of the pertinence of the Confucian model in the modern sense.

The Leader as Educator

The most conspicuous phenomenon throughout the Chinese history is that the state leaders were learned scholars and educators who advocated the value of education and manifested such value at their paramount position.

The rationale for the leader as educator is convincingly stated in the first paragraph of the Essay on Learning, which reads:

- Being able to inspire thinking and to seek for kindness may help one to become well known within a limited community, but such ability is not powerful enough to move the masses in a large scale.

  (发虑宪, 求善良, 足以謏闻, 不足以动众)

- Teaching a group of wise men and embodying what is held as importance may move the people, but such endeavor is not powerful enough to cultivate the masses in their internal locus.

  (就贤体远, 足以动众, 未足以化民)

- In order to cultivate his people and to form up new conventions, the only way for him is through learning initiated and undertaken by the leader himself.

  (君子如欲化民成俗, 其必由学)

This reveals the Chinese traditional perception that the leader must be a learned scholar himself, which enables him to mobilize his people. The best examples given by Confucius are the leaders from the first dynasty onward in the history of China, who are *Yu* (禹, 夏 Xia dynasty), *Tang* (汤, 商 Shang dynasty), *Wen Wang* (文王, 西周 Western Zhou dynasty), *Wu Wang* (武王), *Cheng Wang* (成王), and *Zhou Gong* (周公; see the Essay of *Li Yun* 礼运 in Liu, p. 278, 2006). It has become a granted idea that learning itself certifies a person's teaching and qualifies the person as a leader. Hence, two most prevailing values stem from the Chinese culture, namely, "education" and "learning."

Institutionalizing Education

Being aware of the importance of learning, ancient Chinese leaders felt it imperative to structure systematically nationwide education. The original thinking was thus generated on institutionalizing education, which is specified in the second paragraph of the Essay on Learning:

- Without learning, a person would not know the way things work. Henceforth, ancient leaders established the country and reigned their people through the approach of setting teaching and learning as the priority of governance.

  (人不学, 不知道。古之王者建国君民, 教学为先)

- The various formats of teaching took place as (1) private learning centers known as *shu* (塾), to provide fundamental tuition in the neighborhood; (2) primary schools known as *xiang* (庠), to provide intermediate schooling in the community; (3) vocational schools known as *xu* (序), to provide skill training and advanced learning in professional settings; and (4) the educational institution known as *xue* (学), to provide higher learning at the capital of the state, namely, "Great Learning" (大学).

  (古之教者, 家有塾, 党有庠, 术有序, 国有学)

- According to the statement in the Passage of *Yue Ming*, one must always keep in mind to persist in continual learning.

  (《兑命》曰: "念终始典于学")

An insight of this primitive thought tells us that the leader as teacher would equip him with the authority of governing; that provisions of places for learning would facilitate learning to take place, both in private and public formalities; and that at the national level, there must be a central institution for learners to engage in advanced study, which is equivalent to "higher education" in the modern sense. A philosophy of learning from *Yue Ming* (Passage 3) in the *Book of History* (《尚书 • 商书》说命下) is quoted in the third point of the above statement, which is the nature of learning as a continual process (see Li & Wang, p. 177, 2007).

The main themes of this chapter involve:

- The general guidance for educational management
  - The principle of readiness
  - The ethos across the school
  - The desired qualities and achievements
- The philosophy of teaching and learning
  - The principle of building reciprocally beneficial relationships
  - The general principles for curriculum management
  - The then issues in instructions
- The pedagogical principles
  - The principles of instructions
  - The principle of teacher facilitation
  - Understanding students
  - The methods of inquiries
  - The principle of balanced learning

- The quality teacher
  - The value of respect for the teacher
  - The standards set for quality teachers
  - The principle of congruence

The General Guidance for Educational Management

The Principle of Readiness

In managing institutional operations for effective educational processes, the principle of readiness is set for the administrators, the teachers, and the learners:

- The administrators and teachers of educational institutions must get all ready before the learners are admitted.
- The learners must make up their minds of what they intend to achieve prior to the start of their learning journey.

  (凡学, 官先事, 士先志)

The above statement is from a managerial perspective that the administrators of educational institutions must ensure that they are prepared in all aspects, from the environment, the facilities, and the curriculum to the pedagogies/learning objectives. Yang Kuan, a famous historian, in his book entitled *The History of Western Zhou Dynasty* (2003), depicts the then educational site which was on top of a hill surrounded by water, and the hill looked like a jade (丘如璧). In the ancient Chinese perspective, water stands for danger or hurdle. With water encircling the educational place, the learners would have a sense of safety, hence being able to concentrate on their study. Schools in the ancient times were called *Pan Gong* (泮宫), while the institution at the imperial level was *Bi Yong* (辟雍), named after a poetic description of a learning environment (辟雍泮水) in Western Zhou dynasty. The latest one is still standing in the Imperial Institution of Inspection (also, of Great Learning 国子监) in Beijing, built for the emperors of *Qing* dynasty (清代) to conduct imperial addresses on the Chinese classics to the élite of scholars nationwide.

The first task for the teachers is to check on the learners' setting of determinations and to aspire those who have not made up their minds yet. Aspiring students' determination is an area the teachers exercise their influence on students for a mastery of what they learn. This is in line with the humanistic psychological view which holds that people have the inner drive for a search of the direction to advance to. It is a will of achievements, accompanied by courage to realize the will.

There is a close correlation between the quality of determination, i.e., the feasibility of it, and the age cohort. In the Confucian teachings, the determination that is realistic appears at the age of 15. Confucius himself reflected that his determination of quality was made when he reached 15 ("吾十有五而志于学," in the Passage of *Wei Zheng* 为政, *Confucius Analects*《论语》; see Zhang, Xiong, & Ma, p. 21, 2014).

The Ethos Across the School

In order for institutional functions to be in place, ethos setting and order observation were established since the early stages of educational history in China. In the Essay on Learning, the "Seven Orders to Observe" (教之大伦) are specified for the forming up of a prevailing ethos across the school:

1. Before classes start, the students must be dressed up with formal attires and go through a memorial ritual of esteeming the sages and teachers of the older generations, to show their respect.

   (大学始教, 皮弁祭菜, 示敬道)
   This is the earliest record of the educational values of "respecting teachers" and "paying reverence to the rationale of learning."

2. The students are asked to read the three poems from the *Book of Poems* (《诗经》) to remind themselves of their purposes and determinations of learning.

   (宵雅肄三, 官其始)
   It is understood that the three poems are *Lu Ming* (鹿鸣), *Si Mu* (四牡), and *Huanghuang Zhe Hua* (皇皇者华) in the Volume of *Xiaoya* from the *Book of Poems* (《诗经 • 小雅》; see Ma, pp. 206–209, 2006). The themes of the poems are the necessity of officials observing rituals and ministers remembering their parents and the hardships along the learning journey.

3. Hearing the drums, the students start to open their school bags and begin their learning in an obedient and respectful manner.

   (入学鼓箧, 孙其业)

4. The teacher holds two kinds of canes (*jia* 夏and *chu* 楚), which are meant for creating an awe-inspiring atmosphere and for the students to be discipline oriented.

   (夏楚二物, 收其威)

5. Officials, before attending memorial ceremonies, do not inspect the schools in order not to interrupt the learning process.

   (未卜禘, 不视学, 游其志)

6. The teacher checks on the students' learning but holds back from providing feedback until teachable moments turn up. This is to let the students play an active role in their learning, e.g., taking their initiative according to their own learning rates.

   (时观而弗语, 存其心)

7. Younger learners are encouraged not to ask questions frequently. This is because they have to follow the learning process step by step instead of overstepping

certain stages. In the modern sense, managing various age cohorts in the same class requires the teacher's ability of conducting differentiated instructions.

(幼者听而弗问, 学不躐等)

The features of the above set of seven orders are as follows: (1) education is provided in a formal setting, from the ritual, attire, and atmosphere to manner; (2) learning is a value-driven practice – learners value study and respect their teachers – and (3) through attending procedural formalities and following well-designed lessons, the students build up good learning habits and essential characters of being respectful and well disciplined and valuing learning.

The Desired Qualities and Achievements

Along their learning journey, students are expected to make progress and achieve desired outcomes at two stages. Such expectations are articulated at the 7-year stage of "Intermediate Achievement" and the 2-year stage of "Great Achievement," respectively.

Intermediate Achievement (小成)

During the seven years of learning, students' gradual progress is monitored and the essential qualities they possess are examined, as shown in Table 4.1.

This stage is the characteristic of (1) regarding knowledge of the classics as the essentials; (2) emphasizing a building up of the personal qualities as independent thinking, people orientation, teacher affiliation, and friendship among scholars; and (3) taking "friendship" as a value. The qualities of the people one is close to, especially the teachers and scholars, may determine the qualities one grows to possess.

Table 4.1 The qualities to be examined at Stage 1

Development	Quality
Year 1	Knowledge of the classics
	Determinations
	(离经辨志)
Year 3	Admiration and appreciation of learning
	Socialization and people orientation
	(敬业乐群)
Year 5	All-around learning
	Affiliation with the teacher
	(博习亲师)
Year 7	Independence in learning
	Friendship with learned scholars
	(论学取友)

Great Achievement (大成)

At this stage of year 9, the components of qualities examined are as follows:

- Being knowledgeable, with the capability of categorizing the knowledge (知类通达)
- Being persistent, with a good remembrance of the teacher's guidance (强立不反)

At the end of this stage, students are expected to be able to explore new learning, based on the knowledge accumulated, to possess a sound mind, and thus to think independently. One characteristic is worth mentioning, that is, students' memory of their teachers' instructions as guidance of their progress and accomplishments all along. This has been typical in the Chinese learning culture.

In our opinion, with regard to the successes and achievements of students in East Asia, where certain values about education are shared, the above qualities of learning and teaching are referential. Taking some Special Assistance Plan (SAP) schools in Singapore, for example, the government identified this type of schools preserving the Chinese medium in 1979 to promote the learning of Chinese language and culture. One of the SAP schools upholds this notion of "persistence" to have continuous improvement to ever higher levels. The principal emphasizes to the teachers and students to strive to do even better, no matter how well the school has performed. In doing so, the school looks out to learn from the best practices in other local schools and industries and even go beyond Singapore in collaboration with institutions in Australia, Europe, China, and the USA.

With regard to the notion of "a good remembrance of the teacher's guidance," which is closely related to "respecting teachers" (see section The Ethos Across the School), the aforementioned principal has an understanding of it with new implications – "it does not mean blind obedience, as is the past." He encourages the teachers to gain respect from their students by their words and deeds, namely, by providing quality guidance in their interactions with the students. We believe that it is the teachers' embodiment of the spirit of "persistence" and exemplification of "quality guidance" that will transform their students' diligence and conformity to yield even higher performance.

Great Learning (大学)

Beyond the second stage, those students who are capable to continue to advance reach the tertiary level, namely, Great Learning. The way of Great Learning is demonstrated in the capabilities of:

- Cultivating people
- Improving the conventions
- Influencing younger generations coming from near and far to admire and learn wholeheartedly

(大学之道, 化民易俗, 近者说服, 远者怀之)

Such qualities of the scholarship are reflected in the five components of the Confucian teachings: (1) gaining a wide range of knowledge, (2) asking detailed questions, (3) reasoning thoughtfully, (4) distinguishing things clearly, and (5) practicing genuinely (博学之, 审问之, 慎思之, 明辨之, 笃行之; see Passage 20 of *The Doctrine of the Mean*《中庸》, in Xu, p. 35, 1991). Cai Yuanpei, one of the renowned scholars of the twentieth century, highlighted this Confucian teaching as the motto for Peking University during his term of presidency.

The Philosophy of Teaching and Learning

The Principle of Building Reciprocally Beneficial Relationships

One of the cores of the philosophy of teaching and learning in the Essay on Learning is the emphasis on the rationale of the mutual benefit relationship between teaching and learning. The highlights of the third paragraph of the Essay read:

- At the sight of a good dish, you will never know its delicious flavor without tasting it.
- At the presence of a perfect argumentation, you will not get its essence without learning to gain the insights of it.

(虽有嘉肴, 弗食, 不知其旨也; 虽有至道, 弗学, 不知其善)

- The more you learn, the more you realize that what you have learned is not enough.
- Only when you teach can you notice that your knowledge is insufficient.
- Hence, there is a reciprocally beneficial relationship between learning and teaching.

(学, 然后知不足; 教, 然后知困。知不足, 然后能自反; 知困, 然后能自强。故曰: 教学相长)

The reciprocally beneficial relationship of learning and teaching is one of the powerful cores of the Confucian philosophy which displays the heart of learning and teaching. This is especially so when it refers to one person's capability to learn and to teach. An example below illustrates the benefits of a person with such dual abilities. The case below elucidates such a relationship.

Case 1 Siming Zhang, a renowned first-rate teacher of mathematics at the Affiliated School of Peking University, endorses a philosophy of "students growing together with their teachers." In carrying this out, he upholds a belief that instead of regarding the teacher, the source of knowledge, as a pail of water (一桶水), the teacher himself should be able to open up a fountainhead (一眼泉). In order for his students to be interested in, earnestly like, and learn mathematics well, he has experimented a new method of Mathematical Modeling Teaching since 1993 (Zhang & Bai, 2003).

The mathematical projects his students selected were often related to the frontiers of the subject, which entailed cross-disciplinary knowledge, thus requiring both the supervising teacher and the students to search for new data and to expand perspectives for new knowledge of the latest development in the pertinent domains. During the years of experimentation, Mr. Zhang undertook new learning and earned a master's degree and then a PhD, becoming a well-known educational specialist all over China in teaching mathematics. Meanwhile, his students had their interest and abilities in mathematics elevated tremendously, to yield illustrious achievements.

The Philosophy of Learning in Teaching and of Teaching in Learning

The Essay quotes the *Book of History* (《尚书》), saying "being able to teach what one has learned helps one to reach half of the mastery level of the knowledge" (《兑命》曰: "教学半"). The character "教" (in the old form, "斅") when pronounced as *xiao* means "teaching" and when pronounced as *xue* means "learning." Found in *Yue Ming* (Passage 3) of the *Book of History* (《尚书 • 商书》说命下), this has been the philosophy since the *Shang* dynasty (c. 1250 B.C.).

Giving dual sounds and meanings of the character reveals ancient Chinese thinkers grasping the crux of effective learning and good teaching: as a learner, one demonstrates what he has learned by teaching his peers. As a teacher, one carries on new learning, to deliver good teaching. This has become a principle ever since the *Shang* dynasty, a tremendous contribution to learning and teaching theories.

Case 2 Yangsi Middle School in Jiangsu Province adopts an instructional model of "learning followed by teaching and exercising within the class." The students are given opportunities to teach their classmates. This is practiced in the way that the students do self-study first, while the teacher takes a facilitative role by providing guidance on the problems surfacing from the self-study and on the pitfalls appearing in their exercises. Those students who have understood the key concepts are asked to teach those who are not there yet. During this process, the teacher supervises and then assumes the functions of assessing, complementing, and moderating.

This method of "teaching in learning" is not only carried out in class but also implemented outside the class in a "buddy-buddy" activity where a high achiever and a low achiever are paired up. The difficulties faced by the students with poor performance are effectively dealt with, and meanwhile the students with high abilities are geared up. The rationale behind simply tells that the student teacher, in order to be helpful to the others, must have understood and be able to articulate the piece of knowledge. Hence, it is a "forcing-and-reinforcing" process.

The General Principles for Curriculum Management

Principles are set for teaching at the tertiary level to ensure Great Learning. Firstly, it must go with seasons; secondly, teaching must be based on a regular curriculum; and thirdly, tasks on study must be assigned to students to complete during their vacations (大学之教, 时; 教必有正业; 退息必有居).

Principles of logical connectedness between two seemingly separate types of contents are set for students as guidance of correlative learning in the Essay on Learning:

1. Without learning verses for singing, one cannot play with a musical instrument.

   (不学操缦, 不能安弦)

2. Without the skills of using various types of metaphor, one cannot write poems.

   (不学博依, 不能安诗)

3. Without basic life skills, one cannot practice appropriate rites.

   (不学杂服, 不能安礼)

4. Without an interest in the "six competences" (六艺, in Confucianism), one cannot enjoy learning.

   (不兴其艺, 不能乐学)

Case 3 "Two-Basic" Mathematics Teaching (双基教学) is the characteristic of Chinese mathematical teaching, which emphasizes mastery of basic knowledge and basic skills, while attending to the inherent logic of the curriculum.

Take the learning of quadratic equation of one unknown, for example, which entails the skill of integral operation as the prerequisite. Only on this base can students gradually form up the knowledge chain of the concept of equation, formula method of solving equation, and Vieta's Theorem. Through numerous drills on variants, students learn how to solve problems of diverse quadratic equations of one unknown, including the ability of comprehending the equation of $x^2 + mx + 2 = 0$ with a parameter. In the process of exploring the relations between the roots and coefficient m of an equation, students are able to construct the knowledge networks of quadratic equation of one unknown.

This set of principles is the first record in the Chinese history of education, to rationalize the connections of contents to be learned, with the first of each of the four pairs as the prerequisites. The principles reveal logical linkages between basic life skills with cognitive acquisition and encourage learners to go in for correlative learning. The Essay further suggests a general method of learning and an attitude of learning, if a learner means to present scholarly work:

- He has to concentrate wholeheartedly, cultivate himself, take a break, and go on touring sometimes (君子之于学, 藏焉, 修焉, 息焉, 游焉).

- He has to take an attitude of humility and aptitude of industry, which will lead to his scholarly attainments, as quoted from the Passage of *Yue Ming* (《兑命》曰: "敬孙, 务时敏, 厥修乃来," refers to the phrase of *weixue xunzhi* 惟学逊志 in the *Book of History*).

The curriculum management in a regular institution makes features of (1) scheduling seasonable contact time; 2) providing well-structured programs; (3) pointing out correlative learning, including interest, so as to enjoy learning; and (4) relaxation after a heavy-going period of study. Meanwhile, such personal characters of diligence and humility are expected to be built up, to ensure the learners' success in their learning journey. These characters have become values in the Confucianism.

Applied to science education, these principles may help students to go through more constructive learning experiences. As observation, inquiry, experimentation, and process are central to science learning, schools may also emphasize dialectical thinking and correlative learning in the delivery of their curriculum, for students to build up balanced learning habits and desired personal characters.

"Current" Issues in Instructions

The Essay on Learning addresses some issues at that time existing in instructions which are ineffective for learning and unfavorable for students "enjoying learning":

- Currently, there are teachers who instruct by reading the text on the bamboo slips.

  (今之教者, 呻其占毕)

- They try to give the students a hard time by asking difficult questions, and their concern is only on completing the lesson plan.

  (多其讯言, 及于数进)

- They do not make efforts; the teaching is not authentic; and they do not care much whether they make themselves understood.

  (不顾其安, 使人不由其诚, 教人不尽其材)

- Inevitably there are numerous errors in their instructions, and the objectives are merely opposite to the students' purposes.

  (其施之也悖, 其求之也佛)

- The above unfitting attitudes and inappropriate pedagogies will consequently result in students' suffering from the learning processes.
- As such, the students are unable to comprehend the contents and turn to dislike their teachers simply because of the painful experiences without gaining any benefit.

  (夫然, 故隐其学而疾其师, 苦其难而不知其益)

- They forget upon completion of the term what they have been taught.

  (虽终其业, 其去之必速)

These results are because of the teachers' inappropriate and ineffective instructions (教之不刑, 其此之由). As aforementioned, "enjoying learning" is a principle and a value in the tradition of Chinese education. Lessons drawn from these issues are that teachers should not teach the books but teach the students and that the ultimate purpose is for the benefits of the students. In dealing with these issues, teachers' work ethics, i.e., their authentic alignment with the contents and their willingness to engage with the students, counts the most.

The Pedagogical Principles

The Principles of Instructions

The Essay on Learning sets out four principles of instructions for the Great Learning (大学之法):

1. Be preemptive. In planning their lessons, teachers anticipate possible misconceptions from certain students and are prepared to handle when this occurs in class or try to avoid students' misconceptions by planning for precise language to articulate the key concepts of the lesson.

   (禁于未发之谓豫)

Case 4 Ping Yu, an outstanding teacher in Beijing Primary School, is fully aware of the pitfalls her students tend to make, that is, adding up the numerator and the denominator in the calculation of addition, as in $\frac{3}{4} + \frac{4}{5} = \frac{7}{9}$. Before the lesson on addition of fractions with different denominators, she probes the cause of such misconception and realizes that her students are easily affected by the negative transfer of integer additive operations. With this knowledge, she designs a series of exercises for the purpose of helping her students to understand correctly through connections and comparisons:

$12 + 24 = 36$ Integers can be added up straightly, because digital alignment makes counting units the same.

$0.5 + 0.8 = 1.3$ Decimals can be added up straightly, because decimal point alignment makes counting units the same.

$\frac{1}{5} + \frac{3}{5} = \frac{4}{5}$ For fractions with the same denominator, numerators can be added directly because the fractional units are the same.

$\frac{1}{3} + \frac{1}{2} = \frac{2}{6} + \frac{3}{6} = \frac{5}{6}$ For fractions with different denominators, numerators cannot be added directly because the fractional units are different. Each fraction must be first converted to an equivalent fraction with the same denominator.

2. Be appropriate: catching teachable moments. As the momentum goes in the class, teachers should be observant to catch psychological moments by reading the students' responses to the instructions and to inspire further thinking or to clarify their confusions.

   (当其可之谓时)

Case 5 H is a secondary 1 student in a school in Beijing. Lacking interest in any subject, he produces poor performance. Once he appeared a different person in a lesson of mathematics, presenting extraordinary enthusiasm and excellent mathematical thinking. His study took a turnaround afterward.

What happened then to result in such a change? It was like this: the topic of Mounting invited his attention *a*nd he was fascinated with the idea of exploring the methods of laying plane polygon to a plane graph. The teacher started with an approach of laying bricks on the floor. As soon as the lesson started, H surprisingly answered the teacher's first question. This might have been occasional, but the teacher realized his interest and potential in further exploration. Taking this teachable moment as a turning point, the teacher provided him with opportunities to demonstrate his independent thinking, assessed his results positively, and encouraged him with affirmative words. It was the joyful feeling stemming from active thinking and participation, with the teacher's timely reinforcement, that enkindled H's yearning to learn (Bai & Gao, 2013, pp. 16–20).

3. Follow progressive sequences. In students' cognitive development, the earlier steps are the preconditions of the next ones. The implementation of the curriculum should follow such a progressive sequence, to avoid overstepping students' learning stages. This is pertinent to both the contents and students' development from basic to higher orders of thinking. According to the theory of the nature of human learning, the growth-readiness model, "certain growth patterns must occur before learning can be beneficial" (Gredler, 1992, p. 128).

   (不陵节而施之谓孙)

Case 6 Primary pupils' thinking is characteristic of image and visual. They naturally encounter difficulties with abstracts and deductive reasoning. For instance, when learning the concept of representing numbers with alphabets, they are often puzzled. This is because of the duality of an alphabet which designates both certainty and arbitrariness. Taking a letter as an abstract form of a number, and also perceiving the arbitrary feature of a letter representing certain value, involves pupils' arithmetic thinking to be transformed to algebraic thinking. Obviously, this is a difficult task (Bai, 2012, pp. 15–26).

In the book entitled *Mathematics Curriculum Standard of Compulsory Education* (2011 Edition), it is stipulated that the determination of pupils' ability to learn the concept of representing numbers with alphabets is set as "being able to use alphabets to represent numbers in a particular scenario." The textbooks of various versions, compiled on the principle of spiral development, arrange this part of contents to be learned during the second term (of primary 4–6), to cater for pupils' comprehension.

4. Encourage sharing among students. Learning is not only a behavior of individual listening to the teacher. Rather, it will be more beneficial when students share among themselves what they are receiving from the teacher, how they are interpreting the contents, and where puzzles may possibly surface.

   (相观而善之谓摩)

The above four pedagogical principles, when followed closely, are the derivations good teaching stems from (此四者, 教之所由兴). Perceiving the principles from the negative side, one can see the consequences if the principles are not in place, as reiterated in the Essay on Learning:

1. After what is unexpected forms up, like possible misconceptions stated in the above first principle, it will be difficult to overcome.

   (发然后禁, 则扞格而不胜)

2. When the high time elapses, like the teachable moment in the above second principle, even greater diligence cannot help for effective learning.

   (时过然后学, 则勤苦而难成)

3. If learning occurs as collection of odds and ends, rather than following a progressive sequence, like what is pointed out in the above third principle, it will end up as disorder and chaos in the mind.

   (杂施而不孙, 则坏乱而不修)

4. If learning takes place in isolation instead of learning together with friends who share and learn from each other, like the above fourth principle, one will obtain limited knowledge and scanty information.

   (独学而无友, 则孤陋寡闻)

5. Making friends with bad people, one will overlook the earnest words from his teacher.

   (燕朋逆其师)

6. Bad habits will direct one's study to a wasteland.

   (燕辟废其学)

These six kinds of consequences serve as admonitions for teachers to remind their students to refrain from, because they are the factors of failures of education (教之所由废). The Essay on Learning recapitulates at this point that a gentleman who is fully aware of the derivations of good teaching, and of the factors of failures, may qualify as a *ren shi* (人师), which literally means "human teacher" (now called "teacher"). *Ren shi* denotes explicitly a gentleman undertaking teaching qualifies this title if he is able to cultivate the younger generations to grow. Therefore, the essence of education has long been a characteristic of being humane.

The Principle of Teacher Facilitation

In the process of guiding the younger generations to grow, the teacher is suggested to take the role of facilitator with enlightening and inspiring skills (教喻). The approaches to taking the role and practicing the skills are specified in the Essay on Learning:

- Guide without forcing (道而弗牵).
  This approach encourages learners' active adaptation to the teacher's guidance and helps to build a harmonious relationship between the teacher and the learners, thus avoiding learners' possible passiveness.
- Be forthright instead of coercing (强而弗抑).
  This way suggests an engaging manner from the teacher, with affirmation in correcting misconceptions or misconduct from certain learners, to ensure acceptability and to avoid possible resistance.
- Inspire rather than providing all the answers by frontal talking all the while (开而弗达).
  This way requires the teacher to be precise and concise in explaining the key concepts during the period of lesson, leaving some time for the learners to conceive (see Zhang, Xiong & Ma, p. 20, 2014).

Only by reaching the effects of rapport, acceptance, and conception can a teacher claim that he is an enlightening and inspiring facilitator. The psychological logic lying in the trilogy is that when a harmonious relationship is built, the students tend to accept the teacher; and while following the teacher, the students are allowed or given time to think, they will then take initiative to internalize the instructional materials; thus, real learning takes place. Forcing, coercing, or top-down talking in a domineering manner is a pitfall for teachers. Also teachers are reminded not to be involved in argument in class or to confidently win their students. As Eggen and Kauchak (2009) put it, do not try to win students, simply because even they win, resentment is likely built up in the students in a sullen silence.

Case 7 The daughter of the second author of this chapter, a primary one pupil, brought home a mini project to complete during the winter vacation – measuring the width of the rim of a cup.

Daughter: "I think it is 8 centimeters. I imagine a ruler on the rim and have mental counting of the mark deal on the ruler."

Mother: "A very good idea. But that may not be accurate. How are you going to reach the accuracy?"

Daughter: “Well, I can put a ruler on the rim, to ensure that the 2 semicircles are equal. But the ruler, with its width, covers part of the rim. So I am unable to see if the two semicircles are of the same size.”
Mother: “Then how will you solve the problem? Is there a handy way to use the ruler?”
Daughter: “Put the cup upside down on a piece of paper and draw a circle around the rim with a pencil and then cut out the round.” (She then drew a line across the round and measured the length to be 6.4 cm.)
Mother: “How do you know that the line is the diameter – dividing the two halves equally?”
Daughter (joyfully): “Fold it up, to see if they overlap.”

In our view, this aspect of eastern philosophy and cultural notion is expedient for educators in the western culture. In a typical western cultural setting as in a classroom where the instruction being student centered, teachers being encouraging, activities being interactive, and teacher-student relationships being characteristic of mutual respect, the classic Chinese notion can be reinforcing to learning and teaching. “Guiding without forcing, being forthright instead of coercing, and inspiration instead of providing answers” may help to transform the learners and teachers in a way for teachers to be more facilitative. The transformation may be a little different: while Chinese students are encouraged to become more independent in thinking, students in the west, with such facilitation from their teachers, may practice more critical thinking to learn more from the process rather than to find out the answers only.

Understanding Students

Knowing the students is a process accompanying classroom instructions. There are four types of pitfalls in students’ learning, which teachers need to be aware of (学者有四失, 教者必知之):

1. In learning, learners tend to take more than the amount they are able to digest. Such behavior will lead to their reading without in-depth comprehension within a given period of time.

   (人之学也, 或失则多)

2. They may feel satisfied with inadequate knowledge at certain stage of learning.

   (或失则寡)

3. They may be easily distracted or inclined to withdrawal when faced with difficulties.

   (或失则易)

4. They may stop making further endeavors once they think that they have understood already, which will lead to "shallow learning," using Marshall's (2006, p. 40) term.

   (或失则止)

Case 8 Ms. Zhang is a well-known teacher transferred from outside Beijing. The school principal assigned her to be the form teacher of a secondary 2 class which was the worst of the level. Prior to taking over the class, she planned and did three things: first, she had a discussion with the teacher of mathematics in charge of that class to gain an initial insight of every student; second, she wrote to all the parents to have a basic understanding of the students' backgrounds; and then, she observed the class, checked on the students' assignments, and talked with them one by one, to find out the causes of their learning difficulties.

After that, Ms. Zhang categorized the students into several groups and started teaching them according to their cognitive abilities and personal characteristics. One of her pedagogies was differentiating the amount and level of assignments in accordance with the group each of them belonged to and providing specific feedback to each individual student. In one and half years, when the students were promoted to senior high 1, the class became the best among the 12 classes of the school.

The psychological factors causing the above four pitfalls vary, and only through teachers' counseling on the students concerned can they categorize and analyze the pitfalls before taking pertinent measures to help the students to overcome the pitfalls (知其心, 然后能救其失). In doing so, teachers are advised to help overcoming the students' misbehavior through enlarging and emphasizing their merits, which is what cultivation is about (教也者, 长善而救其失).

The Methods of Inquiries

Questioning Techniques

A skillful method in asking questions is like tackling with a piece of solid wood – starting from the soft part and then the joints. Gradually, the wood will be broken naturally.

(善问者, 如攻坚木, 先其易者, 后其节目, 及其久也, 相说以解)

An unskilled way appears the opposite. What works is the approach. This method is applicable during counseling sessions with problem learners who are not open to reveal the internal locus where the crux of the problem lies. Teachers doing such counseling may take the approach of asking easy questions first and then touching on the difficult part.

Responding Techniques

A skilled way of answering questions is like a bell reacting to the knocking. When the knocking is gentle, the bell gives a soft sound; when the knocking is hard, the bell gives a loud sound. Until the person knocks in a calm manner, the bell sounds as it normally does.

(善待问者, 如撞钟, 叩之以小者则小鸣, 叩之以大者则大鸣; 待其从容, 然后尽其声)

Those who are unskilled at answering questions take the opposite way. These methods are likened to the ways of engaging in scholarship (此皆进学之道). *Jin xue* (进学) in Chinese means scholarly learning, which is also expressed as *xue wen*, literally meaning "learning" and "questioning" (学问). A teacher who has engaged in such learning is able to demonstrate questioning and responding techniques with the above resonant effects.

During the contact time, a teacher who instructs on the basis of what he has learned by rote is not a qualified teacher (记问之学, 不足以为人师). Rather, a qualified teacher is one who starts by probing into his students' current status of abilities. The Essay on Learning articulates with a question on the necessity of the teacher getting answers from a student in particular.

- When a student is puzzled but unable to ask questions for clarifications, the teacher should instantly provide enlightenment and explanations.

  (力不能问, 然后语之)

- When seeing that the student is still unclear after providing illuminations, the teacher has to demonstrate patience instead of requiring the student to have a grasp of the meanings all at once and to consider it alright to postpone the learning process till a later time.

  (语之而不知, 虽舍之可)

The Principle of Balanced Learning

Balanced learning is advocated in the Essay on Learning with analogies:

"The son of a good blacksmith must learn the skill of tailoring leather coats on top of his skills passed on from his father. The son of a good archer must learn how to use a winnowing basket besides the skill he picks from his father. A colt, for the first time to be harnessed, is put behind the carriage."

(良冶之子, 必学为裘。良弓之子, 必学为箕。始驾马者反之, 车在马前)

The Essay further expresses that a gentleman who is able to examine and perceive some insights of the three scenarios may make determinations for scholarly study.

(君子察于此三者, 可以有志于学)

The beliefs in the above statements are as follows: (1) a learner must master both hard skills like those of a blacksmith and those of an archer and soft skills like tailoring and winnowing; and (2) practicum is a necessary experience, like putting a colt behind a carriage for the first time, to have a feel of the mature horses drawing the carriage in the front. The philosophy lying in the statements is "balanced learning" – being good at merely one thing may not help a person go far; but cross-boundary learning will ensure a person to become useful and helpful in his life.

The Quality Teacher

The Value of Respect for the Teacher

The value of respect for teachers was set at the earliest educational institutions in the history of China, as mentioned in Section II B. In the traditional educational philosophy, it has been realized that inculcating a respect for the teacher is never an easy process. Toward the last section of the Essay on Learning, a paragraph reads:

- Respect for the teacher is difficult to be instilled into the learners' minds.

  (凡学之道, 严师为难)

- Once the respect is present, scholarship and the rationale of gaining scholarship are valued.

  (师严然后道尊)

- Only when a reverence for scholarship is held can people regard highly of learning.

  (道尊然后民知敬学)

Two scenarios are provided as good examples in the text following the above statements: the Lord embodies the value of respect for the teacher in (1) treating his minister who takes the role of his teacher not as a minister (君之所不臣于其臣, 当其为师) and (2) practicing the rites of Great Learning by not letting his teacher sit in front of him, facing the north (大学之礼, 虽诏于天子, 无北面, 所以尊师). In the Chinese political philosophy, the Lord (sage) always sits facing the south toward the sun, implying he is the son of the Providence. When the teacher is arranged not to face the north, it implies that the Lord is not sitting on the opposite facing the south when receiving instructions from the teacher who is one of the Lord's ministers. Such a gesture reflects an authentic respect for a scholar who takes and qualifies the position of a teacher.

The Standards Set for Quality Teachers

A skillful teacher, in the Chinese traditional perception, is one who demonstrates capabilities not only in transmitting knowledge, skills, and values to his students but also in successfully letting them procure his will and spirit in pursuing for scholarship. The ability to pass his will onto his students is thus perceived as a strong indicator of a good teacher. The notion of quality teachers is stipulated as a set of skills in the Essay on Learning: a brilliant singer is one who enables his partisans to carry on naturally his expertise in singing (善歌者, 使人继其声), while a decent teacher is one who lets his disciples continue his will willingly (善教者, 使人继其志). In reaching such decency, pertinent criteria are set for quality teachers to reach.

Language Mastery

To reach ideal instructional effects, the teacher must master instructional languages to make his articulation concise and clearly understood (约而达), precise and elegant (微而臧), and sparing the use of metaphors (罕譬而喻). Such language skills will cater for students' comprehension, yielding outcomes in the students following the teacher naturally and willingly.

Vigorous and Rigorous Scholarship

A gentleman intending to become a teacher is one who is aware of the reality that students differ in learning abilities (君子知至学之难易) and in virtuous qualities (而知其美恶). With these well-informed backgrounds, he is able to:

- Deliver differentiated instructions for his students at all levels (然后能博喻).
- Make himself understood by all kinds of students, thus warranting the qualification as a teacher (能博喻然后能为师).
- Possess the potential of advancing to the level of a senior minister in the court (能为师然后能为长).
- Find the paramount position befitting him (能为长然后能为君).

The criteria are perceived to be extraordinary that are difficult to reach without high level of vigor in efforts and rigor in scholarship. This is in accordance with the statement made in Section I A of this chapter that in the Chinese history, national leaders were learned scholars and educators.

Becoming a teacher entails a growing process of capabilities to learn and to earn the potential of making advancements all the way to be the top leader (故师, 所以学为君). This is the highest requirement pronounced as a lofty mission for a gentleman to pursue, throughout his vigorous and rigorous scholarship. While it is impossible for all to advance up to the top leadership, the requirements are set as standards for teaching capabilities to build up and meanwhile for judicious selection of teachers.

The Principle of Congruence

Ancient scholars were able to make comparisons in their processes of observations in order to categorize them and further synthesize to reach congruence as their new knowledge and practice (古之学者, 比物丑类). The Essay on Learning lists four comparisons for congruence:

1. The sound of a drum does not fit in the (ancient Chinese) five-tone scale. But a melody without the role of a drum may not produce a concord.

   (鼓无当于五声, 五声弗得不和)

2. Water, colorless in nature, is not related to the five colors. But without water, the five colors will not be displayed of themselves.

   (水无当于五色, 五色弗得不章)

3. Learning is separate from the duties assumed by officials at various positions. But without continuous learning, the officials will not be able to fulfill their functions well.

   (学无当于五官, 五官弗得不治)

4. Teacher, as a profession, does not belong to any of the relationships in clans and tribes. However, people are unable to maintain close affinities among themselves if they do not learn from teachers.

   (师无当于五服, 五服弗得不亲)

The rationale of congruence stems from the interdependence among those seemingly unrelated factors. It is the factors of influence, involvement (e.g., Ballantine, p. 237, 1989), and systems' environment that educators need to pay attention to, for a greater effectiveness to produce in teachers carrying out their functions.

Conclusion

At the end of the Essay on Learning, the results of learning are reflected in the qualities possessed by a learned gentleman, as follows:

- Great virtues, which endow him with no restriction to any single official position
- The *Tao*, the way things work, which grants him with no parochial mind
- The trustworthiness, which assures others that there is no need for getting a solemn promise from him when working with him
- The sense of time frame, which allows him to control the pace, leaving no anxious feelings in others about his meeting the deadline to accomplish his work

(大德不官, 大道不器, 大信不约, 大时不齐)

Those who engage with Great Learning are illuminated that when seeing the brilliant attainments, they may be determined to make conscious and concentrated endeavors in their scholarly study on the origins of knowledge, namely, the classics (察此四者, 可以有志于本). The Essay on Learning concludes with an inspiring metaphor:

The ancestors, when holding memorial ceremonies for rivers, set the sequences, first for the streams across the land and second for the seas gathering the waters from the streams. The streams are the sources of the waters gathered in the seas; and the seas are the accumulations of the sources. Seeking for the roots of values and the origins of knowledge in the Chinese classics bears an analogy to searching for the sources of the accumulations (源委). The continual concentration on such search and research is thus known as *wuben* (务本).

(三王之祭川, 皆先河而后海, 或源, 或委, 此之谓务本)

In search of the Chinese classic philosophy of education, we gain such findings as (1) articulation of the philosophical base of education in the Chinese tradition, (2) revelation of principle setting that effectively guides the management and provision of learning and teaching, and (3) the prominence of the value of respect for teacher in the Chinese culture that fosters quality teachers who in turn cultivate students with desired achievements. With this overarching philosophy examined, we identify the main components of the Confucian model of education in which the mission of teaching is to cultivate people, carried out by the educator; and the ultimate accomplishment of learning is values and scholarship, as displayed in the figure below.

Values, as one of the four components in the Confucian model in Fig. 4.1, are identified in our exploration of the Essay on Learning. There are ten values specified, as shown in Table 4.2.

Illustrated in Fig. 4.1 are the five key components of the Confucian model, namely, "principles, methods, and skills" which lead to "scholarship" and "values." The practices of this model in teaching and learning explain convincingly the fundamental reasons why Chinese students outperform their counterparts in international tests like the Program for International Student Assessment (PISA) in the recent consecutive years (see Wikipedia, Feb. 2 2014).

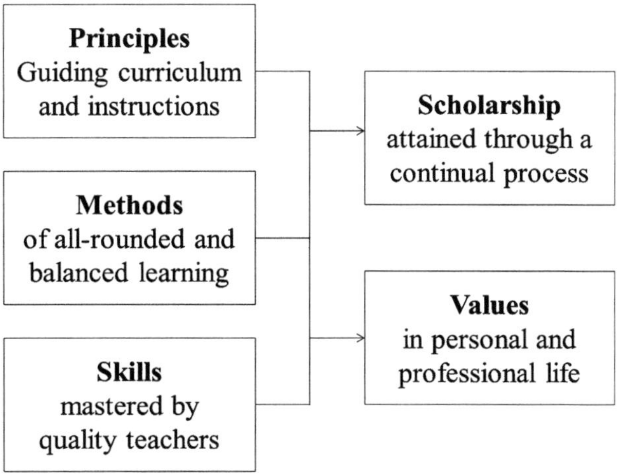

Fig. 4.1 The key components of the Confucian model of education

Table 4.2 Personal characters and values found in the Chinese scholarship

Category	Value
Personal characters	Industry (勤勉)
	Humility (谦逊)
Educational values	Study (学习)
	Enjoying learning (乐学)
	Respect for the teacher (尊师)
	Respect for learning (敬业)
	Balanced learning (兼修)
	Friendship with scholars (学友)
Life/professional values	Lifelong learning (终身学习)
	Seeking for the origins of knowledge in the classics (务本)

Readers are reminded that this chapter addresses the Chinese learners from classical perspectives and maybe Chinese learners today still hold such values and concepts that pass along generation to generation, like "lifelong learning" in Table 4.2. At Nanyang Technological University, this value is highlighted by Prof. Kwok, Associate Provost (Student Life). Furthermore, there are new implications of certain values. Take "respect for the teacher," for example, which does not imply one-way reverence or absolute obedience from the students. Rather, it helps to form up students' positive attitudes toward learning throughout their academic life.

References

Bai, Y. X. (白永潇, 2012). A case study based on 'PCK theory': Taking 'the number represented by the letter' as an example (基于'PCK理论'的课例分析--以'字母表示数' 一课为例"). (J). In *Elementary Mathematics Teachers* (《小学数学教师》), (2), 15–16.

Bai, Y. X., & Gao, Q. P. (白永潇, 高起萍, 2013). A case study on junior middle school mathematical underachievers' transformation (初中数学学困生转化的个案研究) (J). *Journal of Beijing Institute of Education (Natural Science Edition)* (《北京教育学院学报》自然科学版), (2), 16–20.

Ballantine, J. H. (1989). *The sociology of education: A systematic analysis* (2nd ed.). Englewood Cliffs, NJ: Prentice Hall.

Eggen, P.; & Kauchak, D. (2009). *Educational psychology: Windows on classrooms* (8th ed.). Upper Saddle River, NJ: Pearson.

Gredler, M. E. (1992). *Learning and Instruction: Theory into practice*. New York, NY: Macmillan Publishing.

Li, M.; & Wang, J. (李民, 王健, 2007). *The book of history* (《尚书》译注). Shanghai, China: Shanghai Classics Press.

Liu, J. H. (刘剑宏, Ed., 2006). *The Four Confucian Books and the Five Chinese Classics* (《四书 • 五经》). Beijing: Beijing Press.

Ma, Z. K. (马祝恺 Ed., 2006). *The Book of Poems* (《诗经》). Beijing, China: Beijing Press.

Marshall, S. P. (2006). *The power to transform: Leadership that brings learning and schooling to life*. San Francisco, CA: Jossey-Bass.

Mathematics Curriculum Standard of Compulsory Education (2011 Edition). Beijing, China: Beijing Normal University Press.

Wikipedia. (2014, February 2). Program for International Student Assessment (PISA): http://en.wikipedia.org/wiki/Programme_for_International_Student_Assessment

Xu, J. Z. (许景重, 1991). *The Four Confucian Books • The Doctrine of the Mean* (《四书读本 • 中庸》). Taiwan, China: Da Xing Press.

Yang, K. (杨宽, 2003). *The History of Western Zhou Dynasty* (《西周史》). Shanghai, China: Shanghai People's Press.

Zhang, S. M., & Bai, Y. X. (张思明, 白永潇, 2003). *The practice and exploration of mathematics task-based learning* (《数学课题学习的实践与探索》) (M). Beijing, China: Higher Education Press.

Zhang, Y. M.; Xiong, Y., & Ma, Q. M. (2014). A humanistic approach to enhancing teacher Influence. *International Review of Contemporary Learning Research. 3*(1), 20.

Dr. Zhang, Yenming is a senior faculty at the National Institute of Education (NIE), Nanyang Technological University, Singapore. He received Master of Science from Connecticut State University, Master of Education, and Doctorate of Education from Harvard University. Currently, he is working in the Academic Group of Policy and Leadership Studies, NIE, involved in the Doctor of Education, Master of Arts in Educational Management as the coordinator, and Master of Educational Administration (MEA) as the coordinator. He supervises and teaches doctorate and master's candidates. His research interests and publications in books and journal articles are in the areas of "Philosophies in the Chinese classics applied to leadership," "Educational leadership," and "Curriculum leadership from the principal's perspective." He is an editor of the international journal of *Chinese Management Studies*, Emerald, UK.

Prof. Bai, Yongxiao is an associate professor in the Faculty of Teacher Education and Mathematics, Beijing Institute of Education (BIE), Beijing, PR China. She received Master of Education (Mathematical Curriculum & Pedagogy) from Beijing Normal University and is currently working in the Department of Mathematics, BIE. Her research interests and publications are in the areas of mathematical curriculum and teaching, assessment in mathematical education, and teacher education. She has undertaken various research projects, coauthored books, and published numerous journal articles in such domains as "Exploration and practices in mathematical project study" and "Investigations in primary school pupils' high mathematical abilities and their evaluations."

Chapter 5
The 'Why' and 'How' of Engaging Parents in Their Children's Science Learning in Informal Contexts: Theoretical Perspectives and Applications

Ai Noi Lee and Youyan Nie

Introduction

Parents can play instrumental roles in facilitating their children's science learning outside the formal classrooms. A number of scholars have asserted that children's motivation towards school activities is influenced not only by their school experiences but also by their home lives in which parents play a crucial role (e.g. Crowley, Callanan, Jipson, Galco, Topping, & Shrager, 2001; Deci, Vallerand, Pelletier, & Ryan, 1991; Epstein, 1987; Hall & Schaverien, 2001; Ostlund, Gennaro, & Dobbert, 1985). In fact, science education research has reported that parental involvement in their children's science learning outside school not only could positively impact on their children's motivation and academic achievement in science but may also affect their children's inclination towards science-related careers in the later years (e.g. Ainley & Ainley, 2011; Dabney, Tai, Almarode, Miller-Friedmann, Sonnert, Sadler, & Hazari, 2012; Hall & Schaverien, 2001; Harris & Goodall, 2008; Sonnert, 2009; Zuzovsky & Tamir, 1989). It is thus imperative to empower parents to play a more active role to support and enrich their children's science learning experiences outside the school.

Besides the immediate home environment, science education research has reported that many informal learning environments outside the home and school, such as science museums, bird parks, zoos, botanical gardens, nature parks, farms, reservoirs, food factories or even supermarkets in the neighbourhood, can provide good opportunities and useful resources for parents to engage their children in deep and meaningful learning outside the formal science classrooms (e.g. Bamberger & Tal, 2007; Bultitude & Sardo, 2012; Gerber, Marek & Cavallo, 2001; Hofstein & Rosenfeld, 1996; Lee, 2012; Palmquist & Crowley, 2007). As children are naturally curious about their natural surroundings and how things work in their everyday

A.N. Lee (✉) • Y. Nie
National Institute of Education, Nanyang Technological University, Singapore, Singapore
e-mail: ainoi@yahoo.com; youyan.nie@nie.edu.sg

M.S. Khine (ed.), *Science Education in East Asia*,
DOI 10.1007/978-3-319-16390-1_5

lives (e.g. Crowley & Galco, 2001; Rowe, 2004), their knowledge about science could be greatly enhanced if parents are aware of how they could seize 'teachable moments' in their everyday lives to effectively facilitate their children's sense-making and intellectual growth (e.g. Crowley & Galco, 2001; Gerber et al., 2001; Tizard & Hughes, 2002).

In order to develop a better understanding of parental involvement in children's science learning in informal contexts, this article aims to draw upon perspectives from the sociocultural theory, constructivism theory and self-determination theory to discuss the 'why' and 'how' of involving parents in their children's informal science learning process. It will use examples to further illustrate how parents could enhance their children's science learning through engagement in intellectually stimulating science-related conversations or activities. It will also highlight some of the main challenges of engaging parents in their children's informal science learning outside the school and discuss the implications for practice.

Theoretical Perspectives on Engaging Parents in Their Children's Science Learning in Informal Contexts

A major goal of science education is to develop learners' science literacy through acquisition of scientific knowledge and scientific reasoning skills so that they can gain a better understanding of how things work in the natural world and make contributions to the sustainability of the natural environment and improvement of the quality of human life (e.g. Ainley & Ainley, 2011; Crowley & Galco, 2001; Gerber, Cavallo & Marek, 2001). Given that scientific phenomena can easily be observed in our everyday lives, the immediate home environment and many other informal learning contexts outside the school can in fact provide ample authentic learning opportunities and valuable resources for parents, which many science teachers could not easily offer in their formal science classrooms, to engage their children in deep and meaningful science learning (e.g. Anderson, Lucas, & Ginns, 2003; Lee, 2012). Particularly, Crowley et al. (2001, p. 18) noted that parents 'who involve children in informal science activities provide an opportunity for children to learn factual scientific information and to practice scientific reasoning, but they also provide an opportunity for children to participate in a culture of learning about science'.

Parental involvement can be conceptualised as the extent to which the parent is interested in and knowledgeable about and takes an active part in the child's life (e.g. Anderson & Minke, 2007; Grolnick & Ryan, 1989). Given the fact that science education research has widely recognised the importance of parents' involvement and influence on their children's intellectual–cognitive development in science (e.g. Callanan & Jipson, 2001; Crowley & Callanan, 1998; Crowley & Galco, 2001; Gauvain, 2001; Szechter & Carey, 2009), it is imperative to give more attention to understand the 'why' and 'how' of engaging parents more actively in their children's science learning, especially their roles and the kinds of support they could provide to their children in their children's science learning process in informal contexts (e.g. Hoover-Dempsey, Walker, Sandler, Whetsel, Green, Wilkins, & Closson, 2005).

Importantly, a number of scholars have asserted that learning is a cumulative process which involves making connections and reinforcement between the learning experiences a child encounters in school and at home (e.g. Anderson et al., 2003; Crowley & Galco, 2001; Dierking & Falk, 1994; Domina, 2005; Falk & Dierking, 1997; Lee, 2012; Rennie, Feher, Dierking & Falk, 2003). This thus means that in order to provide a child with a more holistic science education, there is a need to take into greater consideration the child's learning experiences in the formal classrooms as well as his or her learning experiences outside the school by which parents could play a fundamental role (e.g. Callanan & Jipson, 2001; Crowley & Callanan, 1998; Gerber, Cavallo, & Marek, 2001; Schauble, Beane, Coates, Martin, & Sterling, 1996).

In the following sections, we will discuss the 'why' of engaging parents in their children's science learning in informal contexts from three theoretical perspectives: sociocultural theory, constructivism theory and self-determination theory.

From the Perspectives of the Sociocultural and Constructivism Theories

The sociocultural theory posits that a child's learning is strongly influenced by the interactions which take place in the child's social contexts (e.g. Vygotsky, 1978). From the sociocultural perspective, a learner can develop his or her intellect through internalising concepts based on his or her own interpretation of an activity or in communication with more knowledgeable or capable others in his or her social contexts (Roschelle;, 1995; Vygotsky, 1978). In other words, in a child's learning process, meaning-making is not an isolated mental activity, but more often, this effort is usually played out within a physical context mediated by the actions of other significant others in the child's social context (e.g. Clark, 2011; Falk & Dierking, 1997; Roschelle;, 1995). Specifically in the context of informal learning, effective parent–child conversations and interactions in everyday life can contribute significantly to enhancing the quality of children's learning experiences outside the school (e.g. Crowley et al., 2001; Gottfried, Fleming & Gottfried, 1998).

In addition, the sociocultural theory emphasises the importance of the 'zone of proximal development' which is the difference between what an individual can do based on his or her own ability and what he or she can do with the help from more knowledgeable others (Vygotsky, 1978). This also suggests that meaningful learning is more likely to take place if the knowledgeable others could provide support structures to help the learner achieve the next stage or level of learning. This process is thus referred to as 'scaffolding instruction' (Jaramillo, 1996; Vygotsky, 1978).

However, scaffolding instruction is usually temporary because the scaffolding support can be gradually reduced or withdrawn once the learner becomes more self-directed and competent in his or her own learning (Jaramillo, 1996; Jonassen, 1998; Vygotsky, 1978).

In the context of informal science learning, it is thus important for parents to provide appropriate levels of scaffolds to guide their children's learning. Given that parents and their children share many everyday life experiences together, science education scholars have noted that parents are in fact in the best position to acquire knowledge about their child's prior learning, current interest and level of motivation to enable them to provide more effective scaffolding instruction and relevant resources to support their children's science learning process at home and in other informal contexts (e.g. Crowley & Galco, 2001; Dierking & Falk, 1994; Glasersfeld, 1992; Jaramillo, 1996; Martin, 2004; Ostlund et al., 1985; Palincsar, 1998; Schauble et al., 1996).

While the sociocultural theory offers valuable theoretical insights regarding the importance of significant others in supporting an individual's learning process, the importance of the interdependence of social and individual processes in the individual's construction of knowledge has also been noted by several scholars (e.g. Cobb, 1994; John-Steiner & Mahn, 1996; Leach & Scott, 2003; Palincsar, 1998). Many science education scholars see the need to incorporate the constructivist approaches of learning in children's science learning to better understand the dynamic nature of knowledge construction. Essentially, constructivist learning is often viewed as a form of active learning which advocates the need for learners to engage in meaningful construction of knowledge from one's prior and current knowledge or experiences (e.g. Chaille & Britain, 1991; Cobb, 1994; Kirschner, Sweller & Clark, 2006). It also recognises that learning can be gradual and assimilative at certain times and at other times significant and rapid (e.g. Anderson et al., 2003) and that direct experience is central to one's intellectual development (e.g. Gerber, Marek & Cavallo, 2001; Tamir, 1990).

Fundamentally, science learning is a process of inquiry and discovery (e.g. Crowley et al., 2001; Jonassen & Rohrer-Murphy, 1999). Moreover, authentic science activities are often socially and contextually bound (e.g. Leach & Scott, 2003; Lee, 2012; Tamir, 1990). As such, effective science learning cannot be solely based on textbook-based knowledge as the acquisition of deep knowledge requires active construction through critical thinking practices and direct experiences (e.g. Crowley et al., 2001; Hawkins & Pea, 1987; Lee, 2012). From a constructivist perspective, it thus suggests that parents play the role of a 'facilitator of inquiry' to provide opportunities for children to engage in inquiry-based, hands-on activities instead of a 'provider of content' who provides direct instruction and spoon-feeding of scientific content knowledge (e.g. Lee, 2012).

For instance, parents can create opportunities for their children to engage in science-related conversations or simple investigative science activities at home. Meaningful science-related conversations could often be initiated when parents and children spend quality time together to engage in daily activities such as when watching a science programme on television, taking a walk in the garden, playing a game or even preparing meals in the kitchen. In fact, studies have found that meaningful parent–child interactions can be a possible mechanism for cognitive change in children's learning process (e.g. Gauvain, 2001; Szechter & Carey, 2009) and everyday parent–child conversations or discourse can contribute significantly to children's scientific understandings if parents are able to make use of 'teachable' moments at home or in other informal settings to enhance their children's science learning experiences (e.g. Tenenbaum & Leaper, 2003). With the advancement of information technology (IT) and the availability of Internet, parents could also make use of IT tools and online resources to allow their children to gain a deeper understanding of the scientific issues and topics they may have encountered in their textbooks or in their everyday lives (e.g. Clark, 2011; Plowman, McPake, & Stephen, 2008).

Parents could also create opportunities for their children to engage in simple investigative science activities or provide experiential learning opportunities through outdoor activities or field trips to educational learning places (e.g. Szechter & Carey, 2009). Empirically, a number of studies have examined how parents might engage their children in home-based science-related activities and how parents could make use of more structured educational places outside the school (i.e. science museums, earth observatories and zoos) to support their children's science learning. For instance, Hall and Schaverien (2001) explored families' engagement with their children's science and technology learning at home over a 6-month period and found that families engaged with children's inquiries at home in many ways, such as by providing resources, conversing and investigating collaboratively with children. Their findings also indicated that when families pursued inquiries together and when children conducted their own sustained intellectual searches, children's ideas deepened. In another study, Cumming (2003) utilised parent diaries and questionnaires to elucidate the experiences of children aged four to seven which might contribute to their knowledge about the origin of food and its destiny after being eaten. The findings indicated that children learn more scientifically correct information with friends and family than teachers in the formal science classrooms might realise and further suggested that children's prior knowledge acquired in informal learning settings can be utilised in classroom discourse to promote science learning in the formal classrooms. On the other hand, Szechter and Carey (2009) examined the nature of parent–child conversations at an informal science education centre housed in an active gravitational-wave observatory. They observed that parents employed a variety of strategies to support their children's understanding of unfamiliar topics, including describing evidence, giving direction, providing explanation, making connections and eliciting predictions. In another study, Crowley et al. (2001) videotaped family interactions at an interactive science exhibit in the Children's Discovery Museum in San Jose, California. Their study suggested that when children engaged in shared scientific thinking with their parents, they had greater opportunity to learn than when they engaged in scientific thinking with peers or by themselves. The study also observed that parents talked to their children about identifying, generating and interpreting evidence. Particularly, parents assumed the role of explainer in helping their children to make connections between prior knowledge and new experiences or introducing abstract principles to their children.

In essence, parents can play instrumental roles in facilitating their children's science inquiry outside the school. When parents are involved in their children's learning process outside the school, they may gain more accurate information about their children's abilities as well as how and what their children are learning in the school. Such information would be useful for parents to provide more relevant guidance for their children to maximise their children's cognitive and metacognitive skill developments (e.g. Pomerantz, Moorman, & Litwack, 2007). More importantly, if parents could spend more quality time together with their children in learning and 'doing' science together, they not only could help their children learn science more meaningfully but could also strengthen the parent–child relationship when they form partnerships with their children by learning together (e.g. Chaille & Britain, 1991; Crowley et al., 2001; Lee, 2012).

As many science education scholars who adopted the sociocultural and constructivist views of learning tended to examine parental influence on the intellectual–cognitive development of children, it is observed that parental influence on children's psychological functioning and development in the science learning process is often underexplored and poorly understood. As asserted by Falk and Dierking (1997, p. 216), 'learning is the process of applying prior knowledge and experience to new experiences; this effort is normally played out within a physical context and is mediated in the actions of other individuals. In addition, learning always involves some element of emotion and feeling'. In fact, research has found that children are likely to function competently and confidently in contexts in which they experience significant others as being both caring and autonomy supportive (e.g. Grolnick & Ryan, 1989; Harris & Goodall, 2008; Joussemet, Landry, & Koestner 2008). Hence, it is imperative to develop a better understanding of the role of parents in supporting their children's psychological needs and functioning in their children's science learning process.

From the Lens of the Self-Determination Theory

Self-determination theory (SDT) posits that humans have three universal psychological needs for optimal growth and adjustment: the need for autonomy, the need for competence and the need for relatedness (Deci & Ryan, 1985; Ryan & Deci, 2000). The need for competence involves understanding how to attain various external and internal outcomes and being efficacious in performing the requisite actions; the need for autonomy refers to being self-initiating and self-regulating of one's own actions; and the need for relatedness involves developing secure and satisfying connections with others in one's social contexts (e.g. Deci et al., 1991). According to SDT, contextual conditions which provide opportunities to satisfy or support any of these three needs are likely to facilitate an individual's motivation, performance and development (Ryan & Deci, 2000; Grolnick et al., 1991). On the other hand, contextual conditions which are controlling or coercive can interfere with an individual's self-determination or autonomy and are likely to alienate the individual from his or her inner motivational resources (e.g. Jang, Reeve, Ryan, & Kim, 2009). Generally, autonomy-supportive environments are more likely to generate self-determined behaviours which are autonomous or intrinsically motivated such that they are usually endorsed by one's sense of self to act volitionally and the regulatory process is choice (Deci & Ryan, 1985; Ryan, & Deci, 2000), whereas controlling environments are more likely to generate extrinsically motivated behaviours compelled by some interpersonal or intrapsychic force and the regulatory process is often compliance or defiance (Deci & Ryan, 1985; Ryan, & Deci, 2000).

Similarly, in the context of children's learning, the satisfaction of these three psychological needs is also crucial to enhancing a child's intrinsic motivation and self-determined behaviours to take charge of his or her own learning and development (e.g. Deci & Ryan, 1985; Grolnick, 2012; Vansteenkiste, Sierens, Soenens, Luyckx, & Lens, 2009). This also suggests that, in the informal science learning contexts, it is important for parents to create a need-supportive environment which is characterised by the provision of autonomy support, structure and involvement. As noted by Grolnick (2012), '... environments that (a) support autonomy (vs. controlling and pressuring behavior), (b) provide structure (clear rules, expectations, and consequences), and (c) provide involvement (resources, tangible and intangible to children) satisfy children's needs for autonomy, competence, and relatedness, respectively'. (p. 60). In other words, it is essential that parents provide support for competence (e.g. providing intellectually stimulating tasks, giving praise, encouragement and constructive or positive feedback), autonomy (e.g. allowing choice, minimising controlling language, giving meaningful rationale) and relatedness (e.g. parents' interpersonal involvement and showing concern, consideration and respect for their children) when engaging their children in science learning in informal contexts. An autonomy-supportive rather than a controlling, learning environment/climate would be more effective for parents to facilitate their children's intrinsic motivation and self-determined behaviours towards the learning of science (e.g. Vansteenkiste et al., 2009; Joussemet et al., 2008).

The positive influences of parents' autonomy support on children's learning outcomes such as children's motivation, academic achievement, creativity and psychological well-being have been noted by several scholars (e.g. Dai, Tan, Marathe, Valtcheva, Pruzek, & Shen, 2012; Grolnick & Ryan, 1989; Hill & Tyson, 2009; Soenens & Vansteenkiste, 2005; Vansteenkiste, Zhou, Lens, & Soenens, 2005). For instance, a study by Grolnick and Ryan (1989) reported that mothers' and fathers' provision of support for competence, relatedness and autonomy with respect to their children's homework predicted the children's maintaining intrinsic motivation for school work and internalising the importance of school-related activities, which in turn predicted the children's school performance and adjustment. In another study, Grolnick, Gurland, DeCourcey and Jacob (2002) reported that elementary school children whose mothers interacted with them in a more controlling manner wrote less creative poems when alone. Their study also highlighted the importance of context, children's competence levels and mothers' styles in determining levels of autonomy support. Although many studies of parental autonomy support have been conducted using samples from the more cultural individualistic western societies, more recently, Liu, Zhang, Zhang, Lee, Wang and Brownell's (2013) study which used a sample of 550 Chinese high school students indicated that students' autonomous motivation positively predicted their creative thinking, and this relationship was moderated by parental involvement characterised by autonomy support. Importantly, Liu et al.'s study also provided support for the cross-cultural validity of SDT and the three psychological needs for a better understanding of Eastern Asian parents' autonomy support on their children's intellectual and psychological development.

In essence, when parents foster a nonthreatening, autonomy-supportive learning environment which supports openness, questioning, reflecting and experimenting, it is likely that their children's intrinsic motivation to engage in science inquiries and discoveries would be enhanced. In addition, children's innate curiosity and creativity is more likely to flourish if there is autonomy support from parents by providing sufficient space and time for them to explore freely and discover about how things work around them. More importantly, when parents foster an autonomy-supportive informal learning climate to support their children's optimal learning, especially by forming learning partnership with their children to learn collaboratively together, it is likely that parent–child interactions could be enhanced and family bond could also be strengthened (e.g. Lee, 2012; Hoover-Dempsey et al., 2005).

How Do Parents Support Their Children's Science Learning in Informal Settings?

Parental involvement in their children's informal science education can generally comprise three kinds of support: (1) *behavioural*, (2) *psychological* and (3) *intellectual–cognitive*. Behavioural support requires parents to form learning partnership with their children and to act as learning role models for their children. It is important that parents could show enthusiasm and commitment in learning together with their children, especially by participating actively in 'doing' science together with their children and providing appropriate guidance to their children whenever necessary. Psychological support requires parents to provide a nonthreatening, autonomy-supportive environment/climate to support their children's autonomous motivation and self-regulation in learning. It is important for parents to take into consideration their children's psychological needs for competence, autonomy and relatedness in order to effectively facilitate their children's interest towards the learning of science. Intellectual–cognitive support requires parents to engage their children in intellectually or cognitively stimulating activities so as to develop their children's intellectual–cognitive abilities in science learning. It is especially important that parents could provide learning opportunities for their children to engage actively in creative and critical thinking in their problem-solving process as well as self-directed learning in their acquisition of scientific knowledge and skills.

To further elaborate on each kind of parental support, we will provide examples/activities to illustrate what parents can do to better facilitate their children's science learning at home or in other informal learning settings.

Behavioural Support

Behavioural support from parents can include engaging together with their children in home-based science activities using simple materials or easily available resources at home. If parents are aware that the study of science is about an understanding of how things work in our physical and natural surroundings, many everyday life scenarios or activities can be used to engage children in learning science meaningfully. In essence, parents can play a more active and instrumental role to provide meaningful learning experiences for their children if they are willing to invest and spend more quality time with their children by participating in their children's science learning process in informal settings outside the school.

In fact, there is no need for parents to acquire sophisticated science equipments or instruments to engage their children in investigative, hands-on science activities at home. Many household items which we use in our daily lives (e.g. electrical appliances or utensils in the kitchen) or even food items we consume (e.g. vegetables and fruits) can enable parents and children to discuss and 'do' science together. Figure 5.1 shows some everyday life examples/scenarios by which parents could engage their children in science inquiry within the immediate home environment.

In addition, children's familiar objects at home, such as childhood or family photos, toys and stationery, can also be used to trigger simple science discussions or explorations within the home among the family members. Figure 5.2 shows examples of how parents could make use of children's everyday familiar objects such as photos or toys to engage their children in simple science conversations and explorations.

Nowadays, as more and more families have personal computers and other IT gadgets in their homes, parents could easily leverage on the Internet for online educational resources to facilitate their children's science learning at home (e.g. Plowman et al., 2008). Many educational technological tools or platforms as well as online teaching aids such as video clips, animated demonstrations and interactive online learning platforms frequently provide pedagogical support structures (e.g. Hall, 2007; Scrimshaw, 1993; Sefton-Green, 2003) and could be easily accessible from various Internet websites. As such, they can serve as valuable resource support for parents to enhance their children's science learning, especially for helping children to understand more abstract science concepts or complex

Fig. 5.1 Examples of parents' participation in home-based science-related activities with their children (Sources: Lee 2008, 2009)

Fig. 5.2 Examples of how parents could engage their children in home-based science inquiry using children's everyday familiar objects (Sources: Lee 2008, 2009)

experiments outside the formal science classroom. Essentially, the IT tools and online educational resources can serve as 'scaffolds' (Vygotsky, 1978) to support children's understanding of abstract science concepts as well as provide interactive learning experiences to enhance children's learning motivation (e.g. de Jong & van Joolingen, 1998; Hall & Schaverien, 2001). Figure 5.3 shows an example of an IT-related science activity which parents could use or adapt to facilitate their children's science learning in informal contexts.

Behavioural support from parents can also include bringing children on a walk in the gardens in the neighbourhood, field trips/excursions to educational learning places (i.e. zoo, bird park or science museum) or even a picnic at the reservoir or beach to allow children to learn more about the natural environment. Generally,

Learning points:
1. Every day, we use electricity to help us do work.
2. Power stations can generate electricity for homes, offices and factories.
3. Power stations generate electricity by burning fossil fuels. This causes pollution and adds to environmental problems.
4. We can help to conserve energy and reduce environmental problems by cutting down on the use of electricity in our daily lives.
5. We need to use electrical appliances safely to prevent electric shocks, fires and burns.

Process skills: Observing, comparing, communicating, information-gathering

This activity aims at increasing your children's awareness of electricity usage and energy saving.

Get your children to make a list of electrical appliances that they can find around the house. They should also take note of the energy usage of each electrical appliance.

First, discuss with them the use of each electrical appliance.

Ask:
How can these electrical appliances help us in our daily tasks? Can you carry out the tasks without them? Can you use these appliances without electricity?

Next, focus on the energy usage of each electrical appliance.

Ask:
Does each electrical appliance use the same amount of electrical energy? How do you find out? Can you rank them according to the amount of electrical energy used?

Encourage your children to search the Internet for the information needed. The idea is to get them to be aware that different electrical appliances use different amounts of electrical energy. For instance, a fan uses less electricity compared to an air-conditioner. You may highlight to your children that electricity is measured in watts. (Refer to the list of suggested websites.)

After your children have done their research, get them to share with you their findings.

Ask:
1. *Where does electricity come from?*
   [Power stations]

2. *How many power stations are there in Singapore? How is the electricity in the power stations usually generated?*

Your children may not know the answers to these questions but allow them to search the Internet and to share with you their findings. The idea is to get your children to understand that power stations generate electricity by burning fossil fuels such as coal, gas or oil. These are non-renewable resources which are limited in supply and are running out. They also cause pollution and add to environmental problems such as the greenhouse effect and global warming. (Refer to the list of suggested websites.)

Prompt your children to think about energy conservation. Get them to suggest how we can reduce the amount of electricity we use daily. Try to instil in your children that we can help to conserve fossil fuel and reduce environmental problems by cutting down on the use of electricity in our daily lives.

Ask:
1. *Why is it important to conserve energy?*
2. *What can you do to help save electricity and conserve energy?*

Suggested extended activity:
Get your children to state as many ways as possible how electricity or electrical appliances can be used safely at home. Discuss with them and highlight the importance of the safe use of electricity.

Fig. 5.3 An example of IT-related science activity which parents could use or adapt to facilitate their children's science learning children can acquire broader and deeper experiences when they can be engaged in experiential learning in outdoor activities or more structured learning from visits to educational learning places (e.g. Dierking & Falk, 1994; Schauble & Bartlett, 1997; Crowley & Callanan, 1998). Parents can thus seize these teachable moments to inculcate in their children positive attitudes, values and responsibilities to protect and ensure the sustainability of our living environment. Such learning opportunities not only could expose children to the diversities of flora and fauna in different habitats but also allow them to have a better understanding of how science and technology may positively or negatively impact on our quality of living. Figure 5.4 shows examples of outdoor activities or excursions/visits to educational learning places which parents could provide for their children's informal science learning.

Psychological Support

Psychological support from parents entails empathic responding to children's feelings and expressions to allow them to be autonomous, effective agents who take charge of their own learning (e.g. Froiland, 2011; Joussemet et al., 2008). Psychological support from parents can thus include creating a noncontrolling, autonomy-supportive learning environment to facilitate their children's intrinsic motivation in science learning. Children who are intrinsically motivated tend to show more appreciation and enthusiasm towards and in pursuing their interests (e.g. Froiland, 2011; Ryan & Deci, 2000). Parents who are autonomy supportive tend to give choice, praise and encouragement to their children more often in their children's learning process, and they also tend to encourage their children to be more innovative and creative, resilient and self-directed in problem-solving (e.g. Tenenbaum, Alfieri, Brooks, & Dunne, 2008).

Particularly, children are often curious about their immediate environments especially about the 'why' and 'how' certain things work or happen. It is important that parents could try to encourage and cultivate their children's learning curiosity instead of discouraging their questions and inquiries (e.g. Soenens & Vansteenkiste, 2005; Sonnert, 2009). Parents could provide learning opportunities to encourage their children to give their views and explanations on why certain things work or do not work. This is because children can gain more confidence to express their ideas or thinking when they perceive that their parents show a genuine interest by listening to them and acknowledging their learning efforts. Sometimes, it does not matter so much whether the children have completed a science task or investigation successfully, as long as they have put in their best effort and could offer logical explanations for their findings or observations; it is important for parents to openly praise or acknowledge their children's efforts on the given task so as to motivate them to do better in future tasks (e.g. Lee, 2012; Liu et al., 2013).

It is highly important that parents could inculcate in their children a mindset which allows them to see failures or imperfections as part and parcel of scientific exploration and that there is always room for improvement in every scientific

Taking a walk in a nearby park or garden

A visit to the zoo

A picnic at the reservoir

Fig. 5.4 Examples of informal science learning opportunities from outdoor activities or visits to educational learning places (Sources: Lee (2008, 2009) investigation (e.g. Crowley et al., 2001). As such, it is also equally important for parents to adopt a mindset which views themselves as learning partners of their children. Especially, parents need to accept that they may not always have the answers to each of their children's questions, and therefore, sometimes it is alright for them to say 'I'm not sure too. Let's find it out together!' (e.g. Lee, 2008, 2009). In fact, this may be a better approach to get their children to think more critically and further seek the answers in a self-directed manner instead of relying on their parents to provide direct answers or explanations all the time.

Most importantly, autonomy-supportive verbal and nonverbal behaviours from parents can critically affect their children's learning motivation and confidence. According to many scholars of self-determination theory (e.g. Grolnick & Pomerantz, 2009; Kenny-Benson & Pomerantz, 2005), controlling parent communication as a form of psychological control may include pressuring children to comply with demands (e.g. 'You must....' or 'You should...') and often disregarding children's views by interrupting children's thinking or explanations (e.g. Froiland, 2011; Grolnick & Pomerantz, 2009; Grolnick & Ryan, 1989). Thus, it is essential that parents are aware of their children's level of learning readiness and interest in order to provide more relevant tasks or appropriate scaffolds to guide their children's learning more effectively and refrain from forcing their ideas to their children or coerce their children in performing tasks/activities which their children may not feel comfortable in.

Taken together, if parents' involvement in their children's learning process is autonomy supportive, it can likely create positive changes in their children's learning behaviours and attitudes. This is because children may internalise the value of learning such that their learning motivation is intrinsically driven rather than extrinsically driven over time (e.g. Grolnick & Ryan, 1989; Grolnick & Slowiaczek, 1994). When children experience intrinsic interest, they would be more willing to take charge of their own learning which could enhance their academic achievement. On the other hand, when parents' involvement is controlling, they may deprive children of feeling that they are autonomous, effective learners which may lead them to be less willing to be engaged in active learning to solve challenges on their own (e.g. Pomerantz et al., 2007). Figure 5.5 presents a home-based activity by which parents could easily conduct for their children to enable them to learn simple science concepts in a fun and meaningful way.

Intellectual–Cognitive Support

Intellectual–cognitive support from parents includes providing intellectual stimulation in their children's science learning process. This may thus include providing intellectually challenging tasks as well as open-ended or guiding questions at appropriate learning intervals to stimulate children to think further on their own and be more self-directed in their own learning. As science learning is basically a

Learning points:
1. A force is a push or a pull.
2. Objects cannot move unless force is applied to them.
3. Friction or frictional force is caused by two different objects moving against each other.
4. Friction can slow down or stop the movement of an object.

Process skills: Observing, comparing, inferring, communicating

Children are usually excited when they watch magic tricks. In this activity, you will demonstrate a classic trick that will make their learning about motion and friction more fun.

For this 'magic trick', you will try and pull a tablecloth out from under a heavy book such that the book does not move much from its original position.

Prepare the following things:
- Tablecloth (one with a smooth edge without any ridge along its perimeter)
- Table
- Heavy hardcover book or encyclopedia

Note:
You may need to practise on your own a few times so that you will be more skilful and confident to perform in front of your children. Practice makes perfect!

Steps:
1. Place a sheet of tablecloth on a table.
2. Place the book on the tablecloth, about 30 cm from the edge of the table. (Remember to remove any breakable or fragile objects from the table.)
3. Pull the tablecloth from under the book downwards and as quickly as possible.

There you go! The book moves but still remains almost at its original position when the tablecloth is removed.

Follow-up activity:
Further arouse your children's curiosity and excitement by asking them some questions.

Ask:
Why does the book still remain at its original position when the tablecloth is removed totally from underneath it? Is this magic? How does it work?

Let your children make some guesses and get them to perform the same trick for you.

Your children might or might not be able to perform the trick successfully. Nevertheless, praise them for their courage in trying. Then explain to them how the trick works.

Ask:
Does your trick work? Why or why not?
[Possible explanation: The book will not move when it stays at rest unless a force is being applied to make it move. By applying an external force such as pulling the tablecloth away, we will expect the book to move because there is contact between the book and the tablecloth. So if the tablecloth is not pulled quickly enough, the book will move together with the tablecloth because of friction between the book and the tablecloth. However, if the tablecloth is pulled quickly enough to overcome the friction between the book and the tablecloth, the book will only move slightly.]

Suggested extended activity:
Get your children to think further or research for more tricks to demonstrate concepts about forces. Encourage them to perform their 'magic tricks' for the whole family.

Fig. 5.5 An activity which illustrates how parents might create an autonomy-supportive informal learning climate to facilitate their children's science learning (Source: Lee 2009)

process of discovery and exploration, it is important that parents play the role of a facilitator of inquiry rather than solely a provider of content knowledge.

In fact, parents could play a critical role to facilitate the extension of science discourse beyond the formal classroom settings. Studies have found that when parents and children engage in meaningful conversations characterised by depth, richness and variety, it may trigger a cognitive change in their children (e.g. Callanan & Jipson, 2001; Callanan & Oakes, 1992; Gleason & Schauble, 2000; Tizard & Hughes, 2002; Shrager & Callanan, 1991; Snow & Kurland, 1996). For instance, parents could engage their children in science-related conversations over the dinner table (e.g. the edible parts of plants we eat, food nutritional value, how food is digested) or even discovering simple science knowledge or concepts when doing simple household chores together (e.g. what kinds of materials make up the parts of electrical appliances, how much energy is consumed when using a vacuum cleaner as compared to using a broom, etc). Sometimes, children's toys can also be a good trigger for an intellectually stimulating science conversation. Figure 5.6 presents an activity to illustrate how parents could enhance their children's science learning through intellectually stimulating parent–child conversations.

Another important goal of science education is to enable learners to develop higher-order thinking skills such as critical and creative thinking in problem-solving skills. Hence, parents could provide opportunities to develop their children's intellectual–cognitive abilities by guiding and encouraging them in planning and carrying out a scientific investigation independently. Such opportunities not only could allow children to construct new knowledge and apply integrated science process skills in problem-solving but also could develop their positive values and attitudes towards science.

For instance, creative toy-making inevitably requires children to apply a number of scientific concepts and skills in an integrated manner in the different phases: design, material selection and construction. This could thus provide authentic learning opportunities for children to apply their high-order problem-solving skills and make creative connections across topics and disciplines. Figure 5.7 presents an activity to illustrate how parents could engage their children in using higher-order science process skills such as in creative toy-making.

As informal learning environments are inherently multidisciplinary, they can provide parents with the additional advantage and opportunities, which teachers in schools seldom have, to encourage children to go beyond specific topics and subject areas to think in a more interdisciplinary manner when solving real-life problems (e.g. Foss & Pinchback, 1998; Lee, 2012). Particularly, real-world problems seldom fit neatly into the confines of a single discipline, and the integration of topic areas and subject matter content can provide greater relevancy and authenticity for the learner's understanding of certain scientific phenomena or issues explored (e.g. Foss & Pinchback, 1998; Lee, 2012; White, 1985). It is thus important for parents to provide opportunities for children to engage in interdisciplinary learning by integrating knowledge and skills learnt from other science topics or subject areas (e.g. language arts, role play, IT) into their problem-solving process. Parents could

Learning points:
1. A system consists of many parts. These parts perform different functions.
2. Our body is made up of many organ systems.
3. The organ systems in our body work together to enable our body to function properly so that we are able to carry out our daily activities. If any of these systems is not working well, it will affect the other parts of the body.

Process skills: Observing, inferring, communicating, information-gathering

In this activity, you will help your children to explore whole systems and parts of a system by looking at a toy. This helps them to appreciate that parts come together to make a whole system which can function in a way that each part alone is unable to.

Show your children a simple toy such as a toy car (or any other toy with movable parts which can move with or without batteries). Ask them to identify and name as many parts of the toy as possible. Then get them to select the part which they think is most important to the toy.

Ask:
What does this part help the toy do? What can it do by itself?

Allow your children to separate different parts of the toy in order to find out whether the whole toy is affected if some of its parts are removed.

Ask:
If this part is removed, can the toy still work? Is this part made of even smaller parts?

Lead your children to realise that the toy will not work properly if one of its parts is missing. Each part of the toy needs the other parts for the system in the whole toy to work.

Next, guide them to share what they have understood about whole systems and their parts in living things.

Ask:
Do living things have whole systems made up of parts too? Can you think of some examples of parts in a living system working together?

Let your children talk about the human body systems. Guide them to name and elaborate on the various parts of each human body system.

Ask:
What will happen if one of these parts is removed? Can the whole system still do its job? What do you think?

Allow your children to think, predict and formulate questions at this point of time. Do not spoon-feed or provide any answers yet. This is a good opportunity to stimulate their curiosity and to encourage them to think independently.

Suggested extended activity:
Get your children to do some research or information gathering using the Internet or bring them to a nearby library. Have further discussions with them after they have done some reading and research on their own. At the same time, you can also gather some interesting information about human body systems to make your discussion with your children more fun and engaging!

Fig. 5.6 An activity which illustrates how parents could enhance their children's science learning through intellectually stimulating parent–child conversations (Source: Lee 2009)

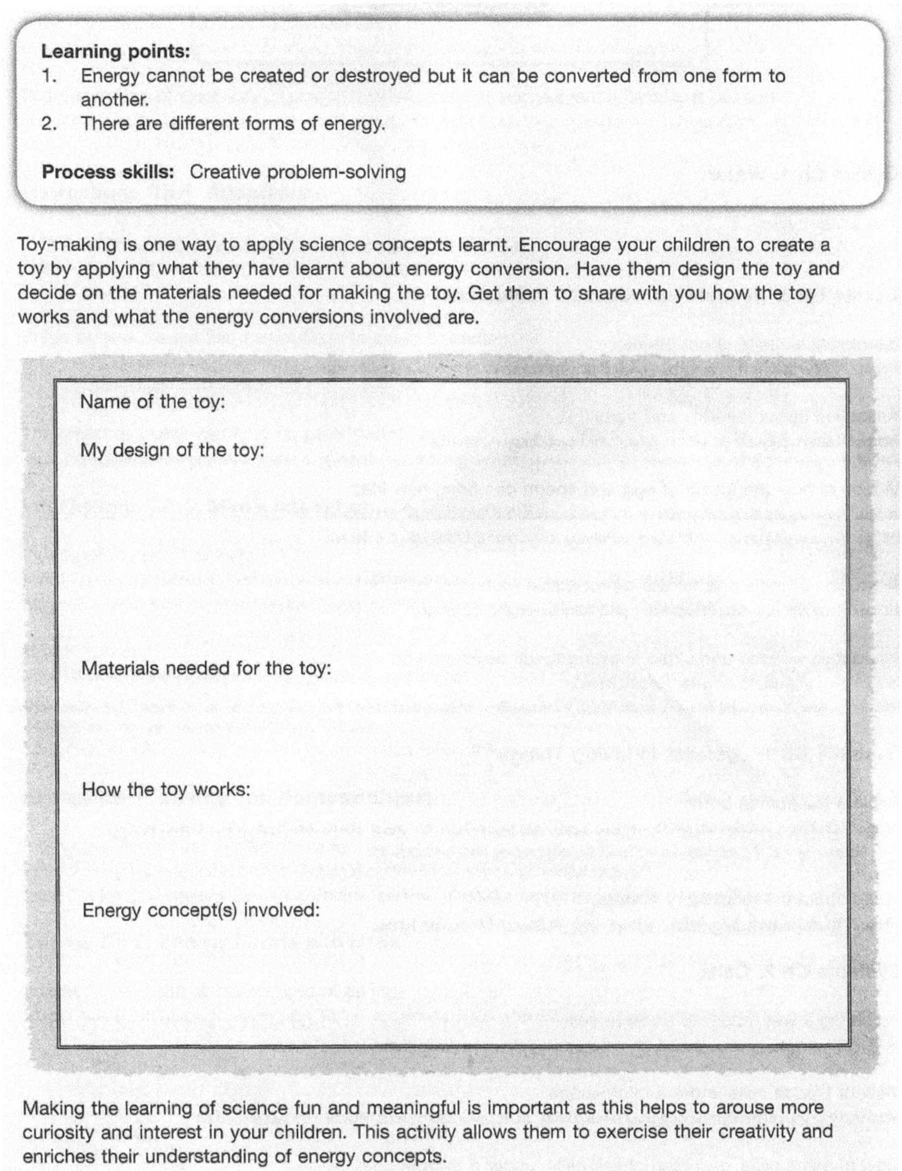

Learning points:

1. Energy cannot be created or destroyed but it can be converted from one form to another.
2. There are different forms of energy.

Process skills: Creative problem-solving

Toy-making is one way to apply science concepts learnt. Encourage your children to create a toy by applying what they have learnt about energy conversion. Have them design the toy and decide on the materials needed for making the toy. Get them to share with you how the toy works and what the energy conversions involved are.

Name of the toy:

My design of the toy:

Materials needed for the toy:

How the toy works:

Energy concept(s) involved:

Making the learning of science fun and meaningful is important as this helps to arouse more curiosity and interest in your children. This activity allows them to exercise their creativity and enriches their understanding of energy concepts.

Fig. 5.7 An activity which illustrates how parents could engage their children in using higher-order science process skills in creative toy-making (Source: Lee 2009) provide learning tasks for children to integrate science with language arts or even with music and drama. The incorporation of interdisciplinary learning into children's informal science learning process can thus provide children with the opportunities to apply their higher-order thinking skills and creativity as a form of integrated processes during their scientific investigations (Lee, 2012). Figure 5.8 presents an activity to illustrate how parents and children may engage in learning simple science concepts of magnetism in puppet-making together and come up with a puppet show for other family members.

Engaging Parents in Children's Science Learning in Informal Contexts: Challenges and Implications for Practice

Although there are many widely recognised benefits of engaging parents in forming learning partnership with their children in informal science learning outside the school, there also exist some challenges which may have important implications for practice. For instance, socio-economic status, parents' educational qualifications or cultural differences have been reported to have possible influences on parents' involvement in their children's out-of-school learning, but the research findings have been mixed and inconclusive (e.g. Anderson & Minke, 2007; Hoover-Dempsey & Sandler, 1995, 1997; Hoover-Dempsey et al., 2005; Weiss, Mayer, Kreider, Vaughan, Dearing, Hencke, Pinto et al., 2003). Often, parents with fewer family resources or lower educational qualifications might tend to experience more difficulties to get themselves involved in their children's out-of-school learning due to limited knowledgeable support system to help address the needs of their children (e.g. Hoover-Dempsey & Sandler, 1995, 1997; Hoover-Dempsey et al., 2005). However, scholars have also argued that it is in fact parents' self-efficacy (especially their perceptions of personal knowledge and skills), time and commitment and value systems which appear to be the main factors affecting parents' thinking about the kinds of involvement and activities they would possibly engage with their children in the informal learning contexts (e.g. Anderson & Minke, 2007; Hoover-Dempsey & Sandler, 1995, 1997; Machida, Taylor, & Kim, 2002; Shumow & Lomax, 2002; Wentzel, 1998).

In the following, we attempt to highlight and discuss three main challenges which parents might face when considering their involvement in their children's science education outside the school. These three main challenges are related to parental awareness, self-efficacy and resource support.

First, many parents are not aware that they have the potential and ability to make a difference in their children's informal science education (e.g. Anderson & Minke, 2007; Hoover-Dempsey et al., 2005. Especially, parents are often unaware about the positive impacts they could have on their children's science learning and are unsure of the roles they could play in their children's science learning process at

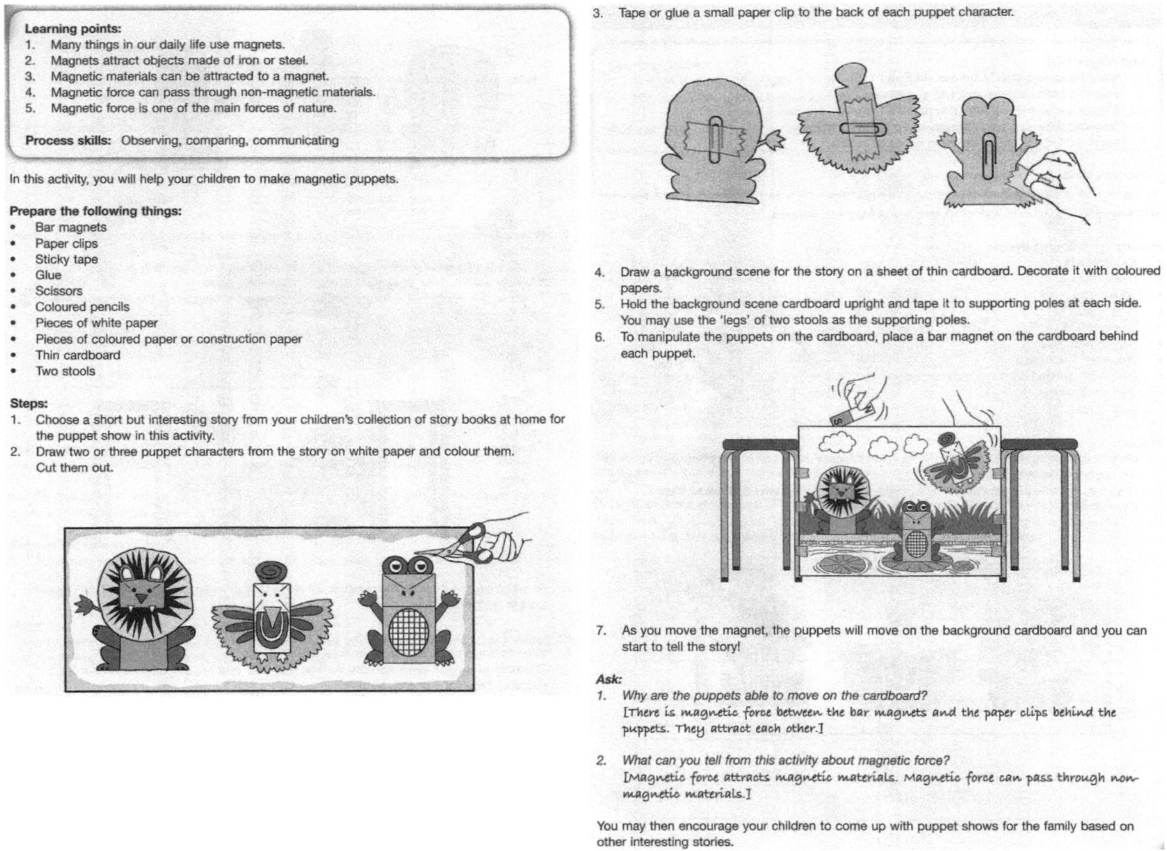

Learning points:
1. Many things in our daily life use magnets.
2. Magnets attract objects made of iron or steel.
3. Magnetic materials can be attracted to a magnet.
4. Magnetic force can pass through non-magnetic materials.
5. Magnetic force is one of the main forces of nature.

Process skills: Observing, comparing, communicating

In this activity, you will help your children to make magnetic puppets.

Prepare the following things:
- Bar magnets
- Paper clips
- Sticky tape
- Glue
- Scissors
- Coloured pencils
- Pieces of white paper
- Pieces of coloured paper or construction paper
- Thin cardboard
- Two stools

Steps:
1. Choose a short but interesting story from your children's collection of story books at home for the puppet show in this activity.
2. Draw two or three puppet characters from the story on white paper and colour them. Cut them out.
3. Tape or glue a small paper clip to the back of each puppet character.
4. Draw a background scene for the story on a sheet of thin cardboard. Decorate it with coloured papers.
5. Hold the background scene cardboard upright and tape it to supporting poles at each side. You may use the 'legs' of two stools as the supporting poles.
6. To manipulate the puppets on the cardboard, place a bar magnet on the cardboard behind each puppet.
7. As you move the magnet, the puppets will move on the background cardboard and you can start to tell the story!

Ask:
1. *Why are the puppets able to move on the cardboard?*
[There is magnetic force between the bar magnets and the paper clips behind the puppets. They attract each other.]
2. *What can you tell from this activity about magnetic force?*
[Magnetic force attracts magnetic materials. Magnetic force can pass through non-magnetic materials.]

You may then encourage your children to come up with puppet shows for the family based on other interesting stories.

Fig. 5.8 An activity which illustrates how parents could engage their children's critical and creative thinking skills through interdisciplinary learning tasks (Source: Lee 2009) home and in other informal settings (e.g. Lee, 2012). There is a need to increase parents' awareness to allow them to realise that they can in fact do much more in enriching their children's learning experiences outside the school if only they are willing to invest more quality time to participate in their children's informal learning process.

Second, parental self-efficacy may affect parents' decision to be involved and engaged in their children's informal science learning process outside the school (e.g. Anderson & Minke, 2007; Hoover-Dempsey et al., 2005). It is not uncommon to find parents often have the mindset that science involves many abstract concepts and complex theories which can only be taught by the qualified science teachers in the formal classrooms (e.g. Lee, 2012). Parents, however, do not realise that they do not need to be a content expert in science in order to guide their children in science learning. Instead, they could act as a facilitator of inquiry to provide a nonthreatening, autonomy-supportive environment outside the school to engage their children in meaningful science learning, and this in turn is likely to have positive impacts on their children's science learning outcomes.

Third, many parents tend to think that science activities and experiments can only be conducted in well-equipped science laboratories using sophisticated science apparatus and instrument (e.g. Lee, 2012). As a result, parents often feel that there is very little they can do at home because of a lack of adequate equipment and available resources to engage their children in science experiments or hands-on activities. With this mindset, it is thus inevitable that many parents feel that science is a difficult subject to 'teach' to their children and they tend to be less enthusiastic to engage themselves in their children's science learning outside the school.

In general, the main challenges of engaging parents to play more active and effective roles in their children's informal science learning are often related to parents' awareness and mindset, pedagogical knowledge as well as resource availability and support. This is mainly because, unlike the science teachers in schools who are trained teaching professional, parents are generally not equipped with pedagogical knowledge and skills. It is therefore unrealistic to expect parents to play a role equivalent to a professional science teacher in their children's informal science learning process outside the school. In fact, there are a number of ways by which science educators and teaching practitioners can do to increase parents' awareness about the multiple roles they could play in their children's informal science learning as well as to change their mindsets to encourage them to participate more actively and effectively in their children's science education outside the school.

For instance, at the school level, regular parental workshops/seminars could be organised to update parents about the new trends or changes in science curriculum and pedagogy. Schools could engage science teachers or science educators outside the schools to conduct parents' seminars or workshops to allow parents to better understand what science learning is about, especially to enable them to realise that many science concepts their children learn in school are linked to our everyday lives so as to allow parents a greater awareness of how they could better support their children's science learning at home or in other informal settings (e.g. Domina, 2005; Epstein, 1987; Wallace & Walberg, 1991). Through such seminars or workshops, parents could acquire some basic science knowledge and facilitation skills in order to effectively facilitate their children's science learning at home and in other informal settings. It is important for parents to realise that they could learn to play the role of a facilitator of inquiry in their children's science learning process by giving prompts or guiding questions at appropriate intervals of their children's earning to support their children's self-discovery and inquiry, instead of spoon-feeding their children by directly giving answers or providing science content to their children all the time. Fundamentally, science learning is a process of inquiry and discovery with a core aim to enable children to gradually take charge of their own learning to initiate their own inquiries and investigations as they become more independent and self-directed in their learning. It is therefore more crucial for parents to facilitate their children's learning by providing opportunities for and encouraging their children to actively engage in scientific inquiry by making observations, asking questions, problem-solving, interpreting data and communicating their findings to others through simple science activities.

Besides, schools could also get parents to be more involved in the school science fairs, field trips or excursions, so as to allow parents to gain a better understanding of what their children are learning in school and at the same participate collaboratively in their children's out-of-school learning with the school. It is imperative for schools to convey the value of parents' active support for their children's science learning outside the school when they invite parental involvement and provide support for guiding skills and strategies that will encourage effective parental involvement (e.g. Hoover-Dempsey et al., 2005; Lee, 2012). Also, it would be helpful if schools or educational learning places (e.g. zoos, botanic gardens, science museums) could provide useful learning resources, such as ready-to-use learning booklets or worksheets and online resources or websites which offer interactive science learning, for parents to use or adapt for their children's science learning at home or when they make a visit to the educational learning places. If relevant educational resources can be made readily available for parents to give them ideas about what kinds of science activities can be conducted in informal settings, it would likely enhance their willingness to engage their children in science learning at home or in other informal settings. In addition, schools could encourage enthusiastic parents to come together to form parents' interest groups or support networks to share and exchange resources and experiences with one another about their support for their children's science learning outside the school. Such collaborative parent–teacher or parent–parent interactions would help to better develop parents' knowledge and skills to support their children's science learning more effectively outside the school.

At the national level, more efforts could also be given to promote science to parents and children as a form of public education through public seminars and talks, exhibitions or television programmes. Besides, public education could promote family bonding and quality parent–child interactions through parent–child engagement in science activities at home or through science field trips organised by community clubs or agencies. In this way, both parents and children are more likely to view science as a subject which is closely and meaningfully linked to the everyday lives. Particularly, if public education could create more awareness among parents for them to be more aware of their roles and responsibilities in their children's science education, they would likely be more encouraged to willingly support and participate in their children's informal science learning process outside the school.

Taken together, parents need to understand that although it is good to gain a deeper understanding of science content knowledge and pedagogical strategies in order to support their children's informal science learning more effectively, it is equally important for them to foster an autonomy-supportive informal science learning environment to facilitate their children's optimal psychological and intellectual–cognitive developments (e.g. Lee, 2012; Liu et al., 2013). Parents need to feel that it is meaningful and worthwhile to engage themselves in learning science together with their children. They need to be aware that their involvement in their children's science education can help not only to develop their children's science knowledge and skills but also their children's positive values and attitudes towards the use of science and technology for the sustainability of our natural environment (e.g. Crowley & Galco, 2001). Perhaps the most effective approach for parents to engage in their children's informal science learning is to be learning partners of their children to learn *from* as well as *with* their children (e.g. Clark, 2011; Lee, 2012).

Concluding Remarks

Learning in the family or with family members is probably one of the most crucial of all educational experiences for an individual (e.g. Crowley et al., 2001; Ostlund et al., 1985; Tizard & Hughes, 2002). Engaging parents in their children's science learning outside the school basically requires a genuine willingness and interest from the parents to commit their time and effort to be engaged as learning partners of their children. When parents begin to participate more actively in their children's science learning process outside the school, they are likely to not only observe many positive influences on their children's academic achievement, motivation and interest in science but, more importantly, the development of a stronger parent–child bonding in the family.

Acknowledgements The authors wishes to express their gratitude to Marshall Cavendish Education Pte Ltd for granting the permission to reproduce some parts of I-Science Parent's Guide (Primary 3 to 6) for the writing of this article.

References

Ainley, M., & Ainley, J. (2011). Student engagement with science in early adolescence: The contribution of enjoyment to students' continuing interest in learning about science. *Contemporary Educational Psychology, 36*(1), 4–12.

Anderson, D., Lucas, K. B., & Ginns, I. S. (2003). Theoretical perspectives on learning in an informal setting. *Journal of Research in Science Teaching, 40*(2), 177–199.

Anderson, K. J., & Minke, K. M. (2007). Parent involvement in education: Toward an understanding of parents' decision making. *The Journal of Educational Research, 100*(5), 311–323.

Bamberger, Y., & Tal, T. (2007). Learning in a personal context: Levels of choice in a free choice learning environment in science and natural history museums. *Science Education, 91*(1), 75–95.

Bultitude, K., & Sardo, A. M. (2012). Leisure and pleasure: Science events in unusual locations. *International Journal of Science Education, 34*(18), 2775–2795.

Callanan, M. A., & Jipson, J. L. (2001). Explanatory conversations and young children's developing scientific literacy. In K. Crowley, C. Schunn, & T. Okada (Eds.), *Designing for science: Implications from everyday, classroom, and professional science*. Mahwah, NJ: Erlbaum.

Callanan, M. A., & Oakes, L. A. (1992). Preschoolers' questions and parents' explanations: Causal thinking in everyday activity. *Cognitive Development, 7*, 213–233.

Chaille, C., & Britain, L. (1991). *The young child as scientist: A constructivist approach to early childhood Science education*. New York, NY: Harper Collins Publishers Inc.

Clark, L. S. (2011). Parental mediation theory for the digital age. *Communication Theory, 21*(4), 323–343.

Cobb, P. (1994). Where is the mind? Constructivist and sociocultural perspectives on mathematical development. *Educational Researcher, 23*(7), 13–20.

Crowley, K., & Callanan, M. A. (1998). Identifying and supporting shared scientific reasoning in parent–child interactions. *Journal of Museum Education, 23*, 12–17.

Crowley, K., Callanan, M. A., Jipson, J., Galco, J., Topping, K., & Shrager, J. (2001). Shared scientific thinking in everyday parent – Child activity. *Science Education, 85*, 712–732.

Crowley, K., & Galco, J. (2001). Everyday activity and the development of scientific thinking. In K. Crowley, C. D. Schunn, & T. Okada (Eds.), *Designing for science: Implications from everyday, classroom, and professional settings*. Mahwah, NJ: Erlbaum.

Cumming, J. (2003). Do runner beans really make you run fast? Young children learning about science-related food concepts in informal settings. *Research in Science Education, 33*(4), 483–501.

Dabney, K. P., Tai, R. H., Almarode, J. T., Miller-Friedmann, J. L., Sonnert, G., Sadler, P. M., & Hazari, Z. (2012). Out-of-school time science activities and their association with career interest in STEM. *International Journal of Science Education, Part B, 2*(1), 63–79.

Dai, D. Y., Tan, X., Marathe, D., Valtcheva, A., Pruzek, R. M., & Shen, J. (2012). Influences of social and educational environments on creativity during adolescence: Does SES matter? *Creativity Research Journal, 24*, 191–199.

de Jong, T., & van Joolingen, W. R. (1998). Scientific discovery learning with computer simulations of conceptual domains. *Review of Educational Research, 68*(2), 179–201.

Deci, E. L., & Ryan, R. M. (1985). *Intrinsic motivation and self-determination in human behavior*. New York, NY: Plenum.

Deci, E. L., Vallerand, R. J., Pelletier, L. G., & Ryan, R. M. (1991). Motivation and education: The self-determination perspective. *Educational Psychologist, 26*(3–4), 325–346.

Dierking, L. D., & Falk, J. H. (1994). Family behavior and learning in informal science settings: A review of the research. *Science Education, 78*, 57–72.

Domina, T. (2005). Leveling the home advantage: Assessing the effectiveness of parental involvement in elementary school. *Sociology of Education, 78*, 233–249.

Epstein, J. L. (1987). Parent involvement: What research says to administrators. *Education and Urban Society, 19*, 119–136.

Falk, J. H., & Dierking, L. D. (1997). School field trips: Assessing their long-term impact. *Curator, 40*, 211–218.
Foss, D. H., & Pinchback, C. L. (1998). An interdisciplinary approach to science, mathematics, and reading: Learning as children learn. *School Science and Mathematics, 98*(3), 149–155.
Froiland, J. M. (2011, April). Parental autonomy support and student learning goals: A preliminary examination of an intrinsic motivation intervention. In *Child & youth care forum* (Vol. 40, No. 2, pp. 135–149). New York, NY: Springer US.
Gauvain, M. (2001). *The social context of cognitive development*. New York, NY: Guilford.
Gerber, B. L., Cavallo, A. M. L., & Marek, E. A. (2001). Relationships among informal learning environments, teaching procedures and scientific reasoning ability. *International Journal of Science Education, 23*(5), 535–549.
Gerber, B. L., Marek, E. A., & Cavallo, A. M. L. (2001). Development of an informal learning opportunities assay. *International Journal of Science Education, 23*(6), 569–583.
Glasersfeld, E. V. (1992). A constructivist's view of learning and teaching. In F. G. R. Duit & H. Niedderer (Eds.), *Research in physics learning: Theoretical issues and empirical studies* (pp. 29–39). Kiel, Germany: IPN at the University of Kiel.
Gleason, M. E., & Schauble, L. (2000). Parents' assistance of their children's scientific reasoning. *Cognition and Instruction, 17*, 343–378.
Gottfried, A. E., Fleming, J. S., & Gottfried, A. W. (1998). Role of cognitively stimulating home environment in children's academic intrinsic motivation: A longitudinal study. *Child Development, 69*(5), 1448–1460.
Grolnick, W. S. (2012). The relations among parental power assertion, control, and structure. *Human Development, 55*(2), 57–64.
Grolnick, W. S., Gurland, S. T., DeCourcey, W., & Jacob, K. (2002). Antecedents and consequences of mothers' autonomy support: An experimental investigation. *Developmental Psychology, 38*(1), 143–155.
Grolnick, W. S., & Pomerantz, E. M. (2009). Issues and challenges in studying parental control: Toward a new conceptualization. *Child Development Perspectives, 3*(3), 165–170.
Grolnick, W. S., & Ryan, R. M. (1989). Parent styles associated with children's self-regulation and competence in school. *Journal of Educational Psychology, 81*(2), 143–154.
Grolnick, W. S., Ryan, R. M., & Deci, E. L. (1991). Inner resources for school achievement: Motivational mediators of children's perceptions of their parents. *Journal of Educational Psychology, 83*(4), 508–517.
Grolnick, W. S., & Slowiaczek, M. L. (1994). Parents' involvement in children's schooling: A multidimensional conceptualization and motivational model. *Child Development, 65*(1), 237–252.
Hall, A. (2007). Vygotsky goes online: Learning design from a socio-cultural perspective. *Learning and Socio-cultural Theory: Exploring Modern Vygotskian Perspectives International Workshop 2007, 1*(1), 94–107.
Hall, R. L., & Schaverien, L. (2001). Families' engagement with young children's science and technology learning at home. *Science Education, 85*(4), 454–481.
Harris, A., & Goodall, J. (2008). Do parents know they matter? Engaging all parents in learning. *Educational Research, 50*(3), 277–289.
Hawkins, J., & Pea, R. D. (1987). Tools for bridging the cultures of everyday and scientific thinking. *Journal of Research in Science Teaching, 24*(4), 291–307.
Hill, N. E., & Tyson, D. F. (2009). Parental involvement in middle school: A meta-analytic assessment of the strategies that promote achievement. *Developmental Psychology, 45*(3), 740–763.
Hofstein, A., & Rosenfeld, S. (1996). Bridging the gap between formal and informal science learning. *Studies in Science Education, 28*, 87–112.
Hoover-Dempsey, K. V., & Sandler, H. M. (1995). Parental involvement in children's education: Why does it make a difference? *Teachers College Record, 97*(2), 310–331.

Hoover-Dempsey, K. V., & Sandler, H. M. (1997). Why do parents become involved in their children's education? *Review of Educational Research, 67*(1), 3–42.
Hoover-Dempsey, K. V., Walker, J. M., Sandler, H. M., Whetsel, D., Green, C. L., Wilkins, A. S., & Closson, K. (2005). Why do parents become involved? Research findings and implications. *The Elementary School Journal, 106*(2), 105–130.
Jang, H., Reeve, J., Ryan, R. M., & Kim, A. (2009). Can self-determination theory explain what underlies the productive, satisfying learning experiences of collectivistically oriented Korean students? *Journal of Educational Psychology, 101*(3), 644–661.
Jaramillo, J. (1996). Vygotsky's sociocultural theory and contributions to the development of constructivist curricula. *Education, 117*(1), 133–140.
John-Steiner, V., & Mahn, H. (1996). Sociocultural approaches to learning and development: A Vygotskian framework. *Educational Psychologist, 31*(3), 191–206.
Jonassen, D. (1998). Designing constructivist learning environments. In C. M. Reigeluth (Ed.), *Instructional theories and models* (2nd ed., pp. 215–239). Mahwah, NJ: Erlbaum.
Jonassen, D. H., & Rohrer-Murphy, L. (1999). Activity theory as a framework for designing constructivist learning environments. *Educational Technology Research and Development, 47* (1), 61–79.
Joussemet, M., Landry, R., & Koestner, R. (2008). A self-determination theory perspective on parenting. *Canadian Psychology/Psychologie canadienne, 49*(3), 194–200.
Kenney-Benson, G. A., & Pomerantz, E. M. (2005). The role of mothers' use of control in children's perfectionism: Implications for the development of children's depressive symptoms. *Journal of Personality, 73*(1), 23–46.
Kirschner, P. A., Sweller, J., & Clark, R. E. (2006). Why minimal guidance during instruction does not work: An analysis of the failure of constructivist, discovery, problem-based, experiential, and inquiry-based teaching. *Educational Psychologist, 41*(2), 75–86.
Leach, J., & Scott, P. (2003). Individual and sociocultural views of learning in science education. *Science & Education, 12*(1), 91–113.
Lee, A. N. (2008). *I-science parent's guide primary 3 & 4*. Singapore: Panpac Education Private Limited.
Lee, A. N. (2009). *I-science parent's guide primary 5 & 6*. Singapore: Panpac Education Private Limited.
Lee, A. N. (2012). Development of a parent's guide for the Singapore primary science curriculum: Empowering parents as facilitators of their children's science learning outside the formal classrooms. *Asia-Pacific Forum on Science Learning & Teaching, 13*(2), 1–11.
Liu, G., Zhang, S., Zhang, J., Lee, C., Wang, Y., & Brownell, M. (2013). Autonomous motivation and Chinese adolescents' creative thinking: The moderating role of parental involvement. *Creativity Research Journal, 25*(4), 446–456.
Machida, S., Taylor, A. R., & Kim, J. (2002). The role of maternal beliefs in predicting home learning activities in Head Start families. *Family Relations, 51*(2), 176–184.
Martin, L. M. W. (2004). An emerging research framework for studying informal learning and schools. *Science Education, 8*(S1), S71–S82.
Ostlund, K., Gennaro, E., & Dobbert, M. (1985). A naturalistic study of children and their parents in family courses in science. *Journal of Research in Science Teaching, 22*(8), 723–741.
Palincsar, A. S. (1998). Social constructivist perspectives on teaching and learning. *Annual Review of Psychology, 49*, 345–375.
Palmquist, S., & Crowley, K. (2007). From teachers to testers: How parents talk to novice and expert children in a natural history museum. *Science Education, 91*(5), 783–804.
Plowman, L., McPake, J., & Stephen, C. (2008). Just picking it up? Young children learning with technology at home. *Cambridge Journal of Education, 38*(3), 303–319.
Pomerantz, E. M., Moorman, E. A., & Litwack, S. D. (2007). The how, whom, and why of parents' involvement in children's academic lives: More is not always better. *Review of Educational Research, 77*(3), 373–410.

Rennie, L. J., Feher, E., Dierking, L. D., & Falk, J. H. (2003). Toward an agenda for advancing research on science learning in out-of-school settings. *Journal of Research in Science Teaching, 40*(2), 112–120.

Roschelle, J. (1995). Learning in interactive environments: Prior knowledge and new experience. In J. H. Falk & L. D. Dierking (Eds.), *Public institutions for personal learning: Establishing a research agenda* (pp. 37–51). Washington, DC: American Association of Museums.

Rowe, A. J. (2004). *Creative Intelligence: Discovering the innovative potential in ourselves and others*. Upper Saddle River, NJ: Pearson Education, Inc.

Ryan, R. M., & Deci, E. L. (2000). Self-determination theory and the facilitation of intrinsic motivation, social development, and well-being. *American Psychologist, 55*(1), 68–78.

Schauble, L., & Bartlett, K. (1997). Constructing a science gallery for children and families: The role of research in an innovative design process. *Science Education, 81*(6), 781–793.

Schauble, L., Beane, D. B., Coates, G. D., Martin, L., & Sterling, P. (1996). Outside the classroom walls: Learning in informal environments. In L. Schauble & R. Glaser (Eds.), *Innovations in learning: New environments for education* (pp. 5–24). Mahwah, NJ: Lawrence Erlbaum Associates.

Scrimshaw, P. (1993). *Language, classrooms and computers*. London, UK: Routledge.

Sefton-Green, J. (2003). Informal learning: Substance or style? *Teaching Education, 4*(1), 37–52.

Shrager, J., & Callanan, M. (1991). Active language in the collaborative development of cooking skill. In *Proceedings of the cognitive science society*. Cambridge, MA: MIT Press.

Shumow, L., & Lomax, R. (2002). Parental efficacy: Predictor of parenting behavior and adolescent outcomes. *Parenting: Science and Practice, 2*, 127–150.

Snow, C. E., & Kurland, B. (1996). Sticking to the point: Talk about magnets as a context for engaging in scientific discourse. In D. Hicks (Ed.), *Discourse, learning, and schooling* (pp. 189–220). Cambridge, MA: Cambridge University Press.

Soenens, B., & Vansteenkiste, M. (2005). Antecedents and outcomes of self-determination in 3 life domains: The role of parents' and teachers' autonomy support. *Journal of Youth and Adolescence, 34*(6), 589–604.

Sonnert, G. (2009). Parents who influence Their children to become scientists: Effects of gender and parental education. *Social Studies of Science, 39*(6), 927–941.

Szechter, L. E., & Carey, E. J. (2009). Gravitating toward science: Parent–child interactions at a gravitational-wave observatory. *Science Education, 93*(5), 846–858.

Tamir, P. (1990). Factors associated with the relationship between formal, informal, and nonformal science learning. *Journal of Environmental Education, 22*, 34–42.

Tenenbaum, H. R., & Leaper, C. (2003). Parent–child conversations about science: The socialization of gender inequities? *Developmental Psychology, 39*(1), 34–47.

Tenenbaum, H. R., Alfieri, L., Brooks, P. J., & Dunne, G. (2008). The effects of explanatory conversations on children's emotion understanding. *British Journal of Developmental Psychology, 26*(2), 249–263.

Tizard, B., & Hughes, M. (2002). *Young children learning* (2nd ed.). Oxford, UK: Blackwell.

Wallace, T., & Walberg, H. J. (1991). Parental partnerships for learning. *International Journal of Educational Research, 15*, 131–145.

Weiss, H. B., Mayer, E., Kreider, H., Vaughan, M., Dearing, E., Hencke, R., & Pinto, K. et al. (2003). Making it work: Low-income working mothers' involvement in their children's education. *American Educational Research Journal, 40*(4), 879–901.

Wentzel, K. R. (1998). Parents' aspirations for children's educational attainments: Relations to parental beliefs and social address variables. *Merrill-Palmer Quarterly, 44*(1), 20–37.

White, A. M. (1985). Teaching as though students mattered. In J. Katz (Ed.), *New directions for teaching and learning* (pp. 39–48). San Francisco, CA: Jossey-Bass.

Vansteenkiste, M., Sierens, E., Soenens, B., Luyckx, K., & Lens, W. (2009). Motivational profiles from a self-determination perspective: The quality of motivation matters. *Journal of Educational Psychology, 101*(3), 671–688.

Vansteenkiste, M., Zhou, M., Lens, W., & Soenens, B. (2005). Experiences of autonomy and control among Chinese learners: Vitalizing or immobilizing? *Journal of Educational Psychology, 97*(3), 468–483.

Vygotsky, L. S. (1978). *Mind and society: The development of higher psychological processes.* Cambridge, MA: Harvard University Press.

Zuzovsky, R., & Tamir, P. (1989). Home and school contributions to science achievement in elementary schools in Israel. *Journal of Research in Science Teaching, 26*, 703–714.

Chapter 6
Characteristics of Chinese Learners as Revealed from Their Affective Domain and Choices of Science Learning in China

Yau Yuen Yeung

Introduction

Being the world's second largest economy, Mainland China has benefitted from her booming economy with huge increase in both the number and percentage of high school students (especially female students) being admitted into universities over the last two decades (National Bureau of Statistics of China, 1994, 2013). During the same period, the USA noticed a troubling trend: 'many of the most proficient students... have been gravitating away from science and engineering toward other professions' as revealed from the President's Council of Advisors on Science and Technology (PCAST, 2010) national report to the president which outlined five overarching priorities to prepare and inspire K-12 education in science, technology, engineering and math (STEM) for America's future. In the European Union (EU), the enrolment rates and percentages of graduates in university STEM programmes have been decreasing steadily in most of its member countries, and in particular female students are underrepresented in those programmes (European Commission, 2004; OECD, 2008). Osborne and Dillon (2008) raised their critical reflections on science education in Europe and proposed corresponding recommendations for the necessary educational reform. In a word, this common concern among the policymakers of those countries has already induced many educators and academics to study the factors influencing recruitment, retention and gender equity in STEM higher education (e.g. Buccheri, Gurber & Brühwiler, 2011; Lyons, 2006). Nevertheless, no similar research study or comprehensive review has ever been published in the Western literature regarding the affective characteristics of Chinese science learners in Mainland China.

Y.Y. Yeung (✉)
Department of Science and Environmental Studies, The Hong Kong Institute of Education, Hong Kong, China
e-mail: yyyeung@ied.edu.hk

M.S. Khine (ed.), *Science Education in East Asia*,
DOI 10.1007/978-3-319-16390-1_6

Therefore, the present work aims to fill in the aforementioned gap by employing the instruments of the following two international research projects to administer reliable surveys in Shanghai and Guangzhou for collecting relevant data from junior secondary three students and university year one students, respectively. Those first-hand research data and findings should be a valuable contribution to inform and provide insights to the national and international science education communities for comparing and contrasting the relative merits and problems of different education systems, including those of China.

The ROSE Project

In the past, individual teams of researchers in specific contexts (and/or countries or regions) developed or adopted different research instruments and methodologies for their own attitudinal research (see, e.g. Blalock et al., 2008; George, 2006). This led to one major problem in common that there is no easy way to undertake a comparative study directly from student responses as collected by different questionnaires or research instruments. Besides, there are often queries on the validity and reliability of individual research instruments and methodologies (Gable & Wolf, 1993). On the other hand, the well-established international comparative studies like TIMSS (IEA, 2011) and PISA (OECD, 2010, 2014), which have been carefully designed to overcome the aforementioned shortcomings, mainly focus on the academic performance or achievements of students (or learning outcomes) to inform policymakers at the national and international levels, with little data collected on the affective domain of science learning. Therefore, in 2002, the ROSE (Relevance Of Science Education) international comparative project (Schreiner &Sjøberg, 2004) was initiated to fill this research gap as based on the previous 'Science-And-Scientists' study by Sjøberg (2000). The study collected data (on students' interest, attitudes, values and career or study plans related to science and technology) from over 40,000 15-year-old students in approximately 40 countries or regions. The findings reveal striking differences between boys and girls and between developed countries and developing countries in the types of interest in science topics, attitudes and values towards science and technology and their aspirations and considerations for pursuing future careers or further study in science and technology (Sjøberg & Schreiner, 2005).

The term 'relevance' in the ROSE project title actually refers to a wide spectrum of factors (including interest and attitude) that lie in the affective domain of science learning in the broadest sense of the definition. In addition, science education in the project also embraces science and technology (S&T) education. Schreiner and Sjøberg (2004) have published a ROSE project handbook to describe the ROSE background, rationale, questionnaire development and data collection, so that all international collaborators can refer to the document to standardize their methods and procedures for administering the questionnaire surveys and inputting the collected data into computer files.

The IRIS Project

The Interests & Recruitment in Science (IRIS) is another international comparative research project which builds upon the experiences, expertise, research findings and network of international researchers from the ROSE project to collect data from university year one students (see IRIS, 2008 for details) on relationship between young people's educational choices and their priorities, considerations, values and experiences with reference to their science, technology, engineering and mathematics (STEM) higher education. It aims to help the improvement of recruitment, retention and gender equity patterns in STEM education at the tertiary level and careers. The IRIS project is funded by the European Union (EU) under the 'Science in Society' category of the Seventh Framework Programme (7FP) for research and technological development and led by consortium of researchers from six universities or institutions in five EU countries (Norway, the UK, Slovenia, Italy and Denmark).

Gender Differences in Students' Attitudes Towards Science

From the educational research over the last three decades (e.g. Brotman & Moore, 2008; Jones, Howe & Rua, 2000; Kahle & Meece, 1994; Smail & Kelly, 1984; Weinburgh, 1995), it is commonly agreed that gender differences in students' interest, attitudes, academic achievements and experience in science learning are significant issues in S&T education. To account for gender differences in science learning, various theories have been put forward. Chang, Yeung and Cheng (2009) recently attempted to relate these underlying differences to the following four key factors:

1. *Societal factors* – Parents and teachers often see science as a male domain, and so they may offer fewer opportunities for science activities to their daughters or female students than to boys (Burkham, Lee & Smerdon, 1997).
2. *Psychological and identity* factors – Science achievement does not naturally enhance self-esteem among females, and some high-achieving women are likely to underestimate their abilities and performance in science subjects (Gilbert & Calvert, 2003).
3. *Curriculum, pedagogy and school factors* – With unconscious gender bias in the teachers' expectations and the teaching of boys and girls, teachers' influence is one of the main causes of the gender differences in students' experiences of and interests in science learning (Skog, 2001). For example, physics teachers often ignore the scope and educational significance of the gender gap in teaching their subject and know little about gender-inclusive practices (Chetcuti, 2009; Zohar & Bronshtein, 2005).
4. *Career factors* – Women scientists are disadvantaged by their underrepresentation as a group and by being overshadowed by high-ability male scientists,

leading to lower self-concept in sciences (Burkham et al., 1997; Buccheri et al., 2011; Crovato & Pellegrini, 2010).

Characteristics of Chinese Learners

Even though Chinese school students in Shanghai outscored their counterparts in dozens of other countries in science and mathematics as revealed by the PISA 2009 study (OECD 2010) and 2012 study (OECD 2014), yet there were little academic literature or rigorous research findings on Chinese school students' affective domain of learning science, which might help educators identify the key factors for explaining (or contradicting) Chinese learners' outstanding performance. It is well known that the two most famous international comparative studies – TIMSS (IEA, 2011) and PISA – have been developed for the sole purpose of comparing students' academic achievement (or cognitive aspects of learning outcomes). Furthermore, Mainland China has long been practising an examination-oriented education system with teacher-centred pedagogies widely adopted to closely follow the examination syllabuses for science subjects in the classrooms and supplemented with plenty of model answer based homework exercises assigned to the students (Gao, 1998; Wang, Wang, Zhang, Lang & Mayer, 1996; Yeung, 2008). Hence, their academic performance as well as review of the relevant curricular documents and classroom pedagogies could not provide a comprehensive and authentic picture of Chinese science learners. Nevertheless, various previous researchers have attempted to explain their success with different characterizations, viewpoints, foci and assumptions. The major findings/conclusions on the characteristics of Chinese learners could be summarized as follows:

(a) Hardworking and outstanding academic performance – Many students, who are from Confucian heritage cultures (CHC) such as Mainland China, Hong Kong, Taiwan, Singapore, Korea and Japan, study much harder and outperform Western students in science and mathematics in the international comparative studies as well as in the colleges or universities where they are studying abroad (Law, 2002; Watkins & Biggs, 2001).
(b) Passive and rote-learning mode – CHC are obediently following or reliant on the teachers' classroom teaching, but they still could achieve high levels of understanding (Gao, 1998).
(c) Highly positive attitudes towards education – Strong support from family, which requires children to respect teachers and for historical texts in accordance with the CHC tradition, overwhelmingly values education as the most important means to climb the social ladder and regards students' achievements as theirs (Ryan & Slethaug, 2010).
(d) Changes of CHC characteristics – Recent researchers attempt to question the abovementioned CHC characteristics as they suggest the need to adopt 'whole culture' explanation, counters to stereotyping of Chinese learners and challenges to the definition of 'Confucian' values (as re-advocated by the People's

Republic of China government in the last decade). They argue that the rapid globalization and modernization processes, which are occurring in Mainland China and other Greater China regions, render the influx of a lot of Western cultures, values and information to dilute and transform the CHC tradition (see, e.g. Biggs, 1996; Ryan & Slethaug, 2010).

Research Methodology

Research Instrument

For collecting student data on their interests, attitudes, views and motivation in the affective domain of science education, Schreiner and Sjøberg (2004)developed the international ROSE research instrument, of which the theoretical framework of the complete instrument and proper procedures of administration were fully documented in their research handbook. They also established the validity, reliability and credibility of their questionnaire instrument, which had undergone one local test survey in Norway and three rounds of international trials in 2002. Based on student feedback and past comments of other researchers on the validity of affective instruments (see, e.g. Gable & Wolf, 1993; Gardner, 1975, 1996), the Chinese version of the English ROSE research instrument was developed by Yeung and Cheng (2008) to include a section for collecting a unique set of socioeconomic data in the Chinese culture and context and has undergone rigorous processes to ensure its validity and reliability. In general, 4-point Likert scale with 1 = 'Disagree' (for views) or 'Never' (for experiences) and 4 = 'Agree' (for views) or 'Often' (for experiences) was assigned in the ROSE instrument for surveying and analysis.

For the IRIS study, the international IRIS consortium has already developed, piloted and validated its set of IRIS questionnaire instruments with detailed guidelines for translation, sampling, data collection and coding (Henriksen & Schreiner, 2010). Hence, the research instrument and sampling methods are concisely described. In essence, the development of the IRIS questionnaire instrument (known as IRIS Q) was based on the literature review conducted in the four work packages on (1) gender perspectives (Crovato & Pellegrini, 2010), (2) priorities and choices regarding higher STEM education programmes (Anastasiou & Dillon, 2010; Bøe, Schreiner & Henriksen, 2010), (3) recruitment initiatives targeted for students in higher STEM education programmes (Jenkins, Jensen & Henriksen, 2010) and (4) students' dropping out or opting out of higher STEM education programmes (Madsen, Ulriksen & Holmegaard, 2010). These packages also provide the theoretical framework and various theories, models, hypotheses, approaches and previous findings to back up the present study. The instrument's English version was finalized in March 2010 after undergoing a pilot run in all IRIS consortium countries, and other language versions will be translated by individual partners. Apart from some questions on personal particulars, the IRIS Q (Henriksen & Schreiner, 2010) contains a large number of 5-point Likert scale items (with responses ranging from 1 = 'Not Important' or 'Strongly Disagree' to 5 = 'Very Important' or 'Strongly Agree') in seven main categories of questions, plus five open-ended questions. In accordance with the IRIS guidelines, the survey was administered to the first-year students in three large universities in Guangzhou. The targeted students were those studying bachelor degree programmes in STEM subjects like biology and biotechnology, physics, chemistry, mathematics (including some statistics), computer science, mechanics and metalwork, electronics and automation or chemical and process engineering.

Research Questions

Based on an intensive review of the relevant literature and the local educational systems and context as well as the original research questions of the ROSE and IRIS projects, the relevant research questions were derived for the analysis of research data collected for the full-scale ROSE study in Shanghai and IRIS study in Guangzhou as follows:

Q1. What are the Chinese students' S&T-related experiences, considerations for career orientation and views on S&T?

Q2. Do the differences in gender and various socioeconomic factors like family incomes, parental education levels and school types have significant effects on the findings obtained for Q1?

Q3. What are the priorities, considerations, values and experiences upon which young people make their choices for STEM higher education? Are there any gender differences?

Sample Populations

Data collection for the ROSE pilot study has been completed in 2007/2008 school years with 2,423 valid questionnaires returned from 70 classes of secondary three students in Hong Kong, Guangzhou and Shanghai plus 251 student interview records. The slight gender imbalance was due to the survey of a few girls' schools in Hong Kong. The preliminary data analysis by Yeung and Cheng (2008) verified that the overall reliability of the ROSE questionnaire instrument is very high with Cronbach's alpha = 0.98. The full-scale ROSE study was administered in Hong Kong and Shanghai with a total of 4411 valid respondents during 2008/2009 and 2008/2009 school year, respectively (see Yeung & Cheng, 2010, 2011 for details). A summary of statistics for the ROSE sample populations is shown in Table 6.1a while Table 6.1b reports the corresponding statistics for the IRIS study of over

Table 6.1 Summary of statistics for the sample populations involved in the ROSE pilot in 3 cities (Hong Kong, Guangzhou and Shanghai) and full-scale study in Shanghai and Hong Kong and IRIS study in 3 Guangzhou universities

		Full-scale study	
(a) ROSE data	Pilot study in 3 cities	Shanghai	Both cities
No. of schools	24	25	53
No. of classes	70	82	139
No. of respondents	2,423	2,419	4,411
Male to female ratio (%)	45.9:54.1	50.1:49.9	50.7:49.3

		Type of university in Guangzhou		
(b) IRIS data	Total	Normal	Technology	Comprehensive
No. of respondents	2,740	760	1,100	880
No. of STEM programmes	27	11	12	10
Male to female ratio (%)	66.3:33.7	42.6:57.4	83.8:16.2	64.9:35.1

2,700 undergraduate year one STEM students in three Guangzhou universities in year 2012 (Yeung, Cheung & Sjøberg, 2013).

Methods of Data Analysis

Taking into account the hypotheses implicitly assumed for the ROSE and IRIS research instruments and data, the SPSS software (Dancey & Reidy, 2002) was employed to carry out exploratory factor analysis (EFA) of the pilot study data of 3 cities to uncover the underlying factor structure of the ROSE instrument (Yeung & Cheng, 2010, 2011). At the same time, this process has rendered data reduction (i.e. simplification of the ROSE instrument) as well as the calculations of factor scores (or scales) for the second-level EFA. In fact, they obtained 25 well-recognized factors which satisfied the scree test and Kaiser's criteria with eigenvalue >1 in the first-level EFA (Yeung & Cheng, 2011). Based on a critical review of the nature of the 25 factors and a second-level EFA, they obtained four distinct categories of factors (i.e. factors of factors) for the ROSE instrument, namely, 13 factors in A (Interest) representing what students want to learn, six factors in B (Experiences) representing students' S&T-related experiences in their daily life and out-of-school activities, three factors in C(Job) representing students' considerations of their career orientations or choices of jobs and another three factors in D (Views),representing students' views on S&T and related issues. For the present analysis, descriptions of those 12 factors in categories B, C and D and their constituent ROSE items, Cronbach's reliabilities and factor scores are reported in Table 6.2. Those 13 interest-related factors in category A on interestingness of S&T topics are excluded as they are being written for publication in another journal paper. Besides, Yeung and Cheng have employed the AMOS software for structural equation modelling (SEM) (Weston & Gore, 2006) to conduct confirmatory factor

Table 6.2 Three categories of 12 noninterest factors (with Cronbach's reliability α) on students' (B) S&T-related experiences, (C) job/career orientations and (D) views on S&T issues as based on the first-level EFA of 160 ROSE items in the pilot studies administered in Hong Kong, Guangzhou and Shanghai and factor scores for the pilot study in those 3 cities and full-scale studies in Hong Kong and Shanghai

			ROSE items involved			Factor scores		
							Full-scale study	
Category	Factor label	Description of factor	Labels[a]	No.	α	Pilot study in 3 cities	Shanghai only	Both cities
B (Experiences)	FB1	Outdoor living experience	H10, H07, H15, H11, H21, H19, H18, H22, H06, H16, H14, H20	12	0.888	1.760	1.674	1.741
	FB2	Hands-on experience of doing transportation	H57, H58, H59, H56, H61, H55	6	0.858	1.740	1.736	1.778
	FB3	Daily life experience of DIY tools and models	H33, H32, H35, H34, H36	5	0.808	1.964	1.940	1.957
	FB4	Students' medical experience	H26, H28, H27, H29	4	0.602	2.243	2.297	2.225
	FB5	Out-of-school experience in learning science	H12, H13, H09, H08	4	0.743	2.747	2.744	2.655
	FB6	Experience of using handy tools and computers	H42, H43, H41, H31, H38, H48, H51, H40	8	0.796	2.944	2.817	2.868
C (Job)	FC1	Jobs of high degree of autonomy and independence	B13, B15, B16, B24, B25, B14, B09	7	0.771	3.368	3.358	3.285
	FC2	Jobs need to get along with others	B02, B26, B01	3	0.640	3.205	3.171	3.118
	FC3	Jobs requiring creativity in S&T	B10, B11, B06, B07, B04, B08, B03	7	0.736	2.734	2.796	2.681
D (Views)	FD1	Favourable views on the nature of S&T	G02, G04, G03, G01, G05	5	0.859	3.415	3.426	3.346
	FD2	Positive views on the effects of S&T	G08, G07, G09, G06, G15	5	0.729	2.501	2.586	2.506
	FD3	Positive views on school science	F12, F10, F11, F07, F15, F09, F08, F04, F02, F13, F05, F06	12	0.904	2.876	2.821	2.777

[a]ROSE items are listed in descending order of factor loadings with cutoff at 0.5, and their full descriptions are given in Schreiner and Sjøberg's (2004) ROSE project handbook

analysis (CFA) or SEM of the full-scale study in Hong Kong and Shanghai. For the absolute measure of CFA fit, the root mean square error of approximation (RMSEA) indicates that the data set for each of their full-scale study should be rated as mediocre to good fit according to the criteria set by various researchers like Mulaik et al. (1989) and Hooper et al. (2008). The factor scores were further adopted as standardized indexes for ease of comparison using various statistical analysis methods (Ferguson & Takane, 1989), like the Student's t-test and ANOVA F-test, in answering the first and second research questions (Q1 and Q2).

To address the third research question Q3, it is quite straightforward to select data of certain relevant questionnaire items of the IRIS study in the three Guangzhou universities for comparison using the Student's t-test and ANOVA F-test.

Results and Discussion

Cross-Region Comparison

From the last three columns on ROSE factor scores of Table 6.2, it is obvious that Shanghai and Hong Kong students are having rather few experiences in outdoor living (FB1, e.g. milked animals, cared for animals on farm, participated in hunting, made dairy products etc.), hands-on activities for transportation (FB2, e.g. used a crowbar, a wheelbarrow and a rope and pulley for lifting, mended a bicycle tube etc.) and using do-it-yourself (DIY) tools and models (FB3, e.g. used an air gun or rifle and made a bow and arrow, slingshot, catapult, a model such as toy plane or boat) as their average scores are all below 2.0 (under the 4-point Likert scale with 1 = 'Never' and 4 = 'Often'). Those findings are much lower than those obtained by Lavonen et al. (2008) on the EFA of ROSE study of Finnish students, revealing six factors (with factor scores of 2.1–3.3) on out-of-school experiences in physics and chemistry-related themes. Likewise, Christidou (2006) obtained factor scores of the values 2.45–3.38 for the five factors on students' out-of-school experiences through the EFA of the responses from 583 ninth-grade Greek students. In all places, students are having richest experiences on using computers. Comparing the factor scores with Hong Kong, the Shanghai students of either genders have statistically less experiences in FB1, FB2 and FB6 (e.g. measured the temperature with a thermometer; used a measuring ruler, tape, stick, a stopwatch and a camera or a computer as a dictionary, encyclopaedia or a word processor; etc.) than their counterparts but more in medical (FB3) and out-of-science science learning (FB5) experiences.

For career orientations, Hong Kong and Shanghai students both put highest preference for jobs of high degree of autonomy and independence, followed by those who need to get along with others and least preference for jobs requiring creativity in S&T (see Table 6.2). The latter finding is quite unfortunate, but this is well expected from the overwhelming teacher-centred pedagogies and excessive drilling exercises adopted in typical Chinese school education. Table 6.3 shows that

Table 6.3 Student's t-test statistics for comparing the factor scores between (a) students from Hong Kong (HK) and Shanghai (SH) and (b) boys and girls in the full-scale study

			(a) Differences between 2 cities (HK-SH)						(b) Gender differences (boys-girls)					
Category	Subcategory	Factor label	All		Boys only		Girls only		Both cities		HK		SH	
B (Experiences)		FB1	7.60	***	4.95	***	5.91	***	5.25	***	3.47	**	3.89	***
		FB2	4.33	***	3.01	**	3.03	**	12.28	***	8.30	***	9.02	***
		FB3	1.71		0.70		1.58		22.03	***	14.29	***	16.76	***
		FB4	−7.67	***	−4.60	***	−6.38	***	2.24	*	2.22	*	1.09	
		FB5	−9.55	***	−5.75	***	−7.90	***	−0.36		0.62		−0.87	
		FB6	5.69	***	3.33	**	4.80	***	−4.03	***	−3.61	***	−2.29	*
C (Job)		FC1	−9.74	***	−6.16	***	−7.75	***	−7.36	***	−4.50	***	−5.85	***
		FC2	−5.61	***	−4.48	***	−3.35	**	−3.43	**	−3.08	**	−1.80	
		FC3	−13.35	***	−8.95	***	−9.99	***	1.48		1.44		0.90	
D (Views)		FD1	−9.30	***	−4.89	***	−8.57	***	−1.23		0.74		−2.25	*
		FD2	−8.74	***	−5.35	***	−7.37	***	7.30	***	5.58	***	4.97	***
		FD3	−4.76	***	−2.95	**	−3.83	***	1.35		1.36		0.63	

Descriptions of individual factors are given in Table 6.2

Remarks: (1) $t > 0$ (or $t < 0$) for HK factor score is higher (or lower) than that of SH (for comparison between the two cities) or the boys' factor score is higher (or lower) than that of the girls (for gender differences). (2) *** for $p < 0.001$; ** for $p < 0.01$; and * for $p < 0.05$ for cases with statistically significant difference

Shanghai students of both genders rated with more importance in those aspects of career orientation than those of Hong Kong students, and the regional differences are all statistically significant. There are no similar types of ROSE factor scores available from other countries for direct comparison.

For views on S&T issues, Hong Kong and Shanghai students have the most favourable views on the nature of S&T, followed by quite positive views on school science, and have the least positive views on the effects of S&T (see Table 6.2). The findings go along with those of the developing countries and are in substantial contrast with those of the developed countries whose children generally hold rather negative or unfavourable views on S&T (Sjøberg & Schreiner, 2005). In particular, Jenkins and Nelson's (2005) ROSE study in England reported that students viewed secondary school science as important but not for them. Lyons (2006) unveiled from some student narratives of their school science experiences as 'boring and very hard'. Atwater, Wiggins and Gardner (1995) revealed from their survey of over 1,400 students (with 91 % of African American) in 3 middle schools that only a quarter of those intending to pursue further study or jobs in S&T had positive attitudes towards science. This is in stark contrast with the largely positive attitudes held by students in many African countries (Sjøberg & Schreiner, 2005). For the Shanghai students, Table 6.3 shows that both genders of them hold more favourable or positive views on S&T than those of Hong Kong students, and the differences are all statistically significant.

Gender Differences

With respect to the three categories of factors given in Table 6.2 and based on the Student's *t*-test results given in Table 6.3b, we highlight the following gender differences for the full-scale ROSE study in Hong Kong and Shanghai:

Experiences related to S&T – For both cities, there are statistically significant gender differences found in factors FB1 'outdoor living experience', FB2 'hands-on experience of doing transportation', FB3 'daily life experience of DIY tools and models' and FB6 'using handy tools and computers'. In all the first three factors, boys were having richer experience than girls. For the last factor, girls were having more experience than boys, and the average factor scores for both genders (of 2.8–2.9 in Table 6.2) are much higher than those for the first three factors. On comparing the findings with those of Science and Scientists (SAS) study by Sjøberg (2000), those gender-specific differences seem to be closer to those of the developed rather than developing countries or regions. Those findings are also similar to Johnson's (1987) analysis of the 1984 APU survey on 11-year-old boys and girls in which boys were more often engaged in activities like 'make models from a kit', 'play pool billiards or snooker', 'play with electric toy sets', 'create models using Lego' and 'take things apart to see inside', while girls got more experience in activities such as 'knit or sew', 'weigh ingredients for cooking' and 'collect/look at wild flowers'. However, the extent and types of gender-specific differences for Chinese learners cannot be exactly compared with those reported by the ROSE study in Finland and Greece because of nonidentical factors extracted from their EFA processes. In particular, the Finnish findings obtained by Lavonen et al. (2008) indicated that boys had richer experiences than girls in three of their six factors (namely, FE1 'S&T-related hobbies or activities', FE5 'camping' and FE6 'use of mechanical tools') while girls had more experiences than boys in another two factors – FE2 'measuring and observing with simple tools' and FE3 'observing natural phenomena and collecting objects'. For the five factors on science-related experiences as obtained by Christidou (2006) in the ROSE data of Greek students, boys had more experiences than girls in the factor 1 'Manual work' and factor 4 'Using a computer' while girls exceeded those of boys in factor 2 'Using instruments and technological devices', factor 3 'Seeking information about nature' and factor 5 'Cuisine and handicraft'. On the other hand, the study of 437 sixth grade students by Jones et al. (2000) gave evidence that more boys than girls had out-of-school experiences in 'a variety of tools and objects, including rifles, batteries, electric toys, fuses, microscopes and pulleys' while more girls than boys had prior experiences in 'bread making, observing birds and stars, knitting, sewing and planting seeds'.

Considerations for career orientations – For factors of significant influence on students' choices or considerations of future career, the factors FC1 (e.g. making my own decisions, working with something I find important and meaningful or fits my attitudes and values, becoming 'the boss' at my job etc) and FC2 (e.g. helping other people, working as a part of a team and working with people rather than things) are considered to be more important by girls than by boys. The gender differences are statistically significant in Hong Kong for both factors and in Shanghai for FC2 only. Those findings are somewhere consistent with and somewhere different from that obtained or commented by Crovato and Pellegrini (2010), Sjøberg (2000) and Warrington and Younger (2000). For example, Sjøberg's (2000) SAS study revealed that boys considered 'control other people', 'become famous', 'make and invent new things' and 'earn lots of money' to be more important for the choice of job than that of girls while girls considered 'work with people instead of things' and 'help other people' to be more important. The previous study by Jones et al. (2000) also led to very similar findings on the gender differences. Warrington and Younger's (2000) survey showed that around a half of girls tended to choose a job dominated by female employees despite of more restrictive income levels and opportunity, probably because of the underrepresentation problem in women's *career factors* (Burkham et al., 1997; Crovato & Pellegrini, 2010). Boys were also found to follow traditionally gendered paths in their career orientation, and around 50 % of them selected male jobs in the fields of science, computing, engineering, mechanics and electricians etc. Less than 10 % of them aimed at jobs in health care, teaching or veterinary science. This kind of gender stereotype is well known to be a crucial determinant of women's career choices (Buccheri et al., 2011).

Views on S&T and related issues – For both cities, boys were holding more 'positive views on the effects of S&T' (i.e. FD2 which contains items like S&T will find cures to diseases; make our lives healthier, easier and more comfortable; give greater opportunities for future generations; and are important for society) than girls. For girls in Shanghai, by contract, they were holding more favourable 'views on the nature of S&T' (i.e. FD1 which contains items like S&T can solve nearly all problems, help to eradicate poverty and famine in the world, are helping the poor, are of greater benefits than harm and scientists are neutral and objective) than boys. The differences are statistically significant. Those gender differences are similar to those found in the developing countries or regions in Sjøberg's (2000) SAS study but contrary to those found in the developed countries and in Francis and Greer's (1999) study of 2129 pupils in Northern Ireland – boys showed a more positive attitude towards science than girls.

Effects of Socioeconomic Factors

As each socioeconomic variable contains three sets of data (labelled as L, M and H with different meanings as given in Table 6.4), it is necessary to apply the ANOVA test (with calculation of the corresponding F statistic) to investigate the effects of four socioeconomic variables, namely, family incomes as perceived by the students themselves, paternal and maternal education levels and school types which are subdivided into high, medium and low profiles in a relative sense. If an ANOVA test shows a statistically significant difference for a particular variable, the profile label for the highest factor score will be marked in front of the F statistic (however, it is sometimes possible for one group to be significantly lower than the other two groups, and so the highest of the other two is not necessarily different from the other). Note that the present article just reports the findings on the Shanghai students without comparison with the Hong Kong students because of the large differences in their socioeconomic background. The four socioeconomic variables do have significant correlation with students' S&T-related experiences. In particular, family income is the most important socioeconomic variable which has a very strong correlation with almost all aspects of students' affective domain of science learning. Under the China's national one-child policy which is strictly enforced in urban areas, it is well known that Shanghai parents would provide all kinds of support and opportunities to their only child whenever they could afford in accordance with the CHC tradition. Hence, this variable would lead to the most significant differences among the Shanghai students. Its importance is closely followed by the parental education levels while the school-type variable plays the least important role. For ROSE factors FB1 'outdoor living experience' and FB2 'hands-on experience of doing transportation', the correlation is sometimes negative (for family income and maternal education levels) and sometimes insignificant (for paternal education levels and school types). In other words, the results indicate that Shanghai students from low-income family or poor mother's education level will tend to have comparatively richer outdoor living experiences. For the remaining four factors in the category B for S&T-related experiences, all variables

Table 6.4 ANOVA test (*F* statistic) for the effects of (a) family incomes, (b) father's education levels, (c) mother's education levels and (d) school types on students' factor scores for the full-scale study in Shanghai

Category	Factor label	(a) Family incomes			(b) Father's education levels			(c) Mother's education levels			(d) School types		
B(Experiences)	FB1	L	7.43	**		1.58		L	3.64	*		1.15	
	FB2		1.72			1.41		L	5.01	**		0.41	
	FB3	H	14.25	***	H	14.32	***	H	8.82	***	H	6.22	**
	FB4	H	16.65	***	H	9.26	***	H	14.85	***	H	5.14	**
	FB5	H	20.07	***	H	12.13	***	H	18.42	***		2.10	
	FB6	H	54.12	***	H	48.11	***	H	49.25	***	H	33.69	***
C(Job)	FC1	H	26.11	***	H	5.83	**	H	9.86	***	H	6.10	**
	FC2	H	4.50	*		0.27			0.49			0.12	
	FC3	H	4.51	*		0.68			0.33			1.19	
D (Views)	FD1	H	27.87	***	H	7.40	**	H	4.58	*	H	3.55	*
	FD2	H	6.86	**	H	5.19	**	H	3.10	*		0.46	
	FD3	H	3.82	*		1.56			2.07			0.87	

Remarks: (1) The group with the highest factor score is labelled as follows:
For family incomes: *L* low-income, *M* ordinary and *H* middle class or rich
For parent's education levels: *L* primary education, *M* secondary education and *H* tertiary education
For school type: *H* high rank, *M* middle rank and *L* low rank
(2) *** for $p < 0.001$; ** for $p < 0.01$; and * for $p < 0.05$ for cases with statistically significant difference

do have a statistically significant and positively correlated effect on Shanghai students' experiences (except school types on FB5). Students from higher-income families, better parental education levels and high-ranking schools are positively correlated with greater extent of S&T-related experiences.

For students' considerations of career orientations and choices of jobs, family income is the key factor which makes statistically significant differences for all the three factors FC1–3, and those students from higher-income families often put greater importance for considerations of those factors. Besides, the factor FC1 is also significantly correlated with the parental education levels and school types in which students with parents of better education levels and high-ranking schools often consider this factor as more important.

For students' views on S&T and related issues, family income is again the key factor which makes statistically significant differences for the all the three factors FD1–3, and those students from higher-income families often hold more favourable or positive views on those factors. Besides, the factors FD1 and FD2 are also significantly correlated with the parental education levels in which students with parents of better education levels always show more favourable or positive views on the nature and effects of S&T. Furthermore, students from high-ranking schools tend to hold favourable or positive views on the nature of S&T.

Considerations Underlying Choices of University STEM Programmes

Table 6.5 reports the key findings on questionnaire items describing students' choices of university programmes from the IRIS study of over 2,700 year one STEM undergraduates in three Guangzhou universities. For the several responses within each question (except Q17), the ANOVA test indicates statistically significant differences in their relative importance. Student's *t*-test has also been carried for every item to reveal the gender differences, if any. For the four factors shown in the IRIS Q6–8 and Q17, students' prior school experiences of both genders are the most important one, significantly higher than those of the other three factors as given in Q7, 8 and 17 according to the ANOVA test (with $p < 0.001$). The three most important items in this factor are (1) students' interest in the subject, (2) using mathematics in lessons and (3) lessons showing the relevance of the subject to society. The findings are more or less consistent with Woolnough's (1994) study of factors affecting students' choice of STEM programmes in England in which the three most important encouraging factors considered by those 1,180 18-year-old respondents were (1) practical nature of the science lessons, (2) supportive math teaching in the school and (3) intellectual satisfaction of doing science. For the IRIS study, there is no statistically significant gender difference found for those three factors and for all items of this factor. However, females rated with higher importance than males in their previous attainment in related subjects, and this provides

Table 6.5 Results of IRIS study in 3 Guangzhou universities on the relative importance of various factors affecting students' choices of university STEM programmes

Questions and possible responses	Male	Female	Both
6. How important were each of the following school experiences in choosing your course?	**3.60(1.33)**	**3.57(1.25)**	**3.59(1.30)**
6a. Interest in the subject	4.01(1.22)	4.03(1.16)	4.02(1.20)
6b. Previous attainment in related subjects	3.54(1.18)	3.69(1.15)	3.59(1.17)**
6c. Experiments/laboratory work	3.59(1.26)	3.43(1.26)	3.53(1.26)**
6d. Field work or excursions	3.26(1.32)	3.16(1.32)	3.23(1.26)*
6e. Lessons showing the relevance of your subject to society	3.72(1.70)	3.68(1.15)	3.71(1.54)
6 f. Using mathematics in lessons	3.86(1.20)	3.82(1.17)	3.84(1.19)
6 g. Lessons showing practical applications of your subject	3.48(1.26)	3.36(1.31)	3.44(1.28)*
6 h. Clear feedback on whether you got the right answer	3.36(1.28)	3.39(1.24)	3.37(1.27)
7. How important were the following persons in choosing your course?	**3.05(1.38)**	**3.12(1.40)**	**3.07(1.38)****
7a. Mother	3.13(1.39)	3.20(1.32)	3.15(1.37)
7b. Father	3.26(1.39)	3.33(1.33)	3.28(1.37)
7c. Good teachers	3.26(1.38)	3.63(1.67)	3.38(1.49)***
7d. Friends (including boyfriend/girlfriend)	2.95(1.34)	2.85(1.25)	2.92(1.31)*
7e. Siblings or other relatives	2.97(1.32)	2.87(1.25)	2.93(1.30)
7 f. Careers advisors in school	2.71(1.36)	2.82(1.33)	2.75(1.35)*
8. How important were each of the following in choosing your course?	**2.91(1.37)**	**2.70(1.33)**	**2.84(1.36)*****
8a. Popular science books and magazines	3.54(1.18)	2.89(1.34)	3.09(1.35)***
8b. Science fiction or fantasy books/films	2.80(1.31)	2.49(1.25)	2.70(1.30)***
8c. Museum/science centre	2.44(1.37)	1.86(1.09)	2.25(1.31)***
8d. Computer games	3.10(1.35)	2.92(1.32)	3.04(1.34)**
8e. Popular science television channels/ programmes (Discovery channels etc.)	3.29(1.32)	3.11(1.31)	3.23(1.32)**
8f. Films or drama on television (CSI etc.)	2.71(1.30)	2.68(1.28)	2.70(1.29)
8g. One specific national item covering competitions etc.	2.75(1.36)	2.76(1.31)	2.75(1.34)
8h. One specific national item covering 'science outreach' of other kinds (festivals etc)	3.01(1.39)	2.92(1.33)	2.98(1.37)
17. How important was the cost of study in your choice of course?	**2.98(1.30)**	**2.88(1.26)**	**2.94(1.28)**

The standard deviations are given in brackets behind the mean scores
Remarks:
*** for $p < 0.001$; ** for $p < 0.01$; and * for $p < 0.05$ for cases with statistically significant difference

another viewpoint on females' underestimation of their abilities and performance obtained by Gilbert & Calvert (2003) for the *psychological and identity factors*. On the other hand, males rated with higher importance than females in (1) the experiments/laboratory work, (2) field work or excursions and (3) lessons showing practical applications of the subject.

For the importance of persons in affecting the students' choices of programmes, good teachers were the most important one as responded by both male and female students in which the latter rated with a significantly higher importance than the former at $p < 0.001$. Statistically significant gender differences are also found in another two items in which more males than females considered friends (including boyfriend/girlfriend) as important persons in affecting their choices of university programmes while more females than males considered career advisors in school as important persons. However, the second and third important persons in this factor were father and mother, respectively. A Student's *t*-test confirmed that the father was rated as more important than the mother, and this is well expected from the paternal dominance in CHC culture as deeply embedded in traditional Chinese families. The present findings are quite different from those obtained by Hazari, Sadler and Tai (2008) in which only females rated teacher's, mother's or father's encouragement towards science as important while males responded with nobody's encouragement. Anyway, the findings confirm that parents and teachers are important *societal factors* which may lead to the gender differences as pointed out by Burkham et al. (1997).

For the eight items on the factor of informal science learning as given in Q8, males rated many of them with higher importance than females, including (1) popular science books and magazines, (2) popular science TV channels/programmes, (3) computer games, (4) science fiction or fantasy books/films and (5) museum/science centre, and the gender differences for those five items and for items as a whole are statistically significant at $p < 0.01$ or 0.001. The results are consistent with the positive affective outcomes resulted from the visit to a science education centre as found by Rennie (1994). The findings are moderately correlated with boys' richer S&T-related experiences as revealed from the ROSE study and consistent with the findings by Hazari et al. (2008) that males reported more outside-of-class experiences than females in the knowledge of physics from hobbies, from media (e.g. TV, Internet) and from reading books/magazines.

For the cost of study as given in Q17, its average rating of 2.94 is slightly below the middle-point value of this 5-point Likert scale, and so it is not considered as an important factor. There is no significant gender difference found. In fact, the local or central government provides the major financial subsidies to all universities in China, and so the university tuition fee is set at a level affordable by most ordinary families. Besides, students could easily earn their living by private tutoring which is very pervasive in most cities of China.

Limitations and Future Work

Despite of the reliable research instruments used for the full-scale collection of data from the ROSE and IRIS study, the present work suffers from the following key limitations and constraints which may affect the reliability and validity of the results and require further work for improvement:

1. Rural region data have not been collected for the ROSE in China, and this will limit the scope of validity or generalizability of the overall results (especially those related to the students' experience of S&T in the wildlife and farming field). Hence, the present findings are applicable to those Chinese students residing in the municipal cities only, and so new surveys should be specifically targeted at those living in the remote rural regions of Mainland China.
2. The sample population of junior secondary three students for the ROSE study in Shanghai, who are mostly urban dwellers, is not compatible with the undergraduate year one students in three Guangzhou universities because the latter came from both urban and rural regions of the Guangdong Province and the whole Mainland China. Hence, the out-of-school S&T-related experiences for those two sample populations may be different, and correlational analysis could be done only after proper identification of the undergraduates' residential regions.
3. Contrary to the ROSE study whose targeted population had undergone the same nine years of compulsory education, the IRIS study collected data from students of STEM programmes only. It cannot reveal the characteristics and considerations of those who did not choose STEM programmes. Another round of IRIS study should be administered to those undergraduates enrolled in non-STEM programmes, and then it may be possible to carry out the comparison between students of different programmes to uncover crucial results to explain why some able students are migrating away from the STEM stream (Anastasiou & Dillon, 2010).
4. The EFA factors on S&T-related experiences as obtained from the ROSE study in different countries like England (Jenkins & Nelson, 2005), Finland (Lavonen et al., 2008; Uitto et al., 2006) and Greece (Christidou, 2006) cannot be directly compared between themselves and with our present study, ruling out the possibility for international comparison. Besides, some S&T-related experiences are not quite applicable to or rarely found in the Chinese communities.
5. The huge amount of qualitative data collected from the third part of the ROSE questionnaire and interview of students in the ROSE and IRIS are to be compared with the quantitative findings for triangulation and explanation of the latter results and to be translated from simplified Chinese into English.
6. Full-scale ROSE study has also been administered in the Xinjiang Autonomous Region and Sichuan Province of China for cross-region comparison, and those findings for interest-related factors have been written for publication in another journal paper (Yeung & Li, 2015)

Conclusions and Educational Implications

Based on the full-scale ROSE study junior secondary three students in Shanghai (as well as Hong Kong) and the IRIS study of university year one undergraduates in Guangzhou, some key characteristics of Chinese learners can be concluded from their affective domain (excluding the interest component) and choices of science learning as follows:

1. Chinese students generally have lesser extent of S&T-related experiences outside their classroom environment as compared with those of the Western students, and the gender-specific differences are closer to those of the developed rather than developing countries (Johnson, 1987; Sjøberg, 2000). It is not possible to make direct comparison with the ROSE findings from other countries because of different factors extracted from the EFA process.
2. Chinese students highly prefer jobs of high degree of autonomy and independence. There are statistically significant gender differences in their career orientation and preferred fields of career in which the gender stereotype and differences are quite similar to those of many other studies (Crovato & Pellegrini, 2010; Jones et al., 2000; Sjøberg, 2000; Warrington & Younger, 2000), confirming the cross-region phenomenon that gender has a greater impact than interest in affecting students' choices of career (Buccheri et al., 2011).
3. Chinese students have very favourable views on the nature of S&T and quite positive views on school science, and there are statistically significant gender differences found in those two factors. In terms of those views and gender differences, Chinese students' characteristics look similar to the students in developing countries but substantially dissimilar to those in developed countries (Jenkins & Nelson, 2005; Lyons, 2006; Sjøberg & Schreiner, 2005).
4. Among the four socioeconomic factors, family income is the most crucial one which positively correlates with greater extent of Chinese students' S&T-related experiences (but negatively with the outdoor living), higher importance for the consideration of jobs and more positive/favourable views on S&T. The underlying reason is likely attributed to the national one-child policy of China. As expected from the CHC tradition, parental influences as reflected from their education levels also play important roles in positively correlating with many types (or FB3-6 factors) of their children's S&T- related experiences, more preference in considering jobs of high degree of autonomy and independence and more favourable/positive views on the nature of S&T and the effects of S&T. The school-types factor has comparatively lesser impact on those three aspects of students' affective domain of science learning.
5. For Chinese students' choices of university STEM programmes, their prior school experiences play the most important roles, especially for their interest in the subject, using mathematics in lessons and relevance of subject lessons to society. There is no statistically significant gender difference found. As expected from the CHC tradition again, good teachers (followed by parents) are the most influential persons for affecting students' choices of study, especially for females. Besides, males are much more influenced than females by their informal science learning experiences, especially by popular science books, magazines and TV programmes.

Educational Implications

The most significant academic merit/impact of this work is to provide the missing data on Chinese learners in the ROSE project to enable the international education community to achieve a more comprehensive and evidence-based understanding of the affective domain and choices of science learning among these students. The characteristics of Chinese learners still look similar to those of developing countries in many aspects but tend to migrate towards those of developed countries, implying a likely decline in their interest of pursuing study in STEM programmes and S&T-related careers in the future. Hence, the Chinese government needs to make a long-term manpower plan to cope with the likely shortage of qualified technicians and engineers for maintaining her huge manufacturing industry in the next few decades or to drastically transform her structure of economy. The large gender differences in many aspects together with this likely trend of unfavourable development should be taken as an early alarm message by (1) the educational policymakers and curriculum planners who should refine the science curricula and related informal science learning opportunities in Mainland China to maintain students' interest in STEM subjects and careers; (2) for science teachers to improve their teaching and learning materials and activities to compensate for the gender-specific and socioeconomic differences of the students with respect to their S&T-related experiences outside their classrooms and to nurture the students' creativity in S&T; and (3) for individual students to reflect on their future choice of university STEM programmes and/or career in the field of S&T, especially for reducing the gender stereotype.

Acknowledgements Financial support from the Hong Kong Institute of Education (HKIEd) and Hong Kong Research Grants Council is gratefully acknowledged. Great thanks are also due to Prof. S. Sjøberg, Prof. M.H. Cheng and Dr. K.L. Cheung for their research collaboration in the ROSE and IRIS projects and to nearly two dozen persons (scholars, research assistants and student helpers) in HKIEd, East China Normal University, South China Normal University, Guangdong University of Technology and Guangzhou University for their help which facilitated the collection and processing of the research data.

References

Anastasiou, P., & Dillon, J. (2010). *Review on young people's (particularly girls') educational choice with focus on reasons for choosing (or not choosing STEM)*. Retrieved 10 June, 2010, from http://iris.fp-7.org/

Atwater, M. M., Wiggins, J., & Gardner, C. M. (1995). A study of urban middle school students with high and low attitudes toward science. *Journal of Research in Science Teaching, 32*(6), 665–677.

Biggs, J. (1996). Western misperceptions of the Confucian-heritage learning culture. In D. A. Watkins & J. B. Biggs (Eds.), *The Chinese learner: Cultural, psychological and contextual influences*. Hong Kong, China: CERC and ACER.

Blalock, C., Lichtenstein, M., Owen, S., Pruski, L., Marshall, C., & Toepperwein, M. (2008). In pursuit of validity: A comprehensive review of science attitude instruments 1935–2005. *International Journal of Science Education, 30*(7), 961–977.

Bøe, M. V., Schreiner, C., & Henriksen, E. K. (2010). *Review on theoretical perspectives for understanding young people's educational choices*. Retrieved 10 June, 2010, from http://iris.fp-7.org/

Brotman, J. S., & Moore, F. M. (2008). Girls and science: A review of four themes in the science education literature. *Journal of Research in Science Teaching, 45*(9), 971–1002.

Buccheri, G., Gurber, N. A., & Brühwiler, C. (2011). The impact of gender on interest in science topics and the choice of scientific and technical vocations. *International Journal of Science Education, 33*(1), 159–178.

Burkham, D. T., Lee, V. E., & Smerdon, B. A. (1997). Gender and science learning early in high school: Subject matter and laboratory experiences. *American Educational Research Journal, 34*(2), 297–331.

Chang, S. N., Yeung, Y. Y., & Cheng, M. H. (2009). Ninth graders' learning interests, Life experiences and attitudes towards science & technology. *Journal of Science Education and Technology, 18*(5), 447–457.

Chetcuti, D. (2009). Identifying a gender-inclusive pedagogy from Maltese teachers' personal practical knowledge. *International Journal of Science Education, 31*(1), 81–99.

Christidou, V. (2006). Greek students' science-related interests and experiences: Gender differences and correlations. *International Journal of Science Education, 28*(10), 1181–1199.

Crovato, S., & Pellegrini, G. (2010). *Gender and feminist perspectives. Empirical findings on scientific careers*. Retrieved 10 June, 2010, from http://iris.fp-7.org/

Dancey, C. P., & Reidy, J. (2002). *Statistics without maths for psychology – Using SPSS for windows* (2nd ed.). Harlow, England: Pearson Prentice Hall.

European Commission. (2004). *Europe needs more scientists*! (Report by the high level group on increasing human resources for science and technology in Europe). Retrieved 9 August, 2014, from http://ec.europa.eu/research/conferences/2004/sciprof/pdf/final_en.pdf

Ferguson, G. A., & Takane, Y. (1989). *Statistical analysis in psychology and education* (6th ed.). New York, NY: McGraw-Hill.

Francis, L. J., & Greer, J. E. (1999). Measuring attitude towards science among secondary school students: The affective domain. *Research in Science & Technological Education, 17*(2), 219–226.

Gable, R. K., & Wolf, M. B. (1993). *Instrument development in the affective domain*. Boston, MA: Kluwer Academic Publishers.

Gao, L. B. (1998). Cultural context of school science teaching and learning in the People's Republic of China. *Science Education, 82*, 1–13.

Gardner, P. L. (1975). Attitudes to science: A review. *Studies in Science Education, 2*, 1–41.

Gardner, P. L. (1996). The dimensionality of attitude scales: A widely misunderstood idea. *International Journal of Science Education, 18*(8), 913–919.

George, R. (2006). A cross-domain analysis of change in students' attitudes toward science and attitudes about the utility of science. *International Journal of Science Education, 28*(6), 571–589.

Gilbert, J., & Calvert, S. (2003). Challenging accepted wisdom: Looking at the gender and science education question through a different lens. *International Journal of Science Education, 25*(7), 861–878.

Hazari, Z., Sadler, P., & Tai, R. (2008). Gender differences in the high school and affective experiences of introductory college physics students. *The Physics Teacher, 46*, 423–427.

Henriksen, E. K., & Schreiner, C. (2010). *Guidelines for IRIS-Q translation, sampling and data collection and coding*. Retrieved 8 September, 2010, from http://iris.fp-7.org/

Hooper, D., Coughlan, J., & Mullen, M. R. (2008). Structural equation modelling: Guidelines for determining model fit. *Electronic Journal of Business Research Methods, 6*(1), 53–59.

IEA. (2011). TIMSS and PIRLS International Study Center. http://www.timss.org/. Accessed 18 Sept 2011.

IRIS. (2008). *Description of work for IRIS – Interests & Recruitment in Science: Factors influencing recruitment, retention and gender equity in science, technology and mathematics higher education*. Unpublished project proposal for the SEVENTH FRAMEWORK PROGRAMME. Retrieved 8 September, 2010, from http://iris.fp-7.org/

Jenkins, E. W., & Nelson, N. W. (2005). Important but not for me: Students' attitudes towards secondary school science in England. *Research in Science and Technological Education, 23* (1), 41–57.

Jenkins, E., Jensen, F., & Henriksen, E. K. (2010). *Recruitment initiatives and choice of STEM higher education: Review of theoretical perspectives and empirical findings regarding recruitment initiatives inside and outside school: Part 1&2*. Retrieved 10 June, 2010, from http://iris.fp-7.org/

Johnson, S. (1987). Gender differences in science: Parallels in interest, experience and performance. *International Journal of Science Education, 8*(4), 467–481.

Jones, M. G., Howe, A., & Rua, M. J. (2000). Gender differences in students' experiences, interests, and attitudes toward science and scientists. *Science Education, 84*(2), 180–192.

Kahle, J. B., & Meece, J. (1994). Research on gender issues in the classroom. In D. Gable (Ed.), *Handbook of research on science teaching and learning* (pp. 542–557). New York, NY: Macmillan.

Lavonen, J., Byman, R., Uitto, A., Juuti, K., & Meisalo, V. (2008). Students' interest and experiences in physics and chemistry related themes: Reflections based on a ROSE-survey in Finland. *Themes in Science and Technology Education, 1*(1), 7–36.

Law, N. (2002). TIMSS science results for Hong Kong: An Ailing Dragon with a British legacy. *Secondary Analysis of the TIMSS Data, 3*, 157–175. doi:10.1007/0-306-47642-8_11.

Lyons, T. (2006). Different countries, same science classes: Students' experiences of school science in their own words. *International Journal of Science Education, 28*(6), 591–613.

Madsen, L. M., Ulriksen, L., & Holmegaard, H. T. (2010). *Literature review on reasons for dropout/opt-out among young people from STM higher education programs*. Retrieved 8 September, 2010, from http://iris.fp-7.org/

Mulaik, S. A., James, L. R., Van Alstine, J., Bennett, N., Lind, S., & Stilwell, C. D. (1989). Evaluation of goodness-of-fit indices for structural equation models. *Psychological Bulletin, 105*(3), 430–445.

National Bureau of Statistics of China. (1994). *China statistical yearbook 1994*. Beijing, China: China Statistics Press.

National Bureau of Statistics of China. (2013). *China statistical yearbook 2013*. Beijing, China: China Statistics Press.

OECD (Organisation for Economic Co-operation and Development). (2008). *Encouraging student interest in science and technology studies*. Paris, France: OECD Publishing. Retrieved 18 September, 2011, from http://dx.doi.org/10.1787/9789264040892-en

OECD (Organisation for Economic Co-operation and Development). (2010). *PISA 2009 results: What students know and can do – Student performance in reading, mathematics and science* (Vol. I). http://dx.doi.org/10.1787/9789264091450-en. Accessed 18 Sept 2011.

OECD (Organisation for Economic Co-operation and Development). (2014). *PISA 2012 results: What students know and can do: Student performance in mathematics, reading and science* (Vol. I). http://www.oecd.org/pisa/keyfindings/. Accessed 10 Apr 2014.

Osborne, J., & Dillon, J. (2008). *Science education in Europe: Critical reflections*. London, UK: The Nuffield Foundation.

PCAST. (2010). *Prepare and inspire K-12 education in science, technology, engineering, and math (STEM) for America's future*. Retrieved 21 September, 2010, from http://science.nsta.org/nstaexpress/PCASTSTEMEdReport.pdf

Rennie, L. J. (1994). Measuring affective outcomes from a visit to a science education centre. *Research in Science Education, 24*, 261–269.

Ryan, J., & Slethaug, G. (Eds.). (2010). *International education and the Chinese learner*. Hong Kong, China: Hong Kong University Press.

Schreiner, C., & Sjøberg, S. (2004). *Sowing the seeds of ROSE. Background, rationale, questionnaire development and data collection for ROSE (the Relevance of Science Education)*. Oslo, Norway: Department of Teacher Education and School Development, University of Oslo.

Sjøberg, S. (2000). *Science and scientists: The SAS study*. http://folk.uio.no/sveinsj/SASweb.htm. Accessed 1 Sept 2005.

Sjøberg, S., & Schreiner, C. (2005). How do learners in different cultures relate to science andtechnology? Results and perspectives from ROSE. *Asia-Pacific Forum on Learning and Teaching, l6*(2), Foreword. [Online] http://www.ied.edu.hk/apfslt/v6_issue2/foreword/

Skog, B. (2001). Girls' avoidance of 'hard' science subjects-protest or a rational choice? *Scandinavian Journal of Educational Research, 35*(3), 201–211.

Smail, B., & Kelly, A. (1984). Sex differences in science and technology among 11-year-old schoolchildren: II – Affective. *Research in Science & Technological Education, 2*(2), 87–106.

Uitto, A., Juuti, K., Lavonen, J., & Meisalo, V. (2006). Students' interest in biology and their out-of-school experiences. *Journal of Biological Education, 40*(3), 124–129.

Wang, W. J., Wang, J. Y., Zhang, G. Z., Lang, Y., & Mayer, V. J. (1996). Science education in the People's Republic of China. *Science Education, 80*(2), 203–222.

Warrington, M., & Younger, M. (2000). The other side of the gender gap. *Gender and Education, 12*(4), 493–508.

Watkins, D. A., & Biggs, J. B. (2001). The paradox of the Chinese learner and beyond. In D. A. Watkins & J. B. Biggs (Eds.), *Teaching the Chinese learner: Psychological and pedagogical perspectives* (pp. 3–26). Hong Kong: CERC.

Weinburgh, M. (1995). Gender differences in student attitudes toward science: A meta-analysis of the literature from 1970–1991. *Journal of Research in Science Teaching, 32*, 387–398.

Weston, R., & Gore, P. A., Jr. (2006). A brief guide to structural equation modeling. *The Counseling Psychologist, 34*, 719–751.

Woolnough, B. E. (1994). Factors affecting students' choice of science and engineering. *International Journal of Science Education, 16*(6), 659–676.

Yeung, Y. Y. (2008). Effective infusion of Chinese culture into the science-related curriculum: An outline and some Exemplars. In May M. H. Cheng, & Joe T. Y. Lo (Eds.), *Integrated learning in secondary schools: Theories and practice* (pp. 67–82) (in Chinese). Hong Kong, China: Marshall Cavendish Education, Times Publishing Ltd.

Yeung, Y. Y., & Cheng, M. H. (2008, February 20–23). *Secondary school students' affective domain of sciencelearning: Comparison of preliminary ROSE findings in Hong Kong, Guangzhou and Shanghai*. Paper presented at the Conference of Asian Science Education, Kaohsiung, Taiwan.

Yeung, Y. Y., & Cheng, M. H. (2010, June 14–18). Factor analysis and Rasch model analysis of the ROSE study on Chinese students' interest of science learning. In B. Lazar & R. Reinhardt (Eds.), *Proceedings of XIV International Organisation of Science and Technology Education Symposium 2010 [CDROM]*, Bled, Slovenia.

Yeung, Y. Y., & Cheng, M. H. (2011, October 25–28). *Redefining the underlying structure of the ROSE instrument and its application to understand Chinese students' affective domain of science learning*. Paper presented in the East-Asia Association of Science Education 2011 International Conference, Gwangju, South Korea.

Yeung, Y. Y., Cheung, K. L., & Sjøberg, S. (2013, July 4–6). *University students' choices of science and mathematics related programmes in Mainland China*. Paper presented in East-Asian Association for Science Education 2013 International Conference, Hong Kong, China.

Yeung, Y. Y., & Li, Y. F. (2015). Chinese students' science-related experiences: Comparison of the ROSE study in Xinjiang and Shanghai. *Research in Science & Technological Education*. doi:10.1080/02635143.2015.1028350 (in press)

Zohar, A., & Bronshtein, B. (2005). Physics teachers' knowledge and beliefs regarding girls' low participation rates in advanced physics classes. *International Journal of Science Education, 27*(1), 61–77.

Professor Yeung is the professor of the Department of Science and Environmental Studies at the Hong Kong Institute of Education. He is a multidisciplinary scholar in two distinct academic disciplines, namely, science and education. The foci of his current research interest include energy technology, crystal field theory, science education and technology-enhanced learning. Currently, he is holding a visiting/adjunct professor title at Sichuan Normal University and Xinjiang Normal University. Besides, he is the chief editor of the open-access journal called 'Asia-Pacific Forum on Science Learning and Teaching' and has played leading role in organising several international conferences in science education.

Part III
Science Curriculum Changes and Reforms

Chapter 7
Science Curriculum Changes and STEM Education in East Asia

Youngmin Kim, Hye-Eun Chu, and Gilsun Lim

Introduction

Every country has its own curriculum for teaching students effectively, which is changing according to both national and international societal needs. The science curricula of East Asian countries have undergone changes over the last about 70 years since the end of the World War II. The science curriculum has been focused on enabling citizens to make reasoned decisions when faced with issues in modern society that are related to science and technology. In the past, the curriculum has emphasized science content knowledge and process skills to enable students to experience what scientists are really doing in their laboratory and in their research areas. But modern science curriculum focuses on scientific creativity and contents related to modern science and technology, as well as scientific concepts and skills.

Curriculum change of countries in East Asia was crucially affected by the World War II. South Korea was independent in 1945 from Japan that was defeated in the war, and subsequently, the curricula of South Korea and Japan were developed under the influence of the USA. Those of Hong Kong and Singapore were developed under the influence of Britain during their colonial period; however, after their independence in 1997 and 1963, respectively, the curriculum of Hong Kong has been influenced by China Mainland and that of Singapore has been influenced by

Y. Kim (✉)
Department of Physics Education, Pusan National University, Busan, South Korea
e-mail: minkiyo@pusan.ac.kr

H.-E. Chu
School of Education, Macquarie University, Sydney, Australia
e-mail: hyeeun.chu@gmail.com

G. Lim
Department of Science, Korea Science Academy, Busan, South Korea
e-mail: glimsun@kaist.ac.kr

M.S. Khine (ed.), *Science Education in East Asia*,
DOI 10.1007/978-3-319-16390-1_7

the British education system. They started to reform the curriculum periodically, embracing the international trend in teaching and learning approaches. A reform movement provokes the public interest and promotes changes in the science classroom and educational policy of science. This sometimes results in imbalance among the important dimensions in science education. Ultimately, identification of all the important dimensions of science curricula and reasoned decision making about appropriate balance among the various dimensions are desirable.

The school systems of four countries above are similar to each other, despite being unique in details. The mainstream is as seen in Fig. 7.1. In the East Asian countries, due to the fact that graduation from a university is essential to get a good career, the entrance examination to a university is the most important for upper secondary school students. Therefore, in Korea, students take a very pragmatic approach to choosing their subjects. For example, if a certain subject being offered to students is not useful for university entrance, students do not study the subject seriously but disregard it instead. In Singapore, there is an examination syllabus in Junior College (School Years 11–12) which is similar to the national curriculum in

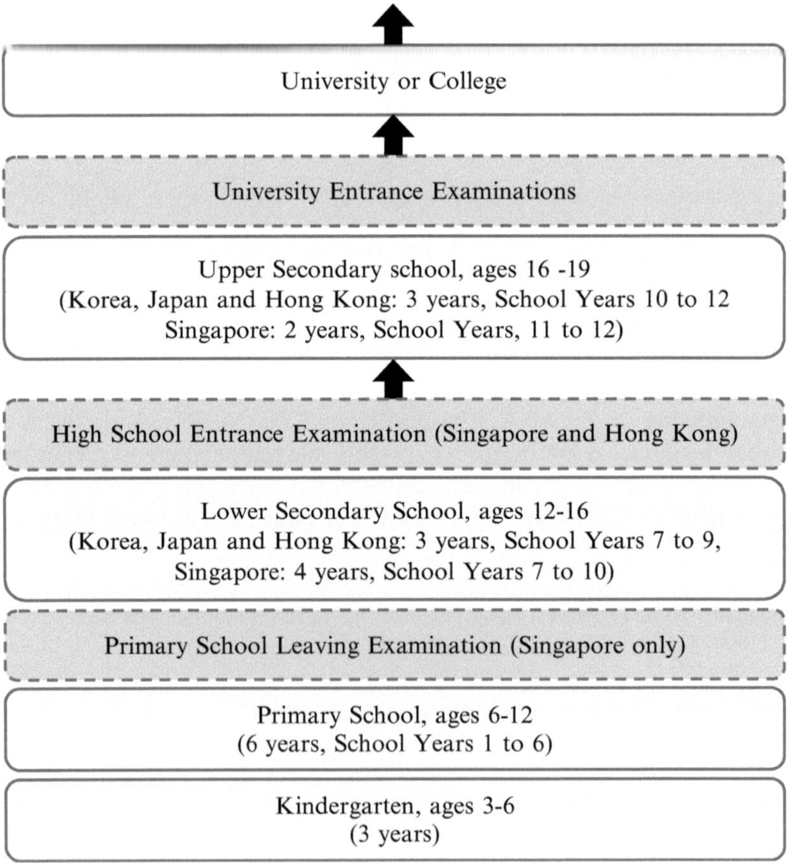

Fig. 7.1 Common school system in East Asia

Korea, but it does not include educational philosophy and framework and teaching and learning approaches. It provides clear explanations and guidelines for the examination at the end of Year 12. Therefore, there is even no time and opportunities during school time to include extra topics/subjects which could be related to their future career.

In this chapter we will discuss changes in the science curriculum mainly in four East Asian countries mentioned above, because of the difficulty in data collection from the other countries in East Asia and the influences of the curriculum reform movements in Europe and the USA. In addition, we will discuss the background of curriculum changes and the need for STEM education to introduce practices relating to students' innovative thinking.

Science Curricula and Educational Practices Before and After the Year 2000 in East Asia

Curriculum is defined usually as all the experiences for learning which are planned and organized by the school (Whitfield, 1971), but in this chapter, only the experiences for learning which are planned were analyzed. In addition, curricula for the secondary school students of the selected countries were mainly introduced. In the East Asian countries, moving into the twenty-first century, substantial change in science education has occurred. In this section, we described science curriculum reforms and educational practices of the countries, dividing the trends into two, before and after the year 2000.

Overview of the Curriculum Reform Trends

The following are the curriculum reform trends of the East Asian countries.

Korea, Republic of

Korea has a centralized education system and the national curriculum is firmly based on the education system. The national curriculum is provided in the form of official document by the Ministry of Education (MOE) of Korea and is mandatory to all schools of K-12. The first formal national curriculum was developed in 1954, and subsequently, it has been reformed regularly by the government. On average, the national curriculum had been reformed every 6–10 years until 2007 and subsequently revised in 2 years. At present, the 2009 National Science Curriculum (MEST, 2011), which was reformed in only 2 years for upper secondary schools and reformed in 4 years for elementary and lower secondary schools after last revision, is being applied to school science education.

The periods of formal curriculum implementation in Korea could be divided into three, which are called experience-centered curriculum period (1954–1973), discipline-centered curriculum period (1973–1981), and humanity-centered curriculum period (1981–present), respectively. The experience-centered curriculum was the first well-formalized curriculum which included objectives, contents, and suggestions for teaching methods. This curriculum was based on American progressive philosophy and emphasized daily life experience and science contents. The discipline-centered curriculum was influenced by the US science education innovation movement. Elementary and secondary science education materials developed in the USA and UK in the 1960s had been translated into Korean, and they had influenced science curriculum reform in Korea. The philosophy, objectives, contents, and methods of teaching of the contemporary experience-centered curriculum were changed almost completely. The new national science curriculum profoundly accepted the philosophy of discipline-centered curriculum, reorganized the contents to reflect the basic concepts of science, changed the methods of instruction from rote memory to inquiry, and emphasized the teaching of science through discovery and problem solving and students' inquiry process skills. At that period, the science contents of this curriculum included the five big ideas: matter, energy, interaction, change, and life. The humanity-centered curriculum was influenced mainly by the movements of "Science for All" and STS. The discipline-centered curriculum had been criticized on the ground that the contents were isolated from real-life situations and problems that students confront in everyday life; they were too abstract and difficult for most students to understand (McCormack, 1992). The attention to "Science for All" and STS increased due to the increased criticisms. However, in the early curriculum of this period, science contents did not overcome the discipline-centered philosophy although new slogans, "Science for All" and STS, were introduced into Korean science education. The STS spirit did not explicitly appear in the objectives until the curriculum was revised in 1987 (MOE of Korea, 1987). Since the curriculum was revised in 1987, "Science for All" and STS spirits had been continuously strengthened even in the revised curriculum in 1997. One of important changes in this period was the development of a new combined or integrated high school science subject. Another important change was the emphasis on scientific literacy including recognition of the relationship between STS and rational decision making in the context of everyday life.

In the 2007 and 2009 science curricula that were reformed in 2007 and 2011, respectively, the terminologies "scientific literacy" and "creative problem solving" were expressed explicitly as the objectives of science education (MEHRD, 2007; MEST, 2011). Actually the 2007 science curriculum was applied to schools with textbooks for a short period and only for elementary and lower secondary students. However, in 2009, a new curriculum, being named as 2009 Curriculum, was developed for upper secondary school students. In addition, a new curriculum for elementary and lower secondary school students, being named also as 2009 Curriculum, was developed in 2011. It aims to help students understand the basic concepts of science through inquiry with interest and curiosity toward natural phenomena and objects and to develop scientific thinking skills and creative problem-solving abilities. Students are expected to be able to develop the scientific literacy necessary for solving the problems of daily life creatively and scientifically (MEST, 2011). The core concepts of science are taught with a close relation to learners' experiences, and students are provided with opportunities to apply science-related knowledge and inquiry skills for problem solving in society and their daily life. For the practice of this curriculum, the STEM or STEAM education is being emphasized (MEST & KOFAC, 2011).

Japan

After the World War II, the national curriculum of Japan has been reformed approximately every 10 years. The first reform occurred in 1951. The national curriculum was influenced by American progressivism that was directed by the occupation army. It was a curriculum based on the usefulness to solve social problems in that period. The second reform occurred in 1958 and implemented from 1961. It stressed systematic learning activities according to the national standard for teaching. The third reform occurred in 1968 and implemented from 1971 onwards. The expectation from the industrial world of Japan gradually increased, and the modern science and technology movement exerted strong effect on the curriculum that the contents of modern science and technology were included into the curriculum. The fourth reform occurred in 1978 and implemented in 1980 to the elementary school level, in 1981 to lower secondary schools, and to upper secondary schools in 1982. At that time, it was argued that excessive content was taught, and the modern contents were too difficult for elementary and secondary students. In the curriculum, humanistic personality, meaningful school life, affluent circumstances, individuality, and understanding ability of basic and fundamental contents were emphasized (Katayama, Takamori, & Kanaizuka, 2004; Nakamichi, 2011).

The fifth reform occurred in 1989 and was implemented to the elementary school level in 1991, while the lower and upper secondary schools adopted it in 1992 and 1993, respectively. Personal individualization and computers in education were gradually valued in society. In this curriculum, basic and fundamental knowledge and skills, enthusiasm for learning independently, ability to cope with societal change and creativity, and qualification of Japanese in international society were emphasized (MESSC, 1989a, 1989b).

The sixth reform occurred in 1998 and implemented to all levels of schools in 2002. The main issue for the change was a five-day school week system. In addition, the problem of diminishing number of students, active use of calculators and computers, and increasing number of students who had disinterest in science and mathematics were debated seriously. In 1996, the CCE (Central Council for Education) White Paper reiterated many of the previous concerns and recommended educational change including the implementation of the five-day school week and integrated curriculum. In this document, the CCE stressed two concepts in Japanese education reform: education with room to grow and zest for living (MESSC, 1989a, 1989b; NIER, 2014). The main goal of this revised curriculum was to provide effective relationship between various curriculum elements and subjects in the school educational activities. According to the goal of the curriculum reform, optional courses were increased, and the students were taught how to think, decide, and organize their thoughts as well as how to gain the skills for learning and how to collect and analyze data (MESSC, 1999). This curriculum cut the content and hours of traditional compulsory subjects at primary and lower secondary levels, in order to allow more hours for elective subjects and a new, cross-disciplinary subject called Integrated Learning. At the lower secondary level, hours for Japanese and social studies was cut by 23 % each, hours for music and art by 34 %, and hours for mathematics and science by 18 % and 17 %, respectively (Cave, 2001; Komatsu, 2002). To promote scientific literacy, science teachers should utilize the period of integrated study in addition to the science lessons. Science teachers needed to make use of facilities outside the school, such as museums and human resources, in order to enrich science education during the school year, a period crucial to the lifelong learning process.

In 2007, Japan piloted a National Assessment of Academic Ability in mathematics and Japanese for students in grades 6 and 9. In 2008 and 2009, MEXT published a reformed version of the national curriculum for primary through upper secondary schools. This new curriculum, which is still applied, places increased emphasis on Japanese, social studies, mathematics, science, and foreign languages, with the hope that students will develop critical thinking ability, rational decision making, good expression, as well as content knowledge. In addition, in the primary school and the lower secondary school, the total number of teaching hours would be increased by around 10 %, and in primary and lower secondary school, the integrated study periods would then be reduced substantially (NIER, 2014).

Hong Kong

Hong Kong was yielded to the British government in 1842 after China was defeated in the Sino-British War (The Opium War). In 1997, Hong Kong's sovereignty was relinquished to China under the "one country, two systems" notion when the lease ended. Under this arrangement, China stepped up its sovereignty over Hong Kong, resulting in Hong Kong having autonomy in all areas except military defense and diplomacy under the arrangement of a Special Administrative Region of China (SAR). In the education aspect, Hong Kong continued to maintain its own system of education under the Education Bureau (EDB) without direct relations with the Ministry of Education in Beijing.

From 1840s to 1940s, education was considered to be a privilege enjoyed by the elite and the rich. Even after the World War II, the Hong Kong education remains as an elite system. Educating the poor was not a priority until universal education was implemented in 1960s. Hong Kong's education system can be divided into two stages – moving toward universal education and education reforms (OECD, 2011; Sweeting, 1990).

Hong Kong announced a 9-year compulsory education plan for primary and secondary schools in 1978. Primary school education has been compulsory since 1965. Even though there was a pressure for social equity in education from outside of Hong Kong, the universal education system was in line with the Hong Kong government's plan. The government needs to maintain this pyramid education system, which means that there can be only 3–4 % of students in the same age cohort studying at Hong Kong University (which was the only higher education institute at the time), after the secondary school education (Cheng, Chow, Yeung, & Cheung, 1998; OECD, 2011).

Due to the 9-year compulsory education, the Secondary School Entrance Examination was abolished. Instead of routine drills and practices from primary four to enter secondary school, the current aim of education is to encourage and help students to think logically (Cheng, 2000)

Hong Kong society has been experiencing rapid changes due to economic developments in the Asia-Pacific area and the political changes in July 1997 due to the transition from the British colony to a Special Administrative Region (SAR) of China. Because of these changes, there were demands for innovative changes in the roles and functions of school education. A comprehensive reform began in 1999, and schools in Hong Kong needed to prepare their students for the workplace (Board of Education, 1997).

Through systematic reform studies about problems in the education system, economic development, and external competitions and outcomes (Cheng, Moc, & Tsui, 2002), there were comprehensive reform progresses from 1999 to 2012 in the education structure, assessment, and curriculum. Firstly, the educational structure and assessment were changed into parallel structure with mainland China and other countries in order to ease students in the process of transferring to schools in Hong Kong. Hong Kong's education structure used to align closely with the system in England – students underwent 3 years of mandatory junior secondary education, 2 years of non-compulsory senior secondary education leading to Hong Kong Certificate of Education Examination (HKCEE; O Level), and 2 years of non-compulsory matriculation course leading to Hong Kong Advanced Level Examination (HKALE; A Level) in the old academic structure as in Fig. 7.2. After the implementation of new academic structure in 2009, where major changes were made on the senior secondary education, Hong Kong Diploma of Secondary Education (HKDSE) replaced both HKCEE and HKALE. The matriculation course was removed. In addition, public assessment after primary schooling was abolished in 2001. This HKDSE examination is the only public examination in the new 3 (secondary)–3 (senior secondary)–4 (university) education system. The reduction in public assessment is well attuned with the new education paradigm that consists of lifelong learning instead of learning for the sake of entering a higher education institute, and it also made reforms possible in primary and secondary curriculum. This new education structure also facilitates higher education program development from 3-year to 4-year education programs.

Secondly, there was curriculum reform focusing on theories of constructivist learning (Curriculum Development Council of Hong Kong, China, 2002). The new

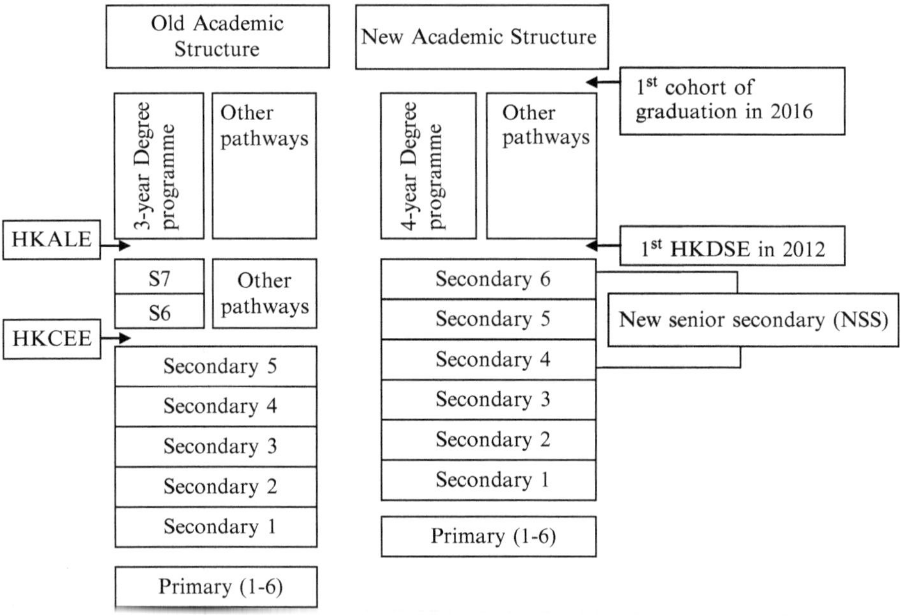

Fig. 7.2 Changes in educational structure and assessment (Chan, 2010)

emphasis was on the process of learning such as "learning is for understanding," "understanding is demonstrated by the effective application of the knowledge and thus constructed," "learning is best experienced in real life with actual effects," "learning is also social action, best achieved in groups," etc.

This is especially evident in the senior secondary level where eight key learning areas were introduced instead of specific subjects: Chinese, mathematics, English, science and technology, social science and humanities, sports and arts/applied learning (real-life workplace experiences), and other learning experiences (service learning, workplace visit, and overseas experiences). In addition, liberal studies was being introduced with a timetabled slot (Curriculum Development Council of Hong Kong, China, 2009). There is no syllabus but only broad guidelines. This learning studies aims to develop higher-order or critical thinking to students through designing their own learning schemes which do not include non-textbook information and from finding directions which are based on analysis, synthesis, and conceptualization-proposing hypotheses or theories (Curriculum Development Council of Hong Kong, 2014).

Singapore

Singapore was under the British colonial rule from 1891 to 1959. During this period, there were many immigrants from India, China, and Malaya. In the last 50 years since Singapore's separation from Malaysia in 1965, Singapore has raised its educational level to be one of the top Asian countries which show outstanding performance in TIMSS and PISA. The current educational system that has achieved high standards has been developed through three broad periods (Goh & Gopinathan, 2008; OECD, 2011), namely, survival driven (1965–1978), efficacy driven (1979–1996), and ability driven (1997–2012).

From the end of British colonization in 1959 to independence from Malaysia in 1965, the Singapore government started to focus on Singaporean's own economic development. There were two considerations about education: firstly, the use of education to solve their pressing conflicts and dilemmas in order to achieve national cohesion and the economic development through restructuring of the society and through expanding educational opportunities and, secondly, the use of education to gain expertise in the areas of science and technology for Singapore's future development. In 1959, through a five-year plan in education by the PAP (the People's Action Party) which has gotten into power from 1959 to date, the standards of education were implemented. This plan emphasized on the study of mathematics, science, and technical subjects. In addition, four language streams of education, namely, Malay, English, Chinese, and Tamil, were treated equally and Malay was established as the national language.

The period from 1965 to 1978 (survival-driven period) was the beginning of a new era involving creating a nationhood instead of statehood after the independence of Singapore. However, there were about two million Singaporean who were illiterate and unskilled because the British rule had not provided education for all people in Singapore. Therefore, expanding basic education and vocational and technical education was emphasized during the survival-driven period. There was a shift of education focus from academic to technical education in order to overcome the lacking manpower during industrialization. By the end of the "survival" period, a national system of public education was created. Also, the bilingual policy and importance of English were emphasized. Evidently, in 1971, the Primary Pilot Project (PPP) (National Commission for UNESCO, 1974) was implemented to pilot curriculum integration for teaching English, mathematics, and science. It means that English is the medium for teaching and learning mathematics and science. The other aim of PPP was to formulate and produce mathematics and science programs for teaching children through guided activities which could help the children to form and develop concepts in order to accomplish the principle objectives of "learning through understanding." In 1976, the government announced that mathematics and science papers in the Primary School Leaving Examination (PSLE) would be set in English. The benefits of the bilingual policy, using English officially, have been recognized since British colonial times. It attracted foreign-owned industries and foreign governments to provide financial and technical support. Indeed, it has contributed to Singapore's fast economic growth in one generation.

The weakness and problems of the one-size-fits-all education approach had been observed and became one of the critical issues in 1979–1996 (efficacy-driven period). Therefore, the New Education System (NES) in 1979 introduced a streaming based on students' learning abilities. It gave flexibility to the education system and introduced many pathways based on students' various academic abilities.

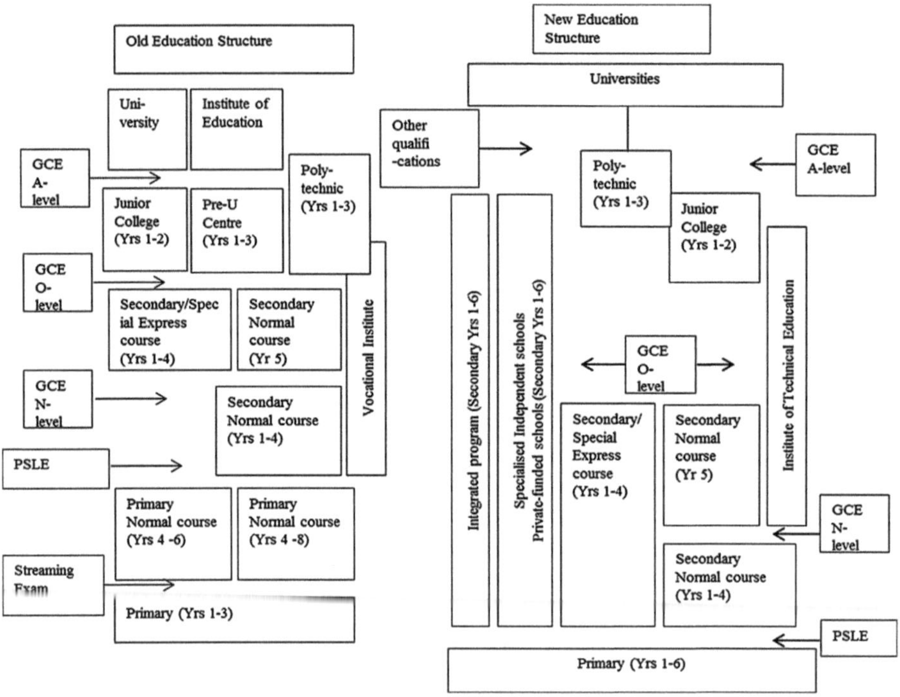

Fig. 7.3 Educational structure and assessment changes in Singapore (Modified from Goh & Gopinathan, 2008) *PSLE: Primary School Leaving Examination

Academically low ability students were able to stay in the primary school for up to 8 years and in the secondary school for up to 5 years (see Fig. 7.3) in order to obtain the General Certificate in Education "Ordinary" Level (GCE "O" Level). Furthermore, a new lower secondary school leaving certificate, the GCE "Normal" Level (GCE "N" Level), was introduced. There were also three types of high schools – academic high schools for college preparation (2 years); polytechnic high schools (3 years) for advanced occupational and technical training in engineering, science, and computer science areas that could also be connected to university studies; and technical institutions (2 years) in order to cater for the lowest 5 % of secondary school graduates for their occupational and technical training that equip students with the basic foundation skills for further studies in polytechnics (see Fig. 7.3). These different types of high school education helped Singapore provide the technical workforce and extended the pool of scientists and engineers in the university. These educational efforts during the efficacy-driven education period produced positive outcomes, for example, students showed excellent performance in TIMSS in 1995 and 1999 (see Table 7.28).

During the financial crisis in Asia in 1997, there were some considerations on preparing young Singaporeans for the growth of the global knowledge economy

(Goh, 1997). The new vision in education was "Thinking School, Learning Nation (TSLN)." It focused on developing active learners with critical thinking skills and creativity. The key strategies in 1996–2012 (ability-driven period) included (1) the explicit teaching of critical and creative thinking skills, (2) the reduction of subject content, and (3) the revision of assessment modes. Based on these strategies, project work and student participation in extracurricular activities were also taken into account together with the result of the General Certificate in Education "Advanced" Level (GCE "A" Level) when students from Junior College apply for higher education institutes (Tan & Gopinathan, 2000). The Ministry of Education (MOE) also announced the master plan for IT in education (MOE of Singapore, 1982) for providing physical infrastructure to schools and for pre- and in-service teachers' trainings.

The ability-driven education provided 10 years of general education including 6 years of compulsory primary school education. Education reform in the educational structure, assessment, and curriculum was implemented during the ability-driven phase. Figure 7.3 shows changes in the educational structure and assessment format. In addition, there was emphasis on maximal development of students' talents and abilities during this period. Therefore, different types of secondary schools were established: the IP (integrated program) schools in which students were not required to take a GCE "O" Level test after 4 years of secondary school education and special independent schools, for example, NUS (National University of Singapore) High School, which is a science and mathematics school for gifted students. However, there were not many changes in the examination system. Streaming examination after primary three was removed from the old education structure, and not only GCE "A" Level but also other qualifications (e.g., International Baccalaureate and NUS high certificate) were recognized after senior secondary school education from IP schools and special independent schools in the new education structure. This flexibility brought about changes in these schools' curricula that are different from Singapore government school curricula. Related to curriculum reform, the Curriculum Committee 2015 in the primary school and the lower secondary school was announced. The new curriculum emphasizes holistic education and balanced assessment between the assessment of learning and the assessment for learning (National Institution of Education, 2010).

The education reform in Singapore is still ongoing. It can be said that the education in Singapore is now in the value-driven phase. There has been an emphasis on education which is rooted in values (Ng, 2008). In addition, the twenty-first-century skills were introduced. It emphasizes roles and directions of school education for the twenty-first-century society. From 2014, MOE initiated the STEM project at the secondary school level. There have been only character development projects, but now schools are keener to develop STEM/STEAM projects for students' knowledge applications.

Science Curricula and Educational Practices Before the Year 2000 in East Asia

East Asian countries developed and have reformed the science curricula since the World War II. To see their change trends, we analyzed their curricula before and after 2000. For the curricula before 2000, the reformed ones after 1980 were investigated and introduced in this chapter. Here we introduce the changes in science curricula of the secondary school level, dividing into the lower secondary and upper secondary schools.

Interval for Science Curriculum Reform

During 1980s and 1990s, each of East Asian countries had reformed its own science curriculum a few times. Korea had reformed it in 1981, 1987, 1992, and 1997; Japan in 1978, 1989 and 1998; Hong Kong in 1985 and 1997; and Singapore in 1979, 1992, and 2012/2013. Therefore, the average interval for science curriculum reform was 7 years in Korea, 10 years in Japan, and more than 10 years in Hong Kong and Singapore.

Science Learning Objectives in Lower Secondary School

The science subject for lower secondary school students was named as "science" in all countries. The objectives for the lower secondary school, however, were different from each other.

In Korea, the science learning objectives for the lower secondary school in the 4th curriculum reformed in 1981 emphasized students' conceptual understanding, improvement of scientific inquiry ability, and interest in science and scientific attitude. Then, STS movement from the USA and UK in the 1980s influenced the next Korean science curriculum reform. From the 5th national science curriculum reformed in 1987, an STS-related objective was included in the main objectives of secondary science curriculum of Korea. The science learning objectives for the lower secondary school in the 5th curriculum were as follows: (1) Students should understand scientific facts, concepts, and principles and apply them for explaining natural phenomena. (2) Students should acquire scientific methods for inquiring nature and should apply them for solving scientific problems. (3) Students should have interests and curiosity about natural phenomena and science learning. (4) Students should experience basic experimental skills and practice necessary for inquiring nature. (5) Students should recognize the impacts of science on the development of technology and progress of society. The last objective was included for the first time in the 5th curriculum. The science learning objectives of the lower secondary school in the 6th and 7th curricula reformed in 1992 and 1997, respectively, were almost same as those in the 5th curriculum.

In Japan, the reformed curriculum in 1989 was called "education to develop students' individual personalities." Therefore, many electives were offered to match students' individual interests and abilities (Nakamichi, 2011). According to the science curriculum reformed in 1989 (MESSC, 1989a, 1989b), during the lower secondary school period, the emphasis shifted more strongly toward the development of investigation and problem-solving abilities in science compared with the curriculum reformed in 1978 (Mayer, 2014). The new curriculum was implemented to primary schools in 1991, to lower secondary schools in 1992, and to upper secondary schools in 1993. The aim of science education for the lower secondary school level was to make students develop the ability of scientific investigation and deepen their understanding of matter and phenomena in nature by arousing an interest in nature and through observations and experiments, thereby developing scientific views and thinking (MESSC, 1989a, 1989b). According to the aim, the specific objectives were to help develop student's ability in solving problems, to foster interest and sensitivity toward nature, and to inculcate a deeper understanding of the phenomena of nature, utilizing various observations and experiments (MESSC, 1989a, 1989b). The "science" subject was divided into mainly two fields, called Field I and Field II. The overall objectives of "science" were to enhance students' interest in nature, to enable them to carry out observations and experiments with identifying clear purpose, and to develop attitudes and abilities to investigate nature scientifically. In addition, the objectives of Field I were: (1) To develop questions about phenomena concerning matter and energy and, through identifying process within these inquiring activities, to enable students to discover regularity and learn the way of explaining the nature phenomena and solving problem. (2) To carry out observations and experiments concerned with chemical phenomena; to learn the observational and experimental skills; to understand familiar substances, chemical change, atoms/molecules, the use of substances and chemical reactions, and so on; and to develop scientific views and thinking about these phenomena. (3) To carry out observations and experiments concerned with physical nature and phenomena; to learn the observational and experimental skills; to understand familiar physical phenomena, electric current, motion and energy, and so on; and to develop scientific views and thinking about these phenomena. (4) To enhance students' interest in phenomena concerned with matter and energy and, through inquiring activities into them eagerly, to develop an attitude of thinking scientifically in relation to everyday life. The objectives of Field II were: (1) To develop questions about living things and the natural phenomena surrounding them and, through investigating the process scientifically, to discover regularity and explain the natural phenomena. (2) To carry out observations and experiments that are concerned with living things and biological phenomena, to learn observational and experimental skills, to develop the ability to give consideration to the results of observations and experiments, and to develop scientific views and thinking about these phenomena. (3) To carry out observations and experiments concerned with the phenomena of earth science; to learn observational and experimental skills; to understand the earth and space, the weather and its change, and so on; and to develop scientific views and thinking about these phenomena. (4) To carry out activities investigating living things and their surrounding natural phenomena, to learn the methods of investigating nature, and at the same time, through the activities, to develop an attitude of respect for life and conservation of the natural environment.

Major emphasis in the curriculum reformed in 1998 was on fostering "zest for living" to children (MOE of Japan, 1999a, 1999b). In this reform, the "science" subject was divided into mainly two fields, called Field I and Field II, as same with the curriculum reformed in 1989. The overall objectives were to enhance students' interest in nature, to enable them to carry out observations and experiments with identifying clear purpose, to develop attitudes and abilities to investigate scientifically, and, at the same time, to deepen their understanding of natural phenomena and to develop scientific views and thinking, to arouse students' interest in nature through experiments and observation, and to help them develop scientific views and thinking. Each field had their own specific objectives. The objectives of Field I were: (1) To enhance students' interest in phenomena concerned with matter and energy and, through identifying problems related with these phenomena and inquiring activities into them, to learn processes of discovering regularity and solving problem. (2) To carry out observations and experiments concerned with physical phenomena; to learn the observational and experimental skills; to develop the ability to give consideration to the results of observations and experiments; to develop and express their own ideas and, at the same time, to understand familiar physical phenomena, such as the nature and use of electric current, the regularity of motion, and so on; and to develop scientific views and thinking about these phenomena. (3) To carry out observations and experiments concerned with chemical phenomena, to learn the observational and experimental skills, to develop the ability to give consideration to the results of observations and experiments, and to develop and express their own ideas and, at the same time, to understand familiar substances, chemical change and atoms/molecules, the use of substances and chemical reactions, and so on. (4) Through investigative activities into phenomena concerned with matter and energy, to develop an attitude of thinking scientifically in relation to everyday life and to see the nature from a comprehensive perspective. And the objectives of Field II were: (1) To enhance interest in living things and the natural phenomena surrounding them and, through identifying problems related with these phenomena and inquiring activities into them eagerly, to learn processes of discovering regularity and solving problems. (2) To carry out observations and experiments concerning living things and biological phenomena; to learn observational and experimental skills; to develop the ability to consider the results of observations and experiments; to develop and express their own ideas and, at the same time, to understand such things as the life and kinds of plants and animals, the cells and reproduction of living things, and so on; and to develop scientific views and thinking about these phenomena. (3) To carry out observations and experiments concerned with phenomena of earth science; to learn the observational and experimental skills; to develop the ability to consider the results of observations and experiments; to develop and express their own ideas and, at the same time, to enable students to understand changes of the earth, the weather and its change, the earth and space, and so on; and to develop scientific views and thinking about these phenomena. (4) To carry out activities to investigate living things and their surrounding natural phenomena; to learn methods in investigating nature and, at the same time, through the activities; to develop an attitude of respect for life and conservation of the natural environment; and to see the natural world from a comprehensive perspective.

Before the year 1997, in Hong Kong, the curricula had been developed centrally by the Education Department at all levels of schooling except forms 6 and 7, which only examination syllabuses existed. Although schools were not compelled to adopt such curricula and this was probably most apparent at grade 9 level (form 3 in Hong Kong), it was the usual practice for schools to follow the guidelines put forth by this Education Department. The intended curriculum for junior secondary science was introduced in 1975. This was a more detailed curriculum with clearly stated objectives for each section and was based on the Scottish Integrated Science course. A curriculum guide produced by the Curriculum Development Committee of the Education Department for each of the 3 years of the course gave aims, teaching suggestions, and resource notes. The curriculum guide for form 1 (grade 7 level) was produced in 1978 and that for form 2 (grade 8 level) was produced in 1982. The curriculum guide for form 3 appeared in 1985 together with a modification of the curriculum at this level (Holbrook, 1989).

In Hong Kong the curriculum guides developed during 1978–1985 had been applied until 1997 without change. The overall aims of the course stated in the science curriculum guide were that pupils should acquire: (1) some knowledge of the empirical world around them, (2) an introduction to the vocabulary and grammar of science, (3) an ability to observe critically, (4) an ability to solve problems and think scientifically, (5) an understanding of the relevance of science to the world beyond school and to the needs of a changing society, and (6) an awareness of the science-related culture. More specific objectives are listed below.

1. Knowledge and understanding related to: (1) facts and concepts concerning the environment, (2) use of appropriate instruments, (3) an adequate scientific vocabulary, (4) use of concepts in familiar situations, (5) selection of relevant knowledge and its application to new situations, (6) analysis of data and the drawing of conclusions, (7) and thinking and acting creatively in science
2. Attitude giving rise to an: (1) awareness of the relationships and contributions of the various sciences to economics and social life, (2) interest and enjoyment in science, and (3) objectivity in observation
3. Experimental manipulative skills

As it was recognized that these traditional areas might no longer adequately reflect science curricula that attempted to relate students' learning experiences in science to their practical experiences in their own environment, hence an additional set of categories was developed to cover areas such as environmental science and health science (Holbrook, 1989).

In Singapore, students in Years 3–6 in primary school and students in Years 1 and 2 in lower secondary schools learn science that consists of these themes –

diversity, energy, interactions, etc. which is integrated science. Science syllabuses are published by the Curriculum Planning and Development Department (CPDD), but each science subject syllabus for Years 9 and 10 in upper secondary schools and Years 1 and 2 in Junior College (Years 11 and 12) is published by the Singapore Examination and Assessment Board (SEAB) in the Ministry of Education and University of Cambridge International Examinations. Most teachers and students recognize this syllabus as the examination syllabus for GCE "Ordinary" Level and GCE "Advanced" Level. There are syllabus aims, contents to be covered, assessment objectives, and schemes of assessment. We will introduce these examination syllabuses as an upper secondary syllabus.

Development of the concept of scientific inquiry has been introduced since the survival-driven phases (1965–1978) in the science syllabus, even though the term was not spelt out specifically. In 1982, the lower secondary syllabus framework for Years 1 and 2 in secondary school included the introduction of inquiry mind using integrated and discovery approaches (MOE of Singapore, 1982). However, science teachers were not familiar with inquiry teaching approaches. Textbooks and teaching resources published by the CDIS (Curriculum Development Institute of Singapore) guided teachers to understand science process skills and to apply these in their science classroom (Singham, 1987) during the efficacy-driven phase (1979–1996). During this period, the curriculum aims (see Table 7.1) expanded and the ideas of science syllabus are similar to the later years' curriculum aims after 2000.

In this science syllabus, we cannot capture how schools have to facilitate inquiry approaches. Some schools are already conducting integrated inquiry-based science projects (Darling-Hammond, 2010; Poon, Lee, Tan, & Lim, 2012) that are incorporated with technology and environmental areas, for example, creating environmental-friendly natural insect repellent, designing effective tachometer considering shapes of blades, mobile learning tasks using mobile devices, etc. These inquiry approaches were more specifically described in the syllabus with the inclusion of the twenty-first-century competencies framework and scientific literacy after 2000.

Science Learning Contents in Lower Secondary School

Science learning contents for lower secondary school students were quite different among countries.

In Korea, the science learning contents of the lower secondary school students in the 4th curriculum reformed in 1981 consisted of scientific concepts according to the discipline-centered curriculum philosophy. There were almost no contents that were related to daily life but mostly basic scientific concepts. In the 5th curriculum reformed in 1987, the contents were not different from those in the 4th curriculum, except the contents of nature environment and our living, which were included according to the objective related to STS. They were taught, in 7th grade, circulation of air and water, plants around us, properties of matter, and force and motion; in 8th grade, construction of matter, structure and function of organisms, change of

Table 7.1 Changes of syllabus aims in 1982 and 1992

<table>
<tr><th>Syllabus aims in 1982</th><th>Syllabus aims in 1992</th></tr>
<tr><td>(i) To develop science concepts and an understanding of our physical and biological environment</td><td>(i) To enable pupils to acquire increasing understanding and knowledge so as:
To become confident citizens in a technological world, being able to take or develop an informed interest in matters of scientific importance
To recognize and appreciate the usefulness and limitations of the scientific method
To be prepared for science studies at the upper secondary level</td></tr>
<tr><td>(ii) To provide the pupils with the essential scientific knowledge and skills that will meet their educational and vocational needs</td><td rowspan="2">(ii) To develop abilities and skills that are relevant to the study and practice of science, are useful in everyday life
To encourage safety consciousness and safe practice</td></tr>
<tr><td>(iii) To develop their ability to use the methods of science</td></tr>
<tr><td>(iv) To provide a science course which is both relevant and meaningful in the technological environment of today</td><td>(iii) To stimulate
Curiosity, interest, and enjoyment in science and its methods of inquiry
Interest in and care for the environment</td></tr>
<tr><td>(v) To enable our pupils to appreciate the humanistic aspect of science</td><td>(iv) To promote an awareness that:
The study and practice of science are cooperative and cumulative activities and are subject to social, economic, technological, ethical, and cultural influences and limitations
The applications of science are generally beneficial; however, the abuse of the application of the knowledge of science may be detrimental to the individual, the community, and the environment</td></tr>
</table>

MOE of Singapore, 1982, 1992

earth surface, and electricity; and in 9th grade, work and energy, continuity of life, change of matters, earth and space, nature environment, and our living. The science learning contents in the 6th and 7th reforms were almost same as the contents in the 5th science curriculum.

Table 7.2 shows the contents of lower secondary school science in the 7th curriculum.

In Japan, science topics in the courses of study reformed in 1989 divided into two fields, Field I and Field II. "Field I" dealt with physics and chemistry concepts and "Field II" with biological and earth science concepts. Students were taught two different topics from each field every year for three years. The topics are shown in Table 7.3.

Table 7.2 Contents of lower secondary school science in the 7th curriculum of Korea

Big idea		7th grade	8th grade	9th grade
Knowledge	Energy	Force	Various motion	Work and energy
		Light	Electricity	Action of electric current
		Wave		
	Matter	Three states of matter	Property of matter	Composition of material
		Motion of molecules	Separation of mixture	Regularity in material change
		Change of state and energy		
	Biology	Structure of living things	Stimulus and response	Reproduction and genesis
		Digestion and circulation	Structure and function of plants	Heredity and evolution
		Respiration and excretion		
	Earth	Structure of the earth	Earth and space	Water circulation and weather change
		Matter of crust	History of the earth and crustal movement	Motion of solar system
		Components and motion of ocean water		
Inquiry	Process	Observation, classification, measurement, prediction, inferring, etc.		
		Problem recognition, hypothesis formulation, control of variables, data transformation, data interpretation, inducing of conclusion, generalization, etc.		
	Activity	Discussion, experiment, investigation, field trip, project, etc.		

During learning contents of the topics, students carried out investigations, experiments, and other types of activities related to process skills. The contents of science for lower secondary school students reformed in 1998 are shown in Table 7.4.

In Hong Kong, the science contents for the lower secondary school were divided into 17 units, six to be studied in each of Years 1 and 2 with the remaining five units allocated to Year 3. Twelve units and subsections for Years 1 (grade 7) and 2 (grade 8) are shown in Table 7.5. The five units allocated to Year 3 (grade 9) are Unit 13, The Earth; Unit 14, Transport System; Unit 15, Electricity and Magnetism; Unit 16, Microorganisms; and Unit 17, Environment and Conservation (Holbrook, 1989).

The curriculum guide, published by the Curriculum Development Committee as a supplement to the 1975 syllabus, gave a more explicit and detailed guideline of the intended topics.

In Singapore, before the year 2000, during the efficacy-driven phase (1979–1996) and at the beginning of the ability-driven phase (1997–2012), the science curriculum consisted of three sciences, physics, chemistry, and biology, even though the science syllabus was considered as an integrated subject using a

Table 7.3 The contents of lower secondary science reformed in 1989 (Japan)

Area	Topic	Subtopic
Field I	1. Substances around us and their changes	(1) Water solutions
		(2) State change of matter
		(3) Generation of gas
	2. Physical phenomena around us	(1) Light and sound
		(2) Heat and temperature
		(3) Force
		(4) Pressure
	3. Chemical change, atoms, and molecules	(1) Chemical change
		(2) Atoms and molecules and use of electric current
	4. Electric current	(1) Electric current and voltage
		(2) Electric current and flow of electrons
	5. Chemical change and ions	(1) Electrolysis and ions
		(2) Acid, alkali, base
	6. Motion and energy	(1) Force and motion
		(2) Motion of object
		(3) Work and energy
		(4) Progress of science and technology and life of human being
Field II	1. Life of plants and their kinds	(1) Life and structure of plants
		(2) Plants and peers
	2. The earth and the solar system	(1) Nearby space
		(2) Planets and solar system
	3. Life of animals and their kinds	(1) Life and structure of animal
		(2) Animals and peers
	4. Weather and its change	(1) Changes in the weather
		(2) The weather in Japan
	5. Linkages among living things	(1) Living things and cell
		(2) Breeding methods of living things and genetics
		(3) Linkages among biological systems
	6. Changes of earth surface and earth	(1) Volcanoes and earthquakes
		(2) Strata and earth's past

thematic approach. For example, the content in the topic "energy" includes only physics topics and "matter" includes only chemistry topics (see Table 7.6).

White space which is 15 % of the free curriculum time was introduced in 1999 due to the TSLN (Thinking School, Learning Nation) initiative in 1997. Highly descriptive contents, not fundamental but advanced concepts for the lower secondary level, concepts less connected to the contents of the syllabus, and topics taught in other subjects/topics were removed from the syllabus. This helped to free up to 15 % of curriculum time as a result of content reduction. Table 7.6 indicates which contents were removed from the existing syllabus.

Table 7.4 The contents of lower secondary science reformed in 1998 (Japan)

Area	Topic	Subtopic
Field I	1. Physical phenomena around us	(1) Light and sound
		(2) Force and pressure
	2. Substances around us	(1) Shape of substances
		(2) Water solutions
	3. Electric current and its use	(1) Electric current
		(2) Use of electric current
	4. Chemical change, atoms, and molecules	(1) Structure of matter
		(2) Chemical change and mass of substance
	5. Regularity of motion	(1) Regularity of motion
	6. Use of substance and chemical reaction	(1) Use of substance and chemical reaction
	7. Science and technology and human being	(1) Energy resources
		(2) Science and technology and human being
Field II	1. Life of plants and kinds of plants	(1) Observing of living things
		(2) Structure and motion of plants
		(3) Kinds of plants
	2. Change of earth surface	(1) Strata and shape of the past
		(2) Volcano and earthquake
	3. Life of animals and kinds of animals	(1) Structure and motion of animals
		(2) Kinds of animals
	4. Atmosphere and its change	(1) Observing of weather
		(2) Change of atmosphere
	5. Cells of living things and reproduction	(1) Cells of living things
		(2) Reproduction process of living things
	6. Earth and space	(1) Motions of celestial bodies and earth's rotation and revolution
		(2) Solar system and planets
	7. Nature and human being	(1) Nature and environment
		(2) Nature and human being

Students are expected to learn each topic based on scientific inquiry approaches by considering science process skills, for example, observing, classifying, using apparatus and equipment, communicating, inferring, applying data, formulating hypothesis, and planning investigations. Therefore, scientific problem solving was integrated in the process skills (MOE of Singapore, 1992): (1) make observations, (2) suggest hypothesis, (3) test the hypothesis, and (4) infer from data. The freed up curriculum time in 1997 aimed to help schools to conduct scientific inquiry-based activities in the lessons such as group discussions, debates, role play, show and tell, model making, etc. The guided teaching and learning approaches in syllabus in the 1992 syllabus were recommended in the lower secondary science syllabus 2013.

Table 7.5 Science contents for lower secondary school (Hong Kong)

Grade	Unit	Subsection
Grade 7	Unit 1: Introducing Science	C.1 Mixing solutions
		C.2 Measuring temperature; weighing; measuring volume
		C.3 Dissolving; filtration; evaporation
		C.4 Using a clock or watch to measure small time intervals
		C.5 Making observation and recording results/E.5 Black box experiment
		C.6 Looking at differences within one kind of organism
	Unit 2: Looking at Living Things	C.1 Looking at snail/earthworm
		C.2 Plants and animals
		C.3 Grouping of living things/E.3 Further classification
		C.4 Use of a simple key/E.4 Construction of key
	Unit 3: Energy	C.1 Different forms of energy
		C.2 Simple energy changes
		C.3 Examples of energy changes/E.3 More daily applications
		C.4 Energy stories
		C.5 Energy converters and daily applications
		C.6 Energy in food
		C.7 Energy and man/E.7 More daily applications
	Unit 4: Matter as Particles	C.1 Sets of things
		C.2 Particles spreading
		C.3 More about particles/E.3 Oil drop experiment
		C.4 Particles moving and spacing/S.4 Particles – more to do
		C.5 A particle model/E.5 Description of particles
		C.6 Gas pressure/E.6 Density
		C.7 Heating substances/S.7 Particles everywhere/E.7
		Applications of heat
		C.8 Joining particles, separating particles
	Unit 5: Solvents and Solutions	C.1 Finding out about water/E.1 More about water
		C.2 Evaporation/E.2 More about evaporation
		C.3 Water purification
		C.4 Dissolving and solubility
		C.5 Crystals
		C.6 Solvents other than water
		C.7 Simple paper chromatography/E.7 Colloids and emulsions
		C.8 Digestion: most foods are complex and insoluble
		C.9 Digestion: the action of enzymes
		C.10 Digestion and absorption

(continued)

Table 7.5 (continued)

Grade	Unit	Subsection
	Unit 6: Cells and Reproduction	C.1 Using the microscope
		C.2 Plants and animal cells
		C.3 Reproduction in animals
		C.4 Reproduction in plants/S.4 Making a model flower/ E.4 Looking at more flowers + grow pollen tubes
		C.5 Growth and development/E.5 Propagation
		C.6 Mammal development
		C.7 Further discussion
Grade 8	Unit 7: Some Common Gases	C.1 Some gases
		C.2 Gases of the air
		C.3 Breathed and unbreathed air
		C.4 Gas exchange between plants and air
		C.5 Formation of starch in green plants
		C.6 Burning food
		C.7 Respiration/E.7 Comparison of respiration and combustion
		C.8 Photosynthesis and respiration
		C.9 Plant and animal interdependence
		C.10 Breathing system in man
	Unit 8: Electricity	C.1 Simple circuitry
		C.2 Conductors and insulators
		C.3 Heating effect
		C.4 At home with electricity
		C.5 Electromagnet and motor effect/E.5 More about electromagnet
		C.6 Chemical effect
	Unit 9: Heat Flow	C.1 Convection in gases
		C.2 Convection in liquids
		C.3 Conduction in solids/E.3 More about conduction
		C.4 Air and water are poor conduction
		C.5 Heat transfer by radiation
		C.6 Application of heat transfer
	Unit 10: Hydrogen, Acids, and Alkalis	C.1 Electrolysis of water
		C.2 Action of metals on water
		C.3 Action of metals on acids
		C.4 Degree of acidity or alkalinity
		C.5 Neutralization/E.5 Conversation and neutralization
	Unit 11: Detecting the Environment	C.1 Perceiving the environment
		C.2 The eye
		C.3 The camera and the eye
		C.4 Limitations of the eye
		C.5 Sources of sound
		C.6 Transmission of sound
		C.7 Ear and hearing
		C.8 Pitch

(continued)

Table 7.5 (continued)

Grade	Unit	Subsection
		C.9 Taste and smell
		C.10 Touch
		C.11 Senses and the brain
	Unit 12: Support and Movement	C.1 What is a force?
		C.2 Friction
		C.3 Weighing
		C.4 Levers
		C.5 Lever systems in human skeleton
		C.6 Work and energy
		C.7 Pulleys
		C.8 Pairs of forces

Science Learning Objectives and Contents in Upper Secondary School

For the upper secondary school students, there are several science subjects, which have their own objectives and contents, respectively. The science subjects for upper secondary school are numerous; therefore, objectives and contents of combined or integrated sciences are mainly introduced in this section.

Korea

In the 5th curriculum reformed in 1987, the subject "science I" (biology and earth science were combined) and "science II" (physics and chemistry were combined) were required for the students who were willing to attend nonscience-bound colleges, while "science I," "physics," and "chemistry" were required for the students who were willing to attend science-bound colleges. And the learning objectives of the science subjects all emphasized same factors. The learning objectives of "science I" were: (1) students should understand basic concepts in biology and earth science and should apply them for explaining natural phenomena; (2) students should develop the ability to inquire about the phenomena of biology and earth science scientifically and should use this ability for solving problems; (3) students should enhance curiosity and interest toward natural phenomena in biology and earth science learning and develop an attitude related to science; (4) students should experience basic experimental skills and practice necessary for inquiring biology and earth science; (5) students should recognize that concepts in biology and earth science are continuously developing; and (6) students should recognize the impacts of biology and earth science on the development of technology and progress of society. The contents of "science I" were: (1) characteristics of living organisms, (2) nutrition of living organisms, (3) homeostasis of living organisms, (4) continuity of life, (5) organisms and environment, (6) our earth,

Table 7.6 Science contents covered in the 1992 and 1999 (Singapore)

Main Theme	Topics and subtopics in 1992	Topics and subtopics 1999
Introducing Science	1.1 What is Science?	1.1 What is Science?
	1.2 Attitudes	1.2 Attitudes
	1.3 Science Process Skills	1.3 Science Process Skills
	1.4 Physical Quantities and Unit	1.4 Physical Quantities and Unit
Matter	2.1 Introduction to Matter	2.1 Introduction to Matter
	Properties	Properties
	Classification	Classification
	Interactions	Interactions
	2.2 Element	2.2 Element
	Properties	Properties
	Uses	Uses
	2.3 Mixtures and Compounds	2.3 Mixtures and Compounds
	Properties	Properties
	Interactions	Interactions
	Uses	Uses
	2.4 Air	2.4 Air
	2.5 Simple Concepts of Atoms and Molecules	2.5 Simple Concepts of Atoms and Molecules
	2.6 Water and Solution	2.6 Water and Solution
	Solutions and Suspensions	Solutions and Suspensions
	Acidic and Alkaline Solutions, pH indicators	Acidic and Alkaline Solutions, pH indicators
	Crystallization	Crystallization
	Water	Water
	Water Consumption and Conservation	***Water Consumption and Conservation (removed)***[CG]

Energy	3.1 What is Energy?	3.1 What is Energy?
	Forms of Energy and Energy Conversion	***Forms of Energy and Energy Conversion (removed)***[L]
	Sources of Energy and Storage of Energy	Sources of Energy and Storage of Energy
	3.2 Mechanical Energy	3.2 Mechanical Energy
	Speed	Speed
	Forces and Pressure	Forces and Pressure
	Work and Energy	Work and Energy
	Turning Effect of Forces	Turning Effect of Forces
	3.3 Heat Energy	3.3 Heat Energy
	Effects of Heat Energy	Effects of Heat Energy
	Transmission of Heat Energy	Transmission of Heat Energy
	3.4 Light	3.4 Light
	The Nature and Behavior of Light	The Nature and Behavior of Light
	Reflection and Refraction at a Plane Surface	Reflection and Refraction at a Plane Surface
	The Spectrum of White Light	The Spectrum of White Light
	Convergence and Divergence	***Convergence and Divergence (removed)***[A]
	3.5 Sound Energy	3.5 Sound Energy
	3.6 Electrical Energy	3.6 Electrical Energy
	Current Electricity	Current Electricity
	Household Circuitry	Household Circuitry
Life	4.1 Diversity of Plant and Animal Life	4.1 ***Diversity of Plant and Animal Life (removed)***[L]
	4.2 Classification of Plant and Animal Life	4.2 Classification of Plant and Animal Life
	4.3 Cells – Structure, Function, and Organization	4.3 Cells – Structure, Function, and Organization
	4.4 Nutrition	4.4 Nutrition
	Food as a Source of Energy in Living Organisms	***Food as a Source of Energy in Living Organisms (removed)***[CH]
	Classes of Food and a Balanced Diet	***Classes of Food and a Balanced Diet (removed)***[CH]
	Digestion in Animals	Digestion in Animals
	Photosynthesis	Photosynthesis

(continued)

Table 7.6 (continued)

Main Theme	Topics and subtopics in 1992	Topics and subtopics 1999
	4.5 Respiration	4.5 Respiration
	4.6 Transport	4.6 Transport
	Water as Solvent in Living Systems	Water as Solvent in Living Systems
	Movement of Substances In and Out of Cells by: Diffusion and Osmosis	Movement of Substances In and Out of Cells by: Diffusion and Osmosis
	Transport in Living Organisms	Transport in Living Organisms
	4.7 Sexual Reproduction in Human Being	4.7 Sexual Reproduction in Human Being
	4.8 Abuses to Life Processes	4.8 Abuses to Life Processes
	Drugs	Drugs
	Alcohol	Alcohol
	Smoking	Smoking
Environmental Studies	5.1 Habitat	5.1 Habitat
	Organisms Living in a Specific Habitat	Organisms Living in a Specific Habitat
	The Physical Environment	The Physical Environment
	The Biotic Environment	The Biotic Environment
	5.2 Simple Concepts of Population, Community, and Ecosystem	5.2 Simple Concepts of Population, Community, and Ecosystem
	5.3 Energy Transfer Process in the Ecosystem	5.3 Energy Transfer Process in the Ecosystem
	5.4 Nutrient Cycles in the Ecosystem	5.4 Nutrient Cycles in the Ecosystem
	5.5 Man's Impact on the Ecosystem	5.5 ***Man's Impact on the Ecosystem (removed)***[CG]

(MOE of Singapore, 1992, 1999a)

Note: the written phrases in ***bold and italic*** were removed from syllabus

L learned topics in primary science, *A* advanced topics, *CG* covered topics in geography syllabus, *CH* covered topics in home economics syllabus

(7) matter of earth crust, (8) changes of atmosphere and ocean, (9) history of the earth, and (10) solar system and universe.

The learning objectives of "science II" were: (1) students should understand basic concepts in physics and chemistry; (2) students should develop the ability to inquire about the phenomena of physics and chemistry scientifically and should use this ability for solving problems; (3) students should enhance curiosity and interest toward natural phenomena in physics and chemistry and their learning and develop scientific attitude; (4) students should recognize that concepts in physics and chemistry are continuously developing; and (5) students should recognize the impacts of physics and chemistry on the development of technology and progress of society. The contents of "science II" were: (1) motion and energy, (2) electromagnetism, (3) light and waves, (4) regularity in the world of matter, (5) chemical reaction, and (6) compounds.

In the 6th curriculum reformed in 1992 (MOE of Korea, 1992), the subject "common science," in which the contents of physics, chemistry, biology, and earth science were combined, was developed and required for all of upper secondary students. The subject "common science" stressed on students' scientific inquiry ability development. In the 6th curriculum, physics I, chemistry I, biology I, and earth science I were required for the nonscience-bound upper secondary students, while physics II, chemistry II, biology II, and earth science II were required for science-bound students.[1] The learning objectives of the subject "common science" were: (1) students should be able to understand the natural phenomena by inquiry and to enhance the ability of solving problems of everyday life scientifically; (2) students should be able to understand basic science concepts comprehensively through inquiry activity and to use them for creative problem solving; (3) students should have attitude to inquire natural phenomena continuously with interests and curiosity about natural phenomena and science learning; and (4) students should recognize the impacts of science on the development of technology and progress of society. The contents of "common science" were divided into two parts, knowledge part and inquiry activity part. The contents of knowledge part were matter (reaction of matter, elements, exothermic reaction, endothermic reaction, reaction velocity), force (description of motion, laws of motion, law of force), energy (heat, solar energy, electric energy, chemical energy, biological energy, flow and conservation of energy), life (nutrition and health, stimulus and response, reproduction, heredity), earth (matter of the earth's crust and crustal movement, geological time, ocean, weather and climate, solar system and stars), environment (greenhouse effect, inversion layer, noise, radioactivity), contemporary science and technology (new material, light communication and semiconductor, life science, space science), and knowledge of inquiry, and the contents of inquiry part were classification, measurement, prediction, experiment, investigation and discussion, and data interpretation.

[1] Subjects I consisted of easy and basic concepts and small amount of contents compared to subjects II.

The learning objectives of the subject "science" for the upper secondary school 10th grade in the 7th curriculum reformed in 1997 (MOE of Korea, 1997) were the same with those for the lower secondary school, because for 3rd to 10th grade, the contents were constructed with sequence and required for all upper secondary students, which were: (1) students should be able to inquire natural phenomena and to solve problems of everyday life through acquiring the scientific inquiry ability; (2) students should be able to acquire basic science concepts through scientific inquiry about natural phenomena; (3) students should have an attitude to inquire and solve problems in everyday life with interests and curiosity about natural phenomena and science learning; and (4) students should recognize the impacts of science on the development of technology and progress of society. In particular, the objective "students should recognize that concepts in science are continuously developing" was not included even in the objectives of physics, chemistry, biology, and earth science learning in the 7th science curriculum. It was considered that even the students who learned upper secondary school level sciences could not recognize conceptual changes in the history of science. The contents of "science" were divided into knowledge part and inquiry activity part as in the 6th curriculum. The contents of knowledge part were energy (force and energy, electric energy, wave energy, energy transformation), matter (electrolyte and ion, reaction of acid and base, reaction velocity), life (metabolism, stimulus and response, reproduction), earth (crustal movement, atmosphere and ocean, solar system and galaxies), and environment (biological concentration, acid rain, greenhouse effect, noise), and the contents of inquiry part were inquiry processes (observation, classification, measurement, problem recognition, hypothesis formulation, control of variables, data transformation, data interpretation, conclusion inducement, generalization, etc.) and inquiry activities (discussion, experiment, investigation, field trip, task research, etc.)

From the 5th to 7th curriculum, the contents of "physics," "chemistry," "biology," and "earth science" had not been changed; almost same contents continued.

Japan

In the science curriculum for the upper secondary school reformed in 1989 (MESSC, 1989b), the overall objectives were "to enhance students' interest in nature, to enable them to carry out observations and experiments with identifying clear purpose, and to develop attitudes and abilities to investigate scientifically." The subjects of science were "integrated science," "physics IA," "physics IB," "physics II," "chemistry IA," "chemistry IB," "chemistry II," "biology IA," "biology IB," "biology II," "earth science IA," "earth science IB," and "earth science II." Each subject had its own objectives for learning. For example, the objectives of "integrated science" were "through carrying out observations and experiments concerned with natural matters and phenomena, to enable students to nurture the comprehensive perspective and thinking, and to enable students to understand

Table 7.7 The contents of "Integrated Science" (Japan)

Topic	Subtopic
Inquiry of nature	Perception of nature
	Planning and conducting of observation and experiment
	Observation, experiment, and clean finish
World of nature and its change	Difference and similarity
	Change, equilibrium, interaction
	Energy and its transformation
Human and nature	Resource, energy, and use of them
	Natural environment and its preservation
	Progress of science and technology and human life
Task research	Observation and experiment about special phenomena
	Survey on the natural environment
	Research of exemplary experiment in the history of science

natural matters and phenomena at the same time and to recognize the linkages among nature and human beings," and its contents are shown in Table 7.7.

In the science curriculum for upper secondary school students reformed in 1999 (MOE of Japan, 1999b), the overall objectives were "to enhance students' interest in nature, to enable them to carry out observations and experiments, to develop attitudes and abilities to investigate scientifically, and at the same time, to deepen their understanding of natural phenomena and to develop scientific views of nature"(Japanese Research Team for US-Japan Comparative Research on Science Mathematics and Technology Education, 2004). The subjects of science in the curriculum were "basic science," "integrated science A," "integrated science B," "physics I," "physics II," "chemistry I," "chemistry II," "biology I," "biology II," "earth science I," and "earth science II." Each subject had its own objectives for learning. For example, the objectives of "basic science" were "to enable students to understand the relationship between science and human life, exploration and revelation of nature, and the development of science through observation and experiments, to enhance their interest and concerns for science, and at the same time, to develop scientific views and thinking" (Japanese Research Team for US-Japan Comparative Research on Science Mathematics and Technology Education, 2004). The contents are shown in Table 7.8.

The objectives of "integrated science A" were "to enable students to understand natural events and phenomena mainly on the formation of matter and energy, through observations and experiments concerned with common events and natural phenomena, as well as to enable them to consider and discuss the relationship between human life and nature, and to develop comprehensive views and the way of thinking of nature." Its contents are shown in Table 7.9.

The objectives of "integrated science B" were "to enable students to understand natural events and phenomena mainly on living things and surrounding environment of them through observations and experiments concerned with natural phenomena, as well as to enable them to consider the relationship between human

Table 7.8 The contents of "Basic Science" (Japan)

Area	Topic	Subtopic
Beginning of science		
Inquiry of nature and development of science	Origin of matter	Investigation of atoms and molecules, how to make synthesis of a substance
	Investigating the life	Discovery of cell and cell theory, thinking of evolution
	Concept of energy	Formation of the concept of energy, use of electrical energy
	Exploring the earth and the space	Geocentric theory and heliocentric theory, establishment of the theory of plate tectonics
Human life and future challenges of science		

Table 7.9 The contents of "Integrated Science A" (Japan)

Area	Topic	Subtopic
Inquiry of nature	Scientific view of nature	
	How to do inquiry	
Resources, energy, and human life	Development and use of resources	Use of energy and resource, the development and use of other resources
	Various energy	Work and heat, transformation, and conservation of energy
Matter and human life	Structure and change of matter	Constituent unit of matter, change of matter
	Use of matter	Matter and daily life, materials within living things
Progress of science and technology and human life		

beings and nature, and to develop comprehensive views and the way of thinking of nature." Its contents are shown in Table 7.10.

When comparing with the former science curriculum, the different points were: (a) in the new subject "basic science," students learn to investigate and solve the problems of nature, contribute to the development of civilization, become aware of the challenges facing science and the relationship between science and human life, and foster a scientific worldview and scientific ways of thinking and (b) "physics I," "chemistry I," "biology I," and "earth science I" have simpler and easier contents than those with "IB," and "physics II," "chemistry II," "biology II," and "earth science II" focused on students' hands-on observation and experimentation (MOE of Japan, 1999a, 1999b).

Table 7.10 The contents of "Integrated Science B" (Japan)

Area	Topic	Subtopic
Inquiry of nature	Scientific view of nature	
	How to do inquiry	
Transition of the earth and life	Transition of the earth	Earth as a planet, changes and movements in the earth
	Transition of life	Transition of living things, genetic regularity
The diversity of living things and balance of nature	The atmosphere and the shape of the earth surface	Diverse landscape, circulation of atmosphere and water
	Living things and the environment	Biodiversity, relationship with the environment and living things
Human activities and change of earth environment		

Hong Kong

In Hong Kong, forms 1 to 3 are lower secondary level, and forms 4 to 5 (grades 10 and 11) are upper secondary level. Holbrook (1990) analyzed the intended curriculum in Hong Kong upper secondary schools. According to his investigation, the major aim of the biology syllabus was to present biology not only as a body of knowledge but also as a field of inquiry. Rather than concentrating on the factual knowledge, students were expected to put more emphasis on the understanding of concepts and principles and their application to both familiar and novel situations. Students should also be able to integrate biological knowledge with other science disciplines. The aims of the sixth form chemistry course were: (1) to develop scientific concepts, especially principles in chemistry through experimentation and observation, and to interrelate diverse phenomena; (2) to understand the terminology and conventions used in chemistry; (3) to appreciate the scientific method and its limitations; and (4) to develop an awareness of the significance of chemical principles in technology and society. And the objectives of the physics examination are to test whether the candidates have acquired: (1) knowledge of the laws, principles and concepts in the syllabus, and their interrelationships; (2) scientific methods and skills; (3) ability to apply materials in the syllabus to both familiar and novel situations; and (4) some understanding of the role of physics in modern society.

As an example of science contents for upper secondary schools, the contents of "biology" for secondary 6 (grade 12) are as follows (Holbrook, 1990): cell structure and function, transport of cellular material, cell metabolism, cell responses, concept of the gene, diversity of life, metabolism of the organism, regulation of the organism, coordination/behavior of organism, reproduction/development of plants, reproduction/development of animals, human biology, natural environment, cycles in nature, natural groups and their segregation, population genetics, and evolution.

Singapore

In Singapore, there are the national examinations for further studies after 4 years of secondary school education and after 2 years of Junior College studies (GCE "A" Level examination). The GCE "O" Level examination and the GCE "A" Level examination include the syllabus of secondary Years 3 and 4 and Junior College Years 1 and 2, respectively. In the secondary science syllabus, there are also two kinds of syllabus for the GCE "N" Level (syllabus for normal technical group) and the GCE "O" Level (syllabus for special/express group). The GCE "N" Level syllabus is designed for students in normal technical groups who will shift to TVE (technical and vocational education) after their secondary school graduation. In this section, we will focus on syllabuses for students who are going for further academic studies such as Junior College or polytechnics and university. Five polytechnics in Singapore have their own course programs, for example, maritime business, science, engineering, technology, and multimedia areas for three-year degree programs. Polytechnics were set up to train professionals to support the technological and economic development of Singapore for a knowledge-based economy. The top 10–15 % of students from polytechnics continue their university studies, and about 90 % of students find jobs after their graduation (Post-Secondary Education, http://www.moe.edu.sg/education/post-secondary/files/post-secondary-brochure.pdf). We will also exclude various polytechnic programs in this chapter.

The educational aims in the upper secondary syllabus are similar to the lower secondary syllabus. In addition, the innovative approaches in lower secondary school science were continuously applied in the upper secondary school syllabus. For example, science contents in the lower secondary science syllabus were reduced in 1999 because of the implementation of TSLN (Thinking School, Learning Nation) in 1997. It was also applied to the GCE "O" Level and the GCE "A" Level syllabuses in 1998. The reasons for deleting contents in upper secondary syllabuses were the same as the lower secondary science syllabus. For example, topics/subtopics/assessment objectives that are mainly descriptive and topics taught in other subjects/topics in the GCE "O" Level physics syllabus were removed. See Tables 7.11 and 7.12 for topics in physics, chemistry, and biology in GCE "O" Level and "A" Level syllabus each. The tables also indicate the deleted contents in the upper secondary syllabus 1999.

The GCE "A" Level syllabuses (Junior College Years 11–12) for physics, chemistry, and biology consist of two parts – core sections I–V and seven options. For example, all students who choose physics had to select the core sections and one option in Table 7.12. Core sections took 90 % of the curriculum time and each option took 5 % of the curriculum time. Students should choose at least two subjects among physics, chemistry, and biology for further studies in a university. Table 7.12 also shows the content summary of these two parts in each subject. Most contents in physics were removed because concepts were covered in the "O" Level syllabus or were not connected to the current syllabus. Most chemistry contents were reduced because topics were learned in the "O" Level syllabus or the concepts were too

Table 7.11 Topics and subtopics in physics, chemistry, and biology for the GCE "O" Level syllabus 1999

Topics in GCE "Ordinary" Level syllabuses		
Physics syllabus	Chemistry syllabus	Biology syllabus
1. General Physics	**1. The Particulate Nature of Matter**	**1. Organization and Maintenance of the Individual**
1.1 Length and time	**2. Experimental Techniques**	1.1 Cell structure and organization
1.2 Speed, velocity, and acceleration	2.1 Apparatus	1.2 Specialized cells, tissues, and organs
1.3 Mass weight and center of mass	2.2a Criteria of purity	1.3 Diffusion and osmosis
1.4 Density (removed)[L]	2.2b Methods of purification	1.4 Enzymes
1.5 Forces	**3. Atoms, Elements, and Compounds**	1.5 Nutrition
1.6 Scalars and vectors	3.1 Atomic structure and the periodic table	1.6 Transport in flowering plants
1.7 Energy, work, and power	3.2 Bonding: the structure of matter	1.7 Transport in man
1.8 Pressure		1.8 Respiration
		1.9 ***Support, movement, and locomotion (removed)***[NFD]
		1.10 Excretion
		1.11 Homeostasis
		11.12 Coordination and response
		1.13 ***Tropisms and toxic responses (removed)***[NFD]
		1.14 The use and abuse of drugs
2. Thermal Physics	**4. Stoichiometry and the Mole Concept**	**2. Diversity of Organisms**
2.1 Simple kinetic molecular model of matter	**5. Electricity and Chemistry**	
2.2 Thermal properties	**6. Energy Changes**	
2.3 Transfer of thermal energy		
3. Properties of Waves, Including Light and Sound	**7. Chemical Reactions**	**3. Relationships of Organisms with One Another and with the Environment**
3.1 General wave properties	7.1 Speed of reaction	3.1 Energy flow
3.2 Light	***7.2 Reversible reactions (removed)***[A]	3.2 Food chains and food webs
3.3 Sound	7.3 Redox	3.3 Carbon cycle
	8. Acids, Bases, and Salts	3.4 Parasitism
	8.1 The characteristic properties of acids and bases	3.5 Effects of man on the ecosystem
	8.2 Types of oxides	
	8.3 Preparation of salts	
	8.4 Identification of ions and gases	

(continued)

Table 7.11 (continued)

Topics in GCE "Ordinary" Level syllabuses		
4. Electricity and Magnetism	**9. The Periodic Table**	**4. Developments of Organisms and Continuity of Life**
4.1 Simple phenomena of magnetism	9.1 Periodic trends	4.1 Reproduction
4.2 Electricity	9.2 Group properties	4.2 Inheritance
4.3 Electric circuits	9.3 Transition elements	
4.4 Practical electric circuitry	9.4 Noble gases	
4.5 Electromagnetic effects	**10. Metals**	
4.6 Introductory electronics	10.1 Properties of metals	
4.7 Electronic systems	10.2 Reactivity series	
	10.3 Extraction and uses of metals	
	11. Nonmetals	
	11.1 Hydrogen	
	11.2 Oxygen	
	11.3 Nitrogen	
	11.4 Sulfur	
	11.5 Chlorine (removed)[NFD]	
	11.6 Silicon (removed)[NFD]	
	11.7 Carbon and carbonates (removed)[NF]	
5. Atomic Physics	**12. Organic Chemistry**	
5.1 Radioactivity	12.1 Names of compounds	
5.2 The nuclear atom (removed)[C]	12.2 Fuels	
	12.3 Homologous series	
	12.4 Alkanes	
	12.5 Alkenes	
	12.6 Alcohols	
	12.7 Acids	
	12.8 Macromolecules	

MOE of Singapore, 1999b

Note: *L* learned topics in lower secondary science, *A* advanced topics, *NC* not connected to current contents, *C* covered topics in chemistry syllabus, *NFD* non-fundamental and descriptive topic

advanced. In addition, non-beneficial activities such as "make, stain, and observe microscopically root tip squash" and non-fundamental objectives that have minimal effect on the overall coverage in the main topics were deleted from the biology syllabus.

Table 7.12 Content topics in physics, chemistry, and biology for the GCE "A" Level syllabus

Topics in GCE "Advanced" Level syllabuses			
	Physics syllabus	Chemistry syllabus	Biology syllabus
Core Sections	**1. General Physics**	**1. Physical Chemistry**	**1. Cell Structure**
	1.1 Physical Quantities and Units	1.1 Atoms, Molecules, and Stoichiometry	
	1.2 Measurement Techniques	1.2 Atomic Structure	**2. Biological Molecules**
		1.3 Chemical Bonding	
		1.4 States of Matter	
		1.5 Chemical Energetics	
		1.6 Electrochemistry	
		1.7 Equilibria	
		1.8 Reaction Kinetics	
	2. Newtonian Mechanics	**2. Inorganic Chemistry**	**3. Enzymes**
	2.1 Kinematics	2.1 The Periodic Table: Chemical Periodicity	**4. Cell and Nuclear Division**
	2.2 Dynamics	2.2 Group II	
	2.3 Forces	***2.3 Group IV (removed)***[C]	
	2.4 Work, Energy, Power	2.4 Group VII	
	2.5 Gravitational Field	2.5 An Introduction to the Chemistry of Transition Elements	
	2.6 Motion in a Circle	2.6 Nitrogen and Sulfur	
	3. Oscillations and Waves	**3. Organic Chemistry**	**5. Genetic Control and Inheritance**
	3.1 Oscillations	3.1 Introductory Topics	**6. Inherited Change and Evolution**
	3.2 Waves	3.2 Hydrocarbons	
	3.3 Superposition	3.3 Halogen Derivatives	
	3.4 Electromagnetic Waves	3.4 Hydroxy Compounds	
		3.5 Carbonyl Compounds	
		3.6 Carboxylic Acids and Derivatives	
		3.7 Nitrogen Compounds	
		3.8 Polymerization	
	4. Electricity and Magnetism		**7. Energetics**
	4.1 Electrostatics (removed)[L]		**8. Ecology**
	4.2 Current of Electricity		
	4.3 D.C. Circuits		
	4.4 Electric Field		
	4.5 Capacitance		
	4.6 Magnetic Fields		
	4.7 Electromagnetism		
	4.8 Electromagnetic Induction		
	4.9 Alternating Currents		

(continued)

Table 7.12 (continued)

Topics in GCE "Advanced" Level syllabuses			
	5. Matter		**9. Transport**
	5.1 Phases of Matter		**10. Regulation and Control**
	5.2 Deformation of Solids		
	5.3 Temperature		
	5.4 Thermal Properties of Materials		
	5.5 Ideal Gases		
	5.6 Transfer of thermal energy		
	5.7 Charged Particles		
	5.8 Quantum Physics		
	5.9 Atomic Structure		
	5.10 Radioactivity		
Optional Section	**A. Astrophysics and Cosmology**	**Biochemistry**	**1. Biodiversity**
	A1 Contents and Scale of the Universe	1. Proteins	1.1 Diversity in Microorganisms
	A2 The Standard Model of the Universe	2. Carbohydrates	
	A3 Techniques of Observation	3. Lipids and Membrane Structure	1.2 Plant and Animal Diversity
			1.3 Animal Adaptations
		4. Nucleic Acids	1.4 Maintaining Biodiversity
	C. The Physics of Materials	**Environmental Chemistry**	**2. Applied Plant and Animal Science**
	C1 Structure of Crystals	1. The Atmosphere	2.1 Plant Metabolism and Productivity
	C2 Microstructure	2. The Chemical Structure of Soil; Processes involving Soils and Water	2.2 Soil Fertility
	C3 Materials Testing	3. The Water Cycle	2.3 Crop Production
		4. Recycling of Materials	2.4 Livestock Production
			2.5 The World Food Problem
	E. Electronics	**Food Chemistry**	**3. Applications of Genetics**
	E1 Analogue Systems	1. Composition and Functions of Food	3.1 Variation
	E2 Digital Systems	2. Food Additives	3.2 Selective Breeding
	E3 Electronics in Society and Industry	3. Food Storage	3.3 Genetic Diversity
		4. Food Processing and its Effects on Food	3.4 Genetic Engineering

(continued)

Table 7.12 (continued)

Topics in GCE "Advanced" Level syllabuses		
		3.5 Human Genetics
F. The Physics of Fluids	***Materials (removed)****	**4. Growth, Development, and Reproduction**
F1 Buoyant Forces	***1. Ceramics***	4.1 Growth and Development
F2 Nonviscous Fluid Flow	***2. Carbon-based Polymers***	4.2 Asexual Reproduction
F3 Viscous Fluids	***3. Composites***	4.3 Sexual Reproduction in Flowering Plants
		4.4 Sexual Reproduction in Humans
		4.5 Control of Growth and Reproduction
M. Medical Physics	**Phase Equilibria**	**5. Human Health and Disease**
M1 Medical Imaging	1. Phase Diagrams	5.1 Health and Disease
M2 Medical Treatment	2. Raoult's Law and Distillation	5.2 Diet
M3 The Physics of Sight	3. Distribution between Phases	5.3 Gaseous Exchange
M4 The Physics of Hearing	4. Separation Techniques for Analysis: Chromatography and Electrophoresis	5.4 Exercise
		5.5 Drugs 5.6 Infectious Disease
		5.7 Immunity
P. Environmental Physics	**Spectroscopy**	
P1 Power Sources	1. Basic Principles of	
P2 Power Consumption	Spectroscopy	
P3 Heat Engines	2. Ultraviolet and Visible	
P4 Pollution	Spectroscopy	
	3. Infrared Spectroscopy	
	4. Nuclear Magnetic Resonance Spectroscopy	
	5. Mass Spectroscopy	
T. Telecommunications	**Transition Elements**	
T1 Communication Principles	1. Complexes	
T2 Communication Channels	2. Chemistry of Transition Elements	
T3 Radio Communication		

MOE of Singapore, 1999c

Note: *L* learned topic in "O" Level syllabus, *C* covered topic in other chemistry topic, * to maintain rigor of each option and to prevent unequal content load in each option

Science Curricula and Educational Practices After the Year 2000 in East Asia

Here we introduce the changes in science curricula of secondary school level after 2000 in East Asian countries, dividing into the lower secondary and the upper secondary schools.

In Korea, the national science curriculum was reformed in 2007 about 10 years since the 7th curriculum reform in 1997, and the reformed curriculum is called the 2007 science curriculum. This one had special features such as inclusion of creativity development and enrichment of STS-related contents. These features were explained in the objectives of the curriculum, which were same from the 3rd to 10th grade. The science curriculum aimed to help students understand the basic concepts of science through inquiry into natural phenomena and objects with interest and curiosity and the development of scientific thinking and creative problem-solving abilities. As a result, students would be able to develop the scientific literacy necessary for solving the problems of daily life creatively and scientifically. The detailed objectives of the science curriculum were: (a) to understand the basic concepts of science and apply them to solve problems in daily life; (b) to develop the ability to inquire about the nature scientifically and to use this ability for solving problems in daily life; (c) to enhance curiosity and interest toward natural phenomena and science learning and develop an attitude to scientifically solve problems in daily life; and (d) to recognize the relationship between science, technology, and society. As was shown in the above objectives, the relationship among science, technology, and society was stressed, rather than impacts of science on the development of technology and the progress of society which was stressed in the 5th to 7th curricula. In the 2007 science curriculum, there were also objectives for learning physics, chemistry, biology, and earth science for upper secondary school level, but they were never applied to actual school science teaching, because in 2009, right after the 2007 curriculum, a new one for the upper secondary school, which is called the 2009 revised science curriculum, appeared, but the one for the elementary and lower secondary students was actually reformed in 2011, even though it is called also 2009 science curriculum.

The objectives of the 2009 science curriculum for the 3rd grade to 9th grade are to enable students to: (a) understand the basic concepts of science through inquiring natural phenomena; (b) develop the ability to inquire about the nature scientifically; (c) enhance curiosity and interest toward natural phenomena and science learning and develop an attitude to solve problems scientifically; and (d) recognize the relationship between science, technology, and society.

A big change in the 2009 curriculum was the introduction of the term "criteria of achievement" and the development of achievement criteria of the contents. In addition, there is no requirement subject in the upper secondary school learning, that is, all subjects are electives. Achievement criteria are divided into key stage achievement criterion, which is equivalent to the objectives of specific stages, and unit achievement criterion, which is equivalent to the objectives of units. Below are examples of key stage achievement criterion.

[*Key stage* (*lower secondary school*) *achievement criteria*]

1. The students have synthetic scientific inquiry ability through integrated inquiring activities based on basic inquiry ability.
2. The students have interest and curiosity through learning the contents in the unit "What is science?" and have positive attitude toward learning science.
3. In the domain of "Matter and Energy," the students should understand the concepts of force and motion, heat and our living, molecular motion and state change, composition of matter, light and wave, property of matter, work and energy transformation, electricity and magnetism, regularity in chemical reaction, and various chemical reactions.
4. In the domain of "Life and Earth," the students should understand the concepts of earth system and change of earth sphere; photosynthesis; structure of water sphere and circulation; atmosphere sphere and our living; digestion, circulation, respiration, and excretion; stimulus and response; solar system; reproduction and genesis; heredity and evolution; exospace; and space development.

In Japan, the courses of study for elementary and lower secondary schools were reformed in 2008 and those for upper secondary schools in 2009. The reformed courses of study continue to aim to nurture students' "zest for life" based on the educational principles expressed in the revisions to the Basic Act on Education. The courses of study enrich the content of education and increases the number of classes, with an emphasis on the balance between acquiring the basic and fundamental knowledge and skills and fostering the ability to think, make decision, and express oneself (MEXT, 2014).

The overall objectives in the reformed science curriculum are to help students to develop basic scientific inquiry ability and attitude by conducting observation and experiment related to objects and phenomena in nature with purposed mind and to help students to develop their own views and way of thinking through deep understanding of objects and phenomena in nature.

The "science" subject is divided into two fields, Field I and Field II, and each field has its own learning objectives. Those of "Field I" are: (1) to discover regularity in matter and energy and to attain the method of task problem solving through the inquiry activity of problem finding and solving related to objects and phenomena in matter and energy; (2) to attain observing and experimental skills through observation and experimentation about physical objects and phenomena and to improve the ability to analyze, interpret, and present the result of observation and experimentation and to have a scientific view or thought about physical objects and phenomena through understanding familiar physical phenomena, electrical current and its use, motion and energy, etc.; (3) to attain observing and experimental skills through observation and experimentation about chemical objects and phenomena and to improve the ability to analyze, interpret, and present the result of observation and experiment and to have scientific view or thought about chemical objects and phenomena through understanding chemical substances around us, chemical change and atoms and molecules, chemical change and ions, etc.; and (4) by having deep perception of the relationship between development of science and technology and human life through the investigation of the objects and phenomena about matter and energy and by attaining attitude to think scientifically, to view nature synthetically.

The learning objectives of "Field II" are: (1) to make connections actively between living things and objects and between phenomena around them, through finding problem among them and inquiring it with enthusiasm, to discover diversity and regularity in them, and to attain the method to solve the task; (2) to attain observing and experimental skills through observation and experiment about living things and biological phenomena and to improve the ability to analyze, interpret, and present the result of observation and experiment and to have scientific view or thought about living things and biological phenomena through understanding the life of living things, kinds of living things, continuity of life, etc.; (3) to attain observing and experimental skills through observation and experiment about geological and terrestrial objects and phenomena and to improve the ability to analyze, interpret, and present the result of observation and experiment and to have scientific view or thought about geological and terrestrial objects and phenomena through understanding the formation of crust and its change, weather and its change, earth and space, etc.; and (4) to develop an attitude to respect wildlife and to contribute for the preservation of nature through the investigation of the living things and their surrounding objects and phenomena and to attain the ability to view nature synthetically.

Hong Kong has become a part of China after 1997. Therefore, it is necessary to enable every student, as a citizen of Hong Kong, to have more in-depth knowledge of modern China and the world (OECD, 2011). In 1999, science education, as one of the key learning areas of the school curriculum, was critically examined as part of the holistic review of the school curriculum. The aims of science education have been redefined to align them with the new aims of education for the twenty-first century. The science education in Hong Kong should build on the above strengths to enhance students' scientific thinking through progressive learning activities. These involve asking questions, hypothesizing, observing, measuring, designing and evaluating procedures, analyzing data, and examining evidence. Learning science will encourage students to learn independently and will enable them to deal with new situations, reason critically, think creatively, make better decisions, and solve problems. Through science activities, students should develop an interest in science, and thus they will be motivated to become active learners in science. Students should also develop an understanding of the interconnections between science, technology, and society (STS). They should be able to make informed decisions based on evidence (Chan, 2010).

In Hong Kong, the arrangement of major learning elements in science into six strands in the science curriculum is needed for the purpose of curriculum planning and organization. And they are interrelated. The six strands are scientific investigation, life and living, the material world, energy and change, the earth and beyond, and science, technology, and society (STS) (Curriculum Development Council of Hong Kong, China, 2002, 2009, 2014).

Through the various stages of schooling, students will acquire the necessary knowledge, skills, and attitudes in science education. The learning targets of science education at these stages are described below. On completion of the lower secondary level, students should: (1) acquire basic scientific knowledge and concepts for living in and contributing to a scientific and technological world; (2) have development of an ability to define problems, design experiments to find solutions, carry out practical work, and interpret the results; (3) apply their understanding of science to technological applications, social issues, and their daily experiences; (4) recognize the usefulness and limitations of science and the evolutionary nature of scientific knowledge; (5) relate their understanding of science to their personal health and cultivate an awareness of safety issues in everyday life and take proper action to control risks; and (6) consider the effects of human activities on the environment and act sensibly in conserving the environment.

The learning objectives at Key Stage 3 (secondary 1–3) for the different strands are shown in Table 7.13.

Generic skills are fundamental in helping students to learn to acquire knowledge, to construct knowledge, and to apply knowledge to solve new problems, thus allowing students to explore and investigate the scientific and technological world. Generic skills should be developed throughout all stages of schooling, across all key learning areas. Students should be able to transfer them from one learning context to another, be it within science or not. The transferability of these skills will help students continue to learn and succeed in personal development and lifelong learning. The nine generic skills advocated in *Learning to Learn* are collaboration skills, communication skills, creativity, critical thinking skills, information technology skills, numeracy skills, problem-solving skills, self-management skills, and study skills.

From 2001–2002 to 2005–2006, priority should be placed on *communication skills, creativity*, and *critical thinking skills*, and specifically for science education, *problem-solving skills* are also one of the focus areas. The science curriculum provides ample opportunities and rich learning experiences for students to develop generic skills as well as scientific process skills. Activities such as scientific investigations, experiments, project work, field work, group discussion, and debates, which allow students to be actively engaged in the learning process, are effective ways to motivate learning and to develop generic skills.

The development of values and attitudes are essential elements of the school curriculum. Values and attitudes should be permeated in the learning and teaching of science to foster the scientific ways of thinking and working, and these include curiosity, perseverance, critical reflection, open-mindedness, appropriately valuing the suggestions of others, caring for the living and nonliving environments, willingness to tolerate uncertainty, respect for evidence, and creativity and inventiveness. From 2001–2002 to 2005–2006, priority should be given to the development of values and attitudes such as responsibility, commitment, perseverance, respect for others, and national identity.

In Singapore, the science curriculum for all students has been developed since their independence from Britain. As mentioned earlier, secondary school students

Table 7.13 The learning objectives at Key Stage 3 (secondary 1–3) for the different strand

Strands	Learning objectives
Scientific investigation	To propose hypotheses and devise methods for testing them
	To plan and conduct scientific investigations
	To evaluate the fairness of tests and draw conclusions based on findings
Life and living	To appreciate the diversity of life and to understand the basic principles of classification systems
	To recognize that cell is the basic unit of life
	To develop a basic understanding of some of the life processes
	To appreciate and understand how a new life is born and be aware of the physiological and emotional changes during puberty
	To recognize the importance of maintaining body health
The material world	To understand physical and chemical properties
	To understand how materials are used and the related consequences
	To investigate some chemical changes and the substances involved
Energy and change	To compare energy options available for particular purposes in the community
	To identify processes of energy transfer and conditions that affect them
	To identify forms and transformations of energy in sequences of interactions
	To relate observed changes in an energy receiver to the quantity of energy transferred
The earth and beyond	To understand that the earth's crust is a useful source of minerals
	To describe the effects of gravity and frictional forces on the motion of an object on the earth
	To understand the basic concepts and conditions for space travel
Science, technology, and society	To understand the development and use of materials and their impact on our living and on the environment
	To recognize the effects of human activities on the environment
	To act responsibly in conserving the environment
	To recognize the limitations of science and technology

in Singapore are streamed into express, normal academic, and normal technical groups depending on their Primary School Leaving Examination (PSLE) results. The recent science syllabus in 2008 aims for the normal technology group include promoting an awareness of the impact of science and technology on society, industry, business, home, and leisure and preparing students for post-secondary technical courses (MOE of Singapore, 2008).

In 1997–2012, during the ability-driven phase, there was innovation in science education. Especially from the 2001 science syllabus, IT education was included for communication purposes not only for data collection tools (MOE of Singapore, 1997). In the current science syllabus implemented in 2013, there are few more changes in the curriculum framework. The twenty-first-century competencies and scientific literacy were emphasized, and it resulted in changes in modes of assessment and teaching and learning approaches. The curriculum aims had not changed, but the previous five science education aims before the year 2000 were grouped reasonably under cultivating students' perception of science; engaging students in science-related issues that concern their lives, the society, and the environment; and helping students develop the domains that are integral to the conduct of science inquiry including knowledge acquisition, developing skills, abilities, and attributes relevant to the study and/or practice science. The recent science syllabus in 2012 aims (MOE of Singapore, 2012) for express and normal academic groups are stated below.

1. Cultivate students' perception of science as a collective effort and a way of thinking rather than just a body of facts: This involves promoting awareness that the study and practice of science are cooperative and cumulative activities. These activities are subject to social, economic, technological, ethical, and cultural influences and limitations. In addition, the applications of science are generally beneficial, but the abuse of scientific knowledge can be detrimental.
2. Engage students in science-related issues that concern their lives, the society, and the environment: This involves stimulating the students' curiosity, interest, and enjoyment in science and matters relating to science and technology as well as developing the students' interest and care for the environment.
3. Help students develop the domains that are integral to the conduct of science inquiry. This includes: (a) acquisition of knowledge and understanding to become confident citizens in a technological world and for further studies, (b) developing skills and abilities, and (c) developing attributes relevant to the study and/or practice of science.

In the syllabus framework, the spirit of scientific inquiry has been emphasized before the year 2000. Comparing with the lower secondary school syllabus of 2008, "decision-making" process skill has been newly included. This is connected to the twenty-first-century competencies and scientific literacy in the science curriculum framework. The twenty-first-century competencies were designed to prepare students to become confident, self-directed learners, by equipping them with various skills and habits of mind that are aligned to the desired in twenty-first-century competencies. On the other hand, scientific literacy is designed to equip students with the skills to be able to utilize scientific knowledge in order to understand and make decisions/communicate about the natural world and the changes made to it through human activities such as doing science research and developing technologies. In addition, the framework influenced the assessment modes in schools. To cultivate and assess the twenty-first-century competencies and scientific literacy in science, performance-based assessments should be conducted in schools. Therefore, 30–40 % of performance-based assessments were indicated clearly through various ways, for example, projects, practical, teacher observations, reflection/journal writing, model making, debates, drama/show and tell, etc. On the other hand, real-world situations-based test questions involving science in daily life, society, and the environment were also recommended for teachers to use as the

Table 7.14 The learning contents for lower secondary students in 2007 curriculum (Korea)

Grade 7	Grade 8	Grade 9
Force and motion	Heat energy	Work and energy, electricity
Static electricity	Light and wave	
Three states of matter	Components of matters	Properties of matter
Molecular motion	Familiar compounds	Electrolytes and ion
State change of matter and energy		
Composition and diversity of organisms	Digestion and blood circulation	Stimulation and reaction
Nutrition of plants	Respiration and excretion	Reproduction and genesis
Matters of earth crust and change	Solar system	Properties of atmosphere and weather change
Earth crustal movement and plate tectonics	Stars and space	Components and movement of seawater

starting points for the construction of questions in order to mirror the application of scientific knowledge in real-life contexts in the lower secondary levels.

Science Learning Contents for Lower Secondary Schools After the Year 2000

The learning contents for lower secondary school students in 2007 curriculum of Korea (MEHRD, 2007) are shown in Table 7.14.

Table 7.14 shows topics of "science" for learning in each grade; in this curriculum subtopics are not presented but instead subtopic statements of objective form are presented. For example, the topic "force and motion" has four content objectives, as follows: (1) to explain various forces such as gravity, electric force, magnetic force, frictional force, elastic force, buoyancy, etc.; (2) to calculate the total force of two forces acting on an object; (3) to represent various motions, such as a motion without change of speed, a motion with change of speed, and a motion with change of direction, as change of position by time; and (4) to predict the motion of an object when a force acts on or not. In addition, it has three inquiry objectives, as follows: (1) to measure total force of two forces acting on an object; (2) to investigate change of speed and direction of an object by force; and (3) to interpret the graph of position by time of an object moving in straight line with constant speed.

In 2011, the science curriculum for the lower secondary school was reformed, which is called the 2009 National Curriculum, and the learning concepts for the lower secondary students in it are shown in Table 7.15 (MEST, 2011).

The two topics, "What is science?" and "Science and human culture," have general contents related to science and inquiry and human culture; therefore, they

Table 7.15 Learning concepts for lower secondary students in 2009 science curriculum (Korea)

Domain	7–9 grades				
Matter and energy	What is science?	Force and motion	Composition of matter	Electricity and magnetism	Science and human culture
		Heat and our living	Light and wave	Regularity in chemical reaction	
		Molecular motion and state change	Property of matter	Various chemical reactions	
			Work and energy transformation		
Life and earth		Earth system and change of earth sphere	Atmosphere sphere and our living	Solar system	
		Photosynthesis	Digestion, circulation, respiration, and excretion	Reproduction and genesis	
		Structure of water sphere and circulation	Stimulus and response	Heredity and evolution	
				Exosphere and space development	

are commonly taught in both domains. Furthermore, all topics have their own content achievement criteria and inquiry activities. For example, the topic "force and motion" has four achievement criteria, as follows: (1) to understand that force is an interaction between two objects and to know that there are interactions between distant objects as well as contact objects; (2) to know the features of various forces such as gravity, elastic force, frictional force, electric force, magnetic force, etc., and to understand how they are used for understanding surrounding phenomena; (3) to draw or calculate the total force of two forces acting on an object and to know the total force is net force; and (4) to explain the motion of an object by interpreting the graphs of distance-time and velocity-time. In addition, it has four inquiry activities, as follows: (1) to measure the strength of an elastic force; (2) to get and represent the total force of two forces; (3) to analyze the motion of a free-falling object using a moving picture; and (4) to induce the net force acting on an object moving with constant change of speed. These contents and inquiry activities are almost the same with those in the 2007 science curriculum.

The learning contents for lower secondary school students in the curriculum of Japan reformed in 2008 are shown in Table 7.16. In this curriculum the subject "science" for the lower secondary school is divided into two parts, "Field I" and "Field II." The learning contents of "Field I" and "Field II" are shown in Table 7.16.

Science education in Hong Kong is implemented through a series of subjects at the secondary levels. At the lower secondary level, the core subject science (S1–3), consisting of topics from various science disciplines, is taught in all schools.

The central curriculum, in the form of an open and flexible framework, sets out what schools are encouraged to help students develop: (1) subject knowledge and

Table 7.16 Learning contents of "Field I" and "Field II" in "science" subject (Japan)

Area	Topic	Subtopic
Field I	1. Familiar physical phenomena	(1) Light and sound
		(2) Force and pressure
	2. Substances around us	(1) Shape of substances
		(2) Water solutions
		(3) Change of state
	3. Electric current and its use	(1) Electric current
		(2) Electric current and magnetic field
	4. Chemical change, atoms, and molecules	(1) Structure of matter
		(2) Chemical change
		(3) Chemical change and mass of substance
	5. Motion and energy	(1) Regularity of motion
		(2) Mechanical energy
	6. Chemical change and ion	(1) Water solution and ion
		(2) Acid/alkali and ion
	7. Science and technology and human	(1) Energy
		(2) Development of science and technology
		(3) Preservation of natural environment and use of science and technology
Field II	1. Life of plants and kinds of plants	(1) Observing of living things
		(2) Structure and function of plants
		(3) Kinds of plants
	2. Structure and change of the earth's crust	(1) Volcano and earthquake
		(2) Stack of strata and shape of past strata
	3. Life of animals and transition of animals	(1) Living things and cell
		(2) Structure and function of animal's body
		(3) Kinds of animals
		(4) Transition of living things and evolution
	4. Weather and its change	(1) Observing of weather
		(2) Change of weather
		(3) Climate of Japan
	5. Continuity of life	(1) Growth of living things
		(2) Regularity of heredity and gene
	6. Earth and space	(1) Motions of celestial bodies and earth's rotation and revolution
		(2) Solar system and stars
	7. Nature and human	(1) Living things and environment
		(2) Grace of nature and disaster
		(3) Preservation of natural environment and use of science and technology

Note: as is shown in Table 7.16, the topics are almost same with the topics in the science curriculum reformed in 1998, although a few of subtopics are changed into other titles or a little changed in their contents skills as embodied in the learning targets and objectives of the six strands such as scientific investigation, life and living, the material world, energy and change, the earth and beyond, and science, technology, and society; (2) generic skills that are developed through learning activities (generic skills such as communication skills, creativity, critical thinking skills, and problem-solving skills are accorded priority positions); and (3) values and attitudes that should be permeated in the learning and teaching of science – curiosity, openness to new ideas, and respect for evidence are accorded priority positions.

In Hong Kong, the arrangement of major learning elements in science into six strands in the science curriculum is needed for the purpose of curriculum planning and organization. And they are interrelated. The six strands are scientific investigation, life and living, the material world, energy and change, the earth and beyond, and science, technology, and society (STS). The goals of six strands are as follows: "scientific investigation" is to develop science process skills and the understanding of the nature of science; "life and living" is to develop the understanding of scientific concepts and principles related to the living world; "the material world" is to develop the understanding of scientific concepts and principles related to the material world; "energy and change" is to develop the understanding of scientific concepts and principles related to energy and change; "the earth and beyond" is to develop the understanding of scientific concepts and principles related to the earth, space, and the universe; and "science, technology, and society (STS)" is to develop the understanding of the interconnections between science, technology, and society. And they are interrelated (Curriculum Development Council of Hong Kong, China, 2009, 2014).

The focus of the science curriculum framework in Singapore is to instill science as inquiry in three domains: (1) knowledge, understanding, and application, (2) skills and processes, and (3) ethics and attitudes. Teaching and learning approaches were also introduced considering inquiry mind. To practice the inquiry-centered approach, the TSLN (Thinking School, Learning Nation) and TLLM (Teach Less, Learn More) were introduced to schools in 2003 by the MOE. It is to help students' active engagement in learning such as peer tutoring, classroom discussions, and more open-inquiry than traditional-inquiry approaches (MOE of Singapore, 2005). However, curriculum implementations sometimes do not really reflect this planned curriculum (Treagust & Rennie, 1993). For example, concept cartoons for class discussion, role play, and drama are welcomed by government schools, but teachers feel that it is difficult to incorporate in their instruction because of time limitation despite the 15 % of free curriculum time (Chu, 2013).

Students in Years 3 to 6 in primary school and students in Years 1 and 2 in lower secondary schools learn science that consists of these themes, diversity, energy, interactions, etc, which were integrated after the year 2000. For example, the theme "interaction" consisted of topic such as effects of heat, chemical changes, ecosystem, etc. The content comparison between years 2008 and 2013 in secondary school is provided in Table 7.17.

Table 7.17 Science content comparison in lower secondary school level between the science syllabuses of 2008 and 2013 (Singapore)

Theme	Curriculum 2008	Curriculum 2013
Science and Technology	Science processes and application	***The Scientific endeavor***
	Scientific inquiry	***Science and technology theme was removed and scientific endeavor was included as an introduction but not as of the themes***
	Science and technology in society	
Measurement	Making measurements	***Measurement skills, knowledge of SI units, and using appropriate units for the respective physical quantities have been infused into the respective topics***
	Use of measurements instruments	
	Physical quantities and units	
Energy	Energy forms and uses	***This energy theme was incorporated into the "Interaction", theme with the exception of electricity that was moved to the "System" theme***
	Energy forms and conversion	
	Light	
	Electricity	
	Photosynthesis and Respiration	
Diversity	(i) Diversity of matter – classification of matter	(i) Exploring diversity of matter by their physical properties
	(ii) Diversity of matter – elements, compounds, and mixtures	(ii) Exploring diversity of matter by its chemical composition
	(iii) Solutions and suspensions	(iii) Exploring diversity of matter using separation techniques
	(iv) Diversity of plant and animal life – classification of plant and animal life	(iv) Understand diversity of living things
Interaction	(i) Interactions of forces and energy – concept of force and pressure	(i) Interactions through the application of forces
	(ii) Moment of a force (removed)	(ii) Energy and work done
	(iii) Work	(iii) ***Transfer of sound energy through vibrations (new)***
	(iv) Effects of heat	(iv) Effects of heat and its transmission
	(v) Transmission of heat	(v) Chemical changes
	(vi) Chemical changes	(vi) Interactions within ecosystem
	(vii) Simple concepts of populations, community, and ecosystem	
	(viii) Energy transfer process in the ecosystem	
	(ix) Nutrient cycles in the ecosystem	
Models and System	(i) Models of cells and matter	***Models***
	Cells – structure, function, and organization	(i) Model of cells – basic unit of life
	Particulate model of matter	(ii) Model of matter – particulate nature of matter
	Simple concepts of atoms and molecules	
	(ii) Plant and human systems	

(continued)

Table 7.17 (continued)

Theme	Curriculum 2008	Curriculum 2013
		(iii) Model of matter – atoms and molecules
	Transport in living things	(iv)***Ray model of light (new) system***
	Digestion in animals	(v) Transport system in living things
	Sexual reproduction in human beings	(vi) Human digestive system
		(vii) Human sexual reproduction system
		(viii) ***Electrical system***

(MOE of Singapore, 2008, 2013)
Note: ***bold and italic letters***, syllabus changes

There were six themes in the 2008 syllabus. As for measurement, it was integrated to each theme, and as for the models and system theme, they were divided into model and system theme in the 2013 science syllabus. In addition, the energy theme with the exception of electricity was moved to the system theme, which was then grouped under the interaction theme.

The upper secondary syllabuses are for students to prepare for the GCE "N" Level, "O" Level, and "A" Level examinations (see Fig. 7.3, p. 158). Most students in the express group in Year 3 of secondary school select two specific subjects from biology, chemistry, and physics. Students have the option of selecting combined science which is not an integrated science, but a combination of two subjects, for example, physics and chemistry or chemistry and biology. Students in the combined science class learn broad, but shallow contents of two combined subjects. See content comparison in Table 7.18 as an example between physics content in the GCE "O" Level physics syllabus and physics content in the GCE "O" Level syllabus in combined science – physics and chemistry.

Specific application of fundamental concepts and formula-based concepts were not included in physics in the combined science syllabus even though the same titles were listed in the syllabus. For example, "Brownian motions" in the kinetic model, "effects of electromagnetic waves on cells and tissues," "the electromagnetic spectrum," "specific heat capacity," and "specific latent heat" in thermal properties of matter were only in the "O" Level physics syllabus, not in the combined science syllabus.

Upper Secondary School

Korea

In the 2009 curriculum reform of Korea, nine subjects, none of them is required, which are "science," "physics I," "physics II," "chemistry I," "chemistry II," "life

Table 7.18 Physics content comparisons between physics and combined science "O" Level syllabuses (Singapore)

Physics area	O Level: physics	O Level combined science: physics
Measurement	Physical quantities, units, and measurement	Physical quantities, units, and measurement
	Physical quantities	Physical quantities
	SI unit	SI units
	Prefixes	Prefixes
	Scalars and vectors	Scalars and vectors
	Measurement of length and time	Measurement of length and time
Newtonian mechanics	Kinematics	Kinematics
	Speed, velocity, and acceleration	Speed, velocity, and acceleration
	Graphical analysis of motion	Graphical analysis of motion
	Free fall	Free fall
	Effect of air resistance	
	Dynamics	Dynamics
	Balanced and unbalanced forces	Balanced and unbalanced forces
	Free-body diagram	Free-body diagram
	Friction	Friction
	Mass, weight, and density	Mass, weight, and density
	Mass and weight	Mass and weight
	Gravitational field and field strength	Gravitational field and field strength
	Density	Density
	Turning effect of forces	Turning effect of forces
	Moments	Moments
	Center of gravity	Center of gravity
	Stability	Stability
	Pressure	Pressure
	Pressure	Pressure
	Pressure differences	
	Pressure measurement	
	Energy, work, and power	Energy, work, and power
	Energy conversion and conservation	Energy conversion and conservation
	Work	Work
	Power	Power
Thermal physics	Kinetic model of matter	Kinetic model of matter
	States of matter	States of matter
	Brownian motion	Kinetic model
	Kinetic model	
	Transfer of thermal energy	Transfer of thermal energy
	Conduction	Conduction
	Convection	Convection
	Radiation	Radiation

(continued)

Table 7.18 (continued)

Physics area	O Level: physics	O Level combined science: physics
	Temperature	
	Principles of thermometry	
	Thermocouple thermometers	
	Thermal properties of matter	Thermal properties of matter
	Internal energy	Internal energy
	Specific heat capacity	Melting, boiling, and evaporation
	Melting, boiling, and evaporation	
	Specific latent heat	
Waves	General wave properties	General wave properties
	Describing wave motion	Describing wave motion
	Wave terms	Wave terms
	Longitudinal and transverse waves	Longitudinal and transverse waves
	Light	Light
	Reflection of light	Reflection of light
	Refraction of light	Refraction of light
	Thin lenses	Thin converging lenses
	Electromagnetic spectrum	Electromagnetic spectrum
	Properties of electromagnetic waves	Properties of electromagnetic waves
	Application of electromagnetic waves	Applications of electromagnetic waves
	Effects of electromagnetic waves on cells and tissue	
	Sound	Sound
	Sound waves	Sound waves
	Speed of sound	Speed of sound
	Echo	Echo
	Ultrasound	
Electricity and magnetism	*Static electricity*	Static electricity
	Laws of electrostatics	Principles of electrostatics
	Principles of electrostatics	Electric field
	Electric field	
	Applications of electrostatics	
	Current of electricity	Current of electricity
	Conventional current and electron flow	Conventional current and electron flow
	Electromotive force	Electromotive force
	Potential difference	Potential difference
	Resistance	Resistance
	D.C. circuits	D.C. circuit
	Current and potential difference in circuits	Current and potential difference in circuits
	Series and parallel circuits	Series and parallel circuits

(continued)

Table 7.18 (continued)

Physics area	O Level: physics	O Level combined science: physics
	Potential divider circuit	
	Thermistor and light-dependent resistor	
	Use of cathode-ray oscilloscope	
	Practical electricity	Practical electricity
	Electric power and energy	Electric power and energy
	Dangers of electricity	Dangers of electricity
	Safe use of electricity in the home	Safe use of electricity in the home
	Magnetism	Magnetism and electromagnetism
	Law of magnetism	Laws of magnetism
	Magnetic properties of matter	Magnetic properties of matter
	Magnetic field	Magnetic field
	Electromagnetism	Magnetic effect of a current
	Magnetic effect of a current	Application of the magnetic effect of a current
	Application of the magnetic effect of a current	Force on a current-carrying conductor
	Force on a current-carrying conductor	
	The D.C. motor	
	Electromagnetic induction	
	Principles of electromagnetic induction	
	The A.C. generator	
	The transformer	

Singapore Examination and Assessment Board and University of Cambridge International Examinations, 2013a, 2013b

science I," "life science II," "earth science I," and "earth science II," are established. "Science" is developed as integrated science, which is totally different from "science" in the 7th curriculum.

The objectives of the subject "science" learning in the 2009 science curriculum for the 10th to 12th grade students are to enable students: (a) to understand the integrated concepts of science, necessary for the understanding of space and life, and modern science and technology and society; (b) to develop the ability to inquire about nature scientifically and to understand the process of the formation and development of scientific knowledge and technology; (c) to enhance curiosity and interest toward natural phenomena and science learning and develop an attitude to scientifically solve problems in daily life; and (d) to recognize the relationship between science, technology, and society and enhance the ability of rational

Table 7.19 Parts of the contents of "High School Science" in the 2009 science curriculum (Korea)

Domain		Elements of content
Origin of the universe and evolution	Origin of universe	Expansion of universe, Hubble's law, line spectrum, age of universe
	Big bang and elementary particles	Elementary particles, proton, neutron, formation of nucleus
	Formation of atom	Atoms of hydrogen and helium, cosmic background radiation
	Stars and galaxies	Beginning of a new star and evolution, combining of heavy element, structure of galaxy, interstellar compounds, covalent bond, reaction velocity
Solar system and the earth	Formation of the solar system	Formation process of the solar system, solar energy, terrestrial planet, Jupiter-type planet
	Mechanics of solar system	Kepler's law, Newton's law of motion, motion of planet, motion of the earth and moon, rotation, revolution
	Atmosphere of planet	Escape velocity, difference between planet atmosphere, structure, and property of molecules
	Earth	Evolution of the earth, the earth systems, element distribution of the earth, terrestrial magnetism

decision making by using the scientific knowledge and inquiry method (MEST, 2011).

In Korea, in the case of lower secondary school science, the contents and structure of science curriculum have almost not been changed since the 1st curriculum, being combined science of which are combined physics, chemistry, biology, and earth science contents as separated units. As for the upper secondary school science in the 5th reform, the subjects "science I," which biology and earth science were integrated, and "science II," which physics and chemistry were integrated, were developed for the students who were willing to go to nonscience-bound colleges. The purpose of the development of those subjects was to reduce the number of subjects which upper secondary school students should learn. The effort of reducing the number of subjects which upper secondary school students should learn has been continued to present science curriculum (2009 revised science curriculum). The subjects "common science" (in the 6th reform) and "science" (in the 7th reform), in which the contents of physics, chemistry, biology, and earth science were combined, were developed and required for all of upper secondary students. The subject "common science" stressed on students' scientific inquiry ability development, while the subject "science" stressed on students' inquiry ability development and scientific concepts understanding both.

The new science curriculum reformed in 2009 for upper secondary school is much different from former ones, particularly in the subjects of integrated science and in physics, as some part of the contents are shown in Tables 7.19 and 7.20. They show the efforts for convergence of science concepts in the "high school science" and "physics I" curriculum, respectively.

Table 7.20 Contents of "Physics I" in the 2009 science curriculum (Korea)

Domain		Elements of content
Space-time and universe	Time, space, motion	Measurement of time and standards, measurement of length and standards, Newton's law of motion, momentum and impulse, mechanical energy conservation law
	New understanding of space-time	Discovery of gravity, relativity theory, black hole and gravitational lens, cosmic model, four kinds of interactions, and elementary particles
Matter and electromagnetic field	Electromagnetic field	Electric field and electric force line, electrostatic induction and dielectric polarization, magnetic field and magnetic force line, induced current and Faraday's law
	Structure and property of matter	Energy level and light emission, energy band theory, semiconductor, new materials
Information and communication	Sound and light	Sound waves and supersonic waves, harmonic sound and noise, microphone and electric signals, photoelectric effect and light sensor, color recognition, and video system
	Information storing	Spectrum of electromagnetic waves, antenna and wireless communication, optic cable, alternative current and control of signal, device for storing information
Energy	Generation of energy	Electromotive force, electric energy, generator, nuclear power, nuclear fusion and solar energy, solar cell, various power systems
	Force and energy use	Transfer of force and torque, equilibrium of forces and stability, laws of fluid, laws of thermodynamics and heat engine, heat transfer, change of states and weather phenomena, use of electric energy

As shown in the above Table 7.20, physics contents in the 2009 science curriculum include high science and technology contents, and traditional contents such as force and motion are reduced much. Korean Ministry of Education is now revising current curriculum again and it is announced officially in 2015. In this curriculum, all high school students can select subjects according to their own interest and aptitude as well as studying common subjects of natural science and social science mandatorily.

Japan

The overall objectives for the upper secondary school science reformed in 2009 are to enhance students' curiosity and inquiring ability in things, to enable them to carry out observation and experiment with identifying clear purpose, and at the same time to deepen their understanding of matter and natural phenomena and to foster a scientific view of nature (MECSST (2009a, 2009b)).

Table 7.21 Parts of the contents of "High School Science (Science and Human Life)" in the 2009 revised science curriculum (Japan)

Domain		Elements of content
Development of science and technology	To allow students to understand how the development of science and technology has contributed to human societies today	
Science in the human life	Science of light and heat	Property of light and its use
		Property of heat and its use
	Science of matter	Resource and reuse of it
		Clothes and food
	Science of life	Living things and light
		Micro-living things and its use
	Science of space and earth	The earth in the solar system and closest to celestial body
		Natural disasters and closest to natural landscape
Human life and the science of the future	Based on the learning of the relationship between nature and human life, and the learning of the influence of science and technology to human life, students understand the interaction between human life and the science of the future	

The subjects of science are "science and human life," "basic physics," "physics," "basic chemistry," "chemistry," "basic biology," "biology," "basic earth science," "earth science," and "science task research."

Their standard teaching units (1 unit means 1 school hour per week for 35 weeks) are as follows: 2 units for "science and human life," 2 for "basic physics," 4 for "physics," 2 for "basic chemistry," 4 for "chemistry," 2 for "basic biology," 4 for "biology," 2 for "basic earth science," 4 for "earth science," and 1 for "science task research." Each subject has its own objectives for learning. For example, the objectives of "science and human life" are "to understand the role of science and technology and the relationship with human life through the observations and experiments for surrounding matter and objects and to enable to nurture the scientific view and the way of thinking scientifically," and parts of its contents are as Table 7.21.

The objectives of "basic physics" are to enhance interest to a variety of energy and movement of the object while making the relationship between society and everyday life; to do the observation and experiment with identifying clear purpose and at the same time to develop the attitudes and the exploring ability physically; to be allowed to understand the principles, laws, and basic concepts; to cultivate ideas and perspective scientific and at the same time to deepen their understanding of matter and natural phenomena; and to foster a scientific view of nature.

The objectives of "physics" are to enhance the inquiring ability for physical matter and phenomenon; to do the observation and experiment with identifying clear purpose; to develop the attitude and the exploring ability physically at the same time; to understand the principles, laws, and basic concepts; and to cultivate ideas and scientific perspective. The contents are: (1) various movements (balance of rigid body and motion in plane, momentum, circular motion, simple harmonic oscillation, gravity, motion of gas molecules, inquiry activities on a variety of exercise); (2) waves (way of transmitted waves, sound, light, inquiry activities on the waves); (3) electricity and magnetism (electricity and current, current and magnetic field, inquiry activities related to electricity and magnetism); and (4) atoms (electrons and light, atoms and nucleus, world building physics, inquiry activities related to atoms).

Hong Kong

At the senior secondary level, biology, human biology, chemistry, and physics are offered as optional subjects at S4–5, while biology, chemistry, and physics are further developed into Advanced Supplementary Level and Advanced Level subjects in the sixth form curriculum (Chan, 2010).

On completion of the senior secondary level, students should: (1) have a fundamental understanding of essential scientific knowledge and concepts for contributing toward a scientific and technological world; (2) have the ability to solve problems by employing scientific approaches and methods; (3) make informed judgments and decisions based on scientific evidence; (4) be able to comprehend and discuss issues related to the nature and development of science and technology; (5) make sensible judgments on their personal health and take responsible actions in safety issues; and (6) acknowledge the effects of human activities on the environment and be committed to act responsibly in conserving the environment (Curriculum Development Council of Hong Kong, China, 2009, 2014).

Learning objectives at key stage 4 (senior secondary) of "scientific investigation" are: to select and follow appropriate methods of investigations for specific purposes; to plan and conduct science experiments for solving problems; and to collect information and draw conclusions for decision making. Those of "life and living" are: to develop a basic understanding of the essential life processes; to understand the basic principles of genetics and its applications; to evaluate the impact of global environmental issues on the quality of life; and to be committed to a healthy lifestyle. Those of "the material world" are: to understand the relation of the uses of domestic chemicals and their properties; to examine classes of chemical reactions and their applications; to investigate the processing of raw materials; and to evaluate the use of materials. Those of "energy and change" are: to describe systems whose purpose is to transfer energy efficiently; to explain the principles of energy input-output devices; to apply ideas of energy conservation and efficiency to sequences of interactions; and to analyze and compare situations to demonstrate the conservation of energy. Those of "the earth and beyond" are: to know ways in which technology has increased our understanding of the universe; to recognize natural changes in the environment, such as seasonal changes, geological changes, and natural disasters; and to describe the impact of human activities on the environment, such as the construction of roads and buildings and the extraction of natural resources. And those of "science, technology, and society" are: to evaluate the impact of science applications on human activities; to analyze ways that scientific and technological development influences our society; to identify the pros and cons of scientific applications for making informed judgments; to develop an awareness of the need to evaluate environmental issues from a variety of perspectives; and to demonstrate responsibility for local and global environments.

The curriculum guides for biology, chemistry, and physics at S4–5 level have also been developed and are scheduled for implementation in September 2003. They reflect the changes in the new science curriculum framework. They are continuations of the science curriculum at S1–3 level, and the subject curriculum guides share the same common features, such as the emphasis on scientific investigation, the inclusion of core and extension parts, and the enhancement of STS connections.

Existing subjects of the Science Education KLA for upper secondary schools are biology (secondary 4–5), human biology (secondary 4–5), chemistry (secondary 4–5), physics (secondary 4–5), biology (Advanced Supplementary Level and Advanced Level), chemistry (Advanced Supplementary Level and Advanced Level), and physics (Advanced Supplementary Level and Advanced Level).

Singapore

During the ability-driven period (1997–2012), there were also changes in the GCE "A" Level syllabus in physics, chemistry, and biology in 2007. Students were able to choose "H1 (standard subjects)" or "H2 (high-level subjects)." Most students, who intend to pursue major science or engineering-related areas in the university, choose H2 physics and H2 chemistry or H2 chemistry and H2 biology at least. The content comparison between H1 and H2 physics "A" Level syllabuses is provided in Table 7.22. Even though the content in the H2 "A" Level syllabus in physics, chemistry, and biology includes more advanced concepts than H1, the syllabus aims are the same. Examples of aims in the H1 and H2 physics syllabuses are listed below (Singapore Examination and Assessment Board and University of Cambridge International Examinations, 2013c, 2013d):

1. To provide, through well-designed studies of experimental and practical physics, worthwhile educational experience for all students, whether or not they decide to study physics beyond this level, and, in particular, to enable them to acquire sufficient understanding and knowledge to:

1.1 Become confident citizens in a technological world and able to take or develop an informed interest in matters of scientific importance
1.2 Recognize the usefulness, and limitations, of scientific method and to appreciate its applicability in other disciplines and in everyday life
1.3 Be suitably prepared for studies beyond A Level
2. To develop abilities and skills that:
2.1 Are relevant to the study and practice of science
2.2 Are useful in everyday life

Table 7.22 Content comparison between "H1" and "H2" physics "A" Level syllabuses 2013

Section	Topics in "A" Level syllabus	Subtopics in A Level (H1)	Subtopics in A Level (H2)
Measurement	Measurement	SI Units	SI Units
		Errors and Uncertainties	Errors and Uncertainties
		Scalars and Vectors	Scalars and Vectors
Newtonian Mechanics	Kinematics	Rectilinear Motion	Rectilinear Motion
		Nonlinear Motion	Nonlinear Motion
	Dynamics	Newton's Laws of Motion	Newton's Laws of Motion
		Linear Momentum and Its Conservation	Linear Momentum and Its Conservation
	Forces	Types of Force	Types of Force
		Equilibrium of Forces	Equilibrium of Forces
		Center of Gravity	Center of Gravity
		Turning Effects of Forces	Turning Effects of Forces
	Work, Energy, and Power	Work	Work
		Energy Conversion and Conservation	Energy Conversion and Conservation
		Potential Energy and Kinetic Energy	Potential Energy and Kinetic Energy
		Power	Power
	Motion in a Circle		*Kinematics of uniform circular motion*
			Centripetal acceleration
			Centripetal force
	Gravitational Field		*Gravitational field*
			Force between point masses
			Field of a point mass
			Field near to the surface of the earth
			Gravitational potential
	Oscillations		*Simple harmonic motion*
			Energy in simple harmonic motion
			Damped and forced oscillations: resonance
Thermal Physics	Thermal Physics		*Internal energy, Temperature scales*
			Specific heat capacity, Specific latent heat
			First law of thermodynamics
			The ideal gas equation
			Kinetic energy of a molecule

(continued)

Table 7.22 (continued)

Section	Topics in "A" Level syllabus	Subtopics in A Level (H1)	Subtopics in A Level (H2)
Waves	Wave Motion	Progressive Waves	Progressive Waves
		Transverse and Longitudinal Waves	Transverse and longitudinal waves
		Polarization	Polarization
		Determination of Frequency and Wavelength	Determination of Frequency and Wavelength
	Superposition	Stationary Waves	Stationary Waves
		Diffraction	Diffraction
		Interference	Interference
		Two-Source Interference	Two-Source Interference
		Patterns	Patterns
			Diffraction Grating
			Interference Patterns
Electricity and Magnetism	*Electric Fields*		*Concept of an electric field*
			Force between point charges
			Electric field of a point charge
			Uniform electric fields, Electric potential
	Current of Electricity	Electric Current	Electric Current
		Potential Difference	Potential Difference
		Resistance and Resistivity	Resistance and Resistivity
		Sources of Electromotive Force	Sources of Electromotive Force
	D.C. Circuit	Practical Circuits	Practical Circuits
		Series and Parallel Arrangements	Series and Parallel Arrangements
			Potential Divider, Balanced Potentials
	Electromagnetism	Force on a Current-Carrying Conductor	Force on a Current-Carrying Conductor
		Force on a Moving Charge	Force on a Moving Charge
		Magnetic Fields due to Current	Magnetic Fields Due to Currents
		Force Between Current-Carrying Conductors	Force Between Current-Carrying Conductors
	Electromagnetic Induction		*Magnetic flux, Laws of electromagnetic induction*
	Electromagnetic Induction		*Characteristics of alternating currents*
			The transformer, Rectification with a diode

(continued)

Table 7.22 (continued)

Section	Topics in "A" Level syllabus	Subtopics in A Level (H1)	Subtopics in A Level (H2)
Modern Physics	Quantum Physics	Energy of a Photon	Energy of a Photon
		The Photoelectric Effect	The Photoelectric Effect
		Wave-Particle Duality	Wave-Particle Duality
		Energy Levels in Atoms	Energy Levels in Atoms
		Line Spectra	Line Spectra
			X-ray spectra, The uncertainty principle
			Schrödinger model, Barrier tunneling
	Lasers and Semiconductors		*Basic principles of lasers*
			Energy bands, conductors, and insulators
			Semiconductors, Depletion region of a p-n junction
	Nuclear Physics		*The nucleus, Isotopes*
			Mass defect and nuclear binding energy
			Nuclear processes, Radioactive decay
			Biological effect of radiation

(Singapore Examination and Assessment Board and University of Cambridge International Examinations, 2013c; Physics Higher 1, 2013d)

2.3 Encourage efficient and safe practice
2.4 Encourage effective communication
3. To develop attitudes relevant to science such as:
3.1 Concern for accuracy and precision
3.2 Objectivity
3.3 Integrity
3.4 The skills of inquiry
3.5 Initiative
3.6 Inventiveness
4. To promote an awareness:
4.1 That the study and practice of physics are cooperative and cumulative activities and are subject to social, economic, technological, ethical, and cultural influences and limitations
4.2 That the implications of physics may be both beneficial and detrimental to the individual, the community, and the environment
4.3 Of the importance of the use of IT for communications, as an aid to experiments and as a tool for the interpretation of experimental and theoretical results
4.4 That physics transcends national boundaries and that the language of science, correctly and rigorously applied, is universal

5. That stimulates students and creates a sustained interest in physics so that the study of the subject is enjoyable and satisfying

STEM Education in East Asia

Definition of STEM in East Asia

STEM is an acronym that refers to the academic disciplines of science, technology, engineering, and mathematics. But it is very difficult to define STEM because it carries different meaning for many different groups of people. The term is sometimes used when addressing education policy and curriculum choices in schools from K-12 through college in order to improve competitiveness in science and technology development. It also addresses concerns that the subjects are often taught in isolation, instead of as an integrated curriculum. Nowadays, however, a common definition of STEM education is seeing it as an interdisciplinary approach to learning where rigorous academic concepts are coupled with real-world lessons as students apply science, technology, engineering, and mathematics in contexts that make connections between school, community, work, and the global enterprise enabling the development of STEM literacy and with it the ability to compete in the new economy (Tsupros, Kohler, & Hallinen, 2009).

STEM has been emphasized in East Asian countries for the past a few decades. This is not a new emphasis for them. For example, math and science have been main focus subjects in Korea, Japan, Hong Kong, Taiwan, and Singapore from primary school to college. These countries have been always at the top ranks in TIMSS and PISA and have emphasized technology related to IT and the importance of engineering. In this section, we will discuss the modern background of STEM education and provide examples of STEM practices in East Asian countries.

Current Status of STEM Education in East Asia

Korea has built a strong basis generating innovation in science and technology, led by the government and the industry since the 1960s. But Korea faces a problem in science and engineering education, which is students' disinterest in science and math, even though their achievement in science and math is high. Educational experts and researchers claim that it is necessary to look at STEM education from a whole new way of solving problems. Recently in Korea, the keywords for the new perspective for STEM education are "creative thinking" and "academic convergence" (MEST & KOFAC, 2011). It is based on the assumption that if science and technology are integrated with other subjects, even with the arts, it could lead to increased curiosity, interest, and problem solving in students. In the reformed

Table 7.23 School hours secondary students learned or learn

Lower secondary school (teaching hours per week)												
	1987			1992			1997			2009		
Subject	7th	8th	9th	7th	8th	9th	7th	8th	9th	7th	8th	9th
Mathematics	4	3–4	4–5	4	4	4	4	4	3	4	4	3
Science	4	3–4	4–5	4	4	4	3	4	4	3	4	4
Technology[a]	3	4–6		1	2	2	1	2	2	1	2	2

[a]Technology was taught as a separate subject until the 5th National Curriculum Period, but since the 6th National Curriculum Period as a combined subject with home economics, the school hours per week for technology teaching has been reduced

Table 7.24 School hours secondary students learned or learn in Japan

Lower secondary school (teaching hours per week)									
	1989			1998			2008		
Subject	7th	8th	9th	7th	8th	9th	7th	8th	9th
Mathematics	3	4	4	3	3	3	4	3	4
Science	3	3	3–4	3	3	2.3	3	4	4
Technology[a]	1	1	1–1.5	1	1	0.5	1	1	0.5

[a]Technology was taught in the subject, "technology and home economics," and the school hours per week is for technology teaching

curricula in 1987 and 1997, STS education was stressed in science education, but the idea of STS education was not practiced thoroughly in the science classrooms of primary and secondary schools.

In Korea since the 2nd National Curriculum, mathematics, science, and technology have been taught for all of primary and secondary school students as requirement subjects, being not integrated but as separated subjects. Table 7.23 shows how many hours per week they are taught in school.

In 2009 National Curriculum, one integrated science subject for the upper secondary school students was developed as one of selective science subjects. And then afterward Korean educational authority has tried to develop programs (not formal subjects) in which science, mathematics, technology and engineering, and even arts are integrated into STEAM.

In Japan, the prominent difference between 1989, 1998, and 2008 is the change of learning hours per week in science, mathematics and technology. In 1998, MECSST decreased the class hours from those in the 1989 reform but increased again in the 2008 reform as shown in Table 7.24. This was due to the "free education" which led to the student's degradation of learning ability. Actually in 1998, MECSST of Japan decreased the mandatory class hours for fostering the students in the "free" (yutori) education with a focus on autonomous learning (Goto, 2000).

In March 2008, MECSST of Japan established courses of study that served as the fundamental standards for curriculum from kindergarten to high school levels to foster a "zest for life" that emphasizes a balance of academic abilities, richness in humanity, health, and physical strength. The reformed main idea of the curriculum is an education that aims to fully develop the personality of each and every child and is essential for children to lead happy lives in the future. In addition, education has the mission of cultivating the children who will be the future leaders of Japan and society, and the importance of such education remains constant, which will remain unchanged across eras. Further, with factors such as intensifying competition from overseas, as society continues to change at a rapid pace, the education of children is becoming increasingly important (MEXT, 2014).

In the Japanese education system, students at the upper secondary level can choose whether to follow a science course track (with emphasis on mathematics and science) or a humanities course track (with emphasis on Japanese and social studies). The two tracks of study, science and humanities, are based on separate curricula, and only science course track students are required to take advanced mathematics and science classes. According to a survey by the Japanese Youth Research Institute in 2005 (Japanese Youth Research Institute, 2005), the number of students in Japan who belong to the humanities course track is considerably greater than those who belong to the science course track. In addition, the percentage of students who enrolled in STEM classes has steadily decreased over the decades from 1970 to 2000.

In Singapore, science and technology has been the main focused educational area for the nation's future. This led to an increase in the number of students in science- and engineering-related majors in the university in the 1980s. At the beginning of 1980s, the number of university graduates in science- and engineering-related majors was 3180 and 2467 each, but at the end of the 1980s, the number of students' became 4105 and 5005 each (Goh & Gopinathan, 2008). In 2001, the number of students studying in these disciplines increased even further. There were 47,961 (7.6 %) graduates from natural, physical, chemical, and mathematical sciences and engineering sciences were 135,945 (21.4 %). Business and administration was the most common field of study, with 32 % of graduates obtaining their qualification, and the second most common field was engineering sciences (Lin, 2012).

The graduates increased in STEM-related areas in the university were due to the continuous emphasis of mathematics and science subject studies in secondary education (see Table 7.25). In lower secondary school, students have science and mathematics lessons for about 5–6 periods each per week. In upper secondary school, depending on their ability groups, students will study different levels of science and mathematics subjects, for example, most students in the normal academic group choose combined science instead of physics, chemistry, or biology and additional math for normal academic groups. Students in upper secondary school (Years 9–10) will have about 5–6 periods of mathematics lessons per week and at least 6–8 periods for science-related lessons. Students in the normal technical group are preparing themselves for their technical and vocational education during their secondary school education. Science and design and technology are elective subjects, but schools strongly recommend students in the normal technical group to choose these subjects for their future preparation.

Table 7.25 STEM-related subject teaching periods per week in secondary school

Subject	School years in streamed groups											
	Express group				Normal academic group				Normal technical group			
	7th	8th	9th	10th	7th	8th	9th	10th	7th	8th	9th	10th
Mathematics	6^{1}	6^{1}	3	3	5^{1}	5^{1}	3	3	4^{1}	4^{1}	3	3
Additional mathematics[b]			3	3			3	3			–	–
Science	6^{1}	6^{1}	–	–	5^{1}	5^{1}	–	–	4^{1}	4^{1}		
Physics[a]			3–4	3–4			3–4	3–4				
Biology[a]			3–4	3–4			3–4	3–4				
Chemistry[a]			3–4	3–4			3–4	3–4				
Combined science[b] (physics + chemistry or chemistry + biology or biology + physics)			3–4	3–4			3–4	3–4			$3–4^{E}$	$3–4^{E}$
Design and [a], [b]technology	2+	2+	2+	2+	2+	2+	2+	2+	2+	2+	2+	2+
	1/2	1/2	$1/2^{E}$	$1/2^{E}$	1/2	1/2	$1/2^{E}$	$1/2^{E}$	1/2	1/2	$1/2^{E}$	$1/2^{E}$

Note: *E* elective, *1* lesson is for 35–40 min

Each lesson (periods) is for 35–40 min or 60 min

[a]Students choose at least two subjects among physics, biology, and chemistry

[b]Students in express and normal academic groups can choose combined science instead of physics, biology, and chemistry

Table 7.26 East Asian students' science achievements in TIMSS 2011

Place	8th grade	
	Country	Average
1	Singapore	590
2	Taiwan	564
3	Korea	560
4	Japan	558
5	Finland	552

Note: international average, 500. From TIMSS (http://timssandpirls.bc.edu/timss2011/index.html)

At the Junior College level, the subject choices are flexible. Depending on university's requirement, students are able to decide to choose science- and mathematics-related subjects. Most students who want to study science- and mathematics-related areas in the university choose three H2 subjects – 6 periods of mathematics and 12 periods of at least two subjects from physics, chemistry, and biology – and one H1 subject, contrasted with other subjects, for example, humanities or arts. H2 level subjects are designed for students to acquire higher level knowledge and skills for university studies. Therefore, some of students who like to study in arts or humanities in the university choose one H1 subject from physics, chemistry, and biology. There are more curriculum times for students to spend in science-related tasks if they select science-related topics for the project work (H1 subject) in their group. Students need to complete this project work for 28 weeks as a group starting from defining the project focus, analyzing and evaluating the information gathered, preparing an oral presentation for both individual and group assessment, and submitting a written group report within 2,500–3,000 words across any areas. Furthermore, there is an individual component file including the first part of project process such as generation of ideas and analyzing and evaluating ideas.

So far, Singapore does not have integrated science subjects in upper secondary school (including Junior College), and we cannot find any mentioning of STEM education in the science-related subject syllabuses. However, if we look at the Lower Secondary Science 2012 syllabus carefully, it emphasizes students' scientific knowledge and skill application to the real world that make connections between school, community, and workplaces. These emphases continuously influence upper secondary science teaching, learning, and assessment as well as in the Junior College.

East Asian Student Achievements in Science and Mathematics

As shown in Tables 7.26 and 7.27, the students of East Asian countries are high achievers in science and mathematics, as compared the students of other countries of OECD.

Table 7.27 East Asian students' mathematics achievements in TIMSS 2011

Place	8th grade	
	Country	Average
1	Korea	613
2	Singapore	611
3	Taiwan	609
4	Hong Kong	586
5	Japan	570

Note: international average, 500. From TIMSS (http://timssandpirls.bc.edu/timss2011/index.html)

Table 7.28 Trends in mathematics achievement of 8th grade at TIMSS

Country	2011	2007	2003	1999
Korea	613	597	589	587
Singapore	611	593	605	604
Taiwan	609	598	585	585
Hong Kong	586	572	586	582
Japan	570	570	570	579

Note: from TIMSS (http://timssandpirls.bc.edu/timss2011/index.html)

As shown in Table 7.28, students of East Asian countries continue to lead the world in mathematics achievement. Korea, Singapore, and Chinese Taipei, followed by Hong Kong SAR and Japan, were the top-performing countries in TIMSS 2011 at the eighth grade, in which were assessed among the 45 countries and 14 benchmarking participants (IEA, 2014). As the report states (OECD, 2014), "The countries with the greatest increases in average mathematics achievement at the eighth grade included Korea, Lithuania, Taiwan, and Chile (more than 20 points), as well as Italy, the United States, Hong Kong, the Russian Federation, and Slovenia (10–20 points)." Korea, Taiwan, and Hong Kong in East Asian countries showed the greatest increases in the average mathematics achievement at the eighth grade.

In Tables 7.29 and 7.30, we can see that student achievements of East Asian countries in science and mathematics are ranked at the top level in PISA 2012.

As shown in Fig. 7.4, the students of East Asian countries have ranked at the top level in mathematics continuously since 2003.

East Asian Student Achievements in Technology and Engineering

There is no international comparison data related to the attitude about technology or engineering.

Table 7.29 East Asian students' science achievements in PISA 2012

Place	8th grade	
	Country	Average
1	Shanghai	580
2	Hong Kong	555
3	Singapore	551
4	Japan	547
5	Finland	545
6	Estonia	541
7	Korea	538
8	Taiwan	523

Note: international average, 501. Source: KICE (2013)

Table 7.30 East Asian students' mathematics achievements in PISA 2012

Place	8th grade	
	Country	Average
1	Shanghai	613
2	Singapore	573
3	Hong Kong	561
4	Taiwan	560
5	Korea	554
6	Macao	538
7	Japan	536

Note: international average, 494. Source: KICE (2013)

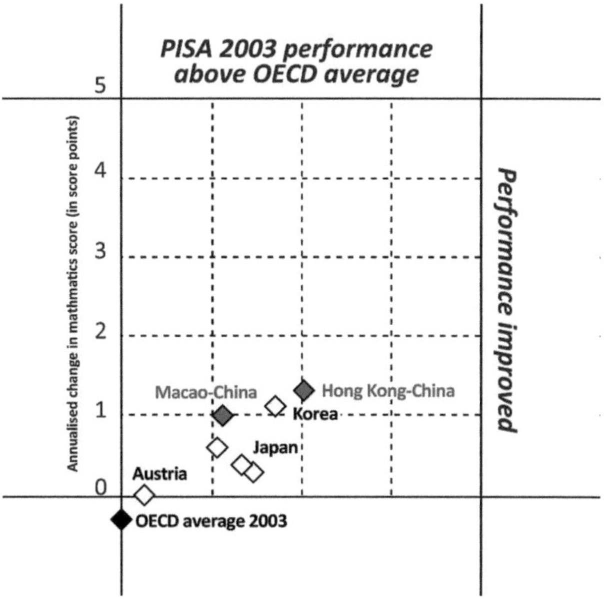

Fig. 7.4 Annualized change in performance between 2003 and 2012 and average PISA 2003 mathematics scores

East Asian Student Perceptions About STEM

Students' Attitudes Toward Mathematics

Table 7.31 shows that more than half of the students in Singapore have a lot or a little confidence and interest, while the students of the other East Asian countries have very low confidence and interest in mathematics, very much lower as compared with Israel and even with OECD average. Except the students in Singapore and Hong Kong who consider mathematics as a relatively more valuable subject, those of the other East Asian countries consider mathematics as somewhat valuable, although it is also much lower than Israel and the OECD average.

Students' Attitudes Toward Science

Table 7.32 shows that more than half of the students of Singapore and Hong Kong have a lot or a little confidence and interest, while the students of other East Asian countries have very low confidence and interest in mathematics, very much lower as compared with Israel and even with the OECD average. More than half of the students of Singapore, Hong Kong, and Korea consider science as a lot or a little

Table 7.31 Students' attitudes about mathematics

	Confidence (%)			Interest (%)			Value (%)		
Country	A lot	A little	No	A lot	A little	No	A lot	A little	No
Singapore	14	46	40	32	44	23	43	47	10
Taiwan	7	26	67	14	33	53	13	41	46
Hong Kong	7	37	56	19	44	37	26	49	25
Korea	3	34	63	8	36	56	14	52	34
Japan	2	24	73	9	38	53	13	50	38
Israel	31	47	22	26	40	35	61	31	8
OECD average	14	45	41	26	42	31	46	39	15

Table 7.32 8th grade (middle school) students' attitude about science (PISA 2012)

	Confidence (%)			Interest (%)			Value (%)		
Country	A lot	A little	No	A lot	A little	No	A lot	A little	No
Singapore	14	48	37	38	46	16	41	43	17
Hong Kong	8	47	45	28	51	21	26	43	32
Taiwan	6	27	67	17	43	40	12	30	58
Korea	4	33	63	11	43	46	14	40	46
Japan	3	28	69	15	47	38	10	34	56
Israel	33	43	24	29	37	34	37	30	32
OECD average	20	49	31	35	44	21	41	33	26

Source: KICE (2013) & PISA Database, http://www.oecd.org/pisa/pisaproducts/

valuable, while more than half of the students of Taiwan and Japan consider mathematics as not valuable, which is relatively much higher than Israel and OECD average.

Singapore is the typical case that shows the positive correlation between students' attitudes and high achievement. One of the attitude items is asking their preferences in science and math. Of the students, 76 % in math and 84 % in science indicated positive preferences. However, students' confidence level is comparatively lower than international average. In TIMSS 2011, there is a question asking "Am I good at science/math?" Only less than 50 % of students in Year 8 showed their confidences as answering "agree a lot" or "agree a little." The students' confidences of science and math are a little lower than international averages.

Students in Japan generally feel less confident about their ability to solve a set of pure and applied mathematics problems than the average student across OECD countries, but have shown improvement over time. Among the 65 countries and economies that participated in the 2012 PISA assessment of 15-year-olds, Japan is ranked seventh in mathematics and fourth in science, with its range of ranks between 6 and 9 in mathematics and between 3 and 6 in science. Matsushita (2010) points out that features of traditional Japanese education such as the high standard and small disparity between levels of academic achievement had been lost by the late 2000s due to the decline of interests and motivations to learn, especially among students who achieved lower benchmarks. Apathetic attitudes toward science constitute a grave problem that plagues Japanese society (Ishikawa, Fujii, & Moehle, 2014).

Poor performances in literacy on PISA and reasoning on TIMSS were attributed to weaknesses of Japanese STEM education and strongly influenced the STEM education policy thereafter. The Curriculum Guidelines of 1998 significantly decreased the number of school hours of STEM subjects in Japanese compulsory education. Promoting the idea of "*yutori* education" ("free education"), designed to reduce the pressure of intensive study and examination anxiety, while enhancing students' motivation to learn, the 1998 Curriculum Guidelines reduced the school hours of mathematics and science curricula to approximately 150 h for mathematics in primary education and 50–70 h for junior secondary mathematics as well as primary and junior secondary sciences. MECSST introduced new special programs for improving literacy (MECSST, 2005), and national surveys targeting primary and junior secondary education were conducted in the latter half of the 2000s (Kudo, 2012; Ishikawa, Fujii, & Moehle, 2014). These reforms ultimately led to the revision of the Curriculum Guidelines in 2008.

Integrated STEM and STEAM

The framework of the twenty-first-century competencies was included in the recent national curriculum in Singapore and the emphasis of communication and decision-making skills also newly emphasized in science syllabuses. Also, sciences in the society and the connections between science and technology have been described in the science curriculum since 1997. Comparing to other East Asian countries such as Korea and Japan, STS program or movement or integrated science waves were not found in syllabuses and activity resources in government schools. However, there is an active adaptation of recent practices from abroad through inviting STEM experts from the USA to train the experienced teachers, researchers, and educators in MOE. The STEM/STEAM framework might be spreading into the government schools. Recently, there is a request from MOE to government schools to design STEM/STEAM-related project or activities as knowledge application projects. These programs will soon be integrated into government school activities and lessons. There has already been a school program in science concept integrated design and technology lessons for students in normal academic group. Amir (2014) indicated that students' creativity, confidence, and motivations increased through integrated programs.

There is not much direct mention about the STEM education in curriculum, syllabus, and classroom practices in Singapore. However, we can often derive the emphasis of STEM-related activities from policymakers' and educators' speeches since 2010. For example, at the official launch of KidsSTOP at Science Center, Mr. Heng, Minister of Education, gave a speech about the roles of the "KidsSTOP" with great emphasis in the importance of interests and learning experience relating to STEM subjects from a young age through "play" and "learning by doing" for creative thinking and innovative skills (Heng, 2014) in order to adapt to a rapidly changing world and to new technologies. Also, he mentioned Singapore's emphasis on the values of STEM education alongside the education system of specialized schools, such as NUS High (NUS High School of Mathematics and Science, http://www.nushigh.edu.sg/special-programmes/da-vinci-programme) for Grades 7 to 12 or Grades 9 to 12 and the SST (School of Science and Technology, http://www.sst.edu.sg) for Grades 7 to 10 (Heng, 2011). Graduates from NUS High do not take the GCE "O" Level and "A" Level examinations. Students receive NUS High diploma that is recognized in world famous universities. Therefore, the NUS High education program is able to incorporate the integrated STEM-related programs from Years 7 to 10 such as Da Vinci program, science presentation, seminar series, research methodology, and design and engineering modules. There is also an arts-oriented school, SOTA (School Of The Arts, http://www.sota.edu.sg/Home.aspx), for Grades 7 to 12 or 9 to 12 that involves taking IB (International Baccalaureate) program in Grades 11 and 12. The learning outcomes of science-related subjects in Years 7 to 10 are to provide opportunities to students by making a connection between science and real-life phenomena and its applications in arts. These science-related subject outcomes are linked to STEAM (science, technology, engineering, arts, and mathematics) which integrate arts and design into STEM subjects. Both schools were established in 2005 and 2008, respectively. SST (School of Science and Technology) was established recently in 2010 for STEM education in secondary school level. The school is offering applied subjects – biotechnology, fundamentals of electronics, and design studies – that are designed, conducted, and assessed by Nee Ann Polytechnic which is one of the tertiary institutions (SST, http://www.sst.edu.sg/curriculum/applied-subjects/).

STEAM education fosters integrative STEM education from artistic perspectives. By merging various subjects and applying it to real life, STEAM intends to develop students' divergent thinking and their excitement about learning. The purpose of this interdisciplinary education is to attract students' understanding, interests, motivation, and potentials in science and technology, as well as to increase their creativity.

In order for STEAM to succeed, the quality and receptiveness of teachers toward the STEAM methods are the most critical factor. In this regard, MEST of Korea is in the process of creating the phased teacher training system for math, science, and technical teachers at all stages of the teaching experience. For example, the government and KOFAC have hosted symposiums, conferences, forums, and events like the STEAM fair in order to spread the concepts and contents of STEAM education among educational subjects.

STEAM education in Korea was started from "the research about direction of STEAM education in Korea," which Back et al. (2012) conducted on 2011–2012 by KOFAC research grant.

The Ministry of Education and KOFAC announced the direction and method about STEAM education in Korea based on the research results including Back et al. (2012). In the research 4C-STEAM education was proposed. 4C-STEAM education refers the STEAM education focused on convergence of knowledge, creativity, communication, and caring, which includes creative design and emotional experience.

At the same time, KOFAC supported the researches for the development of the curriculum including STEAM education subjects for the teachers of primary and secondary school (Back, et al., 2012; Shin, Han, Kim, & On, 2012). KOFAC and MOE hold STEAM education conference annually since 2012. In the conference many of the policies, research results, and action programs related to STEAM education have been presented. In addition KOFAC has supported 10 to 20 STEAM education research and practice teams consisting of both primary and secondary teacher and supported around 20 leading primary and secondary schools for STEAM education practice nationwide. For school STEAM education, two science and arts high schools educating the gifted in science will be established in 2014 and in 2015. In the formal curriculum, integrated STEAM activities are given to the students as requirement subjects (KEDI, 2014). In 2014, "Sejong Science and Arts Gifted High School" was established, and in 2015, another science and arts gifted high school is being established in Korea. In the curriculum for the students of the two schools, five subjects that are consisted of integrated STEAM activities are included as elective subjects (Kim, Kim, & Lee, 2014).

Conclusion

Our society continues to change at a rapid pace, such as the intensifying competition from overseas, and education is an endeavor that aims to fully develop the personality of each and every child and is essential for children to lead happy lives in the future. Science education has the mission of cultivating the children to become future leaders of our society. To improve the educational abilities of schools, reinforcing the curriculum is also a key issue alongside improvements to the quality and abilities of educational environment of each country. In addition, science is deeply related to mathematics, technology, and engineering. Therefore, it is meaningful to consider them together when we investigate educational status of a certain country. According to the change of society, the curriculum of a country also changes. The reform interval of the national curriculum is different from countries of East Asia, averagely about seven years in Korea, ten years in Japan, and more than ten years in Singapore and Hong Kong.

The curricula of East Asian countries, particularly curricula of science and mathematics, have been receiving attention since international evaluations such as TIMSS and PISA began, because the students of those countries have ranked continuous top places in the two international tests. The curricula for science education are similar between Singapore and Hong Kong, in the aspect of separating science subjects that are taught even in the lower secondary level (Years 7 to 9). On the other hand, the curricular for science education between Korea and Japan are similar in the aspect that combined science taught in that level, though there is a slight difference among them.

In Singapore, science curriculum can include higher science contents depending on students' subject choices in upper secondary level. From Year 9 (9th grade), students who are in academically high ability group could move toward specific science subjects. In Years 11 and 12, there are three different levels of science subjects, H1, H2, and H3. H3 is offered by NUS (National University of Singapore), NTU (Nanyang Technological University), and MOE for academically outstanding group who will like to explore science subjects at university level. This means that more difficult science concepts can be taught in Singapore compared with Korea and Japan.

In education systems of Japan and Korea, students at the senior secondary level can choose whether to follow a natural science course track which has an emphasis on mathematics and science or a humanity and social science course track which has an emphasis on literature and social studies. The two tracks of study are based on separate curricula and only science course track students are required to take advanced mathematics and science classes. Up to the 1970s in Japan and up to the mid-1980s in Korea, high school students who aspired to higher education were required to take four science subjects, physics, chemistry, biology, and earth sciences, even though the content levels in science were different between the two tracks. In the two countries, however, in the early or mid-1980s, the Curriculum Guidelines were reformed so that only two science subjects were required, leading to a decline in the proportion of students studying for each science subject. Most of the universities require just one or do not require any science subject for students entering humanity and social science faculties. In addition, reforms of the

Curriculum Guidelines reduced the mandated hours of learning in science and mathematics subjects. The Curriculum Guidelines of the 1990s significantly decreased the number of school hours of STEM subjects in Japanese compulsory education. However, in the 2008 Curriculum Guidelines of Japan, the mandatory hours for teaching science and mathematics were increased because they recognized the decline of student achievement levels in science and mathematics. Even though in the recent curriculum of Japan the mandatory hours for teaching science and mathematics are increased, they are rather decreased in 2013 in Korea, which is on reforming the national curriculum from 2014.

According to the results of international comparison of student achievements in STEM subjects, the lower secondary school students of East Asian countries have been high achievers up to recent years, but their confidence and interest in science and mathematics are very low, except those of Singapore and Hong Kong, comparing with the average of the OECD countries. In Singapore, students are streamed after primary school education, and their learning of science can be customized depending on academic ability levels. There are discussions regarding this early streaming among educators and parents in Singapore. However, it could lead students to different learning paths. For example, some students could take five years instead of four years to sit for their GCE "O" Level examinations. Also, there is a science curriculum for "normal academic students" in secondary school level. Except the streaming in the education system, there is no direct mention of STEM or STS programs in the curriculum of government schools. In the recent curriculum, the twenty-first-century competencies and scientific literacy were introduced. However, it seems that STEM education of Singapore and Hong Kong has been successful in respect of international evaluation test results, which are not shared by Korea and Japan. To overcome depressed situation in Korea, they try to change teaching contents in part, for example, to integrate science contents or to develop STEAM education programs in which STEM subjects and arts are integrated. In Japan, they try to increase formal teaching hours in STEM subjects.

In Singapore, the importance of STEM education has increasingly been mentioned by policymakers. Now there are efforts to bring it down to even government schools not only for specialized schools. However, teachers and schools are facing the dilemma between new initiatives and the rigid examination system (Tan & Leong, 2014). These dilemmas could be overcome by Singapore having high quality of teacher education programs and systematic and structured educational organization. Also, policies can be easily implemented due to its benefits of being a city country. The importance that Singapore placed in STEM-related research and development (R & D) can be seen from the government's increased budget on it (ASTAR, 2013).

Internationally, STEM-related majors are in high demand. Moreover, STEM programs are extended including arts for a more broad-based education. Particularly in Korea integrated STEAM program development and implementation are recommended by education authority guide; however, its effectiveness for increasing academic achievement and confidence and interest in STEM education has not yet been verified.

References

Amir, N. (2014). Showcasing the creative talents in science of the academically less-inclined students through a values-driven toy story-telling project. In L. C. Lennex, & K. F. Nettleton (Eds.), *Cases on instructional technology in gifted and talented education*. Hershey, PA: IGI Global. (in printing). http://www.igi-global.com/contact/

ASTAR (Agency for Science, Technology and Research). (2013). .*National survey of research and development in Singapore 2012*. Singapore: ASTAR. Retrieved September 2, 2014, from http://www.a-star.edu.sg/Portals/0/media/RnD_Survey/RnD_2012.pdf

Back, Y., Kim, Y., Noh, S., Lee, J., Jeong, J., Choi, Y., . . . Choi, J. (2012). *A Study on the Action Plans for STEAM Education*. Seoul, Korea: Korea Foundation for the Advancement of Science and Creativity (KOFAC).

Board of Education. (1997). *Report on review of 9-year compulsory October 1997 Education (Revised Version)*. Hong Kong: Government Printer. Retrieved May 4, 2014, from http://www.edb.gov.hk/en/about-edb/publications-stat/major-reports/consultancy-reports/index.html

Cave, P. (2001). Educational reform in Japan in the 1990s: 'individuality' and other uncertainties. *Comparative Education, 37*(2), 173–191.

Chan, K. K. (2010). *Hong Kong new academic structure for senior secondary education and higher education: A learning journey for all to succeed*. Government of the Hong Kong Special Administrative Region. Retrieved May 3, 2014, from http://www.iie.org/~/media/Files/Corporate/Membership/US-NYIIEabridged.ashx

Cheng, Y. C. (2000). Educational change and development in Hong Kong: Effectiveness, quality, and relevance. In T. Townsend & Y. C. Cheng (Eds.), *Educational change and development in the Asia-Pacific region: Challenges for the future* (pp. 17–56). Lisse, The Netherlands: Swets and Zeitlinger Publisher.

Cheng, Y. C., Chow, K. W., Yeung, S. K., & Cheung, W. M. (Eds.). (1998). *Handbook on educational policy in Hong Kong (1963–1998)*. Hong Kong: Hong Kong Institute of Education.

Cheng, Y. C., Moc, M. M., & Tsui, K. T. (2002). Educational reform and research in Hong Kong: A request for comprehensive knowledge. *Educational Research for Policy and Practice, 1*, 7–21.

Chu, H.-E. (2013). *The effects of course instructions on implementation of assessment for students' learning and pre-service science teachers' beliefs*. Presented paper at the 86th annual international conference of the National Association for Research in Science Teaching (NARST). Rio Grande, Puerto Rico.

Curriculum Development Council of Hong Kong, China. (2002). *Basic education curriculum guide – Building on strengths (Primary 1 – Secondary 3), Booklet 1*. Hong Kong: The Curriculum Development Council. Retrieved May 5, 2014, from http://cd1.edb.hkedcity.net/cd/EN/Content_2909/html/index.html

Curriculum Development Council of Hong Kong, China. (2009). Senior secondary curriculum guide – The future is now: From vision to realisation (Secondary 4–6). Booklet 1. Retrieved May 5, 2014, from http://cd1.edb.hkedcity.net/cd/cns/sscg_web/html/english/index.html, http://cd1.edb.hkedcity.net/cd/cns/sscg_web/html/english/index.html

Curriculum Development Council of Hong Kong, China. (2014). *Liberal studies – Curriculum and assessment guide (Secondary 4–6)*. Retrieved May 5, 2014, from http://ls.edb.hkedcity.net/LSCms/file/web_v2/C_and_A_guide/201401/LS%20C&A%20Guide_updated_e.pdf

Darling-Hammond, L. (2010). *The flat world and education: How America's commitment to equity will determine our future*. New York, NY: Teachers College Press.

Goh, C. B., & Gopinathan, S. (2008). Education in Singapore: Development since. In B. Fredriksen, & J. P. Tan (Eds.), *An African exploration of the East Asian* (pp. 80–108). Washington, DC: The World Bank. Retrieved December 30, 2013, from https://repository.nie.edu.sg/bitstream/10497/15598/1/BC-TWB-80_a.pdf

Goh, C. T. (1997). Shaping our future: Thinking schools, learning nation. In *Speech by Prime Minister Goh Chok Tong at the opening of the 7th international conference on thinking*. Singapore. Retrieved July 5, 2014, from http://www.moe.gov.sg/media/speeches/1997/020697.htm

Goto, M. (2000). Japan. In M. Poisson (Ed.), *Science education for contemporary society: Problems, issues, and dilemmas*. Geneva, Switzerland: International Bureau of Education, The Chinese National Commission for UNESCO.

Heng, S. K. (2011). Keynote Speech at the IBM Centennial Dinner on 2 November 2011, at the Grand Ballroom, Capella Singapore. Retrieved June 30, 2014, from http://www.moe.gov.sg/media/speeches/2011/11/02/keynote-speech-by-mr-heng-swee-keat-at-ibm-centennial-dinner.php

Heng, S. K. (2014). Speech at the official launch of Kidsstop, The Children's Science Centre, on Thursday, 5 June 2014, 7:05 pm, at Science Centre Singapore. Retrieved June 1, 2014, from http://www.moe.gov.sg/media/speeches/2014/06/06/speech-by-mr-heng-swee-keat-at-the-official-launch-of-kidsstop.php

Holbrook, J. (1989). *Science education in Hong Kong: The national report of the Hong Kong science study* (Vol. 1). Hong Kong, Hong Kong: Primary and Junior Secondary Science, Hong Kong National IEA Center, University of Hong Kong.

Holbrook, J. (1990). *Science education in Hong Kong: The national report of the Hong Kong science study*. Hong Kong, Hong Kong: Pre-University Science, Hong Kong National IEA Center, University of Hong Kong.

IEA. (2014, September 9). TIMSS 2011 international results in mathematics. http://timssandpirls.bc.edu

Ishikawa, M., Fujii, S., Moehle, A. (2014). Consultant report securing Australia's future STEM: Country comparisons- Japan. Australian Council of Learned Academies. Retrieved September 1, 2014, from www.acola.org.au

Japanese Research Team for US-Japan Comparative Research on Science Mathematics and Technology Education. (2004). *The courses of study in Japan: General provisions, mathematics, science, art and handcraft, technology, and family and consumer science, information*. Tokyo, Japan: Ministry of Education, Culture, Sports, Science and Technology of Japan.

Japanese Youth Research Institute. (2005), *Survey on senior high school students' attitude toward study and daily lives* (in Japanese), http://www1.odn.ne.jp/youth-study/reserch/index.html, visited on 1 September 2014.

Katayama, N., Takamori, H., & Kanaizuka, Y. (2004). Crisis of biology education in Japan. *Asian Journal of Biology Education, 2*, 75–80.

KEDI (Korean Educational Development Institute). (2014). Development of curriculum for science and arts academy: Interim report, KEDI. Seoul, Korea: Korean Educational Development Institute (KEDI).

KICE (Korea Institute for Curriculum and Evaluation). (2013). TIMSS 2011 Korean students' achievement in mathematics and science, *KICE PIM 2013-1*. Seoul, Korea: Korea Institute for Curriculum and Evaluation (KICE).

Kim, J., Kim, Y., & Lee, M. (2014). *Research on development of curriculum for Science and Arts Gifted High School, KEDI research reports 2014*. Seoul, Korea: Korean Educational Development Institute (KEDI).

Komatsu, S. (2002). Transition in the Japanese curriculum: How is the curriculum of elementary and secondary schools in Japan determined? *International Education Journal, 3*(5), 50–55.

Kudo, M. (2012), PISA and the Educational Subject of Today's Japan, *Bulletin of Shijonawate Gakuen Junior College, 45*, 1–5. Recitation from: Ishikawa, M., Fujii, S., Moehle, A. (2014). Consultant report securing Australia's future STEM: Country comparisons- Japan. Australian Council of Learned Academies. Retrieved September 1, 2014, from www.acola.org.au

Lin, J. S. (2012). Gender differentials in fields of study among university graduates, 2010. *Statistics Singapore Newsletter*. Retrieved May 30, 2014, from http://www.singstat.gov.sg/Publications/publications_and_papers/education_and_literacy/ssnmar12-pg1-5.pdf

Mayer, V. J. (2014). *Developments in Japanese science curriculum*. Retrieved July 23, 2014, from http://earthsys.ag.ohio-state.edu/MayerHP/curriculum.htm

Matsushita, K. (2010, December). What has PISA changed in Japanese education?, *Report of CRET (Center for Research on Educational Testing) symposium* (pp. 3–5) (in Japanese), http://www.cret.or.jp/files/4c2f15b6b31fa47754e2cd22f1f0559f.pdf, viewed 18 January 2013 Tokyo, Japan: Center for Research on Educational Testing. Recitation From: Ishikawa, M., Fujii, S., Moehle, A. (2014). Consultant report securing Australia's future STEM: Country comparisons- Japan. Australian Council of Learned Academies. Retrieved September 1, 2014, from www.acola.org.au

McCormack, A. J. (1992). Trends and issues in science curriculum. In *Science curriculum resource handbook*. New York, NY: Kraus International Publications.

MECSST. (2005), Program for improving reading literacy. http://www.mext.go.jp/a_menu/shotou/gakuryoku/siryo/05122201/014/005.htm, viewed 12 November 2012.

MECSST (Ministry of Education, Culture, Sports, Science and Technology- Japan). (2009a). *The course of study for lower secondary schools*. Tokyo, Japan: Printing Division, Ministry of Finance. (In Japanese)

MECSST (Ministry of Education, Culture, Sports, Science and Technology- Japan). (2009b). *The course of study for upper secondary schools*. Tokyo, Japan: Printing Division, Ministry of Finance. (In Japanese)

MEHRD (Ministry of Education and Human Resources Development). (2007). *MEHRD notification 2007-79 (1): Primary and secondary school curriculum*. Tokyo, Japan: Ministry of Education and Human Resources Development.

MESSC (Ministry of Education, Science, Sports and Culture). (1989a). *The course of study for lower secondary schools*. Tokyo, Japan: Printing Division, Ministry of Finance (in Japanese).

MESSC (Ministry of Education, Science, Sports and Culture). (1989b). *The course of study for upper secondary schools*. Tokyo, Japan: Printing Division, Ministry of Finance (in Japanese).

MESSC (Ministry of Education, Science, Sports and Culture). (1999). *The course of study for upper secondary schools*. Printing Division, Ministry of Finance (in Japanese).

MEST (Ministry of Education, Science, Technology – Korea). (2011). *MEST Notification 2011-361 (9): Science Curriculum*. Seoul, Korea: Ministry of Education, Science, Technology – Korea (MEST).

MEST (Ministry of Education, Science, Technology), & KOFAC (Korea Foundation for the Advancement of Science and Creativity). (2011). *A study on the action plans for STEAM education*. Seoul, Korea: KOFAC.

MEXT. (2014). *Elementary and secondary education*. Tokyo, Japan: Ministry of Education, Culture, Sports, Science and Technology (MEXT). Retrieved July 5, 2014, from http://www.mext.go.jp/english/

MOE of Japan. (1999a). *The course of study for lower secondary schools*. Tokyo, Japan: Printing Division, Ministry of Finance (in Japanese).

MOE of Japan. (1999b). *The course of study for upper secondary schools*. Tokyo, Japan: Printing Division, Ministry of Finance (in Japanese).

MOE of Korea. (1987). *The 5th national curriculum*. Seoul, Korea: Ministry of Education (MOE) of Korea.

MOE of Korea. (1992). *The 6th national curriculum*. Seoul, Korea: Ministry of Education of Korea.

MOE of Korea. (1997). *7th national curriculum*. Seoul, Korea: Ministry of Education of Korea.

MOE of Singapore. (1982). *Lower secondary science syllabus*. Singapore: Ministry of Education.

MOE of Singapore. (1992). *Science syllabus for lower secondary special/express/normal*. Singapore: Curriculum Planning Division and Ministry of Education.

MOE of Singapore. (1997). Master plan for IT in education. Singapore. Retrieved from http://www.moe.gov.sg/media/speeches/2008/09/25/speech-by-dr-ng-eng-hen-at-the-moe-work-plan-seminar-2008.php, http://www.moe.gov.sg/media/speeches/2008/09/25/speech-by-dr-ng-eng-hen-at-the-moe-work-plan-seminar-2008.php

MOE of Singapore. (1999a). *Science syllabus for lower secondary special/express/normal*. Singapore: Curriculum Planning Division and Ministry of Education.
MOE of Singapore. (1999b). *GCE 'O' level physics. Singapore: Curriculum planning and development division*. Singapore: Curriculum Planning Division and Ministry of Education.
MOE of Singapore. (1999c). *GCE 'A' level physics. Singapore: Curriculum planning and development division*. Singapore: Curriculum Planning Division and Ministry of Education.
MOE of Singapore. (2005). Transforming leaning: Teach less, learn more. *The Teachers' Digest Contact, issue 3*. Singapore: Ministry of Education. Retrieved March 3, 2014, from http://www.moe.gov.sg/corporate/contactprint/pdf/contact_oct05.pdf
MOE of Singapore. (2008). *Science syllabus for lower secondary express/normal (academic)*. Singapore: Curriculum Planning Division and Ministry of Education.
MOE of Singapore. (2012). *Science syllabus for lower secondary express/normal (academic)*. Singapore: Curriculum Planning Division and Ministry of Education.
MOE of Singapore. (2013). *Science syllabus – Primary 2014, express/normal academic*. Singapore: Curriculum Planning & Development Division.
Nakamichi, T. (2011). Transition in the course of study for biology education in Japan focusing on lower secondary schools. *Asian Journal of Biology Education, 5*, 26–32.
National Commission for UNESCO. (1974). *Organization of education in 1971–1973/major trends in Singapore educational development in 1971/72 and 1972/73*. Presented paper at the 34th session of the International Conference on Education, Geneva.
National Institution of Education. (2010). Teacher ED: Growing 21st century teaching force. *Singteach.nie.edu.sg, 22*, 1–2. Retrieved March 10, 2014, from http://www.nie.edu.sg/files/SingTeach_Issue22_TE21.pdf
Ng, E. H. (2008). MOE work plan seminar. Retrieved July 29, 2014, from "http://www.moe.gov.sg/media/speeches/2008/09/25/speech-by-dr-ng-eng-hen-at-the-moe-work-plan-seminar-2008.ph, http://www.moe.gov.sg/media/speeches/2008/09/25/speech-by-dr-ng-eng-hen-at-the-moe-work-plan-seminar-2008.php
NIER (National Institute for Education Policy Research). (2014). *Education in Japan: past and present*. Retrieved July 23, 2014, from http://www.nier.go.jp/EducationInJapan/
OECD. (2011). *Strong performers and successful reformers in education: Lessons from PISA for the United States* (Chap 4). Shanghai/Hong Kong: Two Distinct Examples of Education Reform in China. Retrieved from http://www.oecd.org/pisa/46623978.pdfon2January2014
OECD. (2014). PISA 2012 results in focus: What 15-year-olds know and what they can do with what they know. Retrieved from http://www.oecd.org/pisa/46623978.pdfon2January2014
Poon, C. L., Lee, Y. J., Tan, A. L., & Lim, S. S. (2012). Knowing inquiry as practice and theory: Developing a pedagogical framework with elementary school teachers. *Research in Science Education, 42*(2), 303–327.
Shin, Y., Han, S., Kim, H., & On, J. (2012). *Development of curriculum for STEAM education in national university of education*. Seoul, Korea: KOFAC.
Singapore Examination and Assessment Board and University of Cambridge International Examinations. (2013a). *Physics GCE ordinary level*. Retrieved May 4, 2014, from http://www.seab.gov.sg/oLevel/2013Syllabus/5059_2013.pdf
Singapore Examination and Assessment Board and University of Cambridge International Examinations. (2013b). *Science GCE ordinary level: Physics & chemistry, Physics & biology, chemistry & biology*. Retrieved May 4, 2014, from http://www.seab.gov.sg/oLevel/2013Syllabus/5116-5117-5118_2013.pdf
Singapore Examination and Assessment Board and University of Cambridge International Examinations. (2013c). *Physics higher 1*. Retrieved May 4, 2014, from http://www.seab.gov.sg/aLevel/2013Syllabus/8866_2013.pdf
Singapore Examination and Assessment Board and University of Cambridge International Examinations. (2013d). *Physics Higher 1*. Retrieved May 4, 2014, from http://www.seab.gov.sg/aLevel/2013Syllabus/8866_2013.pdf

Singham, J. K. (1987). *An investigation of the science process skills in the intended and implemented PSP of Singapore.* Unpublished PhD thesis. University of Liverpool, UK.
Sweeting, A. (1990). *Education in Hong Kong, pre-1841 to 1941.* Hong Kong: Hong Kong University Press.
Tan, A.-L., & Leong, W. (2014). Mapping curriculum innovation in STEM schools to assessment requirements: tension and dilemmas. *Theory Into Practice, 53*(1), 11–17.
Tan, J., & Gopinathan, S. (2000). *Education reform in Singapore: Towards great creativity and innovation?* (pp. 5–10). Retrieved June 6, 2014, from http://www.apecknowledgebank.org/resources/downloads/SingaporeCurriculumReformCreativity.pdf
Treagust, F. D., & Rennie, J. L. (1993). Implementing technology in the school curriculum: A case study involving six secondary schools. *Journal of Technological Education, 5*(1). Retrieved August 15, 2014, from http://scholar.lib.vt.edu/ejournals/JTE/v5n1/bigtrea1.html
Tsupros, N., Kohler, R., & Hallinen, J. (2009). *STEM education: A project to identify the missing components.* Intermediate Unit 1 and Carnegie Mellon, Pennsylvania.
Whitfield, R. C. (1971). *Disciplines of the curriculum.* London, UK: McGraw-Hill.

Chapter 8
Science Education in Shanghai, China: What Does It Look Like Other Than PISA Score?

Xinying Yin, Xiaoli Guo, and Lan Wang

Introduction

In the 2009 and 2012 Program for International Student Assessment (PISA) conducted by the Organization for Economic Co-operation and Development (OECD), Shanghai teenagers' stunning performance put the Chinese in the spotlight of the international education arena. While some are praising and envying the achievement of Chinese education demonstrated by this international test, some have raised questions about how well the test performance really represents these Chinese teenagers' learning, given that Chinese education is well known for its strong emphasis on examinations and rote learning. As science educators and researchers, we are concerned with such questions as: What is the "real" quality of science education in Shanghai? What do the scores mean with regard to Shanghai teenagers' science learning? What implications can we draw from the science education in Shanghai for the larger science education community? In order to understand these questions, an overview of the distinctive educational context of Shanghai is first provided to put the PISA test scores in proper perspective. Then the reform efforts in science education in Shanghai are reviewed, including curriculum standards, textbooks, assessment, and teaching innovations. Finally, Shanghai science teachers' conceptions and practices of teaching and students' learning are discussed before we draw conclusions and implications.

X. Yin (✉)
Department of Teacher Education and Foundations, College of Education, California State University, San Bernardino, CA, USA
e-mail: xyin@csusb.edu

X. Guo • L. Wang
College of Chemistry, Beijing Normal University, Beijing, China

M.S. Khine (ed.), *Science Education in East Asia*,
DOI 10.1007/978-3-319-16390-1_8

The Educational Context of Shanghai and PISA Score

While deeply sharing the cultural traditions of education in China, Shanghai has its own distinctive educational context comparing to the rest of China. Shanghai is a province-level municipality and has historically attracted the nation's elites. It is also China's largest city and principal business center, with a population of 20.7 million. Shanghai's per capita GDP is more than twice that of China as a whole, and Shanghai's parents invest heavily in their children's education outside of school. An estimated four out of five children in Shanghai attend after-school tutorial groups to help them prepare for exams (OECD, 2010). With its strong economic and political commitment to raising educational standards and student performances, Shanghai has long been a leader in educational reform and indeed an education hub in China. Shanghai was one of the first Chinese cities to achieve universal primary and junior secondary education in 1980s, and more than 80 % of Shanghai high school graduates go to college, compared to 24 % nationally (OECD). The economic and intellectual resources set Shanghai apart from many other areas in China.

Shanghai was the first local government to set its own exams for the college entrance examination (CEE) in 1985 (National Center on Education and the Economy [NCEE], n.d.a). This promoted major changes in curriculum and pedagogy, since the CEE is considered as the "baton" directing teaching among Chinese teachers. The exams are being reformed to move away from a heavy emphasis on memorization and rote learning to an emphasis on students' real-world problem-solving abilities, deep understandings, and interdisciplinary knowledge. For example, some examination questions provide students with new information and require their abilities in applying what they know to new problems, and interdisciplinary papers are used to test students' capacity to apply their knowledge to real-life problems.

The deviation of CEE in Shanghai from China's central national entrance examination heralded a trend in decentralization of education, which allowed Shanghai to have a comprehensive platform in reforming its curriculum. Since 1988, Shanghai has been making curriculum reforms to improve the quality of education and emphasize the development of individual students (Cai, 2002; Yang, 2005). In 1998, the curriculum was reformed to integrate natural science with the humanities, the national curriculum with school-based curriculum, and knowledge acquisition with active inquiry. The curriculum is organized around eight "learning domains," which are meant to encourage active inquiry and interdisciplinary understanding. These domains are language and literature, mathematics, natural science, social sciences, technology, arts, physical education, and a practicum. The curriculum is also organized into three separate components: a core curriculum that is compulsory for all students, an enriched curriculum that allows students to choose their own electives, and a research-based curriculum, which is implemented mainly in extracurricular activities where students conduct individual research projects of their own interests supported by teachers. With these domains and components in mind, schools are encouraged to adapt the government's curriculum frameworks to meet their students' needs. Since 2008, the new curriculum has been implemented throughout the city. Overall, the curriculum reform involves enriching students' learning experiences, enhancing the relevance of subjects by relating them to broader human and social issues, and focusing on the development of students' creativity and application of knowledge. These are reflected in the reform of examinations as well as reform in pedagogy.

Along with the curriculum reforms, teaching practices, teacher education, and professional development have also been changing. Classroom practices are moving toward student-centered teaching, and the evaluation of teacher performance also emphasizes more on students' activities and engagement. In order to improve teaching practices, Shanghai schools require teachers to attend 360 h of professional development every 5 years throughout their career (NCEE, n.d.b). As well, teachers teaching the same subject usually have weekly meetings to discuss lesson plans, and they often observe each other's classes for mentoring and advising. Shanghai is also the pioneer in overcoming inequity issues in education by innovative strategies. For example, a major project is to strengthen weaker schools through "commissioned administration," where the government commissions a "good" public school to send a team of leaders and experienced teachers to help improve the management and teaching in a weaker school (OECD, 2010).

Despite the affluent educational resources in Shanghai, examination pressure persists just as it does in other parts of China and is still a major item on the reform agenda, since examination is consistently seen to be "scientific," "reliable," and hence "fair" way to go to higher education. Under such pressure, students from Shanghai are fully occupied and engaged in learning in and out of school, as the most of Chinese students. The intense concentration on schoolwork is considered a student's responsibility in Chinese culture. Homework is an essential part of their learning activities and in a way governs their lives at home after school. Meanwhile, parents devote their time to students' studying at home. Beyond schoolwork, an estimated 80 % of parents send their children to tutorial schools for exam preparation, and many students also learn nonacademic subjects out of school such as music, sports, and painting (OECD, 2010). Such learning is being critiqued for the little choice left to students and the students' rare direct encounters with nature and social life. There exists the tension between the intense engagement and genuine learning in a broader sense.

In such an educational context, it is not so surprising that Shanghai teenagers can perform well on PISA, given the educational resources in Shanghai, the dedication to education by the society, and the students' hard work toward taking public exams. Their top performance in the recent two PISA tests nevertheless shocked the world. However, it would be unwise to just celebrate the success of education in Shanghai by only looking at the high scores on a test.

On one hand, the stunning PISA scores indeed to a certain extent indicate the excellence of Shanghai's K-12 education. PISA aims to provide cross-nationally comparable evidence on student performance as new basis for policy dialogue and collaboration in defining and implementing educational goals. It is claimed to be the most comprehensive and rigorous international program to assess student performance within an internationally agreed common framework. The PISA assessment takes a broad approach to measuring knowledge, skills, and attitudes and focuses on competencies that 15-year-old students will need in the future such as applying their knowledge and experience to real-life issues and abilities to continue learning. The relevance of the knowledge and skills measured by PISA has also been confirmed by studies tracking young peoples' academic development in several countries (OECD, 2013). Thus, the scores should be able to reflect students' capabilities rather than mere rote learning or test-taking skills. In addition, the scores are representative of the quality of education in Shanghai given that the sample is reprehensive rather than biasing elite schools (Lu & Zhu, 2011). Local experts also believe that this is a piece of evidence of the reform success, whereby students are now exposed to a much broader knowledge base and are taught to integrate their knowledge and tackle real-life problems (OECD, 2010). Students have also become used to identifying questions of interest to themselves and to carrying out open-ended explorations. All these changes are markedly different from the traditional pattern of Chinese education in which students learn by heart and regurgitate on exams.

On the other hand, only test scores cannot represent the whole picture of the quality of education. Educators in Shanghai are not satisfied with the quality of Shanghai's education system, worrying that the dictates of the examinations have left students with little time and room for learning on their own developing creativity and preparing for future real world life and work (OECD, 2010). Thus, we do need to carefully examine the various aspects of K-12 science education in Shanghai, especially the middle grades who were the group being tested in PISA, to better understand the implications of PISA score for science education in Shanghai. This is also an opportunity for science educators to reflect on and discuss how to further improve the quality of science education in the international community.

Science Education Reform in Shanghai

In this section, the recent science education reforms in Shanghai are first reviewed, including curriculum, textbooks, and assessment. Then teaching innovations promoted in the reform are reviewed. And the focus is on secondary school, more specifically junior high school, when students participate in PISA.

Science Curriculum Reform in Shanghai

Since 1988, Shanghai has been leading the K-12 curriculum reform in China for "quality education" (*Sui Zhi Jiao Yu*), to overcome the long-standing examination-oriented education. The first phase of the curriculum reform, from 1988 to 1997, established the primary goal of K-12 education as improving the "quality" (*Su Zhi*, meaning the well-rounded development of a person rather than only test scores) of individual students, as well as a curriculum framework founded on a triangle of societal needs, disciplinary systems, and the development of students. The second phase further established the primary goal of K-12 education as "education based on students' development," emphasizing the developing nature of individual students (Cai, 2002). The second phase of the curriculum reform also emphasizes developing students' creativity and abilities of practices for quality education (Zhang, 2009). In such waves of curriculum reform, science curriculum, as an important part of the K-12 curriculum, also reflects these ideas.

During the first phase of the curriculum reform, the Shanghai Municipal Education Commission adopted an integrated science curriculum for grades 7–9 that aims for cultivating and improving students' scientific literacy, along with which middle school "integrated science" textbooks were developed and adopted. During the second phase of the curriculum reform, a coherent science curriculum was created from a holistic and developmental perspective for grade 1 through grade 12. Integrated science curricula were used for grade 1 to grade 7; physics, chemistry, and life science were studied separately during grade 8 to grade 11; and an integrated science curriculum was used again at grade 12. To be noted, during grades 8–11, students learn all the three subjects on each grade level. The curriculum is consisted of five domains including life science, physical science, earth and space science, science-technology-society (STS), and the history and nature of science. It emphasizes the longitudinal development of the subject matter knowledge through grade 1 to grade 12 as well as the interconnectedness among the subjects. A distinctive feature of the new curriculum is its heavy emphasis on inquiry-based learning to develop students' understandings of the scientific concepts and the nature of science as well as the skills of doing scientific inquiry (Shanghai Municipal Education Commission, 2004). Consistent with the general curriculum framework, the science curriculum also has three components: core curriculum, enrichment curriculum, and research-based curriculum.

Science Textbooks Adopted in the Curriculum Reform in Shanghai

Textbooks have important places in Chinese K-12 education, and they are usually uniformly adopted in a large municipal area, such as a province. The teaching and learning are very much organized around the textbook. For grades 6 and 7, two sets of textbooks are currently adopted for the integrated science curriculum; one is published by *Shanghai* Educational Publishing House, and the other is adapted from Hong Kong *New Integral Science* and published by Shanghai Far East Publishers and Oxford University Press. Both textbooks are aligned with the curriculum reform goals, which are the integration of scientific concepts from different disciplines, the connection of science to technology and society, and the holistic development of scientific knowledge, scientific inquiry, and attitudes toward science (Wang, 2006). As well, both textbooks are organized around activities, such as "observation and experience," "discussion and activity," and "inquiry and research," and have rich texts and visuals to encourage students' reading.

On the upper secondary level (grade 8 to grade 11), the textbooks for separate disciplines also reflect the curriculum reform goals, placing significant focuses on STS and scientific inquiry, as well as connections to other disciplines. For example, the 8th grade *Physics* (Shanghai Educational Publishing House, 2007) used a lot of technologies in students' experiments, such as probes and computers. This textbook also has various activity components in each unit, including (1) observing and describing, thinking and discussing, and communicating and collaborating; (2) student experiment and extended inquiries; (3) reading materials such as "STS" and "science and humanity"; (4) practice items and tasks; and (5) summary of concepts including a concept map for the unit. Particularly, this textbook has a separate "student activity" book for teachers to organize the class activities based on it. Zhang (2010) conducted a survey regarding teachers' opinions about this physics textbook. A majority of teachers favored the separate activity book which can help them change to student-centered, activity-based classroom, but they thought it might contain too much content for the current class time. They also regarded the experiments in this textbook reflecting the reform goals of science education and effective for the development of students' inquiry skills. A survey on students' opinions about the new textbook also showed that majority of the students appreciated the rich texts and visuals that can attract their attentions (Li, 2012).

Assessment of Science Education in Junior High Schools in Shanghai

As public exams have directing effects on K-12 education in China, assessment has to be reformed in order to sustain the curriculum reform. According to *Shanghai Junior High School Science Curriculum Standards* (experimental draft) (Shanghai Municipal Education Commission, 2004), assessment of students' learning should address the goal of improving students' scientific literacy in the reformed curriculum. First, the purpose of assessment should be more on promoting all students' development rather than selection and competition. The assessment should measure different aspects of learning to provide evidence for improving teaching, and both formative and summative assessment needs to be implemented. As well, different forms of assessment should be used, such as portfolios, projects, and performance of real-world tasks other than paper-and-pencil tests. To this end, science assessment has been changed to measure different aspects of science learning aligned with the curriculum goals including scientific inquiry, knowledge and skills, and attitudes toward and values of science. The test questions are also de-emphasizing memorization and including more open-ended, inquiry-based, problem-solving questions. For example, in the 2014 Shanghai high school entrance exam, students need to take a performance test for experiment skills in physics and chemistry. Li (2010) compared the Shanghai high school entrance exam on chemistry from 2004 to 2009 against the chemistry curriculum standards and found that the exam was generally aligned with the standards in different aspects, including content covered and the targeted depth of knowledge, but the part on experiments had higher requirements than the standards. This indicated the emphasis on scientific inquiry in the public exams.

Teaching Innovations Promoted in the Science Education Reform in Shanghai

On the institutional level, the curriculum standards, textbooks, and assessment have been transformed to a new vision of science education – improving scientific literacy among all students. On the classroom level, various teaching innovations have also been promoted in Shanghai's science classrooms. Aligning with the reform goals of cultivating individual students' development, the science classrooms are moving from teacher centered and exam oriented toward student centered and learning oriented (Gu, 2014). This trend is reflected in several aspects: (1) Give students more autonomy in learning and recognize that teachers' role is to facilitate learning. (2) Enhance teachers' conceptual understanding and practice of constructivist-based teaching and learning. (3) Create student-centered teaching models to encourage students learn to learn. For example, Shanghai Zhabei No. 8 Middle School developed classroom models that provide more class time for students' learning and activity, flipped the traditional "teach-learn" sequence to "learn-teach," and emphasized students' classroom discourse (Author Unknown, 2011). In the following, three major teaching models promoted in Shanghai science classrooms are reviewed and discussed, including inquiry-based teaching, STS education, and the use of "learning plan."

Inquiry-Based Teaching

Inquiry-based teaching has become the core pedagogical strategy in the curriculum reform that emphasizes students' abilities of doing scientific inquiry. Many teachers in Shanghai recognize that inquiry-based teaching is student centered, can enhance students' conceptual understandings, and can enrich students' learning experience (Peng & Wang, 2007). Peng and Wang compared traditional approach and inquiry-based approach of teaching a chemistry class on the same topic to explicate how inquiry-based teaching can trigger students' interests to inquire and make the subject more relevant to students' life. In their example of teaching "test and purification of gases," they found that when the teacher only used decontextualized questions from the preparation of chlorine, which is from the textbook, students did not show interest to answer or inquire about those questions. In contrast, when the teacher started the lesson with a real-world problem of the higher price of vitamin C effervescent tablets compared to regular vitamin C tablets, students were very eager to find out the components of the effervescent tablets by using knowledge of testing and purifying the gases produced by the tablets. They also commented that different strategies can be used to enhance the effectiveness of inquiry-based teaching, such as group work, ongoing performance assessment, and using interactive whiteboards.

STS Education

While inquiry-based teaching is being promoted, the substance of the inquiry needs to be relevant to the real world to trigger students' interest in doing scientific inquiry. In the science curriculum, STS education includes three themes: the history of science and technology, the technological design, and the contemporary important STS issues in the society. The goals of STS education include knowledge about the development of science and technology, the interdependence of science and technology, the interactive relationships between science and society, and the technological design process. Science educators argued for the importance of implementing STS education in science teaching, especially in the enrichment curriculum, such as enhancing students' interest in learning science, facilitating students' analytical and creative thinking in solving real-world problems, and promoting students' application of science knowledge and participation in the public discourses about STS issues (Chen & Ding, 2002; Jiang, 2011). Particularly, localized school curricula are used to achieve STS education goals. For example, the No. 2 High School of East China Normal University developed their local school chemistry curricula for the enrichment science courses that have distinctive STS components (Jiang, 2011). Chen and Ding (2002) also provided an example of local STS lesson on energy issues in the modern society. As energy shortage is a worldwide issue as well as a critical problem in Shanghai, this lesson aimed to develop students' knowledge about energy in relation to the society, personal awareness of this issue, and social responsibilities for energy use. This lesson used a variety of authentic learning activities, including doing background research, interviewing experts in the field, classroom debate on the use of nuclear energy, experiencing the clean energy powered city bus, visiting energy research labs, making posters and writing essays on energy use, and carrying out projects to modify home appliances to save energy.

Use of "Learning Plans"

Similar to the activity book accompanying a textbook, learning plans are used by teachers to organize students' classroom activities when the activity books are not available. Learning plans are being widely used in Shanghai secondary school science classrooms (Ye, 2012). The learning plans are created by teachers in local schools. Ye pointed out that this process promoted teachers' collaboration and their investigation of creating school curricula and improving teaching quality. And most importantly, the learning plans helped with students' learning activities in class as well as review after class. For example, learning plans are effectively used for inquiry-based lessons, where students design their experiments, record data, and analyze data under teachers' guidance. As well, when doing review lessons, learning plan is used for students to prereview the major concepts before the class, discuss and collaborate with peers in class, and reflect on their learning after class. Teachers can pre-assess students' knowledge through the prereview, provide specific feedback during the lesson to focus on students' problems, and review students' after-class reflections to ensure students' learning.

Science Teachers' Conceptions and Practices in Shanghai Secondary Schools

In the efforts of science education reform, curriculum, textbooks, assessment, and pedagogy have been significantly renovated to align the socio-constructivist theory of learning and the goal of developing students' scientific literacy. The ultimate goal of improving students' science learning can only be realized in the classroom, and teachers are the changing agents. In this section, we will review and discuss Shanghai science teachers' current conceptions and practices of teaching.

Shanghai Science Teachers' Conceptions of Science

In order to effectively teach inquiry-based science, teachers need to have high levels of knowledge of science, scientific inquiry, and the nature of science. Overall, science teachers in Shanghai secondary schools have fairly good understandings about science and conceptions of the research-based pedagogies promoted in the reform. Zhang (2006) conducted a survey about Shanghai secondary biology teachers' knowledge and views about science, and these teachers scored an average of 14.385 out of 17 points. This study also showed that these teachers have stronger understandings about scientific concepts and scientific practices but weaker in the views of the nature of science. For example, only less than 10 % of the responding teachers perceived scientific research as building theories and using evidence to verify the theories. It is pointed out that these teachers recognized the importance of scientific practices in science learning, as promoted in the new curriculum; however, their lack of experience of doing scientific research, lack of deep understandings about scientific practices, and insufficient understandings of the nature of science prevented them from implementing inquiry-based science teaching.

Science Teachers' Practices in Shanghai

Since the curriculum reform, Shanghai science teachers' teaching has moved a long way from the traditional lecture-based classroom to student-centered and activity-based classroom aligned with the reform goals. Many teachers have also been doing a wide variety of assessments for students' learning in formative ways (Li, 2012). According to a study investigating K-12 teachers' classroom teaching (Shanghai Pudong District Committee on K-12 Teachers' Classroom Teaching, 2009), 81 % of the teachers participating in this study reported that they made intentional effort to achieving the multi-dimensional goals addressed in the new curriculum including knowledge and skills, process and methods, and attitude and values. A majority of these teachers' classrooms had student-centered features, such as providing more time for students' activity and discussion, adjusting teaching based on students' responses, and using more inquiry-based teaching. For example, more than half of the teachers devoted more than 50 % of class time to students' activities instead of lecturing all the time. More than 90 % of the teachers were responsive to students' learning, modifying their teaching based on students' responses and learning from their colleagues. Most teachers would teach students how to learn science and scientific inquiry skills. When students have problems, more than 70 % of the teachers would use interactive teaching to help students, rather than just telling students the right answer. 93.6 % of these teachers had taught inquiry-based lessons, and about 40 % of the teachers had directed students' inquiry projects on real-world issues. As well, a majority of the teachers also tried to make the subject matter relevant to students' lives and the society and supplement the textbook with extra activities and projects. These teachers' responses are validated by students' responses. This indicated that Shanghai K-12 teachers indeed are establishing a new classroom culture where students are active learners and taught to "learn how to learn."

Nevertheless, there are challenges in implementing the curriculum reform in the classrooms. Although many teachers indicated that they had some knowledge about the new curriculum through professional development meetings, such as STS education and inquiry-based teaching, they felt that they were not fully prepared to implement the new curriculum in their classrooms (Chen & Ding, 2002; Yu, 2010). In the survey conducted by Shanghai Pudong District Committee on K-12 Teachers' Classroom Teaching (2009), there were still about 20 % of the teachers who used lecture as the main teaching strategy, especially during the last year of the school when students are going to take the exams to go to the next level of education. In guiding students' inquiry and discussions, very few teachers were able to facilitate students in constructing conceptual understandings. And only 3 % of the teachers in this study reported they were actually promoting students' critical and creative thinking in their classrooms. In STS education, although a majority of teachers considered they used various contexts, such as real-world examples and discrepant events, to engage students in new topics, fewer students perceived the same way (Yu, 2010). Further, technology education is largely ignored in STS education in these studies as well in science classrooms.

In the study surveying about teachers' opinions about the physics textbook for grade 6 and grade 7 in Shanghai, teachers indicated that they had difficulties with implementing inquiry-based activities and the research projects in the textbook (Zhang, 2010). Li (2012) found that many students indicated that they did not have many opportunities to actually carry out the experiments in this physics textbook; instead, the experiments were mainly demonstrated by the teachers. Chen and Ding (2002) also found that 90 % of students in their study would like to do more experiment in chemistry classes and 80 % of the students would like to know more about the application of chemistry in society. Several obstacles in science teachers' implementation of inquiry activities are teachers' insufficient pedagogical knowledge about inquiry-based teaching, lack of subject matter knowledge outside of their major subject in teaching integrated science, lack of resources for science experiments, problems with the activity design in the textbook, and lack of time (Wang, 2006).

In summary, in the curriculum reform in Shanghai, science teachers' conceptions and practices have been changing to align with the reform goals. At the same time, they need more substantial support to gain deeper understandings about scientific inquiry and inquiry-based pedagogy in order to implement the reformed science curriculum. In the following section, we will review and discuss students' science learning that occurred during the education reform in Shanghai in recent years.

Students' Science Learning in Shanghai

Chinese students are known for their test-taking skills and rote learning in the long-standing examination-oriented educational system. Their top performances in 2009 and 2012 PISA tests have spurred educators' interest in knowing whether this is largely a result of their test-taking skills and rote learning, or this indeed demonstrates the top quality of science education in Shanghai. The answer is not that straightforward. In this section, we will first discuss the PISA science framework and test questions and compare it to the Shanghai junior high school science exam. Then we will review studies on different measures of students' science learning, including their views of the contemporary socio-scientific science issues, attitudes toward science, and inquiry skills.

Comparing PISA Science and Public Science Exam in Shanghai

The 15-year-olds participating in PISA are at the age of existing junior high school and entering high school in Shanghai. Comparing PISA science with the public science exam in Shanghai could help us better understand these students' learning outcomes measured by paper-and-pencil tests.

The knowledge and skills at the heart of the PISA science assessment are identifying scientific issues, explaining phenomena scientifically, and using scientific evidence (OECD, 2013). These competencies require students to demonstrate, on the one hand, knowledge and cognitive abilities and, on the other, attitudes, values, and motivations, as they encounter and respond to science-related issues (OECD, 2013). For each item on the assessment, there is a real-world context such as acid rain issues, genetically modified crops, and physical exercise. There are a variety of types of questions under each context, including multiple-choice or closed-constructed response items as well as more open-ended questions requiring more extended and complex answers. Figure 8.1 shows an example of PISA science items on acid rain. These three questions about acid rain assess students' knowledge about common science facts related to the society (question 1), understandings about the basic science principle of chemical reactions (question 2), and students' scientific inquiry skills (question 3).

In 2012 *Shanghai Junior High School Exit Exam-Science* (Shanghai Municipal Education Commission, 2012), other than multiple-choice and closed-constructed response items that measure students' knowledge about basic scientific facts,

Below is a photo of statues called Caryatids that were built on the Acropolis in Athens more than 2500years ago. The statues are made of a type of rock called marble. Marble is composed of calcium carbonate. In 1980, the original statues were transferred inside the museum of the Acropolis and were replaced byreplicas. The original statues were being eaten away by acid rain.

Question 1:
Normal rain is slightly acidic because it has absorbed some carbon dioxide from the air. Acid rain is more acidic than normal rain because it has absorbed gases like sulfur oxides and nitrogen oxides as well.Where do these sulfur oxides and nitrogen oxides in the air come from?

Question 2:
A marble chip has a mass of 2.0 grams before being immersed in vinegar overnight. The chip is removed and dried the next day. What will the mass of the dried marble chip be?
A. Less than 2.0 grams
B. Exactly 2.0 grams
C. Between 2.0 and 2.4 grams
D. More than 2.4 grams

Question 3:
Students who did this experiment also placed marble chips in pure (distilled) water overnight. Explain why the students included this step in their experiment.

Fig. 8.1 An example of PISA science item (From *PISA 2012 assessment and analytical framework mathematics, reading, science, problem solving and financial literacy*, OCED, 2013)

concepts, principles, and skills (e.g., identifying the solvent and solute in sugar water, using the molecular kinetic theory to explain how smoke spreads in the whole room), half of the test consists of open-ended questions called "experiment and inquiry" and "analysis and synthesis." In these open-ended questions, real-world contexts are used, and students are required to understand the context, develop hypotheses, design or evaluate experiments, analyze provided data, and interpret graphs to draw conclusions. Figure 8.2 shows two examples of the items from the exam. We can see that the science assessment items in Shanghai junior high school exit exam are comparable to the PISA items.

PISA is the most comprehensive and rigorous international program to assess student performance (OCED, 2013). We can assume the assessment items have

Example 1:

On the sunny afternoon, Xiao Hong saw a lot of bubbles were being produced on the plant leaves in her fishbowl. In order to find out what gas the bubbles contained, she gathered the bubbles in a container and did the following tests: (1) she added the gas into a clear lime water , nothing happened. (2) she put a barely burning wooden stick into the container of the gas, the wooden stick started to burn again.

Question 1: What question Xiao Hong wanted to solve by doing the two tests?

Question 2: What gas is it in the bubbles according to the tests?

Example 2:

This task is about finding out the factors influencing the evaporation rate in the context of air-drying clothes.

1. Here are some phenomena in everyday life, please write your hypothesis about what factor is influencing the evaporation rates following each phenomenon. The first one is already given.

    (1) It is faster to dry clothes if it is flattened out.

    Hypothesis: surface area is a factor influencing the evaporation rate of liquid.

    (2) On cloudy days, we can use electric fans to dry the clothes faster.

    Hypothesis:

    (3) It is faster to dry clothes in the sun.

    Hypothesis:

2. Design an experiment to test the first hypothesis. List the procedures and predict the results.

Fig. 8.2 Example items from 2012 *Shanghai Junior High School Exit Exam-Science* (Shanghai Municipal Education Commission, 2012) high validity and reliability. From this perspective, Shanghai students indeed demonstrated high levels of science competencies measured by PISA. On the other hand, however, since the public exams taken by these Shanghai teenagers have similar items as PISA science, it is very likely that they have taken many practices for such exams. This eludes the question of whether the high PISA scores are the result of high-quality science education in Shanghai or students' test-taking skills and rote learning. Then we need to look at other measures of students' science learning in order to better understand this question.

Other Measures of Shanghai Students' Science Learning

Zhang (2006) surveyed about Shanghai middle school students' knowledge about some of the important scientific facts and theories in life science (e.g., evolutionary theory, gene mutation, cell as the basic functional unit of life, etc.), their views about some large-scale research projects (e.g., climate change research, spaceship research, genetically modified organism research, etc.), and attitudes toward science. The study shows that students have fairly good understandings about the important scientific knowledge in life science, with an average correct rate of 67.7 %. With respect to students' views about the large-scale research projects, they generally showed clear opinions toward different research. For example, they strongly agreed with the necessity of doing climate research and strongly disagreed with research on chemical weapons. This also indicated their knowledge about these important socio-scientific issues. In addition, the majority of the students showed positive attitudes toward science, recognizing the importance of science in the society and people's lives. But a few students indicated that it was not important to know science in everyday life, alarming Chinese science educators about the need to make school science more relevant to students' lives.

Shanghai junior high school students also showed improved learning strategies and inquiry abilities with the reform efforts on inquiry-based teaching and students' scientific inquiry skills (Li, 2012; Yang, 2005). Li found that a majority of the students regarded their inquiry skills had improved and they could apply their knowledge learned in physics classes to solve real-life problems. Meanwhile, teachers also reported students had stronger communication skills, more active learning habits, and enhanced inquiry abilities (Yang, 2005). Further, studies show that a majority of Shanghai teenagers are interested in learning science and doing science as well as science-related careers (Li, 2012; Yang, 2005). In Li's study about junior high students' learning experience with the new physics textbook, about 68 % of the students said they like physics, about 57 % of the students indicated they would like to do physics-related jobs in the future, and 80 % of the students indicated that the physics textbook enhanced their interest in physics. However, Gu (2014) also found that Shanghai 8th grade students' inquiry abilities such as analytical and problem-solving skills have not been improved significantly, a lot of students still do not know how to learn, students still have very heavy workload and pressures from exams, and many students do not have genuine interest in learning.

In general, the various measures of Shanghai teenagers' science learning reflect the reform goals of developing students' scientific literacy and support the students' science competencies measured by PISA. Despite the fact that preparing for exam is still an important part of their learning, Shanghai teenagers are not only test takers who use rote learning and test-taking skills to get high scores on paper-and-pencil tests; they are growing into scientifically literate citizens and future workforce for science-related careers.

The Limitations of the Literature

Due to the relative small amount of evidence-based educational research in China, especially on science education in Shanghai, there are some limitations of this review. For example, there is a lack of research on science teacher conceptions and practices, especially research based on interviews and classroom observations other than teacher self-reported surveys. As well, measurement of students' science learning primarily relies on surveys. Students' achievement of the Shanghai junior high school performance assessment of physics and chemistry experiment skills was not available; thus, we are missing an important piece of information on students' science learning. Further, no data are available to show the actual percentage of students in Shanghai who choose a college major and eventually choose a career in science/engineering-related fields. Nevertheless, the current review provides us with a fuller picture of science education in Shanghai other than PISA scores. In the following section, we will answer the three questions raised at the beginning.

Discussions and Implications

1. What is the "real" quality of science education in Shanghai?

Shanghai is no doubt a leader in Chinese science education and also in the top rank of the world. It has made great advancement in reforming science education from teacher centered and knowledge based to student centered and inquiry based. The K-12 science curriculum features an "integration-separation-integration" structure, with grades 8–11 teaching different subject areas separately. In such a curriculum structure, the K-7 integrated science emphasizes cultivating students to be scientifically literate citizens, the separate science subject in grades 8–11 provides students with deeper science learning for college preparation, and the integration in the last year of high school intends to help students make deeper connections and applications of the knowledge and skills learned in the previous years. As well, the core curriculum, enrichment curriculum, and research-based curriculum provide teachers with a structure for differentiated education, ensuring all students' learning of the core curriculum and allowing a flexible agenda for teachers and students to enact the enrichment and the research-based curriculum. Meanwhile, textbooks are developed to support teachers and students to achieve the curriculum goals of students' abilities to do scientific inquiry and application of science to real-world problems. The activity-based science textbooks help teachers organize their teaching and also aid students' learning in and out of classrooms. The relatively uniform adoption of high-quality textbooks throughout the city also can help teachers provide quality learning activities to students, while they could make local adaptations and create enrichment curriculum to fit students' needs.

In the science education reform efforts, if assessment of students' learning did not change, then teachers would not be motivated to change their teaching as their evaluation is mainly based on students' achievement on the public exams. The public examination in Shanghai has come a long way in recent years to reflect the reform goals. Not only does the curriculum standards clearly state that the assessment should measure different aspects of students' science learning especially inquiry skills, the public exam has been changed significantly in terms of the form and content of the assessment. This can be seen from the comparison between the 2012 *Shanghai Junior High School Exit Exam-Science* (Shanghai Municipal Education Commission, 2012) and PISA science items. As well, all students need to take experiment skill performance assessment when they exit junior high school and apply for high schools.

Educational reform eventually comes down to the classroom teaching, and the reform goals can only be realized in the classroom. Along with the changes in the curriculum, assessment, and textbooks, different teaching models are being promoted in Shanghai science classrooms for students' active learning. Inquiry-based learning, STS education, and learning plans have been widely accepted by science teachers, and science classrooms are moving toward student centered. In teachers' classroom practice, most teachers are intentionally addressing scientific inquiry, application of science to real-world issues, and students' attitudes toward and values of science. A majority of teachers are also using more interactive teaching and being responsive to students' learning. As well, many teachers are using various forms of classroom assessment for students' learning formatively.

Meanwhile, teachers face various challenges of implementing the new curriculum with full fidelity. First of all, the high-stakes exams are still the golden standards to evaluate teaching, and teaching innovations are restricted by the pressure of preparing students for the exams (Li, 2012). The enrichment curriculum and especially the research-based curriculum are not implemented to its full potential yet. Another major reason is teachers' lack of deep understanding of scientific inquiry and inquiry-based pedagogy. Although these science teachers commonly have a solid foundation of scientific knowledge and have understandings about the reformed curriculum and reform-based pedagogical strategies, they often lack the experience of doing scientific inquiry and the support for them to develop practical knowledge for the implementation of inquiry-based teaching. Thus, although teachers have conceptually accepted the inquiry-based teaching methods, students' perception of the lack of hands-on activities indicates that the "inquiry skills" perceived by teachers may be probably more about the cognitive processes, but not so much about *doing* science yet. This could also be due to the large class size, limited class time, and lack of resources for students to conduct experiments.

2. What do the scores mean with regard to Shanghai teenagers' science learning?

With respect to students' learning in science, PISA scores and other measurements of students' learning showed that Shanghai teenagers' high level of knowledge and skills in science and positive attitudes toward science and science-related careers. The achievement of students' learning can be attributed to the investment in educational reform by the Shanghai government, the high level of teacher qualification and requirement for professional development, parents' tremendous support for children's education, and the students' engagement in learning and significant amount of hard work. In addition, Chinese schools and students put more emphasis on science compared to the Western countries, as there is an old saying "If you learn math and science well, you can go anywhere" meaning that it is easier for math and science-related majors to find good jobs. All students are required to take science courses throughout K-12; when science subjects are taught separately from grades 8 to 11, students learn three subjects, physics, chemistry, and life science, on each grade level.

On the other hand, students' science knowledge and skills have not gone too far beyond paper-and-pencil tests yet. Students are still weak in creativity and real-world problem solving. One major reason is still the heavy emphasis on preparing for exams and students' lack of time participating in social activities outside of schoolwork. Although the reformed exams have been redirecting teachers to teach more inquiry skills, many teachers are still at the stage of training students to take exams and using large amount of practices. Thus, the students' score on PISA can be also partially attributed to their practices for exams that are similar to PISA items.

As well, Chinese students do not have much autonomy in choosing what to learn as their Western counter partners do, as they devote most of their time to the subjects on the college entrance examination. This could limit their intrinsic motivation to learn. However, this does not mean Chinese students lack the motivation to learn. While research suggests that in the West, education should be its own end and students striving for external goals such as high marks often report less intrinsic motivation toward learning and mastery, Chinese students' extrinsic motivation from exam performance, meeting family expectations, and gaining social recognition tends to co-occur with intrinsic motivation of mastery goals (Tweed & Lehman, 2002). Such achievement approach among Chinese students comes with competition and using strategies that maximize cost-effectiveness of time and effort, but it does not mean only surface learning (Gao, 1998). Many Chinese students are self-disciplined and systematic in their work habits; emphasize the understanding of the basic concepts, principles, and laws; and appreciated activities related to learning skills (Gao, 1998). As well, while in the Western culture inherent ability is perceived more important for achievement, Chinese culture attributes success to effort and failure to lack of effort (Lee, 1996). Therefore, Chinese students can still achieve at least a certain level of deep learning under the pressure of high-stakes testing.

3. What implications can we draw from the science education in Shanghai for the larger science education community?

As the top performer in the international benchmark test PISA, Shanghai science education provided some important lessons for the science education community. First, Shanghai indeed has had success in its educational reform including science education. Although Shanghai is one of the most affluent areas in China, its GDP per capita ($14,547, in 2013) is much lower than some of the other countries, such as Singapore ($55182, in 2013) and the United States ($53142, in 2013). The resources for all the students in Shanghai are still relatively limited; the class size is still big (around 50), and labs are still insufficient to accommodate students' regular hands-on activities. With the limited resources, the policy makers and science educators in Shanghai have developed new curriculum along with new textbooks, public assessment, teacher evaluation methods, and reformed teaching models that are applicable to the science classrooms in Shanghai. This indicates that systematic change is key to the success of educational reform. Particularly, public assessment and evaluation of teachers have helped to encourage teachers to make the changes in the classroom. Also, it is important that Shanghai government has been paying attention to educational equity in the reform, ensuring the balanced development in different districts and for different student populations in Shanghai.

However, we could also see the challenges of enacting educational reform. Under the high-stakes testing context in China and the social and parental expectations for students' high test scores, it is extremely difficult for teachers to divert their teaching away from "teaching to the test" even though teacher evaluation has already put more emphasis on classroom teaching and learning. More in-depth teacher professional development is needed to support teachers' more fundamental change of their beliefs, knowledge, and practices of teaching, and this is a challenge faced by science teacher educators from all over the world.

Another important lesson we can learn from Shanghai science education and the Shanghai teenagers' top performance on PISA is the challenges in the evaluation of students' science learning. Even with an international benchmark assessment such as PISA that has high validity and reliability, it is still difficult to capture an accurate picture of students' learning and, in turn, the "real" quality of science education received by the students. No matter how well the assessment is constructed, students still can be "trained" to take the test, although they could indeed learn something different as the assessment changes. It is particularly hard to evaluate students' creativity and real-world problem-solving abilities through paper-and-pencil test. PISA is claimed to assess students' competencies such as applying their knowledge and experience to real life, but the educators and students themselves in Shanghai consider they still lack the experience of applying knowledge to real life. Even with test questions that require students to comprehend and then apply new information to a new context, teachers still can teach students test-taking skills or "tips" to earn points.

This issue comes back to teachers' beliefs, knowledge, and practices of teaching. Teachers need to believe that students can still achieve good scores on tests even if they are not "teaching to the test," and they need to have the necessary strategies to ensure students' learning of the required knowledge and skills when they are "teaching for learning." Previous studies did show that, even in the particularly high-stakes testing context in China, teachers could relieve the conception of "teaching to the test" and focus on students' learning if they are given enough support (Yin & Buck, 2015). Thus, in the science education reform, the investment should be balanced between creating new curriculum, textbook, and assessment and supporting teachers' classroom implementation.

References

Author Unknown. (2011). 25 exemplary K-12 education curriculum reform projects. *Shanghai Education, 2011*(6), 54–67. (in Chinese)

Cai, Z. (2002). Promote quality education for the development of students: A review of the second phase K-12 curriculum reform in Shanghai. *Journal of Jimei University, 3*(3), 81–84 (in Chinese).

Chen, S., & Ding, S. (2002). Infusing STS concepts into chemistry teaching. *Chemistry Teaching, 2002*(5), 14–17 (in Chinese).

Gao, L. (1998). Cultural context of school science teaching and learning in the People's Republic of China. *Science Education, 82*, 1–13.

Gu, L. (2014). Teachers and their mentors with the vision of classroom. *Curriculum Textbook, Pedagogy, 2014*(1), 27–29 (in Chinese).

Jiang, H. (2011). Achievement, problems and possibilities in localizing high school chemistry enrichment curriculum: Centered on introducing STS education. *Contemporary Basic Education Research, 2*, 98–103 (in Chinese).

Lee, W. O. (1996). The cultural for Chinese learners: Conceptions of learning in the Confucian tradition. In D. A. Watkins & J. Biggs (Eds.), *The Chinese learner: Cultural, psychological and contextual influences* (pp. 25–41). Hong Kong, China: CERC/ACER.

Li, N. (2010). *A study on the alignment of chemistry curriculum standards and Shanghai high school entrance examination-chemistry*. Retrieved from China master thesis full-text database (CNKI). (in Chinese)

Li, X. (2012). A survey on middle school students' satisfaction of physics published by Shanghai Educational Publishing House. *Journal of Inner Mongolia Normal University (Educational science). 2012*(4), 91–94. (in Chinese)

Lu, J., & Zhu, X. (2011). How to perceive Shanghai PISA 2009 assessment results. *Shanghai Educational Research, 2011*(1), 17–19 (in Chinese).

National Center on Education and the Economy. (n.d.a). Shanghai-China overview. In *Shanghai-China*. Retrieved from http://www.ncee.org/programs-affiliates/center-on-international-education-benchmarking/top-performing-countries/shanghai-china/

National Center on Education and the Economy. (n.d.b). Teacher and principal quality. In *Shanghai-China*. Retrieved from http://www.ncee.org/programs-affiliates/center-on-international-education-benchmarking/top-performing-countries/shanghai-china/shanghai-china-teacher-and-principal-quality/

Organization for Economic Co-operation and Development. (OECD). (2010). Shanghai and Hong Kong: Two distinct examples of education reform in China. In *Strong performers and successful reformers in education: Lessons from PISA for the United States* (Chapter 4). Retrieved from http://www.oecd.org/countries/hongkongchina/46581016.pdf

OECD. (2013), *PISA 2012 assessment and analytical framework mathematics, reading, science, problem solving and financial literacy*. Paris, France: OECD publishing. Retrieved from http://dx.doi.org/10.1787/9789264190511-en

Peng, X., & Wang, P. (2007). Practice and reflection on inquiry-based teaching in review lessons in junior high school chemistry. *Chemistry Teaching, 2*, 1–3 (In Chinese).

Physics. (2007). Shanghai, China: Shanghai Educational Publishing House. (In Chinese)

Shanghai Municipal Education Commission. (2004). *Shanghai junior high school science curriculum standards* (Experimental draft). Shanghai, China: Shanghai Educational Publishing House. (In Chinese)

Shanghai Municipal Education Commission, (2012). *Shanghai junior high school exit exam-science*. Retrieved from http://wenku.baidu.com/view/026f6fc40c22590102029d72.html (in Chinese)

Shanghai Pudong District Committee on K-12 Teachers' Classroom Teaching. (2009). A survey on K-12 teachers' teaching in the context of new curriculum. *Contemporary Teaching, 2009* (Z1), 22–29. (in Chinese)

Tweed, R. G., & Lehman, D. R. (2002). Learning considered within a cultural context. *American Psychologist, 57*(2), 89–99.

Wang, Q. (2006). *Investigating activity-based teaching in Shanghai junior high school science*. Retrieved from China master thesis full-text database (CNKI). (in Chinese)

Yang, S. (2005). Judgment based on commonsense and rationality: A report on the second phase curriculum reform in Shanghai. *Shanghai Educational Research, 2005*(10), 16–20 (in Chinese).

Ye, P. (2012). A survey on the use of learning plans in middle school chemistry teaching. *Chemistry Teaching, 2012*(5), 19–22 (in Chinese).

Yin, X., & Buck, G. (2015). There is another choice – An exploration of integrating formative assessment in a Chinese high school chemistry classroom through collaborative action research. *Cultural Studies of Science Education*. Advance online publication, doi:10.1007/s11422-014-9572-5

Yu, J. (2010). Optimizing science experiment teaching through STS education theories. *Contemporary Educational Science, 2010*(2), 12–13 (in Chinese).

Zhang, H. (2009). *Study on senior high school integrated science curriculum*. Retrieved from China doctoral dissertations full-text database (China National Knowledge Infrastructure, [CNKI]). (In Chinese)

Zhang, W. (2006). *A study on Shanghai biology teachers' scientific literacy based on new curriculum standards and solutions*. Retrieved from China master thesis full-text database (CNKI). (in Chinese)

Zhang, Y. (2010). *A comparative study on the previous and current Shanghai junior high school physics textbooks*. Retrieved from China master thesis full-text database (CNKI). (in Chinese)

Part IV
Pedagogical Innovations in Science Education

Chapter 9
The Implications of Science Teaching and Practices on Educational Neuroscience

Chia-Ju Liu, Wen-Wei Chiang, Chin-Fei Huang, and Ming-Hsun Shen

Introduction

An interdisciplinary approach is essential to introducing educational neuroscience to common learning. It requires collaboration from cognitive neuroscience, developmental cognitive neuroscience, educational psychology, educational technology, education theory, and other related disciplines (Liu & Chiang, 2014). Scientists seeking to understand brain behavior and educators using cognitive neuroscience to enhance education must overcome existing mindsets and fixed perceptions related to these issues. Determining whether particular traits increase or decrease with age could help researchers to analyze the relationship between behavioral changes and the human brain. Despite the limitations of current scientific knowledge, public expectations with regard to the partnership linking neuroscience and educational learning have already far exceeded existing scientific data (Fischer, Immordino-Yang, & Waber, 2007).

Educational neuroscience is essentially the integration of neuroscience and education, with the main aim of exploring the underlying physiological mechanisms that are at work during students' learning processes and of trying to establish a permanent link between theoretical assumptions and practical use. Scientific researchers hope that through educational practice, using some of the valuable findings in neuroscience, they can be helped to validate all the research findings in neuroscience in the past 10 years. Moreover, from the perspective of many science educators, they hope to find solutions to solve the numerous problems encountered in the classroom that have persisted for many years. The field of neuroscience education needs to be both theoretical and practical to be used by education institutions. The interplay between them can become beneficial when the

C.-J. Liu (✉) • W.-W. Chiang • C.-F. Huang • M.-H. Shen
Graduate Institute of Science Education and Environmental Education, National Kaohsiung Normal University, Kaohsiung, Taiwan
e-mail: chiaju1105@gmail.com

M.S. Khine (ed.), *Science Education in East Asia*,
DOI 10.1007/978-3-319-16390-1_9

integration of different disciplines is successful. Its sustainable development not only lies in drawing inspiration and practical use from the educational neuroscience research but also the outcomes of many significant early childhood education studies, to reveal the secrets of lifelong learning. The wide application of the results of many modern neuroscience researches has changed the way the public look at science learning.

Some of the most dominant issues in the educational sector require the assistance of neuroscience to reveal their origins and how they can lead to misuse in classroom contexts (Geake, 2011). Moreover, while the research is fascinating in its own right, the results only have practical use if educators actually apply what is learned in their classrooms.

Purpose

First, we will consider the information yielded by the analyses of how the brain functions and reacts to daily learning routines and how they have been approached by modern educational neuroscience. Following that, we will review the development of educational neuroscience in the past few decades by examining the empirical research conducted by some of the most acclaimed researchers in the field. Finally, we will consider potential educational neuroscience trends as they might apply to the future of science education.

According to Selden, Sherrier, and Wooters (2012), the purpose of education in modern times is to cultivate what we now consider "full humans" who are more sensitive to developments in all areas. More specifically, the learning concerned with curriculum, memory, emotions, creativity, society, and other brain-based elements needs to be well coordinated and managed under some precise scientific mechanism. Judging from both of the developments of brain-based science and neuroscience, most of the modern technologies such as PET, MRI, and EEG can be used to gain more insights into how the brain functions in the process of learning and solving problems.

Theoretical Background

Using Neuroscience Technology to Enhance Education

Traditional Western views of the mind and body, such as that of Descartes, divorced high-level and rational thought from what were considered basal, emotional, or instinctual body processes (Damasio, 2003; Immordino-Yang, 2011). However, these bodily processes are important contributors to rational thought. From a psychological perspective, learning experiences can be deepened using cognitive enhancement, which increases mental capacity with external stimuli (Chiang & Liu, 2014; Liu et al., 2014).

Table 9.1 Levels of development: optimal and functional

Level	Optimal level	Functional level
Single attribute	2 years old	2–5 years old
Attribute mapping	4 years old	4–8 years old
Attribute system	6 years old	7–12 years old
Single abstract	10 years old	13–20 years old
Abstract mapping	15 years old	17–30 years old
Abstract system	20 years old	23–40 years old (or permanent inability in many areas)
Single rule	25 years old	30–45 years old (or permanent inability in many areas)

Functional age levels vary and this table provides only rough estimates

When researching natural linguistic development in Dutch children, Ruhland and van Geert (1998) observed accelerated development in specific categories of vocabulary among most children aged approximately 24 months. Cognitive development passes through ten phases from age four months to adolescence. Table 9.1 summarizes the phases that are closely related to education from childhood to adulthood (Dawson, 2006; Dawson, Fischer, & Stein, 2006). Research on tasks, such as arithmetic, self-conceptualization, reflective judgment, moral reasoning, classification, and conservation, has shown that development in these areas is associated with the formation of the skills indicated in the following table.

Noonan et al. (2004), Since the 1960s, one of the most popular theories of neural science is that the brain is not a stand-alone organ—but separated by both left and right brain hemisphere. The left brain is rational and logical and mainly responsible for language processing, deductive reasoning, abstract thinking, and other functions. In the meantime, the right brain is more emotional and creative and so is solely responsible for pattern recognition, spatial perception, and other functions.

Liu et al. (2012) found that the left and right brain division theory holds that there are some theoretical assumptions about the function of the whole brain in science learning which lead to the misconception that the two hemispheres operate independently. Indeed, the whole brain theory looks at the left and right parts of the human brain as an integral organ. Brain functions are supposed to compensate for each other. On a wide view, left and right brain functions are both crucial to develop logical thinking and solve problems with the designed teaching activities. However, we must be clear that the left and right brain are not independent, as it is a premise for dissecting the true relationship between the right and left brains. The two hemispheres in the brain are active at the same time, regardless of the activity being undertaken. In short, they are cooperating together to accomplish tasks.

Liu et al. (2013) emphasized that there are two types of messages related to the optimal level of development in the left brain which can go through the department of neurons. Vidyasagar and Pammer (2010) found that the volume of the external stimulus threshold needs to be reached to form a new connection. Mostly, the distribution of ions inside and outside the axon will generate transient action potential which later would spread all around the axon terminals. One of the more obvious effects is that it immediately prompts the release of neurotransmitters in accordance with the proliferation of postsynaptic receptors. At times, it would cause the learners to be excited or accustomed to the difference in materials. The continuation of neuron messages is delivered through the qualification process on behalf of many neurons which have been mutually joined together to release the biological and synapses between nerves. Ach et al. (2013) thought that the basic consequence of this kind of neuron reaction is that it forms a weak stimulus a few minutes later, affecting the short-term memory, which is essentially stored in the synaptic deformation.

Discovering individual potential of their human brain on all levels has now become a major task for many educators and neuroscience researchers. On the whole, this includes research in combination with the whole brain development theory on the molecular, cellular, and cognition levels. The known activity in cognitive neuroscience is intended to answer the fundamental proposition which is whether the continuous development of cognitive neuroscience for educators will prove to be of practical use from a new perspective. Many students interested in science, with the help from school's psychological department, are able to reap many benefits of the cognitive neuroscience. All the results point to the fact that relevant theories and methods used in educational research are increasingly compared to those before. For instance, gifted students have greater interconnectivity between different areas of their brains, the coordination and integration of which is supported by Duncan and colleagues' study in which neural activations in response to difficult IQ test items were contrasted with activations in response to easier IQ test items from a positron emission tomography (PET) study (Duncan & Owen, 2000). Thinking about difficult IQ test items, whether spatial or verbal, involved the bilateral PFC. Consistent results were found by Lee et al. (2006) in a functional magnetic resonance (fMRI) study which measured experimentally induced changes in the neural vasculature, the BOLD (blood oxygen level dependent) response which compared gifted and age-matched adolescents on difficult and easy IQ test items; in the gifted group, difficult tasks increased activity in the bilateral PFC regions. This supports a suite of high-level neurocognitive abilities including a relatively enhanced executive capability, with a more efficacious working memory (Geake, 2011).

Current studies showed that the reaction of long-term memory which is affected and reconstructed by the retention system of learning tends to lead to positive learning outcomes (Ho, Chou, et al., 2012; Ho, Huang, et al., 2012). However, brain science has confirmed that memory is not a passive storage system, which leads us to the discussion of whether an individual accepts the external environment stimulation when he tries to learn new knowledge. Normally, the brain will continually change or adapt to changes (neural plasticity), and more often than not, the restructuring process goes on in parallel with the network system between neurons, thus giving form to the continuous learning behaviors and reactions triggered by the strengthened ties stored in the neural network. For instance, in order to stop learning, one must weaken or even "disconnect" nerves connected together (neural ties), which is impossible. Estevez, Lindgren, and Bergethon (2010) claimed that as far as whole brain science is concerned, the key to effective learning of new knowledge is to identify what lies in between neurons which produce effective communication. That is, effective learning is associated with repeated contact with curricula and materials over time because in so doing, the synaptic function will become more "familiar" with all the neurons and ties. On a more physiological basis, the function and formation of memory indicate that the learning organizations in human brains rely more on linkage (memory) than on individual stimulation.

Memory is the foundation on which human beings comprehend and adapt to everyday life. fMRI is commonly used in research on the mechanisms underlying memory by revealing the working principles of the human brain, with the hope of one day enhancing memory capacity:

- *Short-term memory*. Short-term memory refers to the storing of data in easily accessible areas within a short amount of time and is an essential condition for functions such as decision making, reasoning, and comprehension. D'Esposito et al. used event-related fMRI to conduct in-depth research on working memory in the regions of the prefrontal cortex during the performance of delayed-response tasks (D'Esposito, Postle, Ballard, & Lease, 1999; D'Esposito, Postle, Jonides, & Smith, 1999).
- *Long-term memory*. Long-term memory is categorized into explicit, implicit, and episodic memories. Busatto et al. (1997) used fMRI to explore the anatomical aspects of the encoding and retrieval processes involved in episodic memory. They found that repeatedly recognizing previously presented words did not effectively stimulate the right prefrontal cortex. Encoding processes were associated mainly with the left prefrontal cortex, which is located in Broca's area rather than the dorsolateral prefrontal cortex. Semantic memory refers to the recollection of ordinary knowledge and events and is unrelated to the time and location of their occurrence.

Kraut, Hart, Soher, and Gordon (1997) used fMRI in experiments involving object association tasks. Subjects were shown the words "desert" and "hump" and were then asked to think of a word associated with these two words. The answer was "camel." Their results showed that the thalamus was activated during this process only when associating the paired words with an object. In this light, Ho, Huang, et al. (2012) speculated that the thalamus was not activated if the paired words did not facilitate the association with the object, or there was only a simple semantic association, or the words belonged to the same category. Memory is a complex cognitive function that is not restricted to a single neuron or area of the brain.

Nehdi (2007) claimed that the test procedure is simple, safe, and easy to detect. This is why many neural researchers use it to explore the variety of sensory, mental, and physiological changes in whole brain activities such as perception, cognition, judgment, exercise, language, learning, and other brain functions. As far as science learning is concerned, the fMRI technology still presents many technical difficulties, but it is noninvasive, provides high spatial resolution, does not require the injection of a contrast agent, and displays brain activity both at the surface as well as deep within (Bandettini, 2012).

Making Science Education Evidence-Based Practice

If the structure or content of new knowledge is something learners have been in touch with before, either by hearing or seeing, it takes less time to integrate the new knowledge into their cognitive system (Baron-Cohen, Knickmeyer, & Belmonte, 2005). Depending on the course that is to be taught to the students, the proposition, disciplines, and the ideas attached to induce more meaning in their minds are of great importance if the curriculum is examined in relation to the structure of disciplines. For instance, the core application of whole brain theory learning emphasizes the objective of absorbing knowledge and relation to other past experiences (strengthened neurons). It is believed that if the curriculum is meaningful enough, most students will be able to relate the logical value of the body of knowledge resulting in learning efficiency.

In Ach et al.'s (2013) research, the EEG results showed that we have to be familiar with the different brain waves involved in the phenomena that most educators try to explain. The framework of a curriculum needs to be of practical use to the students. After all, regardless of the structure or content of the curriculum, it is the students' cognitive framework and tendency that are the most important.

Perception, Attention, and the Senses

One of the classical theories about perception regarding education includes several different paradigms: (a) indirect perception theory, which suggests that personal perception or interpretation of meaning is driven by both inner elements such as emotion and past experience and direct interaction with the external stimulus; (b) direct perception theory, which suggests that the messages the visual environment provides are sufficient to allow people to interact directly with the environment without involving inner processes and characterization; and (c) integration theory, which suggests that the relative importance of uplink and downlink processes depends on various factors, such as the clarity of the situation (Eysenck & Keane, 2000).

A current study conducted by Liu and Liu (2012) involved performing event-related potential (ERP) tasks designed separately using 2D and 3D images. Offline analyses of collected brain wave data of the experimental subjects were performed as participants were doing the task. The participants were 20 senior high school and vocational school students. The results of this study suggest significant differences between the amplitude from N1 and P300 in the central midline sites and from P1 in the occipitotemporal cortex, when subjects were shown 2D and 3D images (see Fig. 9.1).

In formal theories of categorization, attention is typically modeled with weights that selectively bias the processing of stimulus features. Blair, Watson, Walshe, and Maj (2009) investigated university students who had normal or corrected-to-normal

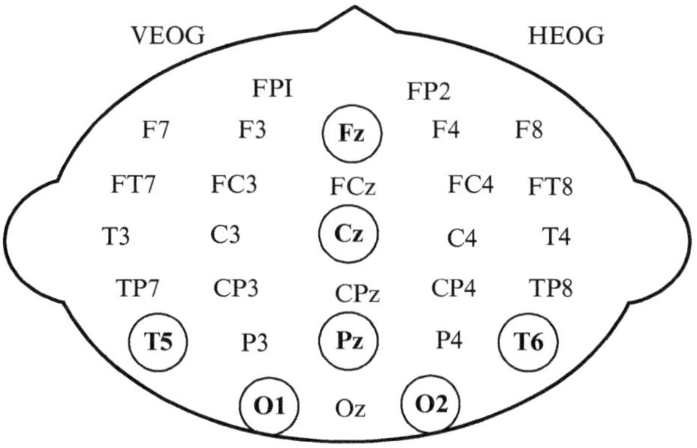

Fig. 9.1 In the image recognition process, N1 and P300 mainly induce the Fz, Cz, and Pz electrode points in the midline district. P1 mainly induces the O1, T5, O2, and T6 electrode points in the occipital temporal lobe

vision. For instance, when the same teaching method was used for different curricula, the result showed that the stimulated areas were not shown in the same region. This is because individual interpretation of which subjects to perceive and sense is different.

The phenomenon of many brain activities can be understood in various ways, but in order to compare the effectiveness and applicability of each temporal gyrus activation method, it is necessary to examine how each temporal gyrus activation is triggered. In addition, it is found that the stimulations of the brain tend to result in distinctive developments in the poststimulation periods. For instance, in chemistry learning, some are combined with the results of experiments and some with the knowledge imparted during classes by the educator. Lehky and Sereno (2007) put forward some concrete methods of learning using memory strategies and discovered that when students are discouraged to use their memory (such as memorizing equations), their results will be wrong more often. On the other hand, when students are encouraged to use perception and trust their feelings, most of them would correctly answer non-memory-based questions.

The experiment consisted of a series of categorization trials in which a fixation cross appeared. When the participants clicked on the cross, a stimulus was displayed. The participants identified it as belonging to one of the four categories, and response feedback was given. The results show that humans can rapidly learn to differently allocate attention to members of different categories. However, they were not asked to focus on remembering the actions in the clips since they did not need to recollect or reproduce them, as shown in Fig. 9.2.

The actions included two parts: (1) the subject learned and operated the assembly, beginning with parts in a standard arrangement and ending with a completed object; and (2) the subject learned and operated the decomposition, beginning with a fully assembled object and ending with the disassembled parts in the standard arrangement. By MRI, the subjects' brain activation parts were observed to realize their perceptual control condition and learning condition, as shown in Fig. 9.3. In the perceptual control condition, in the subjects' brain, their parietofrontal mirror system was activated, including the inferior frontal cortex (IFc) and the inferior

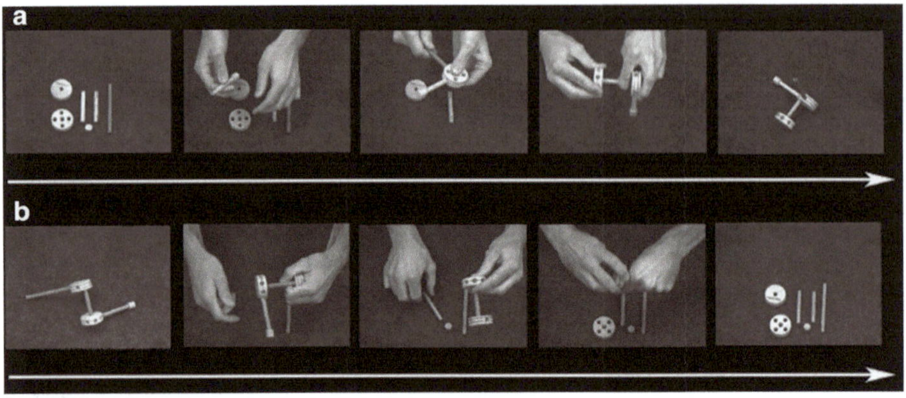

Fig. 9.2 The actions of composition and decomposition

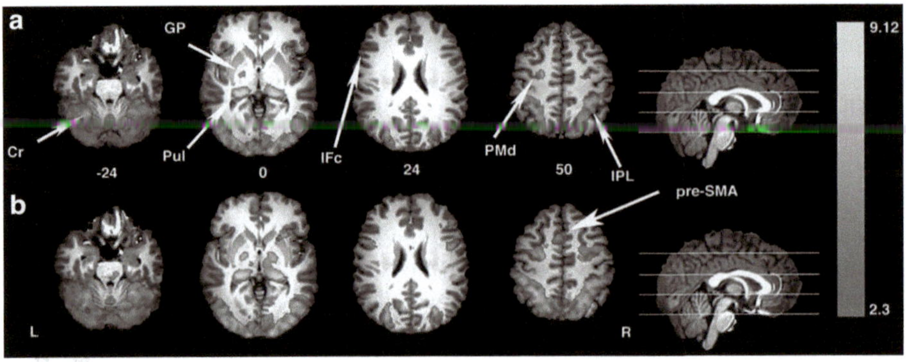

Fig. 9.3 The activation parts of the subjects' brains showing in MRI

parietal cortex (IPl). The results of this study showed that the activation parts of the perceptual control condition were similar to those of the learning condition in the subjects' brains (Gilbert, 2013).

One of the reasons why educational neuroscience is developing rapidly is that the functional brain imaging techniques have reached a mature state with which neural researchers can test out their theories. The aforementioned research with figures show that educational activities influence the intensity of the brain activity when the brain mechanisms of learning required by functional brain imaging techniques are presented. Ethical issues, moral constraints, and other experimental practices in the field of brain science research mainly target specific functions at best, which means that the emergence of functional brain imaging techniques can to some degree help general teachers to improve class performance.

As far as the brain imaging techniques are concerned, some researchers use more advanced techniques, which are more risky, to achieve their practical goals. For instance, through the observation of cerebral blood flow, glucose metabolism and receptors shown in the EEG reflect different levels of brain magnetic signals. Many researchers think that, in order to understand the changes in neuronal activity, one can get the images of the whole brain function while the observed subject is conducting certain tests that require the use of both the left and right brains. With this technique, researchers are more likely to understand the changes in cognition, information flow, and time and spatial variation.

The Visualization of Science Concepts

Various relations exist between the mental and physical aspects of an individual. For instance, approximately 80 % of human knowledge related to the external environment is obtained through visual means. Ungerleider and Mishkin (1982) proposed that the cerebral cortex has two pathways for visual functions: (1) the occipitotemporal pathway, which is located in the anterior region of the brain and is thus termed the ventral stream, and (2) the occipitoparietal pathway, which is located in the upper regions of the brain and is thus termed the dorsal stream. Vidyasagar and Pammer (1999) proposed a model of the interaction between these two pathways. The dorsal stream first conducts pre-attention positioning of visual stimuli by identifying the object for processing.

In this light, course design needs to rely on visual stimulation which leads to relevant theories and outcomes. It is the way to achieve educational goals, of course, but for the most part, it expands the usage of many seemingly unimportant educational resources. Vidyasagar and Pammer (2010) found that when visual concepts and physical props are used in physics and chemistry classes, there will be fewer time restrictions, thus empowering both the teacher and the students. The visual arrangements to achieve neural stimulation to form connections with the neural ties are important in that the courses would not facilitate studies of early childhood due to stress. As evident in many cognitive programs using visual aids, the teaching of music, games, and other types of logic-based functions is essentially the regulation of both the body and the mind. In some nonmainstream curricula, neuroscientists report that the visual connotation of brain imaging techniques such as the positron emission tomography will decrease the abstract meaning and complexity in patterns, thus forming the activities in the cerebral cortex. In some cases, the neurotransmitters are powered to stimulate the spatial conception in relation to the groups of neurons affected by the visual stimulation.

Researchers tend to use conflict tasks such as stroop tasks to research the control component of attention. In stroop tasks, subjects must react to the color of ink (such as red) while disregarding the name of the color (Bush, Luu, & Posner, 2000). Solving conflict in a stroop task chiefly activates the frontal midline region (anterior cingulate) and lateral prefrontal cortex (Fan, Flombaum, McCandliss, Thomas, & Posner, 2003). Eye-tracking equipment can track a person viewing a visual representation of the dynamic trajectory of attention and thinking, so we can understand the distribution of the subjects' attention, their viewing area, as well as looking at an object using a strategy (Poole & Ball, 2005). Prior knowledge will also affect the span of attention students have in life and in class. When they have relatively adequate amount of contact with a new knowledge in the past, chances are they would understand certain concept more quickly and react to teachers' questions more promptly.

Past studies such as that of Liu et al. (2013) have clearly pointed out that human attention, especially that affecting the role of learning and efficiency, is especially important in raising the functions to another level so that the effectiveness of training is more memorable. Furthermore, after doing repetitive exercises, all the critical functions in both the left and right brains will be more sensitive to performance, and their study showed that even when the students had no prior experience of doing similar tests or were not able to perform the said functions, it is considered that the training of attention will be spread equally over patterns and symbols rather than equations.

However, when problem solving needs to extract characterization, the prior knowledge of the context would not matter much, even if the visualization is to make the acceptance of new knowledge easier (Mathai & Ramadas, 2009). Liu and Hou (2009) explored the influence of prior knowledge on eye movements while participants looked at scientific diagrams concerning plate tectonics (see Fig. 9.4). From the study of the brain working to deal with the qualification of the neural process, it was found that the same brain region will not start to engage or participate in multiple similar activities unless the single stock of prior knowledge is in relation to the qualification process. For instance, the brain areas responsible for logical thinking, which is most important in science learning, will participate in dealing with the logic, particularly when using the nonverbal brain region. It is found that all the nonverbal logics can be sorted out when one can improve the performance of the neural district. The study verified that prolonged practice in time

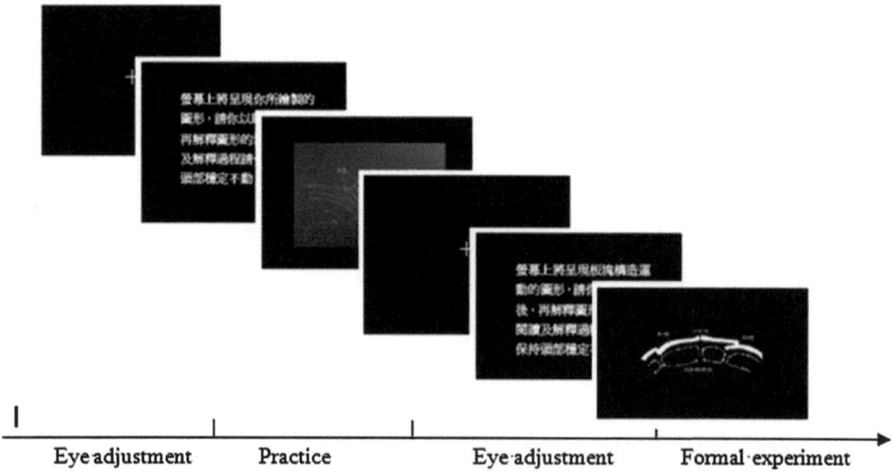

Fig. 9.4 Eye movement tracking experiment

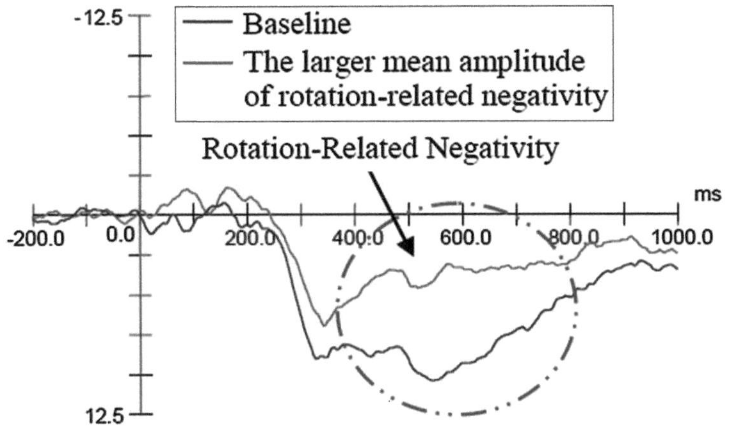

Fig. 9.5 Rotation-related negativity in ERP analysis

and space can improve reasoning ability and the diversity of learning, which is largely contributed by the junction between the simulated fields and cognition.

An alternative that is attracting attention is the use of electroencephalography (EEG) signals, which has progressed considerably, to the point where the user can wear a lightweight recording unit that transmits data wirelessly for analysis (Croft, Chandler, Barry, Cooper, & Clarke, 2005). Furthermore, there are many methods for eye movement correction in EEG recordings reported in the literature, with varying degrees of success (Kierkels, van Boxtel, & Vogten, 2006; Shipulina & Campbell, 2009). Stevens, Galloway, Berka, Johnson, and Sprang (2008) compared novices and experts as they solved a series of chemistry problems and reported that the two groups showed distinct patterns of visual attention that correlated with problem-solving time and accuracy.

The main set of components was rotation-related negativity with latency between 400 and 800 ms. The negative peak that occurs between the latencies of 400 and 800 ms is called rotation-related negativity (Milivojevic, Johnson, Hamm, & Corballis, 2003; Nunez-Pena & Aznar-Casanova, 2009). As Fig. 9.5 shows, the larger mean amplitude of rotation-related negativity indicates that the participants used more mental rotation to complete the task than the baseline. The high activity of the brain, such as memory, cognition, and science learning, which require substantial logical thinking, over time strengthens the relationship between neuron sets resting in different brain areas. Moreover, when they participate in the activities of dynamic intensity, the brain activities are more visible in the ERP.

An ongoing study is investigating how mental rotation strategies affect the identification of 2D and 3D geometric figures and chemical structural formulas (Huang & Liu, 2012). Sixty university students majoring in science were recruited for the Chemical Structure Conceptual Questionnaire (CSCQ), the scores of which were used to divide 26 students into a high-achievement group (HSG) and a low-achievement group (LSG). In the ERP experiments, the participants were

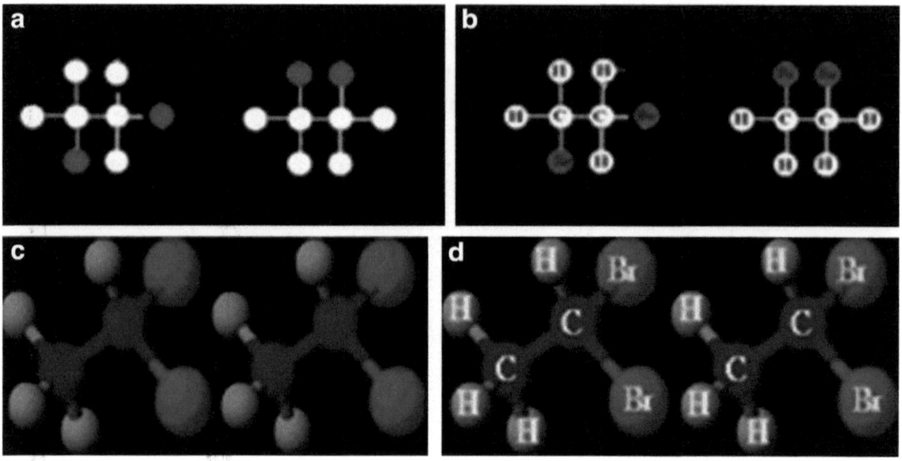

Fig. 9.6 Examples of 2D and 3D pictures and 2D and 3D chemical structural formulas

required to identify 2D and 3D figures and 2D and 3D chemical structural formulas (see Fig. 9.6). The results of this study illustrate that the high-achieving and low-achieving students spent a similar amount of time and got similar scores on identifying 2D figures and they used similar mental rotation, inferred from the analysis of the ERP data from identifying 2D figures.

The Learning Sciences/Technology

Perani et al. (2001) explored what happens to an object in free fall in one's brain mechanism. In the experiment of demonstrated component actions in their study, the subjects observed the videos of objects which were constructed, but they were not asked to learn the sequences of the actions. Their intention of recreating items manually was examined with MRI scanning, with the results as in Fig. 9.7, which shows the increases in activity. The conclusion was that, in the intention to copy, the used parts of the brain were the same as those when learning by physically doing the activity.

To realize if the processes of live and video observations are similar, in this study, teaching and reeducation/rehabilitation settings were used to know the practical implications (Perani et al., 2001). The results showed that the activation parts in the brain are different, as shown in Fig. 9.8. Besides, in typically developing children, live observation better improves learning than video observation (Gilbert, 2013).

Trueswell, Medina, Hafri, and Gleitman (2013) state that, as far as the brain mapping function is concerned, the current results are still far from proving the positive connection of cognition and conscious absorption of knowledge. Current

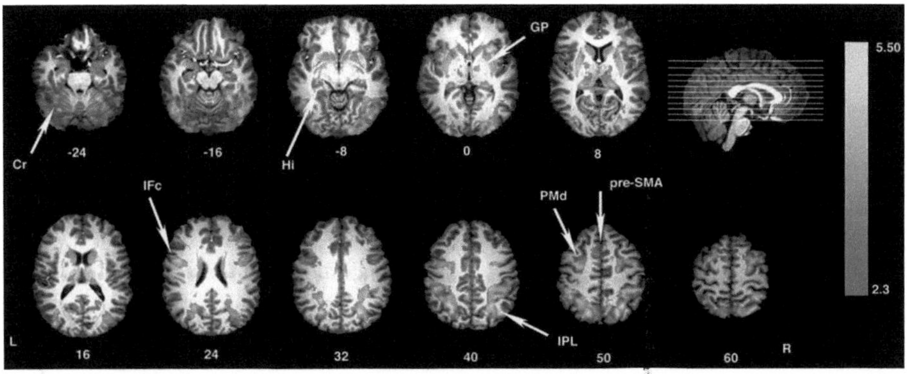

Fig. 9.7 The MRI detection in areas of the sequence learning system

brain research from the perspective of the brain mechanism can further illustrate the science relation in terms of a very close relationship with the spatial arrangement of objects. For instance, the temporary interruption of the left angular gyrus function can result in a decreased visual recognition of patterns. Basic mathematical cognitive processes of the neural mechanisms of fMRI study reveal that the number of abstract representations of neural loops positioned on the region of space neural circuits is more sensitive to the representation of symbols.

In specific contexts, the use of two different nonverbal characterization systems can be regulated to increase the number of memories (Fig. 9.9). Research shows that, with this brain science, digital and spatial processing both enable the whole brain to overlap in the parietal region, characterized by the number and spatial links taught in particular conditions. This explains the number of teachers using structured pedagogy to fulfill the class objectives in science teaching. In the field of in-class cognition of a material, the left brain processing is divided based on gender difference, as males have more active regions in the left brain, thus increasing their ability of logical thinking. On the contrary, females have comparatively larger areas that are related to the size of the mapping scale (Liu et al., 2015).

The growth of technology-based instruction over the last two decades has led to an increased interest in estimating the extent to which students are trying to learn while at the computer. Lin, Liu, and Liu (2012) applied cognitive load theory and employed eye-tracking technology while six classes of fifth graders ($n = 192$) were surfing a website offering teaching resources, interesting experiments, inquiry experiments, virtual animations, multi-assessments, and supplementary materials. Three units, "Air and Combustion," "Heat Effects on Substances," and "Healthy Diet," from the Science Education Website set up by the Ministry of Education (Taiwan) were chosen to assist students' learning.

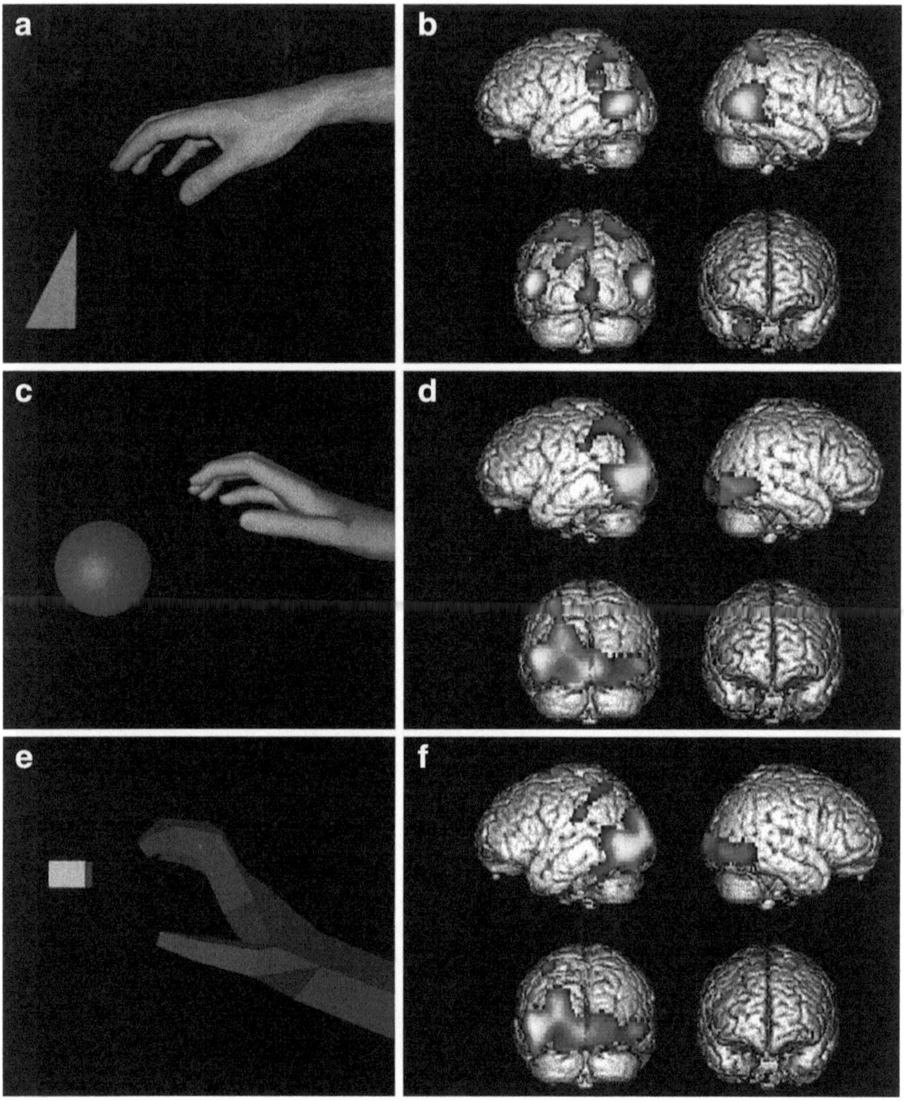

Fig. 9.8 The differences of the activation parts in the brain among live (first hand), 3D high virtual reality, and 2D low virtual reality

The Role of Emotions in Academic Engagement

Research on emotions in science education might profit from the rapid progress in psychophysiological and neuroscientific research on psychophysiological approaches to education (Ellsworth & Scherer, 2003). Tobin and Ritchie (2012) illustrated a methodology of event-oriented social science, covering video

Fig. 9.9 Example sequence of two learning trials for the word "zud," meaning "bear." In the first learning trial, only two alternatives are displayed; in the second, five alternatives are displayed

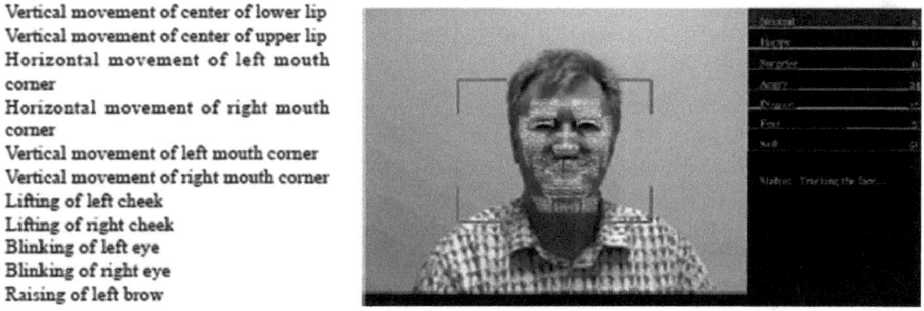

Fig. 9.10 Action units used in eMotion software

ethnography, narrative, conversation analysis, prosody analysis, and facial expression analysis. Moreover, the role of emotions in teaching and learning was examined with eMotion software to analyze the facial expression with different emotions, as shown in Fig. 9.10.

Human cognition and control in the objective world are a wide-ranging process. Physiologists have found that the muscles and joints have proprioceptors that feed real-time reactions to the brain, which then issues commands to perform appropriate actions. This should be seen as the second phase of human reaction to the objective world and can be termed the feedback stage. Basmajian (1982) described biofeedback as the use of instruments to display visual and auditory signals in order to reveal physiological activity (regular or irregular) in the human body.

Medicine has proposed that the concept of the brain "trinity" in relation to science learning strongly intensifies the brainstem – the limbic system – which in turn affects the accumulation of emotions over time. Although the brainstem survival instincts dictate most human beings' behavioral patterns and control the limbic system, the emotional management of learning behaviors caused by experience and response to timely stimulation can be seen as the conscious perception of the human brain. Eagleton and Muller (2011) pointed out that people are always reacting to positive learning environments regardless of the content and meaning—as long as it is beneficial to their survival. Indeed, cognition and behavior control can be changed with science, and it is possible to concentrate on the cerebral cortex for precise thinking and logical arrangement. The clear operation to solve problems in math, physics, and chemistry will be intensified by more participation in the same learning action and memory function. The long-term growth in the positive environment for children relies on the distribution of resources across both parts of the brain. This is largely beneficial when it comes to logic such as the relationship between an individual and his environment and the people around him.

In some ways, invention means creating or discovering a meaningful relationship between objects previously thought to be unrelated. Creativity is the ability to identify such a relationship and use it to accomplish specific objectives. Liu (2005–2008) investigated the elements of scientific creativity using electroencephalograms (EEG), a scientific questionnaire, and interviews. Six junior high school students participated in the study. The results showed that tiny differences in scientific creativity could not be distinguished from the questionnaire results. However, the EEG results showed these differences. The study also found that positive emotion promotes scientific creativity (see Fig. 9.11).

Davidson and Begley (2012) advised teachers to develop a reinforcing strategy that enables students to feel their progress, such as designing a competitive list that keeps the score each time they memorize a new series of numbers. Emotional interaction is important for science learning as well. A case study by Ritchie, Tobin, Hudson, Roth, and Mergard (2011) explored how a new science teacher produced and reproduced positive emotional interactions with her students in her first year of teaching.

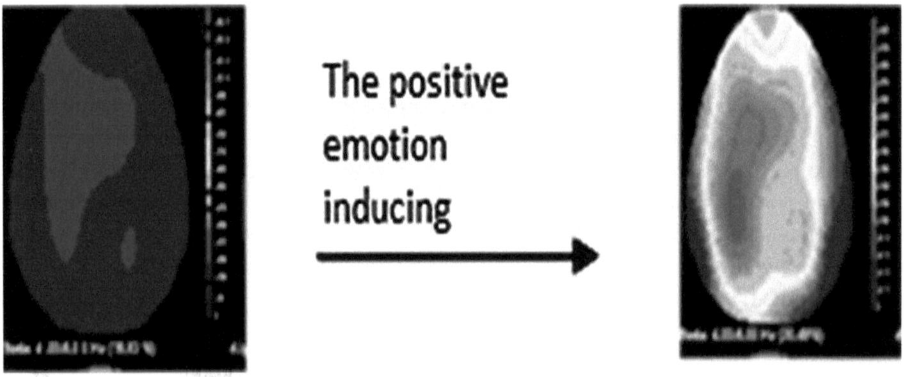

Fig. 9.11 Activation of θ wave in the brain of subjects before and after experimental processing

Conclusion

The findings of neuroscience research in the past few decades reviewed in this chapter center on the study of whole brain, brain-based curricula, building new neural pathways, and how these factors interact with each other to produce optimum learning results. However, neuroscience-assisted learning is far from pervasive in schools due to its lack of coverage and the high costs of formulating individualized lesson plans for each student. Teachers with some training of neuroscience (such as how human brain works while absorbing or coming to contact with new concept) generally are more able to transform the learning process into a fun one. First, preexisting concepts are often deeply and firmly rooted in student behavior and understanding. Second, the brain requires restructuring in order for personal experiences to enhance neural pathways. This is seen by the fact that "you have decided to expand your research to include aspects of learning and education, which are specifically human activities," as stated by Pope John Paul II in his address to the Pontifical Academy of Science on November 10, 2003.

Although exchanges between education and neuroscience have become more frequent, due to large differences in academic background, many educators and scientists suffer from a lack of practical use of the theories stated in this research. For teachers to use the findings of neuroscience and apply them in their pedagogy, the professional training is also very different, so much so that most science teachers lack the research exchanges and cooperation with neuroscience.

As far as the theoretical knowledge is concerned, it is difficult to design a more humane, more targeted curriculum for many teaching frameworks. That is, not all neuroscience frameworks can be applied. The collaboration between neuroscience and education should continue and try to apply a more bidirectional communication so that the results of the application of neuroscience can be more effective in the classroom.

References

Ach, E., Gerhardt, C. A., Barrera, M., Kupst, M. J., Meyer, E. A., Patenaude, A. F., & Vannatta, K. (2013). Family factors associated with academic achievement deficits in pediatric brain tumor survivors. *Psychooncology, 22*(8), 1731–1737. doi:10.1002/pon.3202.

Bandettini, P. A. (2012). Twenty years of fMRI: The science and the stories. *NeuroImage, 62*, 575–588.

Baron-Cohen, S., Knickmeyer, R. C., & Belmonte, M. K. (2005). Sex differences in the brain: Implications for explaining autism. *Science, 310*(5749), 819–823.

Basmajian, J. V. (1982). Clinical use of biofeedback in rehabilitation. *Psychosomatics, 23*(1), 67–73.

Blair, M. R., Watson, M. R., Walshe, R. C., & Maj, F. (2009). Extremely selective attention: Eye-tracking studies on dynamic attentional allocation to stimulus features. *Journal of Experimental Psychology. Learning, Memory, and Cognition, 35*, 1196–1206.

Busatto, G. F., Pilowsky, L. S., Costa, D. C., Ell, P. J., David, A. S., Lucey, J. V., & Kerwin, R. W. (1997). Correlation between reduced in vivo benzodiazepine receptor binding and severity of psychotic symptoms in schizophrenia. *The American Journal of Psychiatry, 154*, 56–63.

Bush, G., Luu, P., & Posner, M. I. (2000). Cognitive and emotional influences in anterior cingulate cortex. *Trends in Cognitive Sciences, 4*(6), 215–222.

Chiang, W. W., & Liu, C. J. (2014). Scale of academic emotion in science education: Development and validation. *International Journal of Science Education, 36*(6), 908–928.

Croft, R. J., Chandler, J. S., Barry, R. J., Cooper, N. R., & Clarke, A. R. (2005). EOG correction: A comparison of four methods. *Psychophysiology, 42*(1), 16–24. doi:10.1111/j.1468-8986.2005.00264.x.

D'Esposito, M., Postle, B. R., Ballard, D., & Lease, J. (1999). Maintenance versus manipulation of information held in working memory: An event-related fMRI study. *Brain and Cognition, 41*, 66–86.

D'Esposito, M., Postle, B. R., Jonides, J., & Smith, E. E. (1999). The neural substrate and temporal dynamics of interference effects in working memory as revealed by event-related functional MRI. *Proceedings of the National Academy of Sciences of the United States of America, 96*, 7514–7519.

Damasio, A. (2003). *Looking for Spinoza: joy, sorrow and the feeling brain*. New York, NY: Harcourt.

Davidson, R. J., & Begley, S. (2012). *The emotional life of your brain: How its unique patterns affect the way you think, feel, and live – And how you can change them*. New York, NY: Hudson Street Press.

Dawson, T. L. (2006). Stage-like patterns in the development of conceptions of energy. In X. Liu & W. Boone (Eds.), *Applications of Rasch measurement in science education* (pp. 111–136). Maple Grove, MN: JAM Press.

Dawson, T. L., Fischer, K. W., & Stein, Z. (2006). Reconsidering qualitative and quantitative research approaches: A cognitive developmental perspective. *New Ideas in Psychology, 24*(3), 229–239. doi:10.1016/j.newideapsych.2006.10.001.

Duncan, J., & Owen, A. (2000). Common regions of the human frontal lobe recruited by diverse cognitive demands. *Trends in Neurosciences, 23*, 475–483.

Eagleton, S., & Muller, A. (2011). Development of a model for whole brain learning of physiology. *Advances in Physiology Education, 35*(4), 421–426. doi:10.1152/advan.00007.2011.

Ellsworth, P. C., & Scherer, K. R. (2003). Appraisal processes in emotion. In R. J. Davidson, K. R. Scherer, & H. Goldsmith (Eds.), *Handbook of affective sciences* (pp. 572–595). New York, NY: Oxford University Press.

Estevez, M. E., Lindgren, K. A., & Bergethon, P. R. (2010). A novel three-dimensional tool for teaching human neuroanatomy. *Anatomical Sciences Education, 3*(6), 309–317. doi:10.1002/ase.186.

Eysenck, W. M., & Keane, T. M. (2000). *Cognitive psychology: A student's handbook* (4th ed.). New York, NY: Psychology Press.

Fan, J., Flombaum, J. I., McCandliss, B. D., Thomas, K. M., & Posner, M. I. (2003). Cognitive and brain consequences of conflict. *NeuroImage, 18*(1), 42–57.

Fischer, K. W., Immordino-Yang, M. H., & Waber, D. P. (2007). Toward a grounded synthesis of mind, brain, and education for reading disorders: An introduction to the field and this book. In K. W. Fischer, J. H. Bernstein, & M. H. Immordino-Yang (Eds.), *Mind, brain, and education in reading disorders* (pp. 3–15). Cambridge, UK: Cambridge University Press.

Geake, J. (2011). Position statement on motivations, methodologies, and practical implications of educational neuroscience research: fMRI studies of the neural correlates of creative intelligence. *Educational Philosophy and Theory, 43*(1), 43–47. doi:10.1111/j.1469-5812.2010.00706.x.

Gilbert, M. (2013). *Cognitive function: Mechanisms underlying learning and memory*. Kaohsiung, Taiwan (R.O.C.): Graduate Institute of Science Education and Environmental Education. National Kaohsiung Normal University.

Ho, M. C., Chou, C. Y., Huang, C. F., Lin, Y. T., Shih, C. S., Han, S. Y., ... Liu, C. J. (2012). Age-related changes of task-specific brain activity in normal aging. *Neuroscience Letters, 507*, 78–83.

Ho, M. C., Huang, C. F., Chou, C. Y., Lin, Y. T., Shih, C. S., Wu, M. T., ... Liu, C. J. (2012). Task-related brain oscillations in normal aging. *Health, 4*, 762–768.

Huang, C. F., & Liu, C. J. (2012). An event-related potentials study of mental rotation in identifying chemical structural formulas. *European Journal of Educational Research, 1*(1), 37–54.

Immordino-Yang, M. H. (2011). Implications of affective and social neuroscience for educational theory. *Educational Philosophy and Theory, 43*(1), 98–103. doi:10.1111/j.1469-5812.2010.00713.x.

Kierkels, J. J., van Boxtel, G. J., & Vogten, L. L. (2006). A model-based objective evaluation of eye movement correction in EEG recordings. *IEEE Transactions on Biomedical Engineering, 53*(2), 246–253. doi:10.1109/TBME.2005.862533.

Kraut, M. A., Hart, J., Soher, B., & Gordon, B. (1997). Object shape processing in the visual system evaluated using functional MRI. *Neurology, 48*, 1416–1420.

Lee, J. H., Gurney, P. T., Dharmakumar, R., Wright, G. A., Hargreaves, B. A., Shankaranarayanan, A., ... Pauly, J. M. (2006). *Blood oxygenation (BOX) level dependent functional brain imaging using steady-state free precession.* Paper presented at the 14th ISMRM, Seattle, Washington, DC.

Lehky, S. R., & Sereno, A. B. (2007). Comparison of shape encoding in primate dorsal and ventral visual pathways. *Journal of Neurophysiology, 97*(1), 307–319. doi:10.1152/jn.00168.2006.

Lin, Y. C., Liu, B. Y., & Liu, C. J. (2012). A study of applying cognitive load theory to science education websites. *International Journal of Science and Engineering, 2*(3), 53–58.

Liu, B. Y., & Liu, C. J. (2012). Using ERPs to investigate two-dimensional and three-dimensional image recognition. *Journal of Research in Education Sciences, 57*(2), 1–23.

Liu, C. J. (2005–2008). *Evaluation and development of creativity sub-program 10: the study composed of elements of the scientific creativity researched by the brain signals.* Kaohsiung, Taiwan (R.O.C.): National Science Council.

Liu, C. J., & Hou, I. L. (2009). Explore the influence of prior knowledge on understanding in scientific diagrams through eye tracking. *Bulletin of Educational Psychology, 43*(Special Issue on Reading), 227–250.

Liu, C. J., & Chiang, W. W. (2014). Theory, method and practice of neuroscientific findings in science education. *International Journal of Science and Mathematics Education, 12*(3), 629–646.

Liu, C. J., Huang, C. F., Chou, C. Y., Kuo, W. J., Lin, Y. T., Hung, C. M., ... Ho, M. C. (2012). Age- and disease-related features of task-related brain oscillations by using mutual information. *Brain and Behaviour, 2*(6), 754–762.

Liu, C. J., Huang, C. F., Chou, C. Y., Yu, C. H., Lin, Y. T., Wu, M. T., & Ho, M. C. (2013). The influence of 40 Hz electromagnetic wave induce phase-synchronization on brain. *Applied Mechanics and Materials, 311*, 491–496.

Liu, C. J., Hou, I. L., Chiu, H. L., & Treagust, D. F. (2014). An exploration of secondary students' mental states when learning about acids and bases. *Research in Science Education, 44*(1), 133–154.

Liu, C. J., Huang, C. F., Liu, M. C., Chien, Y. C., Lai, C. H., & Huang, Y. M. (2015). Does gender influence emotions resulting from positive applause feedback in self-assessment testing? Evidence from neuroscience. *Educational Technology & Society, 18*(1), 337–350.

Mathai, S., & Ramadas, J. (2009). Visuals and visualisation of human body systems. *International Journal of Science Education, 31*(3), 439–458. doi:10.1080/09500690802595821.

Milivojevic, B., Johnson, B. W., Hamm, J. P., & Corballis, M. C. (2003). Non-identical neural mechanisms for two types of mental transformation: Event-related potentials during mental rotation and mental paper folding. *Neuropsychologia, 41*, 1345–1356.

Nehdi, H. (2007). Creative and strategic thinking: The coming competencies. *FOCUS, 8–10.*

Noonan, K. J., Farnum, C. E., Leiferman, E. M., Lampl, M., Markel, M. D., & Wilsman, N. J. (2004). Growing pains: Are they due to increased growth during recumbency as documented in a lamb model? *Journal of Pediatric Orthopaedics, 24*(6), 726–731.

Nunez-Pena, M. I., & Aznar-Casanova, J. A. (2009). Mental rotation of mirrored letters: Evidence from event-related brain potentials. *Brain and Cognition, 69*(1), 180–187. doi:10.1016/j.bandc.2008.07.003.

Perani, D., Fazio, F., Borghese, N. A., Tettamanti, M., Ferrari, S., Decety, J., & Gilardi, M. C. (2001). Different brain correlates for watching real and virtual hand actions. *Neuroimage, 14* (3), 749–758.

Poole, A., & Ball, L. J. (2005). Eye tracking in human-computer interaction and usability research: Current status and future prospects. In C. Ghaoui (Ed.), *Encyclopedia of human computer interaction* (pp. 211–219). Hershey, PA: Idea Group, Inc.

Ritchie, S. M., Tobin, K., Hudson, P., Roth, W. M., & Mergard, V. (2011). Reproducing successful rituals in bad times: Exploring interactions of a new science teacher. *Science Education, 95*(4), 745–765. doi:10.1002/Sce.20440.

Ruhland, R., & van Geert, P. (1998). Jumping into syntax: Transitions in the development of closed class words. *British Journal of Developmental Psychology, 16*, 65–95.

Selden, S., Sherrier, T., & Wooters, R. (2012). Experimental study comparing a traditional approach to performance appraisal training to a whole-brain training method at C.B. fleet laboratories. *Human Resource Development Quarterly, 23*(1), 9–34.

Shipulina, O., & Campbell, S. (2009). *Electrooculography: connecting mind, brain, and behavior in mathematics education research.* Paper presented at the brain, neuroscience, and education SIG, San Diego, CA.

Stevens, R. H., Galloway, T., Berka, C., Johnson, R., & Sprang, M. (2008). *Assessing students' mental representations of complex problem spaces with EEG technologies.* Paper presented at the 52nd annual meeting of the Human Factors and Ergonomic Society, New York, NY.

Tobin, K., & Ritchie, S. M. (2012). Multi-method, multi-theoretical, multi-level research in the learning sciences. *Asia-Pacific Education Researcher, 21*(1), 117–129.

Trueswell, J., Medina, T. N., Hafri, A., & Gleitman, L. R. (2013). Propose but verify: Fast-mapping meets cross situational word learning. *Cognitive Psychology, 66*, 126–156.

Ungerleider, L. G., & Mishkin, M. (1982). Two cortical visual systems. In D. J. Ingle, M. A. Goodale, & R. J. W. Mansfield (Eds.), *Analysis of visual behavior* (pp. 549–586). Cambridge, MA: MIT Press.

Vidyasagar, T. R., & Pammer, K. (1999). Impaired visual search in dyslexia relates to the role of the magnocellular pathway in attention. *Neuroreport, 10*(6), 1283–1287.

Vidyasagar, T. R., & Pammer, K. (2010). Dyslexia: A deficit in visuo-spatial attention, not in phonological processing. *Trends in Cognitive Sciences, 14*(2), 57–63.

Chapter 10
Kids Science Academy: Talent Development in STEM from the Early Childhood Years

Manabu Sumida

Introduction

There are rising concerns about what is being referred to as "science and technology phobia" or "science phobia" in junior and senior high school education. While numerous policies are moving toward implementation, few efforts are being made in education to try to commence the development of science literacy in childhood. One of the reasons that science education has not been emphasized in childhood is the influence of Jean Piaget's staged theory of cognitive development. Piaget viewed childhood as the preoperational stage, during which a child is unable to process scientific concepts sufficiently.

However, since the 1980s, studies from all over the world have demonstrated that young children do possess scientific competence (e.g., Wellman and Gelman (1998)). This has urged educators to revise curricula, which had been developed based on the conventional idea that childhood is a period in which scientific exploration cannot be performed. The timeliness of education needs to be examined from a variety of perspectives that take into account not only the child's age but also their personalities and social and cultural context (Copple & Bredekamp, 2006). Some ideas, like the notion that scientific practice must be experiential at this age because of the nature of childhood, and the notion that the introduction of scientific survey methods, such as experimentation and observation, need to wait until the child reaches the next age level, are no longer convincing (Metz, 1997).

Gopnik et al. (2001) argue that young children and scientists are the best learners in the world; they are highly similar in the ways in which they adjust and change existing knowledge as they interact with the outside world, and they can perform computations that cannot be duplicated even by the most advanced computers. Gopnik (2009) also shows that while careful observation of young children will

M. Sumida (✉)
Department of Education, Ehime University, Matsuyama, Japan
e-mail: sumida.manabu.mm@ehime-u.ac.jp

M.S. Khine (ed.), *Science Education in East Asia*,
DOI 10.1007/978-3-319-16390-1_10

reveal differences in their level of precision or competence, this will also reveal that young children and researchers share a sense of curiosity that makes them want to know more about the world through experimentation and observation and that they seem to be born with this characteristic.

Childhood educational curricula used overseas are starting to implement a variety of practices in an effort to promote scientific interest and exploration among children between the ages of 3 and 8. These include the Science Start program in the USA, which aims to position science education at the center of all other subjects (Conezio & French, 2002), the Clarion Program for developing the skills of gifted young children (Center for Gifted Education, 2008), and the Reggio Emilia approach (Katz & Chard, 2000), which focuses on the implementation of projects in which children engage in exploratory activities. However, none of these go any further than the presentation of individual case studies. The formulation of an overarching theory that presents a cohesive model of science education in childhood is an urgent challenge that needs to be addressed in twenty-first-century science and education (AAAS, 1999).

Neuroscientist Hideaki Koizumi et al. (2007) proposes five aspects of a "scientific mind": (1) deep appreciation for the wonders of nature; (2) direct recognition of truth and lack of interest in deceit; (3) absence of bias or prejudice, but rather a concern to make sound judgments and taking appropriate action; (4) an understanding of one's own position within nature and a respect for life; and (5) respect for diversity and compassion for others. He has exchanged ideas with Nobel Prize-winning scientists and developed new practices taking these aspects into account, which are being implemented in kindergartens. In Japan, however, generally speaking, there is a gap between the formal science learning that begins in third grade and the scientific lessons and education presented prior to that time, such that the lessons are not connected in a gradual, cohesive manner. Kindergartens tend to deal with scientific content in an environmental context, while primary school classes tend to focus on science in the context of everyday life (life environment studies). Current practice demonstrates the following trends (Sumida, 2013a):

1. While efforts are aimed at providing hands-on experiences during childhood, it has not been clearly shown what children gain from that experience.
2. The emotional aspects of experience are emphasized during childhood, but scientific exploration is not emphasized.
3. Experiences in the field of biology are emphasized in childhood, but other fields of science are rarely explored.
4. The deliberate pursuit of scientific experiences during childhood is rare.

Thus, it would be difficult to argue that the scientific interest and curiosity of children are adequately promoted through the scientific instruction provided during kindergarten childcare and in primary school life environment studies classes. Nonetheless, our research findings suggest that, even during childhood, the introduction of appropriate teaching materials and experiential learning programs can stimulate and develop intellectual inquisitiveness and curiosity of children in various scientific fields, including water, physical fitness, and biology (e.g., Fukada,

Sumida, et al. (2005); Sumida (2013b); Sumida, Fukada, et al. (2007)). Efforts must be made to think about the best ways to increase children's scientific exploration and thinking based on their voluntary behaviors. As emphasized by Ritz (2007), the editor of "A Head Start on Science," scientific learning activities should strike an appropriate balance between child-led and instructor-driven priorities. Coates and Wilson (2003) created a model that combines the scientific exploration activities of children and provided guidelines regarding what is required of instructors when such activities occur. As was pointed out by Settlage and Southerland (2007), it is not enough simply to offer access to "activities" to all children. Rather, we must build bridges toward exploration by explanations based on information that children have gathered for themselves and by appropriately connecting what they have learned to relevant scientific knowledge. Also, efforts must be made to ensure that, starting in childhood, all children have access to high-quality opportunities to encounter science.

Thus, in this chapter, I introduce an experiential science learning program (Ehime University Kids Academy Science) for kindergarteners and first and second graders that aim to expand, intensify, and develop the simple intellectual curiosity of young children into scientific exploration and scientific ideas. This report presents specific information on the rich scientific learning that young children achieved through the implementation of this program and on the curriculum and worksheet materials developed in conjunction therewith. The report also proposes a sustainable experiential science learning program that offers both breadth and depth of knowledge for science learning starting in childhood and provides opportunities for children to encounter science at an early age and which goes beyond the dichotomy of experience-oriented versus knowledge-cramming-oriented approaches to science education.

Gifted Young Children: Participants of Kids Academy Science Program

Of all the children with special needs, younger gifted children are the group most frequently ignored throughout the world (Koshy & Robinson, 2006). Sumida (2013c) notes that there is an imperative need to reevaluate just what young children are potentially capable of doing. Harrison (2004) investigated characteristics of gifted young children aged 3 to 5 in Australia and found (1) Curiosity, Intrinsic Motivation, and Independent Investigation; (2) Exceptional Memory; (3) Formulating, Researching, and Testing Theories; (4) Advanced Awareness of Literacy and Numeracy Concepts and Processes; (5) Advanced and Detailed Visual Representation; and (6) the Search for Complexity and Connection. She also pointed out the importance of cooperation between the gifted young children, his/her family, and educators. Sankar-DeLeeuw (2004) found a spectrum of abilities of gifted young children in academic areas, social interactions, emotional intensity and sensitivity, perfectionism, gross and fine motor skills, and creative pursuits. She also pointed out parents and educators as keys in helping gifted young children grow intellectually, socially, and psychologically.

However, there is very little study on identification of gifted young children in science. In this study, revising Sumida's gifted behavioral checklist (Sumida, 2010a) in science for primary children, 25-item checklist was developed to identify gifted young children in science and asked the parents to answer the list. Some of the items are shown below:

1. *Is knowledgeable about nature and scientific topics and knows many scientific terms*
2. *Takes care of animals or grows plants with regard to their ecology*
3. *Is persistent in exploring specific topics and tasks*
4. *Comes up with many ideas and answers about a question*
5. *Understands the causal relations of a natural phenomenon*
6. *Is confident about his/her knowledge and understanding*
7. *Asks unanticipated questions to the teacher*
8. *Observes natural events and phenomena in detail*
9. *Shows interest in analysis using numbers*
10. *Tries to do things in his/her own way, not by the instruction*

In 2010, ten gifted young children (one kindergartener, six grade 1 children, and three grade 2 children) were screened from 50 applicants for Kids Academy Science at Ehime University on the basis of the answer for the checklist and written answer by their parents. There were five boys and five girls. The average total points for the checklist among the selected young children was 82.5 that was significantly high with the comparison to those (72.9) among the children who were not selected in the year. I also asked the parents to write some episodes that show uniqueness of their child in science and to specify points for careful consideration if their child would participate in Kids Academy Science. Some responses from the parents whose child participated in Kids Academy are shown below:

Examples of episodes of uniqueness of their child

When he was a kindergartner, he made over 100 paper planes and explored the influence of the shape and size of wings and balance of weight on its flight (grade 1, boy).

He is very curious about many broad things in everyday life, nature, and ecology. He also loves exploring and he remains focused until he is satisfied with reading books. He often raises questions which we cannot answer (grade 1, boy).

When she was a kindergartener, she was mumbling to herself about calculating the nutritional value of many different foods (grade 2, girl).

Examples of episodes of points to be considered

I am worrying that she cannot continue participating with confidence. She looks to be hurt when she is laughed at by someone (kindergartener, girl).

He has many things that he cannot do by himself but he wants to be perfect and stops doing if he will not succeed (grade 1, boy).

Even in the context of science education, gifted education as a kind of special education in which is neither merely a tool of social selection for the elite, nor a tool for the social exploitation of intelligence, is deemed an important perspective. When designing gifted education, the provision of high-quality education that meets individual needs in early grades—while the foundations of all of the children's future learning are being established—is crucial in terms of meeting the goals of the national educational policy.

Curriculum and Lesson Models of the Kids Academy Science Program

Childhood education has traditionally been divided into child-driven, experience-oriented play and teacher-driven, knowledge-oriented learning (Sommer, Samuelsson, & Hundeide, 2010). The Kids Academy Science Program is an early science education program that has been developed to offer experiential learning opportunities to expand, intensify, and develop the intellectual curiosity of young children into scientific exploration and thought, going beyond the dichotomy of experience-oriented versus knowledge-cramming-oriented approaches to science education. This section explains the basic concepts of this program, the model used for curriculum development, and the guiding principles behind its development and implementation.

Stimulating the "Scientific Mind" of Children

I selected themes and scientific concepts that form the foundations of science with which children are familiar, specifically water, the human body, and plants. I conducted project-based activities designed to help children identify their own problems and work on solving them. These themes include important concepts in Japanese science education at the primary, secondary, and high school levels and are commonly addressed by related childhood science education programs implemented elsewhere in Japan and abroad (e.g., Center for Gifted Education, 2008; Chalufour and Worth 2005; Harlan & Rivkin, 2004; Lind, 2005; McIntyre, 1984; Ritz, 2007). By strategically designing the topics and conducting project-oriented activities, we enabled children to approach learning in a way that excites and challenges them.

Deepening the "Scientific Mind" of Children

I included activities that allowed children to actively use measuring equipment and devices, such as digital scales and beakers, and to think analytically as they convert information into numbers. Through these activities, I expected that these children would be able to exercise precision and to learn voluntarily, continuously, and systematically, and we took into consideration the related scientific content that they would be learning in higher-grade levels. By effectively using the measuring equipment and devices, children are able to understand things beyond what they know from their own five senses. Also, their learning becomes more accurate and cumulative. For example, by using a tool known as a thermometer, the expressions used by children of "a little hot," "hot," and "very hot" turn into "XXC," which is much more reliable and can be shared with others. By using a scale, children's concepts of "heavy" or "light" can be turned in a specific number of grams. By using a microscope, children are able to see a world that is inaccessible to them using the naked eye. The use of tools, including computers, and creative thinking practices, like those required for technological innovations, must be required of science education in this new era (Eshach, 2006; National Research Council, 2010).

Expanding the "Scientific Mind" of Children

Experiential scientific learning is extremely well suited to the integrated play that is characteristic of childhood, the learning that occurs through the natural environment as experienced in everyday life, and collaborative learning with others. Once children have learned certain core scientific concepts experientially, Harlan and Rivkin (2004) recommend integrated activities, such as calculation activities and activities involving the expression of shapes, play activities and expressions related to the body, creative thinking activities, and activities involving foods. Lind (2005) also recommends integrated activities that include the expression of shapes, math, and language. In this program, the experiential science lessons were designed to include activities that could be integrated into other subjects, such as language, maths, home economics, and drawing and art, and I took into consideration the need to create a balance between independent and collaborative activities.

Sustaining the "Scientific Mind" of Children

How captivated a young child is by an activity can be used to determine how much time should be spent on that activity. Sumida (2010b) proposed a captivation and curiosity index for scientific exploration activities among young children by focusing on the child's level of concentration and length of time spent working continuously. We developed this program primarily for first graders, such that they would spend about two hours on each program session and would study a single theme (e.g., "water") continuously over the course of three sessions. The attitudes of the children who took part in the activities and the results of their explorations can be viewed as reflecting of their level of captivation and level of curiosity and also can be viewed as something of an evaluation of the program. In this program, we distributed an experiment sheet and a materials file to each child. By having the children record their own lessons and giving them books on the activity contents and information about activities that they could do at home, we promoted the continuation and accumulation of children's learning within the context of their families and communities.

Based on the four approaches outlined above, we designed a program development model for curriculum development. Ten guidelines were also established for the specific development and implementation of the curriculum, and we used them to develop and implement this program.

Ten Guidelines for the Development and Implementation of the Kids Academy Science Program Curriculum

1. Focus on the child's spontaneity and sense of discovery, and conduct integrated project-style activities.
2. Include elements that stimulate the child's creative thinking skills.
3. Use materials that are familiar and inexpensive.
4. Use scientific terms correctly.
5. Include activities in which children use simple measuring equipment and devices.
6. Include both group activities and individual activities.
7. Design activities in a way that is conducive to integration with other fields and subjects.
8. Don't focus solely on the intellectual aspects of an activity, but also focus on the child's feelings and emotions.
9. Incorporate familiar, seasonal, and local themes and materials.
10. Promote partnerships with families and communities.

When developing and implementing a science learning program for young children in Japan, scientific views of nature and uniquely Japanese views of nature need to be taken into consideration (Sumida, 2012). This program placed priority on theme-based experiential learning activities, established the "Work of Scientists" Project, and tried to get children to be more aware of the fact that they were engaging in scientific exploration. Thus, the children learned about scientists and the distinctive characteristics of scientific research. The importance of learning science as a method (process) from a very young age was emphasized in a variety of ways (National Research Council, 2010; Saracho & Spodek, 2008). Focus was also placed on the contribution of scientific method to the development of language and societies (Settlage & Southerland, 2007). The model used for the Kids Academy Program was developed based on the model of scientific research and thinking.

For the Kids Academy Science Program, I developed a three-session curriculum (2 h per session) and worksheet materials for three different themes: water, the human body, and plants. I have been implementing this program at Ehime University continuously since 2010, primarily for first graders, but also for second graders and some kindergarteners, with over 150 children participating in the program to date. All of the classes are open to the public, including parents.

Sample Curricula and Practices of the Kids Academy Science Program

Example 1: Welcome to Kids Academy Science

Goals of the "Work of Scientists" Project

Help children learn about scientists and scientific methods of thinking and research, so that in the future, the Kids Academy Science Program will help raise those children's interest in pursuing scientific exploration.

Activity period: 30 min

Supplies used

White lab coat for children, beakers, microscopes, and other equipment, worksheets, and images of scientists

Lesson flow

(1) Children wear their white lab coats and prepare their beakers and other lab equipment. Next, the instructor says, "From now on, at the Kids Academy, you are all scientists." Ask: "Do you know a scientist?"; "What kinds of things do you think scientists do?"
(2) Show pictures of people in several different occupations and ask the children to identify whether each is a scientist. Then ask, "Why do you think this person is (or is not) a scientist?" and explain the kinds of things that scientists do.
(3) Ask about the various things scientists do. Present a figure that shows methods of scientific thinking and research and compare these to the children's ideas, explaining the meaning of each.
(4) The instructor says, "At the Kids Academy, everyone is a scientist, and everyone will learn to conduct research and think like a scientist." Explain that it is important for a scientist to keep records of what they learn during their research and the ideas they come up with. Enter the date in the logbook. It is important to have someone there to confirm the findings. Tell the students how to use the experiment notebook.
(5) Ask the children to imagine what it would be like to be a scientist and draw a picture of what they envision. Ask them, "What would you want to research and study?" Encourage each child to share their drawing of themselves as a scientist.

New scientific terms children will learn
Science, scientist, and scientific research methods
Try it at home!

- You can learn about scientists and scientific research at home.
- Ask other members of your family what they would research and study if they were a scientist.

There are relatively few opportunities for children between the ages of five and seven to visit a university. For young children, a university laboratory is surely a mysterious world. The children who were brought to the laboratory with their guardians appeared nervous at first. However, their expressions changed when they put on the lab coats that were distributed to them, and the moment each of them hung their membership badges over their necks marked the birth of several young scientists.

There are without doubt children who are highly interested in natural objects and phenomenon but too shy to express this interest. In this class, the children were given an opportunity to develop, announce, and share their own ideas. The highly original pictures of scientists drawn by the children were very appealing.

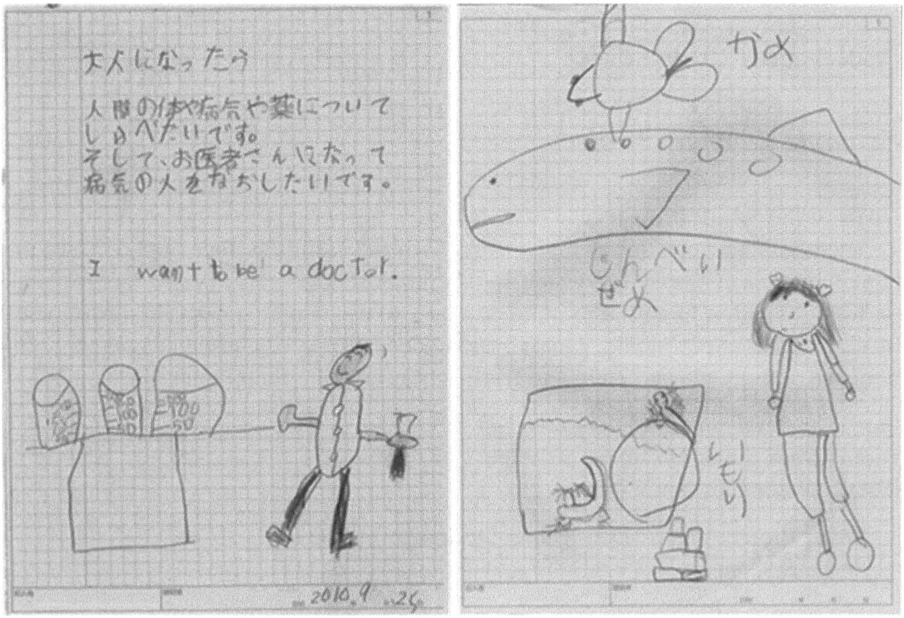

Grade 1 (boy) Kindergarten (girl)

Grade 2 (girl) Grade 1 (boy)

Example 2: Water Project

Goals of the Water Project

The ocean accounts for as much as 70 % of the earth's surface, and 60 % of the human body is comprised of water. In this program, children will conduct research on water, which is essential to life; learn how to use simple measuring instruments by conducting experiments; gain a basic understanding of phase changes, dissolution, and water environments; and learn some scientific thinking skills.

Part 1: Water Transformation!
Part 2: Does Sugar Disappear in Water?
Part 3: Water Flows in Circles

Water Project Part 2: Does Sugar Disappear in Water?
Program Concept

Core Scientific Understanding

- Some substances dissolve in water while others do not.
- When a substance dissolves in water, it permeates all of the water and becomes uniformly distributed.

Enrichment

- Present the predictions and results of the experiment in a table.

- Use standards for the analysis.

  Extension

- Make a prediction before conducting the experiment.
- Use a scale to ascertain results accurately.

  Acceleration

- Even when a substance dissolves in water, the total weight of the water and the substance does not change.
- The amount of a substance that will dissolve in water differs depending on the water temperature and volume.

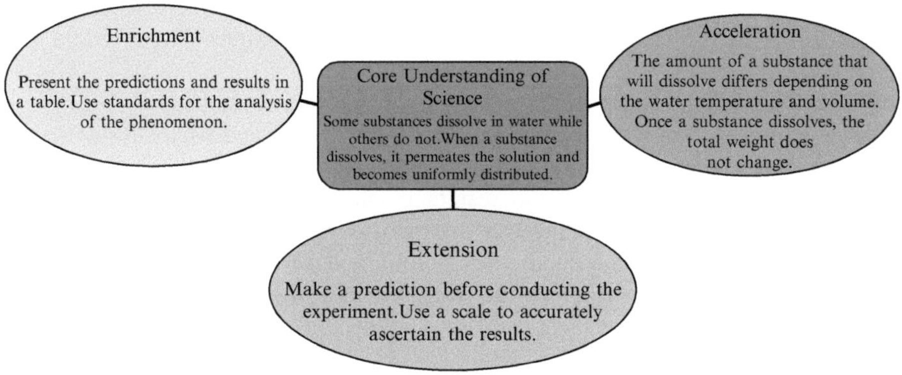

Water project part 2 conceptual diagram
Activity period: 120 min
Supplies needed
Water, sugar sticks, potato starch, chocolate, beaker, glass rod, digital scale, saccharimeter, and worksheet
Lesson flow

Lesson 1

(1) While presenting a panel that shows scientific research methods and ways of thinking, the instructor tells the children that they will think and act like scientists while at the Kids Academy.
(2) Ask the children to recall how the water changed when it was solid, liquid, and gas. Ask how the water changed when various things were added into it.
(3) Divide the children into groups, distribute three substances of different types, and explain that some substances dissolve in water while others do not.
(4) Explain that when a substance dissolves in water, "it permeates all of the water so that it has the same concentration throughout." Note that water in which something has been dissolved is called a "solution," and present several examples of different solutions.

(5) Ask the students to predict whether or not each substance will dissolve.
(6) Distribute a beaker filled with water and a glass rod to all the children, and explain the hazards. Allow the children to perform the experiment and record their findings.
(7) Using a substance that dissolves in water (sugar), the children combine the weight of the water before the sugar is added, and the weight of the sugar, and predict the weight of the resulting solution. Even when a substance has dissolved and is no longer visible, a saccharimeter is used to confirm the presence of the substance throughout the solution. Try the experiment again using a different substance, as needed.

Lesson 2

(1) Predict how much sugar will dissolve in the water.
(2) Add sugar, a little at a time, to the water, experiment with how much can be added before it stops dissolving, and record the findings. Also, record the change in weight, as necessary.
(3) Have the children share their results and discuss their conclusions.
(4) Have the children think about what substances change and what substances do not before and after they are dissolved in the water.
(5) Have the children discuss the results of their experiment, expressing the findings as though they themselves are the water.
(6) Ask the children to think about what should be done to increase the amount of sugar that can be dissolved in the water. Demonstrate several of the options presented.

New scientific terms children will learn
Dissolve and solution
Extending the Lesson: Try it at home!

- You can try and dissolve some other substances in water.
- If you change the substance (water) in which you try to dissolve something, do things dissolve differently?
- Think about how you might be able to extract salt from seawater.

The common Japanese word used to refer to melting, "tokeru," could also refer to dissolving or loosening. In science classes, however, it's uncommon for children to learn the distinctions between these different possible meanings. For example, when a child uses the word tokeru to refer to chocolate, he or she may not be thinking of melting chocolate but may instead be thinking of dissolving chocolate. The children were taught that even when words have the same phonetic spelling in hiragana characters (a set of phonetic characters used in Japanese), the meanings of the words may be expressed using different kanji characters (logographic Chinese characters). The words may even correspond to entirely separate words in other languages (such as English). The children were able to experience the strictness and power of scientific terms, and this experience helped to add depth to their experiments and discussions.

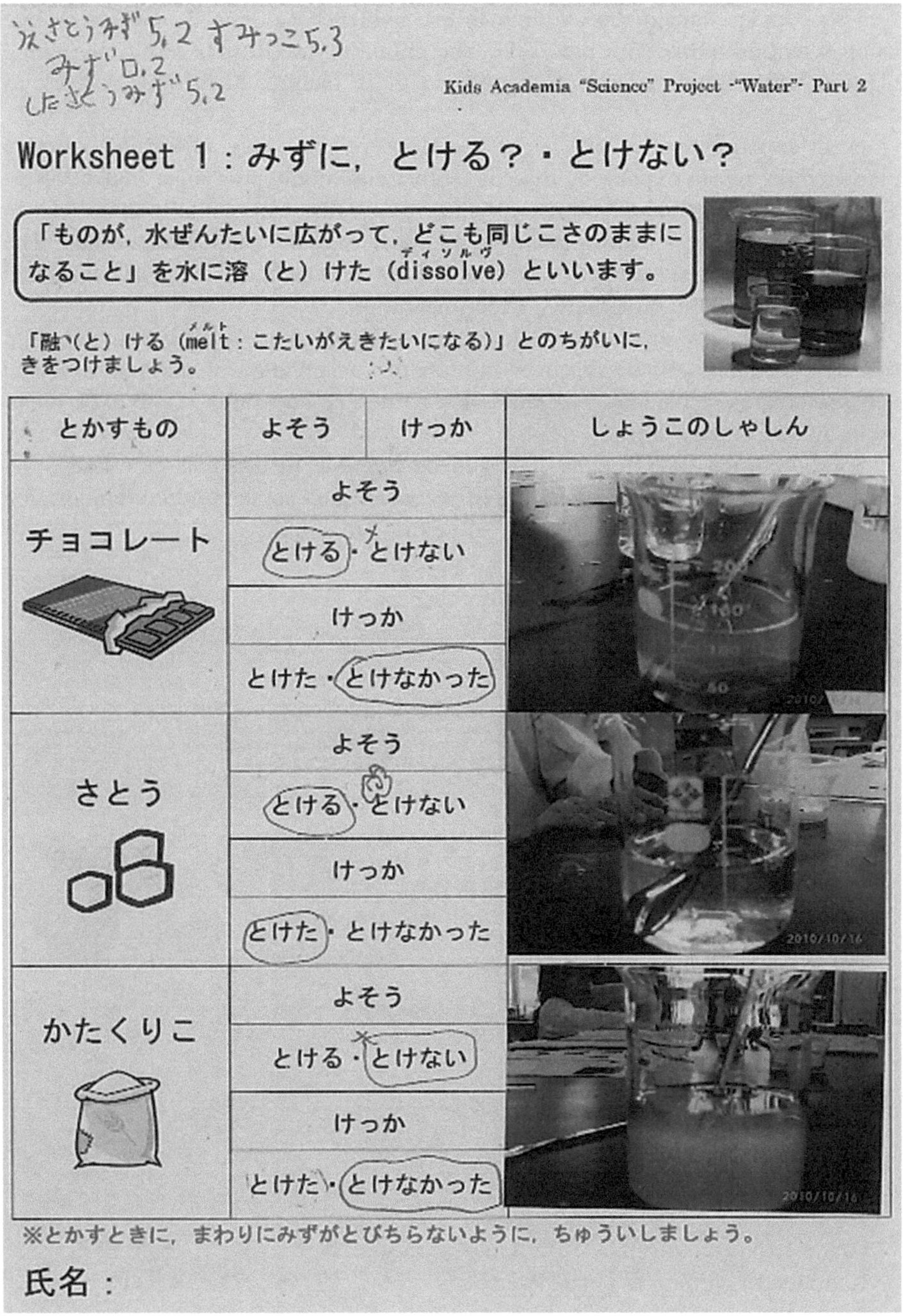

えきとうみず 5.2 すみっこ 5.3
みず 0.2
したさとうみず 5.2

Kids Academia "Science" Project -"Water"- Part 2

Worksheet 1：みずに，とける？・とけない？

「ものが，水ぜんたいに広がって，どこも同じこさのままになること」を水に溶（と）けた（dissolve）といいます。

「融（と）ける（melt：こたいがえきたいになる）」とのちがいに，きをつけましょう。

とかすもの	よそう / けっか	しょうこのしゃしん
チョコレート	よそう：とける・とけない（とける に○）	
	けっか：とけた・とけなかった（とけなかった に○）	
さとう	よそう：とける・とけない（とける に○）	
	けっか：とけた・とけなかった（とけた に○）	
かたくりこ	よそう：とける・とけない（とけない に○）	
	けっか：とけた・とけなかった（とけなかった に○）	

※とかすときに，まわりにみずがとびちらないように，ちゅういしましょう。

氏名：

Grade 1 (girl)

This course enabled the children to use evidence to draw conclusions while using scientific terms with precision. The children shared their results by taking pictures with digital cameras and attaching those pictures to their worksheets as evidence.

In an experiment in which sugar was dissolved in water, one child correctly predicted the result, explaining that his mother commonly puts sugar in her coffee. The child ended up being surprised by how much the sugar actually dissolved. In this experiment, we used 5 g packages of sugar to make calculations easier. The children did not compete to see who could dissolve more sticks of sugar but instead diligently engaged in quantitative experimentation.

The figure below shows the worksheet of a first grader. Even this younger child was able to quantitatively discover the law of conservation of mass and seek to verify increases in volume. The child also wrote that he would like to experiment with salt and other solutes.

Some children thought that it would be possible to dissolve more sugar by changing the quantity or temperature of the water, and some children thought that substances other than sugar might dissolve more readily.

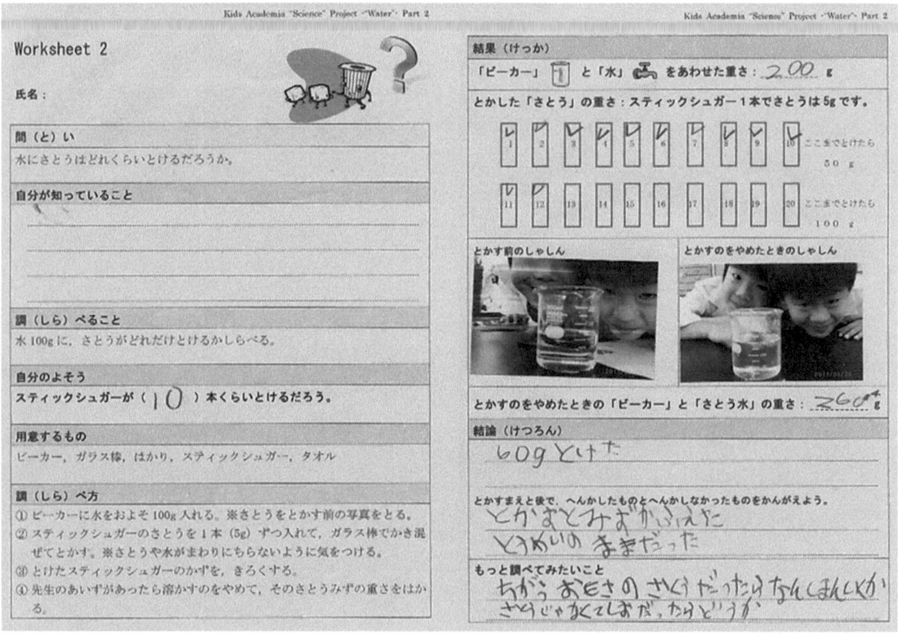

Grade 1 (boy)

Continued Learning at Home 2-2

At the end of each course in the Kids Academy Science Program, we introduced activities that children could engage in at home and in their communities. These activities were not mandatory, and the children were free to choose whatever topics they liked. They used their lab notes and eagerly pursued additional related activities at home.

For example, an excerpt from a second grade girl's lab notes is shown below. According to a detailed plan that she came up with herself, the girl used household ingredients to test whether substances such as salt, flour, and gelatin would dissolve in liquids such as water, vinegar, and carbonated water. Just as she had done in the Kids Academy, she took pictures to use as evidence and attached them to her lab notes. She wrote that in the future, she would like to examine in detail how the temperature of a solvent affects the amount of material that dissolves.

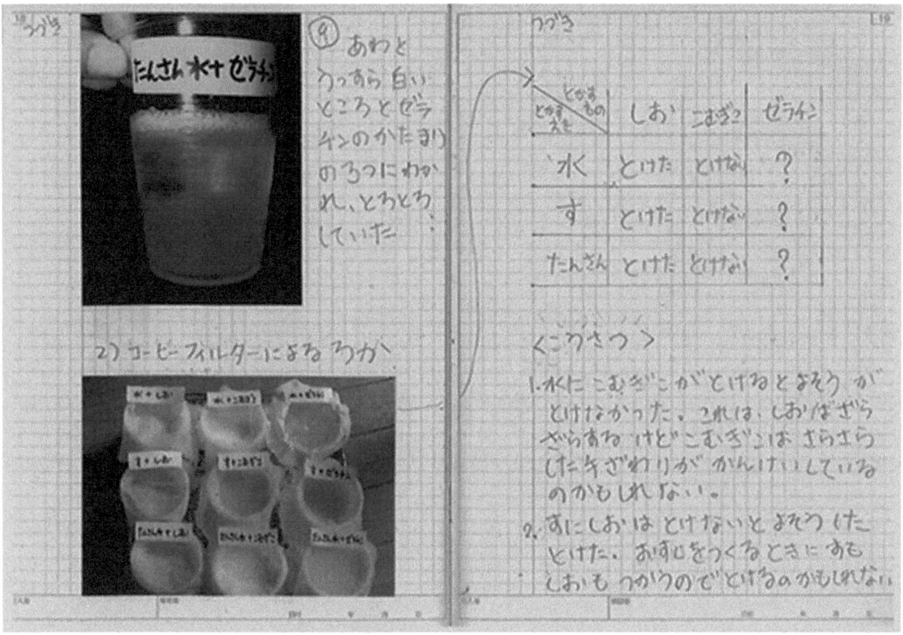

Grade 2 (girl)

An excerpt from a first grade girl's lab notes is shown in the below figure. The girl first dissolved different solutes, such as starch and paint, in a single solvent and observed how layers were formed. She also checked to see whether salt would dissolve in oil and then tried extracting salt from seawater. The child began each experiment and observation by predicting what would happen, and she modified her methods in her own way when things did not go well at first.

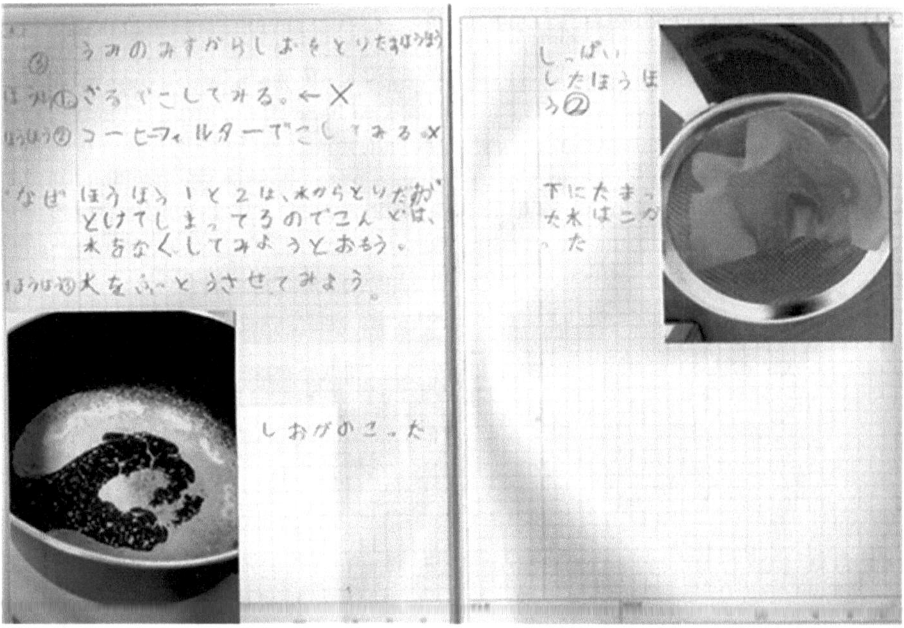

Grade 1 (girl)

For example, in the experiment in which she tried to extract salt from seawater, she first tried to filter out the salt using a strainer. Her notes indicate that she then tried to filter out the salt using a coffee filter. At this point, it appears that she was focused entirely on the salt. However, in her third experiment, she came up with the idea of getting rid of the water and was able to remove the salt by putting the seawater in a pan and heating it up. She even introduced the methods that failed, leaving detailed records using photographs.

During the project "Water," the children produced a variety of original opinions in an activity when they were asked to identify various forms of water they encountered in their daily lives. In the worksheet, a first grader listed more than 20 different forms of water, including rivers, toilets, tears, baths, lakes, pools, running noses, and sweat. The child divided these different forms of water into clean and dirty water and was able to explain and share how she determined which was which. At the end of the project, the children created a filtering device. They came up with the materials, amounts, and procedures on their own and tested the results of their arrangements. They used pictures as evidence and explained how they arranged their devices and what the results were. A K-grade child disassembled the water filter in her yard at home after the program and systematically performed additional experiment. She tried to improve its performance by changing how it was put together and adding different materials and recorded the process and results in detail.

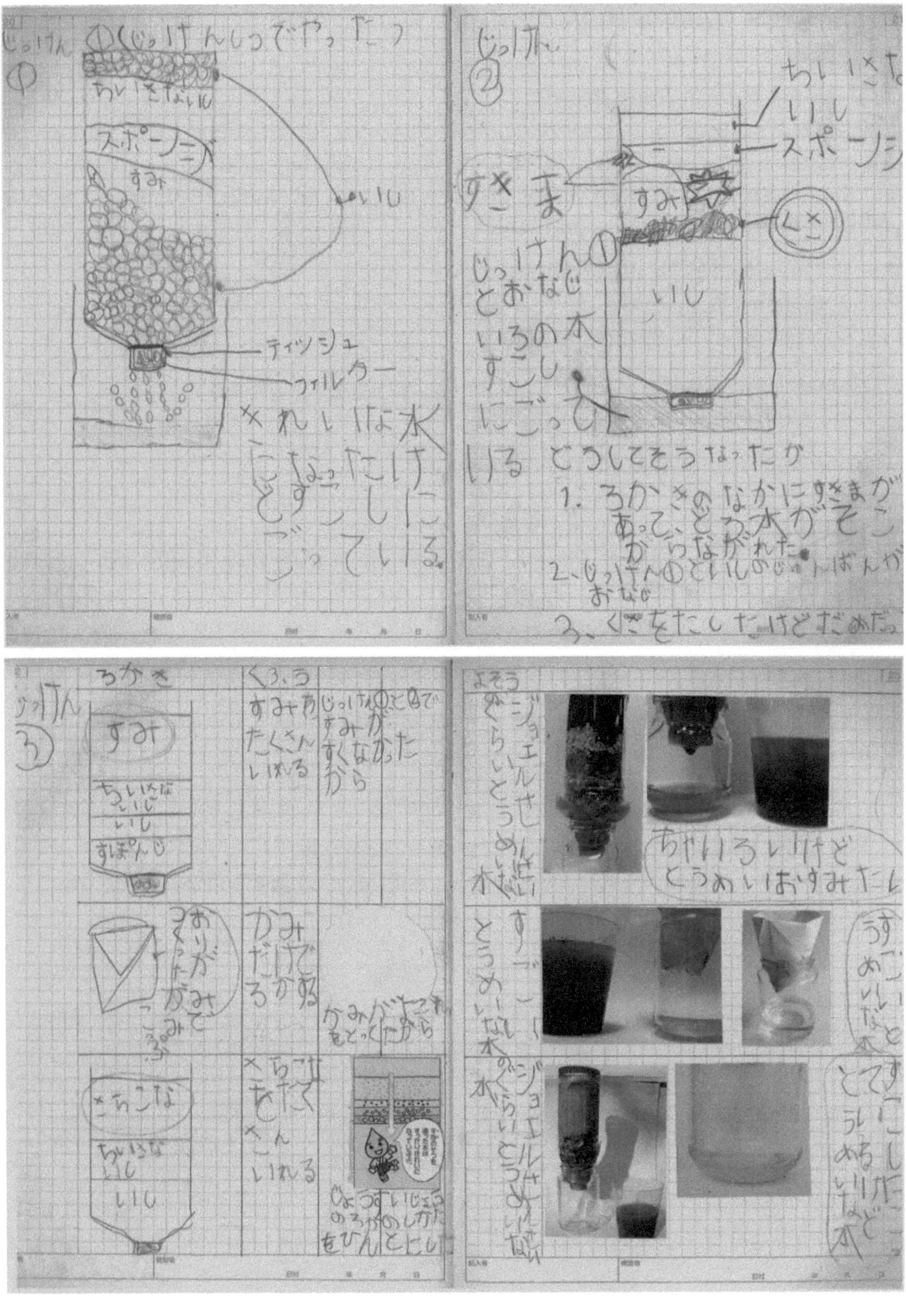

Kindergartener (girl)

Example 3: Human Body Project

Goals of the Human Body Project

Growing up healthy is important for both young children and their parents. In this program, growing children learn to recognize that they themselves are growing, and they learn experientially about the structure and workings of their five senses and their heart, using graphs and tables.

Part 1: Growing Up Healthy and Strong!
Part 2: Sensing Your Body
Part 3: Your Body Is Always Working

Human Body Project Part 3: Your Body Is Always Working
Core Understanding of Science

- Your heart is essential for maintaining life.

Enrichment
- Children will gain a real sense of the strength of their own bodies through the workings of their heart.
- They will listen to the heartbeat of small animals, like marmots.

Extension
- Children will measure the number of heartbeats numerically and will compare the results.

Acceleration
- The heart works to pump blood throughout the body.
- The number of heartbeats changes as a result of exercise and other factors.

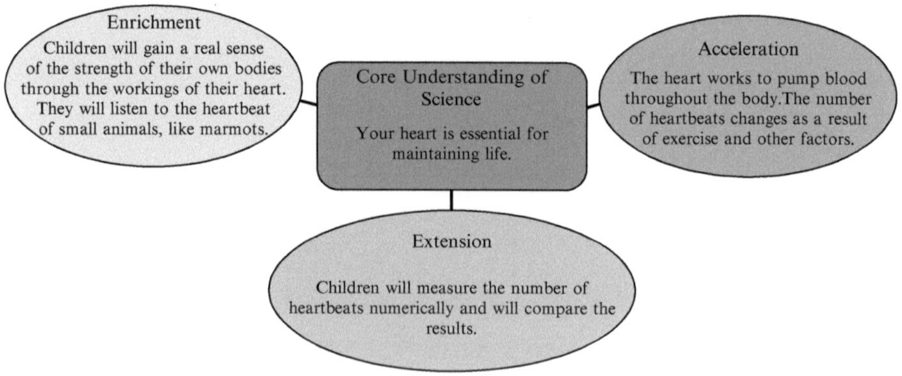

Human body project part 3 conceptual diagram

Supplies needed

Stopwatch, stethoscope, counter, animal (pig) heart, small animal (guinea pig), and worksheet

Lesson flow

(1) While presenting a cardboard panel that shows scientific research methods and ways of thinking, the instructor tells the children that they will think and act like scientists while at the Kids Academy.
(2) Today, we will study everyone's bodies. First, stand up and try to stop all movements in your body. Try to stop blinking and breathing. Is there anything that you just can't make stop? It's your heart! Today, we will learn about your heart.
(3) Ask the children what they know about their heart. Have children listen to different rhythms on a metronome and predict which one comes closest to the beating of their own hearts.
(4) Have the children listen to the rhythm of their own hearts using a stethoscope, and have them listen to their heartbeat for 1 min.
(5) Encourage them to predict the heartbeat rhythm of other animals. Once they share their predictions, encourage them to listen to the heartbeat of a small animal (marmot). Then, try to listen to the heartbeat of a baby while it's still in its mother's womb.
(6) Give the students the opportunity to observe the actual heart of a pig, allowing them to feel its size and ruggedness. Confirm that the heart is a mass of muscle, that it has openings that are surrounded by thick muscle tissue, and that it has thick, rugged arteries.
(7) Ask if a person's heartbeat is always the same. When does a person's heartbeat speed up?
(8) Organize the conditions children have thought of, plan an experiment, and share the results.
(9) Finally, confirm that after a short time, the heartbeat returns to its original rhythm.

New scientific terms children will learn
Heart, heartbeat, and aorta
Try it at home!

- How does your own heart rhythm (heartbeat) change from the time you wake up in the morning until you go to sleep at night?
- Listen to the heartbeats of your family members and compare them to your own.
- Let's examine how the blood pumped from our heart is sent all throughout our body.
- Other than your heart, think about what other kinds of things are in your body and draw a picture of what the inside of your body looks like.

Stethoscopes were used in the conclusion of the human body theme. The children looked like little doctors. As they kept their bodies still, the children learned about the heart, whose movement cannot be kept still. The children listened to a metronome and predicted the rhythms of their own hearts and then investigated what the actual rhythm was. After a little practice using a counter, even the younger children were able to skillfully measure their heart rate. At first, there were children who made a fuss, saying that they could not hear anything, but they started quietly concentrating for the sake of the experiment and seemed very happy when they were able to hear their own hearts.

Many children guessed that a guinea pig would have a slower heart rate than theirs. The children carefully placed the stethoscope on the guinea pig, closed their eyes, and concentrated on the heartbeat. They were shocked to discover that the heart was beating so quickly that the hand they were using for the counter could barely catch up.

Next, when the children observed a pig's heart, they were surprised to discover that it was indeed shaped like a heart. The children were able to experience the heart's weight and thickness, as well as the thickness of its muscles and blood vessels firsthand. Even children who were a little afraid at first and kept a safe distance approached closer and closer to the heart until they were able to touch it themselves. Even after break time came, there were children who wanted to keep touching the heart, and some children wanted to investigate its inner chambers.

In the final experiment, the children came up on their own with different conditions in which they thought their heart rates would increase, and they performed experiments under these conditions. The following figure shows the worksheet of a second grader. The child first came up with 11 different conditions and then discussed the conditions with the friend who would be the child's partner for the experiment. The friends came up with more appropriate conditions and then performed the experiment.

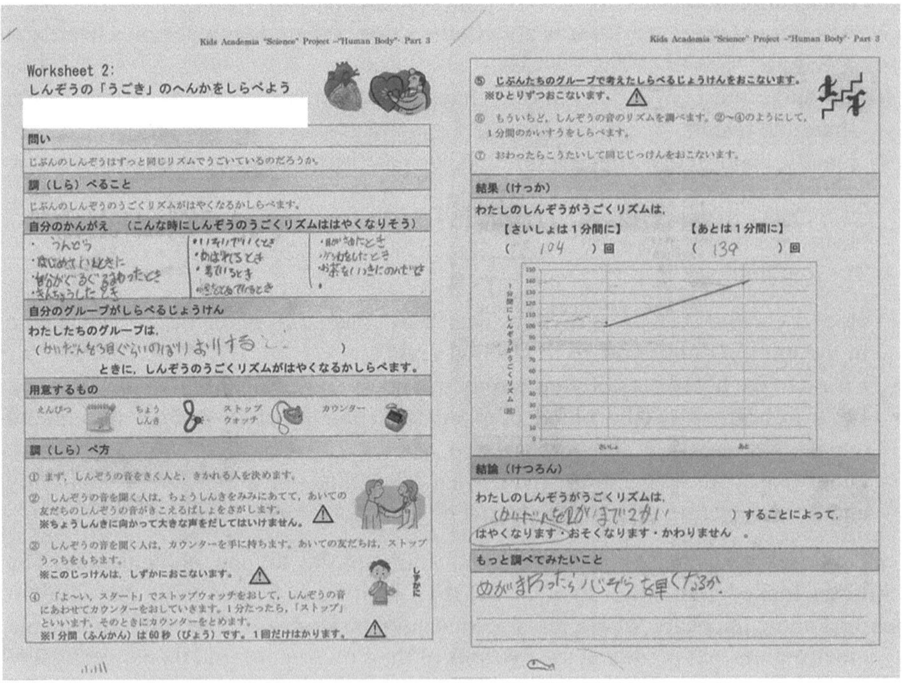

Grade 2 (girl)

Children's Growth Through the Kids Academy Science Program

After completing Theme 1 "Welcome to the Kids Academy Science Program," Theme 2 "Water Project," Theme 3 "Human Body Project," and the research presentation session, we conducted a survey among parents regarding the frequency of program sessions and number of classes offered; activity period; difficulty of, level of interest in, and satisfaction with class content; and the growth in understanding children who had taken part.

None of the respondents indicated that there were too many program sessions or classes or that the activity periods were too long. All of the respondents indicated that they were "just right" or "too few/short (more/longer would have been better)." With regard to the difficulty level of, level of interest in, and satisfaction with class content, virtually all respondents indicated that their participating child had a high level of interest in the program, reported being very satisfied, and indicated that the level of difficulty was appropriate (selected option 3 on a scale of 1 (very difficult) to 5 (very easy)).

In questions that asked which lessons their child seemed to have the most interest in, all of the sessions listed received responses from some parents, with some even indicating that the presentation session had held the greatest interest for their child. When the survey was conducted, we had only developed and implemented programs on the two themes of water and the human body, but both of these themes were familiar and of interest to young children.

The following 15 questions were developed to explore the growth experienced by the children as a result of the Kids Academy Science Program, and respondents were asked to indicate their responses on a scale of 1 (no) to 5 (yes).

The Table 10.1 shows that the parents felt that their children showed progress in all 15 of the items listed. Particularly, strong progress was reported on items (1) the ability to enjoy natural objects and phenomena and to enjoy science, (2) the ability to observe natural objects and phenomena, and (5) the ability to examine things that seem mysterious or that they don't understand. The average value reported on these items was higher than 4.5. Parents' assessments of their children's progress in the following areas were also high, averaging more than 4.0 points: (3) the ability to focus and work on a single topic, (7) the ability to think theoretically and organize their thought, (10) the ability to operate lab equipment and measuring instruments, (11) the ability to convey their ideas and what they have learned, (12) the ability to understand natural objects and phenomena scientifically, (13) the ability to cooperate with friends to solve problems, and (15) the ability to come up with new ideas and recognize other ideas.

Table 10.1 Children's growth through the kids academy science program

	Questions on growth achieved through the program	Average score
(1)	Ability to enjoy natural objects and phenomena and to enjoy science	4.8
(2)	Ability to observe natural objects and phenomena	4.6
(3)	Ability to focus and work on a single topic	4.0
(4)	Ability to maintain focus over a period of time and work diligently on a single topic	3.6
(5)	Ability to examine things that seem mysterious or that they don't understand	4.6
(6)	Ability to create a plan on one's own and solve a problem	3.6
(7)	Ability to think theoretically and organize their thought	4.1
(8)	Ability to carry out calculations and create a graph	3.9
(9)	Ability to think about things in a multifaceted and flexible way	3.8
(10)	Ability to operate lab equipment and use measuring instruments	4.1
(11)	Ability to convey their ideas and what they have learned	4.1
(12)	Ability to understand natural objects and phenomena scientifically	4.1
(13)	Ability to cooperate with friends to solve problems	4.0
(14)	Ability to actively challenge themselves to do new things	3.9
(15)	Ability to come up with new ideas and recognize other ideas	4.1

I asked the children's guardians which learning activities the children particularly found interesting and why. Their answers are organized in Table 10.2. Each course had some children who found it interesting. Some children were most interested by the independent research presentations.

In this class, my child discovered how fun and mysterious science and experiments are and learned scientific methods and procedures. My child also learned how to interact and cooperate with other children and learned methods and approaches for presenting experimental results and summaries.

The preparation of notes and files also gave my child an opportunity to use kanji characters and letters of the alphabet. Now, my child has started looking up kanji in the dictionary whenever she writes something → Japanese, English.

Braille → integrated study graphs → arithmetic

Evaluation of food nutrition → home economics

The source of the water we use → social studies

As indicated above, not only did my child learn about science, she also learned about a wide variety of other topics, so this was a very good experience for her.

These past 4 months have been very enriching for both parent and child alike. Thank you very much. (From a guardian whose child is a boy at grade 1)

After taking the courses about water, my child started paying much more attention to the changes in the water taking place all around him.

(continued)

For example, everyday, he checks the percentage on the hygrometer and says things like "it's dry again today" and "it's humid today, so it might rain." He's trying to think of ways to get rid of the fog on the mirrors in the bathroom. When hail fell, he happily observed: "It's falling as a solid." And when he saw that the pond in our garden had frozen, he took a picture to show how it freezes from the surface of the water.

When we eat or have snacks, he also asks how many calories are in our food, and he's been concerned about our red-colored food consumption. It's not really pleasant to have the meals you prepare scrutinized, but it's clear that the desire to grow healthy and strong has taken root in him. (From a guardian whose child is a boy at grade 1)

I believe that being exposed to science has had a significant effect on how my child thinks about the world around him. He's started thinking more about the reasons behind things and tries to come up with his own solutions. As his parent, I'm very happy to see that he's looking at the same things that he used to look at before in a new way.

I'm very grateful that my child was given this wonderful opportunity at such an opportune time in his life. (From a guardian whose child is a boy at grade 1)

Partially, because she has two brothers who are in elementary school, Sayo has always wanted to go to elementary school as soon as possible. This program gave her the opportunity to skip all the way to college. She was delighted and always looked forward to going to the courses.

Raising her hand to answer questions from the teacher, wearing a lab coat, and using scientific tools were all fresh experiences that filled her with joy.

I think the experience of taking notes in her own childish way and summarizing the results in tables and graphs helped her develop important learning skills. However, I regret not being able to sufficiently develop those skills further at home, and I feel like I've let the teachers down.

For example, I would call to Sayo while looking at a book, but she would reply unenthusiastically: "That's boring." I wasn't able to draw out the enthusiasm from her that she showed in the classroom. However, when I helped Sayo with her research presentation, I was surprised by her ability to observe phenomena and classify and organize results. I believe that her ability to observe phenomena objectively and express her observations in words marked a tremendous improvement. Neither we (her parents) nor her school would have been able to provide her with this opportunity on our own. Even looking at the presentations of the other children, I could see how wonderful they all were, and I was very impressed by the children.

If an opportunity like this comes up again, I would love to participate. I also feel that it would be wonderful if this program could be shared with as many children as possible. Thank you very much. (From a guardian whose child is a girl at kindergarten)

Table 10.2 Guardian responses regarding what courses children found particularly interesting and why

Course	Guardian response
Project 1-1 The Job of a Scientist	This course, which was the first step to teaching the children what the program is all about, really seems to have grabbed my child's attention
Project 2-1 Water Transformation!	My child appears to have been deeply fascinated by the way in which water changes its shape in a variety of different ways and the way water connects with nature. After this class, my child's thinking had broadened, and he began looking at clouds, pots with food cooking in them, and other things more carefully
Project 2-2 Sugar Disappears in Water?	My child was surprised by the amount of sugar that can be dissolved in water. "So how much sugar is dissolved in the sweet drinks at the coffee shop? How much salt is in miso soup?" My child started asking questions like these. This course opened a door to a world of curiosity; realizing that there are many things that are dissolved in many other things, my child started wondering how much of what is dissolved into what
Project 2-3 Water Circulates	My child appears to have been surprised that his predictions were way off. He was also delighted to be able to accurately confirm the truth
Project 3-1 Let's Grow Healthily!	Learning about nutrients and making a menu were apparently a lot of fun for my child. He also said that he'd like for the elementary school menu to be color coded. Up until now, he didn't have any interest in the menus that were passed out by the elementary school. Now, he's interested in the menus, and I can see how the course has had an impact on his daily life
Project 3-2 Let's Feel Our Bodies	My child appears to have been very interested in the investigation of her palm. Examining her own sensations appears to have been a very novel experience for her
Project 3-3 The Body Is Always Moving	In the activity in which the children moved around and confirmed changes in heart rate, my child appears to have enjoyed the act of moving around and measuring in and of her friends and was thus able to participate enthusiastically. Also, hearing the heartbeat of a guinea pig had a big impression on her. She enjoys interacting with animals, so she was very satisfied
Independent Research Presentations	My child enjoyed doing research by herself, and she enjoyed the progression of her research and presenting her research. She also said that it was fun to watch her friends' presentations

Conclusions

The ability to express one's knowledge of natural objects and phenomena and to share that knowledge with others is a unique advantage of being human. Matsuzawa (2002) compared performance on cognitive problems related to social intellect among primitive primates. He then highlighted characteristics that can be seen in chimpanzees and *Homo sapiens* but cannot be seen in monkeys, such as the ability to use social tools and to look others in the eye in a friendly way, noting that "make-believe play, which involves pretending or make-believe," and "teaching, which involves actively educating another being," are distinctive characteristics of *Homo sapiens*. The fact that humans, unlike other living creatures, can create intellectual theories has been extremely advantageous in identifying causal relationships that exist in the world. The ability to express these, share them with others, and accumulate relevant knowledge is a distinctive aspect of human wisdom. The key feature of a culture of ideas is communication based on ternary relationships that include other people and things, is unique to humans, and appears suddenly at about 9 months after birth. An important factor in engaging those young children is the active creation of scaffolding by other people, the daily accumulation of participatory behaviors, and a nurturing human environment that foregrounds education (Syowa, 2010).

Based on the latest trends in education research related to the proficiency of young children in understanding nature, the Kids Academy Science Program is an early STEM education program that has been developed to offer an experiential learning environment that expands, intensifies, and develops the intellectual curiosity of young children into scientific exploration and thought, going beyond the dichotomy of experience-oriented versus knowledge-cramming-oriented approaches to science education. Children who participate in the programs ranging from Theme 1 "Welcome to the Kids Academy Science Program" to Theme 3 "Human Body Project," and who participate in the question and answer session conducted during the presentation session, receive a completion certificate, so increasing their sense of self-accomplishment and their sense of self-efficacy. The program received higher than expected ratings from parents after its initial implementation with regard to the attitudes of children toward the program and its effectiveness, and the programs are still under development. The young children who participate, who are predominantly first graders, participate in an intensive two-hour session, get carried away with their activities, and conduct highly individualized exploration activities with designing experiments and making things. This approach is starting to draw the attention of newspapers, television, and other media outlets.

Our twenty-first-century society is a knowledge-oriented society in which new knowledge, information, and technologies are rapidly increasing in importance because of their role in laying down the foundations of activities in all kinds of social arenas, and science is the driving force behind this trend. As scientific knowledge becomes increasingly globalized, it is also undergoing constant reform and innovation. All people, regardless of gender or age, have some connection with science and technology. It is important not only for future scientists but also for other nonprofessionals to acquire and share sound, basic scientific knowledge, and attitudes starting in childhood. STEM activities are extremely well suited to the integrated play that is characteristic of childhood, the learning that occurs through the natural environment as experienced in everyday life, and collaborative learning with others. STEM education programs for young children, like the Kids Academy Science Program, incorporate such elements as (1) the innate capacity of children for learning, (2) the diversity and creativity of learning, (3) the continuity and expandability of learning, and (4) the collaborative nature of learning. High-quality STEM activities starting in childhood, which include elements of innate learning, are particularly important today as they enable children to engage emotionally and collaboratively in persistent intellectual exploration across a breadth and depth of topics, building on the foundations of physical images from their own experience, which may be more pronounced in potentially gifted young children.

References

American Association for the Advancement of Science (AAAS). (1999). *Dialogue on early childhood science, mathematics, and technology education*. Washington, DC: American Association for the Advancement of Science.

Center for Gifted Education. (2008). *Water works – Project clarion primary science units*. Waco, TX: Prufrock Press.

Chalufour, I., & Worth, K. (2005). *Exploring water with young children*. St. Paul, MN: Redleaf Press.

Coates, D., & Wilson, H. (2003). *Challenges in primary science: Meeting the needs of able young scientists at key stage two*. Oxon, OX: David Fulton Publishers.

Conezio, K., & French, L. (2002). Science in the preschool classroom: Capitalizing on children's fascination with the everyday world to foster language and literacy development. *Young Children, 57*, 12–18.

Copple, C., & Bredekamp, S. (2006). *Basics of developmentally appropriate practice*. Washington, DC: National Association for the Education of Young Children.

Eshach, H. (2006). *Science literacy in primary schools and pre-schools*. Dordrecht, The Netherlands: Springer.

Fukada, S., Sumida, M., Masukagami, M., & Sakata, C. (2005). Learning about water in a Japanese kindergarten (Yochien): Two approaches to teaching science. *International Journal of Early Childhood Education, 11*(2), 65–80.

Gopnik, A., Meltzoff, A. N., & Kuhl, P. K. (2001). *How baby think*. London, UK: Phoenix.

Gopnik, A. (2009). *The philosophical baby*. New York, NY: Picador.

Harlan, J. D., & Rivkin, M. S. (2004). *Science experiences for the early childhood years: An integrated affective approach*. Upper Saddle River, NJ: Prentice Hall.

Harrison, C. (2004). Giftedness in early childhood: The search for complexity and connection. *Roeper Review, 26*(2), 78–84.

Katz, L. G., & Chard, S. C. (2000). *Engaging children's minds: The project approach (2nd ed.)*. Connecticut: Ablex Publishing Corporation.

Koizumi, H., Akita, K., & Yamada, T. (Eds.). (2007). *Developing "scientific mind" in early years*. Tokyo, Japan: Shogakukan (in Japanese).

Koshy, V., & Robinson, N. (2006). Too long neglected: Gifted young children. *European Early Childhood Education Research Journal, 14*(2), 113–125.

Lind, K. K. (2005). *Exploring science in early childhood education* (4th ed.). Clifton Park, NY: Thomson Delmar Learning.

Matsuzawa, T. (2002). The evolution of mind. In T. Matsuzawa & Z. Hasegawa (Eds.), *The evolution of minds* (pp. 11–20). Tokyo, Japan: Iwanami-Shoten (in Japanese).

McIntyre, M. (Ed.). (1984). *Early childhood and science*. Washington, DC: National Science Teachers Association.

Metz, K. (1997). On the complex relation between cognitive developmental research and children's science curricula. *Review of Educational Research, 67*(1), 151–163.

National Research Council. (2010). *Exploring the intersection of science education and 21st century skills*. Washington, DC: National Academy Press.
Ritz, W. C. (Ed.). (2007). *A head start on science: Encouraging a sense of wonder*. Arlington, VA: National Science Teachers Association Press.
Sankar-DeLeeuw, N. (2004). Case studies of gifted kindergarten children: Profiles of promise. *Roeper Review, 26*(4), 192–207.
Saracho, O. N., & Spodek, B. (2008). *Contemporary perspectives on science and technology in early childhood education*. Charlotte, NC: Information Age Publishing Inc.
Settlage, J., & Southerland, S. A. (2007). *Teaching science to every child*. New York, NY: Routledge.
Sommer, D., Samuelsson, I., & Hundeide, K. (Eds.). (2010). *Child perspectives and children's perspectives in theory and practice*. Dordrecht, The Netherlands: Springer.
Sumida, M. (2010a). Identifying twice-exceptional children and three gifted styles in the Japanese primary science classroom. *International Journal of Science Education, 32*(15), 2097–2111.
Sumida, M. (2010b). Science education for enriching children's potentials and enhancing their personalities from early childhood years. *Proceedings of the 60th Annual Conference of the Society for Japan Science Teaching*. (in Japanese)
Sumida, M. (2012). The Japanese and Western view of nature – Beyond cultural incommensurability. In P. Theodora (Ed.), *Debates on early childhood policies and practices: Global snapshots of pedagogical thinking and encounters* (pp. 123–135). Oxon, OX: Routledge.
Sumida, M. (2013a). Science curriculum for enriching children's potentials from early childhood years. *Science Education Monthly, 62*, 34–35 (in Japanese).
Sumida, M. (2013b). Japanese early childhood education: Its view of nature and implications for the teaching of science. In J. Georgeson & J. Payler (Eds.), *International perspectives on early childhood education and care* (pp. 243–256). Berkshire: Open University Press.
Sumida, M. (2013c). Emerging trends in Japan in education of the gifted: A focus on science Education. *Journal for the Education of the Gifted, 36*(3), 277–289.
Sumida, M., Fukada, S., Nakamura, H., Masukagami, M., & Sakata, C. (2007). Developing young children's scientific, technological, and social competency through "Pendulum" play activities at Japanese kindergarten. *Asia-Pacific Journal of Research in Early Childhood Education, 1* (1), 83–100.
Syowa, M. (2010). The evolutional foundation of the development of mind and education. In T. Mastuzawa (Ed.), *What is human being? – Perspectives from studies on chimpanzee* (pp. 232–236). Tokyo, Japan: Iwanami-Shoten (in Japanese).
Wellman, H., & Gelman, S. (1998). Knowledge acquisition in foundational domain. In D. Kuhn & R. Siegler (Eds.), *Handbook of child psychology* (pp. 523–574). New York, NY: Wiley.

Chapter 11
Changes in Collaborative Discussion and Engagement with "Smart Technology" in Science Classroom

Suna Ryu and Seoung-Hey Paik

Introduction

Smart mobile devices (mostly smartphones and tablets—mobile devices with higher computing power and Internet access) are now a familiar part of the everyday lives of most students and teachers in South Korea. 49.2 % of elementary, 85.4 % of middle, and 83.7 % of high school students in South Korea have a smartphone (Ministry of Education Science and Technology, 2013). They use their smartphones for text messages, social network services (SNS), Internet surfing, and entertainment (e.g., games). For educational purposes, students state that they use smartphones to search information using search engines or web 2.0 knowledge communities and to watch educational videos. Most of them indicate that they upload and share their own pictures and movies via a UCC service such as YouTube or a popular SNS service. Thus, today's young students search, collect, and consume information and knowledge, but also produce and create knowledge and artifacts (Seol & Son, 2012).

In science education, a number of researchers have studied the use of simulation, computation, visualization, and/or structured communication and collaboration that support science practices. However, though increasing, relatively little research in science education has addressed how the use of smart technology may facilitate science learning and science practices (Ryu, Han & Paik, 2015).

Since this is still just the beginning of using smart mobile devices for teaching and learning, many challenges currently exist. Some researchers criticize current popularity studies that tend to focus on the general capacity of technologies rather than

S. Ryu (✉)
Graduate School of Education & Information Studies, University of California, 1320 Moore Hall, Los Angeles, CA 90095, USA
e-mail: sunaryu@ucla.edu

S.-H. Paik
Korea National University of Education, Cheongju, Korea

M.S. Khine (ed.), *Science Education in East Asia*,
DOI 10.1007/978-3-319-16390-1_11

their actual learning practices mediated by technology (Pachler, Bachmair, Cook, & Kress, 2010). Thus, they argue that more research needs to be done about how the specific functions of these devices are coordinated with discipline-specific learning practices (conceptual, epistemic, and social) and how using these technologies may or may not support teaching and learning. Increasing numbers of researchers also argue for the need for research about how such educational practices with advanced technology must be considered and reflected based on our current understanding of how people learn and develop (Roschelle, Pea, Hoadley, Gordin, & Means, 2000; Roschelle, 2003; Stahl, Koschmann, & Suthers, 2006). Indeed, a number of existing proposals and reports often provide the hypothetical potential of using smart devices in classrooms, but not necessarily with evidence-based investigations, and often present increased interests or motivations as the only evidence to support learning. Researchers are thus concerned that no widely accepted learning theory that includes the capability of smart devices could hamper the effective use of these technologies for learning as well as for improving the design of a learning environment (for review, see Pachler et al., 2010).

In response to these great potentials but also with challenges, this chapter addresses how specific capacities of using smart mobile devices support students' engagement in science practices and science learning both in an elementary and a high school science classroom. In particular, We focus on how the use of mobile technology can facilitate collaborative learning and argumentation by acting as a resource that supports engagement and participation.

The chapter begins with a background of the use of smart technology in education, including the definition of what "smart technology" really means. Theoretical and pedagogical perspectives on using smart technology in science education are reviewed, from the viewpoints of situated learning perspectives and knowledge-building communities, with a focus on building new agencies and identities in learner-generated content and contexts. Two case studies conducted in South Korea are presented next. One study examines the ways in which incorporating digital technology into science classrooms contributes to improved communication and participation between teachers and students and among students themselves. This issue of unbalanced or limited participation has been a commonly recognized issue in classrooms in Korea, as well as other East Asian countries. For example, by allowing individual students to send anonymous questions to the teacher via text messages, students with low performance scores overcome their fear of embarrassment and may be more willing to ask questions, improving their overall self-esteem. The classroom can also share instructional materials (pictures, video clips, Google docs, and presentations) with other classrooms located in rural areas, and they can share and exchange ideas to improve one another's knowledge; students could also develop reflective thinking as well as collaborative argumentation skills. The second study examines how students in a high school afterschool science program voluntarily use their mobile phones in the context of modeling activities. In particular, the analysis is focused on how searching information can be a catalyst for students to engage in conceptual discussion and argumentation. The chapter concludes with a discussion of the implications of integrating these technologies into science practices.

Defining Smart Technologies: Mobile Technologies with Networked Technologies

The term “smart technology” needs to be explicated within the context of education. Although different researchers have proposed different ideas in Korea, smart technology conventionally refers to the technical capacities of smart devices with higher computational powers, including touchscreens and networking (Seol & Son, 2012). Smart technology also includes functions that support ubiquitous computing and that use artificial intelligence (e.g., Siri, OK Google). Human-like ways of communicating with devices, such as touchscreens, voice activation, handwriting recognition, and multimodal interfaces (gestures and sensory uses), are identified as distinct characteristics. Compared to PDA, some researchers focus on the nature of communication as a distinctive characteristic of smart devices as the origin comes from the conversion that occurs between email and phones. Commonly, iPhones, Android phones, iPads, and Android tablets are referred to as smart devices. Simply speaking, learning that uses these smart devices is called “smart learning.” Smart learning also includes the use of embedded, ubiquitous, and networked capabilities and technologies that enable individually customized learning as well as collaboration. In South Korea, the Ministry of Education Science Technology (2011) synthesized and characterized five important properties of smart technology for education. The properties are (1) self-directed, (2) motivated, (3) adaptive, (4) providing resources, and (5) technology embedded. While mobile learning traditionally focuses on individualized learning on the move, thus learning with movable, portable, and personal technology, recent development of smart devices with advanced computing power and network capacity has moved the locus of smart learning toward collaborative and participatory learning for co-knowledge building.

Smart technology may provide diverse advantages for teaching and learning in science education, which includes the advantages of previously studied mobile and networked technologies. With respect to the advantages of mobile technology, Klopfer and his colleagues (2002) identified five properties of using mobile devices that support learning, such as portability (lightweight to carry around), social interactivity (real-time conversation and data exchange, face-to-face discussion), context sensitivity (gathering data and having conversations in the current location, environment, and time), connectivity (creating and sharing a network), and individuality (customized individual help). For networked technology, Kyza (2013) identified the advantages of network technology for teaching and learning, such as fostering students’ active and reflective engagement, providing opportunities for collaborative learning, and engaging students in authentic contexts and just-in-time feedback. Synthesizing these advantages, smart technology may enable learners to have access to relevant information, promote active socializing, and customize individual scaffolding.

Smart Technologies in Science Education for Fostering Collaboration and Argumentation

In science education, the potential of using Internet-enabled mobile devices for classroom learning has been studied as a means to support students' engagement in authentic science and collaborative learning. Several handheld computer projects, conducted by the center for Highly Interactive Computing in Education (Hi-CE), reported that handheld computers successfully supported students' science learning by providing immediately accessible functions, including note-taking, data collection, and concept mapping (Tinker & Krajcik, 2001). By using these functions, students expressed greater competence with technology and showed enhanced understanding of the relevant scientific concepts. These handheld computers also included designed scaffolds. For example, Pocket PiCoMap, a concept map tool, provides map notes that enable students to give and receive feedback, while the node notes tool allows students to articulate abstract ideas. Research on PiCoMap, which supports students' concept mapping, also found that, through such technological scaffolding, concept mapping—typically done as individual work—becomes collaborative work entailing rich reflection and discussion. For the Hi-CE research, students first created their own maps to show their understanding of weather and then shared the maps with partners. Through the feedback function via map notes, students critiqued one another's concept maps and collectively revised and improved their own. While doing so, students also demonstrated an enhanced conceptual understanding of weather systems. More recently, the MyArtSpace project (Sharples, Arnedillo-Sánchez, Milrad, & Vavoula, 2009) allowed students to send photos, audio recordings, and notes taken at a museum to a website via a mobile phone application. Students shared and later discussed these artifacts in the classroom. In doing so, they connected learning in the classroom with learning at the museum, in addition to having a greater opportunity to communicate with other museum visitors by giving and receiving feedback.

Collaborative discussion and argumentation have been the focus of computer-supported collaborative learning (CSCL). Thus, literature about CSCL could inform the directions and ideas of smart mobile learning. Collaborative discussion and argumentation are viewed as keys for learning critical thinking, elaboration, reasoning, and reflection. Different from the common idea of argumentation (to prevail over an opponent's argument), collaborative argumentation highlights the evaluation and critique of claims with supporting evidence in consideration of counterarguments to reach a consensus for the best argument. This is particularly important practice for science because scientists engage in such argumentation to establish new theories over time. For example, sending anonymous questions or opinions to teachers help students to propose and present their ideas and viewpoints, whether or not all students are comfortable with presenting them. This feature could be beneficial for argumentation because being comfortable with proposing ideas and disagreeing with others are fundamental conditions for engaging students in argumentation. Because students are often afraid of proposing ideas (they worry about their answers being incorrect or about being embarrassed by others), such anonymous engagement could encourage students to step further into the mode of argumentation.

The portability of mobile devices with network availability also supports real-time discussions on topics. The feature of real-time discussion is important because students often forget to keep and record relevant questions or ideas when they return from a computer lab activity. In this sense, their activities in the computer lab and their activities in the classroom are not connected well. When students use a smart mobile device, they can engage in discussion as part of what they do with technology. This also helps students to collaborate to answer questions raised during an activity. For example, in studies that use wearable computers in participatory simulation (Colella, 2000), students were more likely to raise relevant questions followed by active discussion compared to students in a control group who did a similar activity in a computer lab and had the chance to discuss it after returning to the classroom.

The properties of mobile smart technology described above have the potential to support collaborative discussion and argumentation in science classrooms. Clark, Sampson, Weinberger, and Erkens (2007) categorized different functions of these technologies for promoting argumentation, such as accessible knowledge bases, asynchronous communication, enriched representations of subject matter, visualization tools, and monitoring of cognitive moves. Scheuer, Loll, Pinkwart, and McLaren (2010) developed two different categories: individual and collaborative technologies and supportive and educational technologies. Technologies used in individual functions focus on the presentation of the "rules" of argumentation. In contrast, technologies used in collaborative functions focus more on social interactions and promoting dialogical argumentation. Regarding the differences between supportive and educational functions, educational software provides specifically designed functions that help students learn about a specific knowledge domain. From these reviews, three shared focuses are identified: (1) allowing students to have access to information and knowledge, (2) providing structural scaffoldings that guide the rules of argumentation, and (3) promoting engaging social interactions.

Mobile technology also provides individually customized help, which in turn encourages students to participate in collaborative learning more actively. For example, when students visited a butterfly farm in Taiwan, an individual student was able to use a PDA (with a camera) to take pictures and then identify the species of butterfly by comparing their own pictures with the pictures from a database (Chen, Kao, Yu, & Sheu, 2004). Individual students were also encouraged to record their observations by writing a journal. In this process, students independently seek useful information for recording observations and identifying the species of butterflies. After this, students as a group decide which species is the best match based on their journals.

Authoring and Enacting Identity and Agency Through the Mobility of Contexts

Social Interaction with Technology

As described earlier, a growing number of researchers point out the need for elaborated theoretical and pedagogical perspectives in the use of advanced technologies for education. As part of these efforts, increasing numbers of studies in CSCL and ICT education have paid greater attention to the sociocultural aspects of using technology for education (for review, see Pachler et al., 2010). One of the key findings from this line of research is that technology on its own is not sufficient to assist with the building of a knowledge community. Rather, the building of a social infrastructure combined with layers of a technical infrastructure seems to be essential. In other words, how technology is used is influenced by the nature of social practices and participant structures within communities just as technology changes the ways in which people interact and engage in different practices. Thus, the use of smart technology must be involved with a broader understanding of the context of learning. Traxler (2007) pointed out that all that is good about technology becomes a pedagogical tool when the capabilities of different technologies are woven into social practice. Thus, in order to see whether and how students engage in learning with the use of technology, it is essential to observe and trace students' engagement in practice using smart technology.

Those who are interested in sociocultural orientation in learning with technology have shown that the new relationship between knowledge and learners enable them to establish agency, especially with regard to their knowledge. The social theory of learning (Wenger, 1998) and the notion of social practice theory provide an idea to characterize the use of smart technology as social participation and a way to enact agency and identity for knowledge. From this perspective, learning is defined as a meaning-making appropriation process of practice within a community. For learners, changes in participation are interrelated with the changes in who they are (identity) and how they author their acts (agency) as they engage in the practice of community.

Scardamalia (2002) defined epistemic agency as a collective cognitive responsibility for knowledge building. Using a Computer-Supported Collaborative Learning Environment (CSCLE), students coproduced written explanations regarding many different topics. They also shared and utilized the CSCLE as a knowledge resource. Students acknowledge each other's explanations, share and critique opinions, and produce further explanations in a synthetic way. Through the coproduction of knowledge, and through the exchanging of comments or acknowledgments, students show ownership and responsibility for their work. Since these kinds of interactions (knowledge sharing and critiquing) seem to help students to expand their responsibility and ownership, authors argue that epistemic agency emerges socially.

Barton and her colleagues (2010) investigated the development of agency when students participate in an after-school science program involving youth and digital culture. These students produced a short movie that consists of interviews and their own investigations regarding climate change in their community. Through this activity, these students shifted their roles as consumers and as passive recipients to those of producers and critics of scientific knowledge. The students explicitly called themselves "community experts" in their video. These students acted as local scientists. They became involved in conducting literature reviews and scientific investigations on global warming issues, especially on the topic of the UHI issue. However, what seems to be most interesting to students is the opportunity to produce their own movie. Instead of "let's pretend to be scientists," they were in a position to produce a real artifact for the world as community science experts. This changed their learning context and helped them to situate themselves as knowledge producers and local experts on science. Therefore, it seems that the specific contexts and situations in which students engaged themselves were due to the creation and production of YouTube movies. Just as students are positioned as consumers and recipients of science, an individual student usually tends to be taken as a consumer and a passive recipient of digital media. However, since they can now create and publish their own media content, all processes become much easier and more accessible. For example, just a few years ago, one took a picture or recorded a video by using a digital camera or video camcorder. If one ever wanted to edit these, the pictures and movies needed to be transferred to a computer. One also needed relatively expensive software such as Photoshop or Premiere for editing. Finally, the movie or picture was published on an individual website or blog. However, nowadays with smartphones or tablets, these processes can be done on one device almost instantly. It is also possible to share and work together (e.g., with Google docs) to accomplish these tasks. For younger students, the capability to use digital inking and sketching on touchscreens (using a digital pen to draw pictures and take notes) facilitates their ability to take notes, to draw pictures, and to annotate information and share it with others.

This capability not only enables students to create and produce their own content; it enables them to change their learning contexts, which helps them to reconfigure the relationship between them and their learning and ultimately to enact agency and identity. By changing the learning contexts, one may first think about overcoming the limits of physical and time barriers. However, this also enables learners to establish a new social position in which they take full responsibility and authorship for their work (especially by recognizing their potential audience in the world and trying to share not only knowledge but also emotion and empathy). Traxler (2007) argues that mobility is not just a property of devices; rather, mobility is the emergent property of the interactions between people and technologies. In this sense, mobility extends its meaning from the portability of a device toward the mobility of people in physical, conceptual, and social spaces. With this extensive notion of mobility, Traxler argues that mobile learning is "uniquely suited to support context-specific situated learning," meaning that context is everything when people with smart devices overcome the constraints of physical, conceptual, and social space.

The context-aware technology of smart devices, often called "localized computing" in commercial terms, gathers and provides information and measures what is currently available around users and their devices. A smart device recognizes the command "find a Chinese restaurant nearby" and shows restaurants located within a close proximity. In an educational context, a group of students who live in a rural area might take pictures and investigate several characteristics of diverse local plants and animals. Their smartphones diagnose the location and show information about local plants, insects, and animals, along with other related data, so that the students can identify the species of what they observed and then gather further information about them. On the other hand, students who live in a city away from this rural area are able to share these experiences using a web video-conferencing function on their smartphones. They can ask questions, share information, and co-develop a report together as a group. For the local students, their learning activity was more likely a context-aware activity. For the urban students, their learning activity could be called a situated activity because they feel like they experienced and shared the same quality of activity as the local students did and because all experiences took place at the same time.

Smartphones with Mobile Internet as a Relevant Resource

The notion of relevant resources, which support productive disciplinary engagement (Engle & Conant, 2002), seems suitable to explain the benefits of using mobile technology for active argumentation. They suggested four principles to support productive engagement in classrooms, including problematizing, authority, accountability, and relevant resources. Problematizing means that students themselves generate their own questions, and they propose and challenge ideas rather than simply answering questions from teachers and memorizing facts. Authority means that students are empowered in their knowledge-building activities, and so they take an active role. Accountability means that students ensure that their knowledge-building work is responsive to the rules of the relevant disciplinary community (i.e., respecting epistemic criteria). The provision of relevant resources is meant to provide time and materials that support student engagement.

These authors argued that, when students are productively engaged, they are able to provide more evidence-based argumentation. Thus, these principles of productive engagement seem to work as important conditions to enact active argumentation. In this study, students have to use and critically reflect upon conceptual knowledge in order to build a model (problematizing). Through the process of discussing and presenting the model to their peers, and convincing them of its integrity, they bring about their epistemic agency (authority). They want to show others that they are responsive to the epistemic criteria of the scientific model (accountability).

Among these four principles, the authors highlighted that providing resources is the fundamental condition needed to embody and enact other principles. For example, they recognized that securing sufficient time and having access to information were relevant resources for problematizing. The building of a public platform and having access to discipline-specific experts are considered resources for fostering authority and accountability, respectively. This study uses smart mobile devices as relevant resources, as they enable students to access information: students seek information by using web-based search engines. Smartphones may also constitute resources for authority and accountability because real-time discussions with the use of smartphones may encourage students to actively raise and discuss questions, which can play a role in the manner of a public forum. The use of question–answer communities (e.g., Yahoo! Answers) allows students to have access to discipline-specific experts who can be resources for authority and accountability.

Case Study 1: Connecting Three Science Classrooms in a City, a Farm Village, and a Fishing Village

"Classting across three villages" was conducted by three 4th grade elementary school classrooms in South Korea. One classroom was located in a big metropolitan city (24 students). Another classroom was located in a southern farm village (19 students). The third classroom was located in an eastern fishing village (18 students). Because these classes were part of the national SMART science initiatives pilot (i.e., a national digital textbook project), each student had a table PC. Learning activities were fully developed through collaboration among the three teachers as part of their lesson study.

These three classes used Classting (Cho, 2013), a classroom SNS application developed by a former elementary school teacher in Korea. Using Classting, it is possible to share notes, photos, and videos with classmates and parents by using smart devices as well as computers. Students also can send private or anonymous messages to their teacher. Classting consists of a "wall," where members post their opinions or thoughts, a "news board" for announcing classroom news or information, and a "photo album" and a "video room" for sharing pictures and video clips.

A learning group for social studies and science was generated, consisting of 4–6 members. The members of the learning group came from all three classes: for example, 1–2 students each from the city, farm-village, and fishing-village classes. The group worked together on a series of combined-inquiry topics in social studies and science regarding the characteristics of objects, minerals, animals, plants, and places and histories. According to the teachers, the topics were selected by considering local availability, potential interest, and content difficulty. For example, students investigated local animals and reported similarities and differences among them. The group members shared reports and notes, and then they presented their results as a group. The three classes arranged a few opportunities for real-time video conferences. Each class also uploaded videos and shared comments.

Helping Behaviors, Developing Reflective Thinking, and Collaborative Argumentation

Giving and receiving help is an important condition of collaborative learning (Webb, Nemer, & Ing, 2006), which benefits both help providers and receivers. In explaining something to others, help givers have the opportunity to engage in knowledge building, to develop deeper understanding by reflecting their knowledge, and to become more motivated to learn because of its expectation for helping others. It would be a little more complicated for help receivers because help only works when receivers genuinely need help, and the given help needs to be relevant and appropriate.

Students in these three classes benefited from giving and receiving help. Even though there was no explicit demand from teachers, these students spontaneously shared relevant pictures, video clips, and information with explanations. When questions or comments for these resources were posted, answers and other related information were provided in response. Students recognized that sharing information and providing questions and answers were helpful for both their collaborative project and themselves. For example, a student from the city school noted that she could concentrate on searching for relevant information and reading relevant magazine articles with less distractions because she was about to share this information, which she expected to help others. A student from the fishing village said that he could enjoy reading school-related materials because he knew that other friends shared these materials with him, and these materials were relevant for their mission project.

At the beginning of this project, the academic achievement gap between students in the urban and rural areas was a potential concern to the teachers. However, using smart technology and Classting seemed to resolve this issue by filling the gap through helping behaviors such as sharing information and answering questions. Moreover, students' sharing of information and Q&A seemed to help them improve their reflective thinking. Rodgers (2002) pointed out that reflective thinking is more likely to occur when one has a chance to explain ideas to others as they interact with others to achieve the same goal together. Because the students as a group shared the same mission, they tried to search relevant information, share it with others, and provide explanations. In this process, students could think about a topic in greater depth.

Students' helping behaviors (sharing information and Q&A) also contributed to the engagement of active collaborative argumentation. By "collaborative argumentation," the aim of engaging in argumentation is not to win over opponents verbally; rather, the aim is to reach a consensus by examining the quality of the claims, evidence, and justification. For example, scientists constitute and critique each other's arguments by examining evidence and then agreeing upon the best theory.

Sharing new information and discussions about topics provided students with even richer topics to argue about. Although it seems to be taken for granted that students' learning tasks should have something to argue and debate, often the information presented by their textbooks is rather straightforward. However, what students observed from local sites motivated them to engage in further questions and argumentation.

Many ideas and questions are grounds for fostering argumentation. It seemed possible to form a number of interesting questions and ideas because the students were excited about the new and interesting data they collected. Instead of sitting in a science lab or classroom, the students recorded and collected their data from outside the school. They collected a relatively richer set of data (though not all of them were relevant). For example, when they investigated local trees, they not only recorded information about the trees but also investigated the directions of tree branches and sunlight, the temperature, the color of the sky, and the color of the soil. This was possible for them because their tablet PC had apps such as a compass, thermometer, camera, and note-taking/editing capabilities. They also instantly shared their data by uploading their investigations to the group notes section of Classting. Comments and questions regarding these investigations trigger students to engage in argumentation. For example, because students' theories drew upon relatively narrow local data, other students often opposed them and provided a counterargument. For example, a group argued that pine trees in the farm village are taller and stronger than those in the city, based on their investigation. The group provided an interesting theory that the trees were taller and stronger in the farm village because they have been used as firewood. However, the argument was opposed by others because even taller pine trees exist in the city; they were also skeptical about the causal theory between taller trees and firewood. After the teachers saw this interesting argumentation among the students, one teacher asked students to look for Lamarckism, the theory of use and disuse. Later in the semester, the group who originally proposed their theory on taller trees and firewood referred to Lamarckism and reflected on why their original theory was wrong.

It is also important to note the characteristics of their mission to foster argumentation. While big topics such as objects, plants, animals, and comparisons among the three local sites were given, other details were open for students to decide. Different groups worked on different topics and data. Such collaboration on this kind of ill-structured task seems to provide students with more opportunities to make collective decisions and to argue. In addition, the local data that they collected brought high authority and affection (for rural students, the information belongs to the area where they live).

More Opportunities for Participation and Better Communication

Students could send their questions and comments to their teacher. By using this question function, those who were not comfortable (some of them were shy, and some of them were afraid of being laughed at by their friends) could ask questions and provide comments. These became opportunities to provide their ideas to the class because teachers often introduced these questions to the whole class. When these were positively recognized, these students were more confident in asking questions and proposing their ideas, resulting in better participation in class. Teachers often manipulated their participation by encouraging less active students to answer questions or to provide ideas. However, when these students could not answer the questions properly, they could lose their confidence and willingness to participate. Moreover, this kind of encouragement, different from the original intention of the teachers, could be seen as enforcement by students. Instead of being forced to participate, by allowing students to send questions and comments anonymously, they could become more willing to participate while gaining confidence.

Sending questions and comments individually acts as customized assistance and scaffolding and could facilitate better communication. Scaffolding is, in general, defined as instructional assistance given to perform a task that is otherwise difficult to achieve. When students send their questions and comments, some teachers provide hints or send Internet links to useful websites instead of giving them direct answers. In addition, students who have difficult issues related to family or friends could arrange consultations with teachers, which becomes a hint for teachers to facilitate classroom communication and classroom disciplines.

Authoring and Enacting Agency and Identity

In this study, the mobility of smart technology helped students to connect different contexts in two ways. First, they could connect their everyday life with their science class. Second, they could expand such connections to different places across urban and rural areas. Through their collaborative project, students seemed to successfully combine what they directly observed and collected from local sites with information searched using the Internet and textbooks. By doing this, students were able to generate and produce meaningful knowledge for their community. Students also reflected on their theories and improved them. While students in Barton's study explicitly identified themselves as community science experts, these elementary school students did not explicitly call them as scientists or experts. However, modifying theories based on reflections of scientific knowledge is beyond what students usually do in their science classes.

Especially for those who live in rural areas, this project became an opportunity to reconfigure the relationship between their school and their everyday lives. Although students who live in rural areas tend to think that what they learn at school is not relevant to their current or future lives, this project enabled them to develop a new conceptual frame that connects their schoolwork to their lives. By doing so, they seemed to be able to activate their agency for this connected version of knowledge, which might help them to enact their new identity just as students in Barton's study did.

Case Study 2: Using Mobile Internet as a Relevant Resource for Modeling Practices and Argumentation

The second case study ("smartphones for modeling practices and argumentation") investigated how mobile Internet and smartphones are used as relevant resources that facilitate collaboration and argumentation. The study was conducted in an afterschool science/math class of 16 tenth graders (8 boys and 8 girls) on a small college campus in South Korea from early March to late July (a typical Korean semester). The classes took place once a week for a period of 2 h after regular class time. The general mode of instruction was group work, after which each group presented its experiment's designs, models, and results to others. Students generally seemed to enjoy these opportunities, and they became more eager to participate in discussions as the class progressed. Other than oral presentations, students often brought self-built models to facilitate their explanations.

Among 16 students, 12 had smartphones, whereas another four had a previous generation of mobile phone. In a group that consists of four students, at least one or two students have smartphones. While the school provides free Wi-Fi service in the library and the labs, most students prefer to use personal 4G services because of the faster Internet speed. No particular learning software or websites are provided or specified. Students freely use the available Web 2.0 tools to seek necessary information. A representative characteristic of Web 2.0 is openness to participation in which users generate and evaluate knowledge and information. As for mobile applications (apps), in the recent past, game consoles, cameras, presentation tools, graphic or word processors, and independent software needed to be purchased; today, these features are already part of mobile devices (e.g., smart phones). Moreover, many apps are free to use, and they are easily downloaded from Google Play stores. With respect to popular Web 2.0 tools, Naver Knowledge-iN (KiN), the Educational Broadcasting System (EBS) forum, the DC-inside science forum, and the Biological Research Information Center (BRIC) are popular knowledge communities. KiN is a popular question–answer community: a user (questioner) posts a question under a relevant topic or category and the question remains open to receive answers. Any user can answer the question. However, because answerers are given a degree of expertise that is derived from questioners' votes for the best answer, questioners tend to prefer highly ranked answerers. EBS provides a number of educational resources related to science, math, and English, and forums play a role similar to KiN. While BRIC serves as a resource center for biologists, it has also been recently perceived as an expert resource for adults and secondary students.

Mobile Internet as a Relevant Resource for Facilitating Collaborative Argumentation

Students use mobile Internet to identify, screen, and answer questions regarding scientific concepts. They also try to use searched information to explain their ideas and then compare and negotiate ideas together. For these students, just using mobile Internet and searched information does not address their questions, because, when they are unable to explain it well, the information is not yet meaningful to them. However, the searched information at least enables students to continue to explore various concepts and to discuss them in depth. Students discuss what to search, compare different information related to a question, and discuss various interpretations and understandings. In doing so, students can understand a concept in greater depth, which also leads them to try to understand other related concepts. Discussion of which words, phrases, or sentences they want to search helps students to organize their ideas coherently and make their conceptual questions clearer and often helps them to develop more sophisticated questions. On the first try, students tend to try a few keywords or terms without conceptual reflection. However, this gradually becomes a concrete level of curiosity that directly addresses their inquiries about a model. For example, in their modeling activity involving a needle floating on water, they worked on building a model regarding how/why a needle could float. Students first tried "water, needle, relationship" as search words. Searching these words indicated that the students had not developed a concrete question about what to model. Through the discussion phases of choosing words and comparing and discussing information, their questions became more sophisticated such as: "How does a needle float on water?" "How does a tissue help a needle to float?" "How does surface tension help a needle to float?" One reason for more frequent usage of the Internet was in part because there were several scientific concepts and terms that they nominated as relevant concepts for their model making. Typically, when students search for the definition of a term or concept, there are other related terms or concepts that they are also unclear about, which prompts another search. Even for the same information given by a search, students often interpret the information differently. Consequently, students are more likely to engage in negotiating meanings to reach an agreement. Moreover, even when students were successful in searching for the definitions of terms and sharing them with others, this did not mean that students were able to provide successful explanations by connecting the searched information to the phenomenon being studied.

Students have easy access to many relevant concepts, which help them to understand better and to stay engaged in discussions. While modeling practice requires students to build a conceptually coherent relationship between the concepts used and the target phenomenon, it's not often easy to make their thoughts and concepts visible or clear. Moreover, when they face conceptual challenges, it is often hard to resolve such challenges if there is little or no assistance from the teacher. Providing appropriate assistance is difficult because such help needs to be provided at the appropriate time, and the level of it also needs to be appropriate. Not providing enough assistance to aid understanding, or providing too much or too difficult information, hinders students' ability to understand something in depth. When students found some terms or questions difficult to understand, they easily found the necessary information on the web. Though this kind of information-searching behavior seems very easy and simple, the selection of the searched words seemed to reflect students' understanding about their modeling activity. For example, a student said, "How about searching 'relationship, water, needle'?" Another student suggested the question "How does a needle float on water?" instead. The group members agreed upon searching the latter because they thought the latter sentence reflected their question more precisely.

Such continuous student discussions facilitated by smartphones can be viewed as negotiating resources. Engle and Conant (2002) highlighted that the provision of relevant resources does not mean a passive delivery of resources from the teacher to the student. To be "relevant" resources, the learning community must negotiate what constitutes "relevance." However, traditionally, these resources are simply presented to students, who do not necessarily know how these resources must be used to promote problem-solving or to establish authority or accountability. Often, this could be a case for the use of technology because students are sometimes simply asked to use technology without a sufficient reason. They usually spend significantly longer periods of time learning the operations and functions of technology. Then, when they become satisfied with learning the technology, they are less likely to integrate that technology into their learning. When using smartphones, students, in most cases, do not have to learn how to operate the basic functions because this is part of their everyday routine. Thus, instead of spending time learning the actual technology, they can think more about how to use it to facilitate learning. In other words, they negotiate how to use resources.

Using mobile phones also contributed to facilitating students' collaboration by engaging students in discussion more actively. More importantly, such seeking, screening, and critiquing behaviors via mobile Internet helped students engage in argumentation: the negotiation of meanings of concepts in order to reach a consensus on the statements or definitions of ideas. Students were thus more likely to engage in deeper-level conceptual discussions. One concern with using Internet searches for terms in science learning is that students might accept information from the Internet without any screening. Moreover, a common concern is that students simply write down what they find on the Internet without engaging in any discussion or reflection. Despite these common concerns, however, there are actually more opportunities for discussion. Thousands of search results pop up that sometimes suggest different ideas. Students also interpret information in different ways. Thus, students in either case have to decide which information they want to use by deciding why they thought certain information was better than others, which requires conceptual and epistemic reflection and evaluation. Such reflection and evaluation leads students to engage in active argumentation. Moreover, they have to convince others why the searched information is best. In the following example, when a student (S8) found "surface tension" to be a relevant concept that explains a floating needle, other students (S7 and S16) in the group asked for further explanations.

S7:	Can you explain this? You know, I have no idea what you meant by surface tension. Explain as if I never heard about surface tension before.
S8:	Look. It's like this. I found a nice explanation from this website. Listen. "Intermolecular force pulls the water molecules together and inward."
S7:	So why do these molecules pull together? And more importantly, why does this happen?
S8:	See, a piece of tissue could float on the surface, so all the water from this part is absorbed by the tissue. After that it will sink, but during the time that the tissue absorbs water, it will float. So. . . no water on the needle because this piece of tissue absorbs all the water. That's why the needle could float.
S16:	Then what about the heavy one? In your theory, the heavy ones should float. And it still doesn't make sense to me. Is this about surface tension?
S8:	Yeah, definitely. Surface tension eventually means that something could float well when the surface area is bigger. . ..
S16:	Let me say this again. In your theory, heavy ones can float better than light ones. How do you explain that?
S7:	[S16], what is this about? Explain it again in relation to what we are doing now.
S8:	Heavy needles are too heavy to float. What about this? The surface of water becomes smooth. . . so the water molecules. . . look like they are smooth in the center.
S16:	Is that because of buoyancy? Or a difference of density?
S7:	What is that? Why do you say that?
S8:	Let's look at this. . . Let me send you the link. Water molecules inside of the water . . .

Perhaps, one reason that S8 could say "nice explanation" for what he found from a website is because the explanation given by the website made sense to him. Thus, from the example above, S8 did not merely deliver what he searched; rather, he made an argument regarding why "surface tension" could be a relevant concept that explains a needle floating on water. Other students also asked S8 to explain how and why surface tension could explain a floating model.

More Opportunities for Participation and Better Communication

Even in small group discussions, it was common that the more academically advanced students dominated the group discussion, especially for conceptual discussions that required content understanding and previous science knowledge. In addition, at high school age, students are less likely to encourage others to engage in their learning activities when some students do not participate.

However, using mobile phones could encourage less academically motivated students to participate and engage in their science activities. In this project, one student (S10) barely participated in discussions in the beginning. Other students in the group began to encourage him to look up related concepts, given the fact that he had a smartphone with a bigger screen and a faster Internet connection (4G) than others in the group. Although he seemed to be bothered by other members' requests in the beginning, he began to look up the information and shared it with others. As he began to engage in the discussion by sharing information, he began to cross the border from peripheral participation into the core features of this practice. That is, while he read material that he searched upon others' requests, he stayed in tune with the following discussion and began to volunteer for searching other information. In doing so, he occasionally provided his own ideas, searched information, and spontaneously uploaded a question to a knowledge portal site. He also once indicated that he looked up further information after the class because a question occurred to him while he was on his way home. It might be difficult to say whether his conceptual understanding or epistemic knowledge about modeling practice dramatically improved through engaging in these activities. However, it is important to note the changes in his participation, which facilitate active collaboration. Similarly, another student (S6) mentioned that he joined this afterschool program not because he was interested in learning science but rather as a way to keep a promise to his parents that he would study math and science more. By mentioning his low interest in this class explicitly, he seemed to indicate his role as a watch-and-sit member. He was using his smartphone for playing games or texting messages. However, other members encouraged him to look up some information with his phone. While he reluctantly searched information and read it to others, he began to provide his comments and ideas and thus became an active participant.

S6:	I found something. They boiled the compounds of water and ethanol, so it's not pure ethanol like ours. Maybe that's why the ethanol in this condition is boiled at a higher temperature.
S3:	It doesn't really matter. As far as I know, boiling point is the unique nature... so the amount doesn't really matter.

In the example above, S6 reflected on what he found instead of merely delivering it. He found that the ethanol used for their experiment was pure ethanol, but the information from the website deals with a compound of water and ethanol.

Typically, one concern with allowing students to use their personal mobile phones is that they might play games and send text messages instead of participating in the classroom. In this classroom, though some students used mobile phones for personal entertainment at first, such distractions were kept under control, not by the teacher's warnings or evaluations but by implicitly shared norms spontaneously generated by the students. For example, when one student was looking at his text messages, another student told him to use his phone to look up "How to make a needle float." In addition, some practical limitations, such as battery life or limited data usage, prevented students from using the Internet for personal reasons and encouraged them to save their phone usage for searching academic information. This could be seen as a collective negotiation of selecting and using resources because personal resources (such as smartphones) can become collective learning resources.

Conclusion

The use of smart technology for learning is continuing to increase. Many countries are using or planning to use interactive digital textbooks. For example, students in Los Angeles received a free iPad preloaded with digital textbooks. Teachers in Singapore received nationally organized training that integrates online educational resources into local classrooms using tablets.

Many researchers and educators are excited about how the use of smart technology may expand learning opportunities by overcoming physical and cultural barriers. This chapter focused on how smart technology may act as learning resources that promote participation and engagement, which eventually help students to engage in collaborative argumentation. The chapter also reviewed how such participation in collaborative learning, where students connect their school experiences with everyday life, might enact their agency and identity. By using their tablets and Classting as tools for data collection, analysis, presentation, and communication, elementary school students overcome physical and time limits and create a new learning context that connects three different places between school and everyday life. Their co-knowledge-building activity enables them to shift their learning from passive acquirement toward reflection. The connectivity enabled by tablet PCs and Classting helps these students to share social situations, to negotiate and make decisions about activities, and to share knowledge together. For high school students, the case study shows how searching information using mobile Internet can be an opportunity to develop argumentation skills regarding science concepts and how less-engaged students become active participants who begin with performing simple searches and reflecting on the searched information.

As highlighted throughout this chapter, the use of technology as a social and cultural mediator can contribute to learning through changes in participation and engagement in collaborative activities. While new functions and capacities of innovative technology are exciting, and future smart technology may provide even more exciting opportunities for learning (e.g., artificial intelligence, ubiquitous computing), more research is needed to increase the use of smart technology as a means to facilitate collaborative knowledge building and social practice in science education rather than as a method for the simple inputting of new technical functions.

References

Barton, A. C., & Tan, E. (2010). We be burnin'! Agency, identity, and science learning. *The Journal of the Learning Sciences, 19*(2), 187–229.

Chen, Y. -S., Kao, T. -C., Yu, G. -J., & Sheu, J. -P. (2004). *A mobile butterfly-watching learning system for supporting independent learning.* Paper presented at the wireless and mobile technologies in education, 2004. In: Proceedings. The 2nd IEEE international workshop on.

Cho, H. G. (2013). *Method of sharing knowledge and information between learning communities: Google patents.*

Clark, D. B., Sampson, V., Weinberger, A., & Erkens, G. (2007). Analytic frameworks for assessing dialogic argumentation in online learning environments. *Educational Psychology Review, 19*(3), 343–374.

Colella, V. (2000). Participatory simulations: Building collaborative understanding through immersive dynamic modeling. *The Journal of the Learning Sciences, 9*(4), 471–500.

Engle, R. A., & Conant, F. R. (2002). Guiding principles for fostering productive disciplinary engagement: Explaining an emergent argument in a community of learners classroom. *Cognition and Instruction, 20*(4), 399–483.

Klopfer, E., Squire, K., & Jenkins, H. (2002). *Environmental detectives: Pdas as a window into a virtual simulated world.* Paper presented at the Wireless and Mobile Technologies in Education, 2002. In: Proceedings. IEEE international workshop on.

Kyza, E. A. (2013). Networked technologies to foster students' collaboration and reflection. In C. Mouza & N. Lavigne (Eds.), *Emerging technologies for the classroom* (pp. 113–126). New York: Springer.

Ministry of Education Science and Technology. (2011). A strategic plan for implementing smart education

Ministry of Education Science and Technology. (2013). Internet and smart phone use

Pachler, N., Bachmair, B., Cook, J., & Kress, G. (2010). *Mobile learning.* New York: Springer.

Rodgers, C. (2002). Defining reflection: Another look at john dewey and reflective thinking. *The Teachers College Record, 104*(4), 842–866.

Roschelle, J. (2003). Keynote paper: Unlocking the learning value of wireless mobile devices. *Journal of Computer Assisted Learning, 19*(3), 260–272.

Roschelle, J. M., Pea, R. D., Hoadley, C. M., Gordin, D. N., & Means, B. M. (2000). Changing how and what children learn in school with computer-based technologies. *The Future of Children, 10*, 76–101.

Ryu, S., Han, Y., & Paik, S. H. (2015). Understanding co-development of conceptual and epistemic understanding through modeling practices with mobile internet. *Journal of Science Education and Technology, 24*(2–3), 330–355. doi:10.1007/s10956-014-9545-1

Scardamalia, M. (2002). Collective cognitive responsibility for the advancement of knowledge. In B. Smith (Ed.), *Liberal education in a knowledge society* (pp. 67–98). Chicago: Open Court.

Scheuer, O., Loll, F., Pinkwart, N., & McLaren, B. M. (2010). Computer-supported argumentation: A review of the state of the art. *International Journal of Computer-Supported Collaborative Learning, 5*(1), 43–102.

Seol, M., & Son, C.-i. (2012). A survey on teacher's perceptions about the current state of using smart learning in elementary schools. *Korea Association of Information Education, 16*(3), 309–318.

Sharples, M., Arnedillo-Sánchez, I., Milrad, M., & Vavoula, G. (2009). *Mobile learning.* Heidelberg, Germany: Springer.

Stahl, G., Koschmann, T., & Suthers, D. (2006). Computer-supported collaborative learning: An historical perspective. In *Cambridge handbook of the learning sciences, 2006.*

Tinker, R. F., & Krajcik, J. S. (2001). *Portable technologies: Science learning in context* (Innovations in science education and technology). ERIC. New York: Kluwer Academic/Plenum Publishers.

Traxler, J. (2007). Defining, discussing and evaluating mobile learning: The moving finger writes and having writ. *The International Review of Research in Open and Distance Learning, 8*(2), 1–12.

Webb, N. M., Nemer, K. M., & Ing, M. (2006). Small-group reflections: Parallels between teacher discourse and student behavior in peer-directed groups. *The Journal of the Learning Sciences, 15*(1), 63–119.

Wenger, E. (1998). *Communities of practice: Learning, meaning, and identity.* New York: Cambridge University Press.

Chapter 12
A Study of the Internet Resource-Based e-Learning Environments in Hong Kong: Animal Classification in a Primary School

Winnie Wing-Mui So and Fiona Ngai-Ying Ching

Introduction

Internet World Stats (2013) indicated the growth of Internet users from 2000 to 2012 was 566.4 %. Hong Kong is among the areas/countries highly penetrated by the Internet in 2010, with 69.5 % penetration. Nielsen (2011) reported that 87 % of consumers in Hong Kong have used the Internet in the past year, making Hong Kong the top in Internet usage among Singapore, Malaysia, the Philippines, Thailand and Indonesia. Without any doubt, the invention of the Internet has the potential to transform education. The Hong Kong government has been advocating the use of information technology in education since 1998 with the introduction of the First Strategy on IT in Education (Education and Manpower Bureau, 1998), with the goal of turning e-learning into an effective mode of learning (Education Bureau, 2009). The Secretary for Education of Hong Kong accepted the Task Force to Review Learning and Teaching Materials report (2011), which stated that 'this is an opportune time to further implement e-learning'. These initiatives have empowered primary schools in Hong Kong to gradually implement the e-learning approach by increasing the use of information technology in the learning and teaching of different subjects. However, there is no evidence that these resources are used either widely or wisely (Education Bureau, 2009). Hence, this case study of animal classification using Internet resources, a science topic usually taught at the primary level, was conducted to provide insights for learning and teaching with Internet resources in primary classrooms.

W.W.-M. So (✉) • F.N.-Y. Ching
Department of Science and Environmental Studies, The Hong Kong Institute of Education, Hong Kong, Hong Kong
e-mail: wiso@ied.edu.hk; nyching@ied.edu.hk

M.S. Khine (ed.), *Science Education in East Asia*,
DOI 10.1007/978-3-319-16390-1_12

Literature Review

Students' Conceptions of Animal Classification

Classification is a fundamental science practice used across the science domains, but it is especially important in the biological sciences to establish living and nonliving things and taxonomy of diverse living things: Animalia, Plantae, Protista, Fungi and Monera. The classification of the animal kingdom, a high interest topic for primary school students, involves the establishment of a hierarchical structure of categories using a variety of external and internal attributes and processes on the basis of more inclusive concepts (phylum) such as invertebrates and vertebrates, as well as less inclusive concepts (class) such as mammals, birds, insects and others (Braund, 1998; Trowbridge & Mintzes, 1988).

Animal classification has occupied the interest of science educators around the world for many years and revealed thought provoking results. For example, researchers in New Zealand (Bell, 1981), the United Kingdom (Braund, 1998; Ryman, 1974), Israel (Jungwirth, 1971), Germany (Kattmann, 2001), Taiwan (Yen, Yao, & Mintzes, 2007), the United States (Shepardson, 2002; Trowbridge & Mintzes, 1988), Slovakia (Prokop, Prokop, & Tunnicliffe, 2008) and Hong Kong (So & Cheng, 2008) have investigated students' understanding and learning of animal classification. A synthesis of these regional studies shows that students have a variety of concepts about animals and animal classification. Trowbridge and Mintzes (1988) suggested that students often had a highly limited view of animals and tended to label them as familiar vertebrates and common mammals in their daily life. Braund's (1998) noted that students had less knowledge of invertebrates than of vertebrates and that they tended to identify invertebrates as either having small bodies or no legs. Kattmann (2001) found that students had the conception that everything that lived in the water was a fish and everything that flew was a bird. More recently, Yen et al. (2007) discovered that to most students, the concept 'animal' referred to 'vertebrates, especially common, well-known mammals and birds', and the most common attributes used by students to define 'animals' are 'movement' and 'viability'. Classifying animals involved most commonly used cues about external morphology (Kattmann, 2001; Yen et al., 2007), habitat (Yen et al., 2007) and movement patterns (Kattmann, 2001) of the animals.

It is evident from the literature that animal classification has always been a challenging topic for students. The variety of misconceptions that students have about animals and animal classification makes it difficult for teachers to design instructional approaches to improve student understanding.

Student Learning of Animal Classification

According to the TIMSS 2007 International Science Report (Martin, Mullis, & Foy, 2008), on average across countries, teachers reported that they had taught 77 % of fourth grade students about types, characteristics and classification of living things. In Hong Kong, the topic 'animal classification' belongs to the strand 'People and Environment' of the interdisciplinary subject 'General Studies' which includes the Key Learning Area of Science Education (Curriculum Development Council, 2011). The General Studies curriculum adopts a spiral approach, i.e. the idea of animal classification is usually first introduced in grade 3 and is revisited at more complex and deeper conceptual levels in grade 6. Animal classification is one of the central concepts of biological science education (Yen et al., 2007) and is closely related to students' attitudes towards the environment and living things as well as their understanding of nature.

How classification of animals should be taught and learned has also been one of the research interests of science educators. Yen et al. (2007) and Cardak (2009) suggested that providing opportunities for students to interact with animals and to be informed of their characteristics in a zoo or natural environment would be a good technique. Cardak (2009), however, noticed that students living in cities are far away from nature and could only learn about the animals through printed materials. He therefore suggested the use of educational techniques to make the information more concrete. Prokop et al. (2008) found that keeping pets significantly contributes to children's factual knowledge about the anatomy of animals, especially vertebrates, and suggested that teachers encourage children to keep a diverse range of animals. But in Hong Kong, a city with limited land supply, most schools can at most manage to have a small pond, and most families do not have the space to accommodate a pet. It was estimated in 2005 that only 12.6 % of all households in Hong Kong had pets (Census and Statistics Department, 2006), compared with 46 % of households in the United Kingdom.

The old saying, 'a picture is worth a thousand words' may provide some insight into effective learning environment for animal classification. A photograph captures and preserves an instant frozen in time, which makes it a great asset for study (Krauss, Salame, & Goodwyn, 2008), allowing the features and the surroundings of the object captured to be scrutinised over extended time. High-quality photographs of a great diversity of animals can be found easily on the Internet for free. These photographs, if carefully selected by teachers, can become excellent materials for students to learn about different animals. Furthermore, videos, a medium that learners are very comfortable with, can provide clarity and explicitness about movement, context, and other features that are not possible with words or still pictures alone (Williams, 2004). Moreover, videos can facilitate students' cognitive processing (Reiser & Dempsey, 2007). As a result, photos and videos complement each other to become powerful learning materials for students to learn about animal classification such as different categories, the features of different groups of animals and the animals belonging to different groups.

Concept Mapping in Science Learning

Contemporary interpretations of learning involve students constructing their own understandings or changing their conceptions by representing and re-representing ideas. A concept map is defined as a set of propositions mainly focusing on expressing the relationships between concepts in a two-dimensional diagram (Andrews, Tressler, & Mintzes, 2008). It is a visual representation or organisation of knowledge (Novak & Cañas, 2008; Rosas & Camphausen, 2007). Novak and Gowin (1984) suggested that concept maps should be hierarchical with the more general or inclusive concepts being placed at the apex and the more specific or less inclusive concepts arranged at the lower part of the structure.

Concept mapping is the process of organising and representing concepts and the relationships between concepts in visual form (Novak, 1998). The potential advantages of concept mapping have been widely recognised; it has also been employed in large-scale research in science education and classroom assessment (McClure, Sonak, & Suen, 1999; Pinto, Doucet, & Fernández-Ramos, 2010; Yin, Vanides, Ruiz-Primo, Ayala, & Shavelson, 2005) as well as in assessing thinking (So & Watkins, 2005). Concept mapping, when viewed through the contemporary views of interactive-constructive learning, can be seen as a potential asset in promoting conceptual growth and change.

Concept mapping has long been taken as an effective assessment tool of student learning for several reasons. Firstly, it is an explicit presentation of concepts held by the students, making it possible to understand and measure students' prior knowledge (Kinchin, 2000; Novak & Gowin, 1984). Secondly, it is a hierarchical organisation of concepts of a particular topic and hence reveals students' knowledge structures (Kinchin, 2000; McClure et al., 1999; Novak & Gowin, 1984). Thirdly, it provides rich information about the change and growth of students' concept understanding (Pinto et al., 2010), and it helps understand the quality of the learning that has taken place (Hay & Kinchin, 2008). In addition, it documents individual student' learning outcomes in detail as the process of concept mapping expresses personal organising and understanding of particular topics (Hay & Kinchin, 2008). Since the concept of animal classification is hierarchical in nature, it is appropriate to make use of concept mapping or similar procedures as an assessment tool to examine students' conceptions and learning of animal classification.

Resource-Based e-Learning Environments

Twenty-first-century learning environments are no longer only physical places and spaces, they could also be structures, tools and communities that inspire learners to attain the knowledge and skills that this era demands of them (The Partnership for 21st Century Skills, 2004). The advancements of information technologies and rapid expansion of the Internet have influenced learning environments and encouraged teachers to implement various forms or combinations of different forms e-learning resources (Lee et al., 2011).

The resource-based e-learning environments (RBeLEs) framework (So, 2012), an earlier stage of research, was developed based on two related learning environments findings. Hill and Hannafin (2001) proposed the resource-based learning environments characterised by four essential components—resources, contexts, tools and scaffolds. Such learning environments could allow both teachers and learners to take advantage of digital information systems to enhance teaching and learning by expanding the resources used. Blumenfeld, Kempler and Krajcik (2006) suggested that learning environments with features of authenticity, inquiry, collaboration and technology would engage learners in thinking deeply about the content and constructing understand involving the integration and application of key ideas in the discipline.

The RBeLEs framework aims to support teachers in creating e-learning environments with Internet resources more effectively by helping them to think about four framework components: creation of contexts, selection of resources, use of tools and adoption of scaffolds. Contexts can be determined by teachers, generated by learners or negotiated between the two parties. Teachers should create authentic contexts to increase students' interest in the topics and let students understand how the learning tasks are related to their everyday life. When learning contexts are generated or negotiated by students, students' sense of autonomy increases because they make choices and play a significant role in directing their own activities.

Hill and Hannifin (2001) defined resources as 'media, person, place or idea that has the potential to support learning' (p. 38). Resources in the RBeLEs can be categorised as either static or dynamic. Static resources have relatively stable contents that may soon become obsolete or inaccurate; dynamic resources undergo frequent, sometimes continual, change. Online resources can act as 'hooks' that get students to participate, sustain interest and promote cognitive engagement (Blumenfeld et al., 2006).

Tools aid in locating, accessing, manipulating, interpreting and evaluating resources (Hill & Hannafin, 2001). Teachers should select appropriate tools for students to use so as to provide them with engaging experiences when performing learning tasks, increase their perceived competence in succeeding on specific tasks and foster positive interactions among peers (Blumenfeld et al., 2006). Tools can be classified as information processing tools, searching and seeking tools, information and data collecting tools, organising tools, collaborating and integrating tools and communication tools.

Scaffolds provided by teachers are necessary to sustain learners' interest, help them become more willing to approach challenges, construct understanding and have a positive inquiry experience with their peers (Blumenfeld et al., 2006). The four aspects of teachers' scaffolds are asking and discussing, searching and selecting, doing and observing and summarising and conceptualising.

Methodology

This quasi-experiment pretest-posttest design explored the learning about animal classification of two different ability sixth grade classes. The host elementary school was in North District of Hong Kong and serve an area of families of low to middle socioeconomic status.

Participants

Two sixth grade classes in a local primary school were involved in this study, one with higher and the other with lower academic ability in terms of the students' examination results of the previous school year. The higher-ability class, Class H, consisted of 38 students, while the lower-ability class, Class L, consisted of 32 students. Both classes comprised a similar proportion of girls and boys. All of them are ethnic Chinese. The language of instruction was Cantonese and the teaching and learning materials were in Chinese.

Lesson Design Based on RBeLEs Framework

With the support of the research team, the teacher who taught the two classes involved in this study designed a lesson with reference to the RBeLEs framework. Figure 12.1 summarises the lesson design in terms of the four components of the RBeLEs framework.

Students were assigned into heterogeneous groups of 3 to 4 by the teacher, and each group had a laptop computer with a Wi-Fi connection. The lesson developed over three stages a modified learning cycle: inquiry, application and consolidation, taking altogether 60 min.

The Inquiry Stage

This stage required each student group to work on one category of animals from five categories provided (fish, mammals, reptiles, birds and insects). Each group followed two guiding questions to explore a list of animals given:

Which taxonomic group do these animals belong to?
What are the features of this group of animals?

These questions were used to find out the taxonomic groups and the features of the animals provided by viewing the Internet resources including videos, photos, other visuals and text.

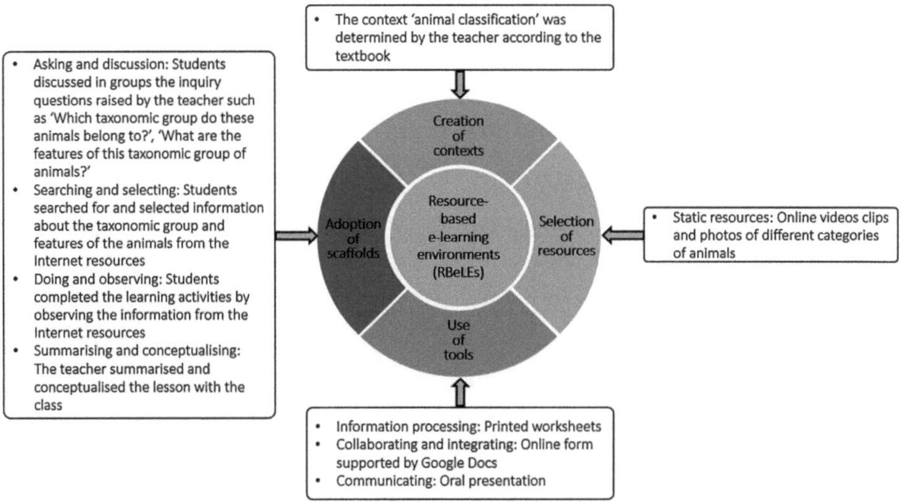

Fig. 12.1 Summary of the lesson design with the RBeLEs framework

Figure 12.2 is an example of the different Internet resources provided by the teacher for the students to learn about fish. Students searched for and selected information about the taxonomic groups and features of the animals from the Internet resources and recorded the group decision of the category on the worksheet. Afterwards, each group of students was invited to report their findings to the rest of the class. This arrangement made reference to the cooperative learning strategy of 'expert group' so that students shared the responsibilities to learn the five different categories of animals within limited class time.

The Application Stage

The application stage required the students to view video clips of two animals, penguins and dolphins, and then discussed and decided which categories these two animals belong to and justified the reasons by what they have learned in the inquiry stage of the lesson. Each group demonstrated their applications by filling in the online form supported by Google Docs to share their views and reasons for the categories of these two animals.

The Consolidation Stage

The consolidation stage of the lesson involved a teacher-led whole class discussion. The discussion established, clarified and summarised the underlying reasons for animal classification central to the lesson.

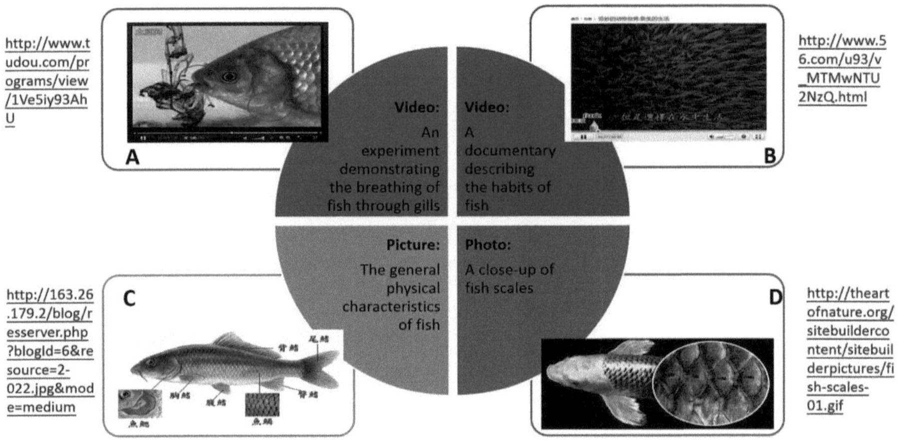

Fig. 12.2 Internet resources for learning about fish

Data Collection and Analysis

Student Word Webs

Word webs drawn by the students 1 week before and *1 week* after the lesson were used to collect data to address the first research question. The pretest and posttest requested that individual students use a modified concept mapping procedure and put down as many ideas as they could around the central concept 'Animal classification', on a blank paper, with the instruction to draw the word web showing three levels: taxonomic group (level 1), features (level 2) and examples (level 3). Figures 12.3 and 12.4 are samples of the student-drawn word webs of the same student's pretest and posttest.

The scoring procedure for the word webs was developed inductively from the context of the lesson and Internet resources. A reference guide was developed which contained the common features of the animals provided in each taxonomic group that students would be able to find from the Internet resources identified (Table 12.1). This reference guide was not meant to be exhaustive, but it contained the specifics expected as learning outcomes of this lesson.

Since all the word webs comprised a central concept and three levels, the items for each level could be systematically organised into a table similar to Table 12.2. Each word web was examined, specific features and examples identified and the number of correct items at each level was counted. An item would be considered correct and given a tick if it matched the reference guide. However, if a taxonomic group was named incorrectly, any features and examples given for that group would be ignored, regardless of their correctness. Table 12.2 illustrates how scoring of the student's pretest word web was done. First, items in Fig. 12.3 were organised into a table with three columns. Ticks indicated the items that matched the reference

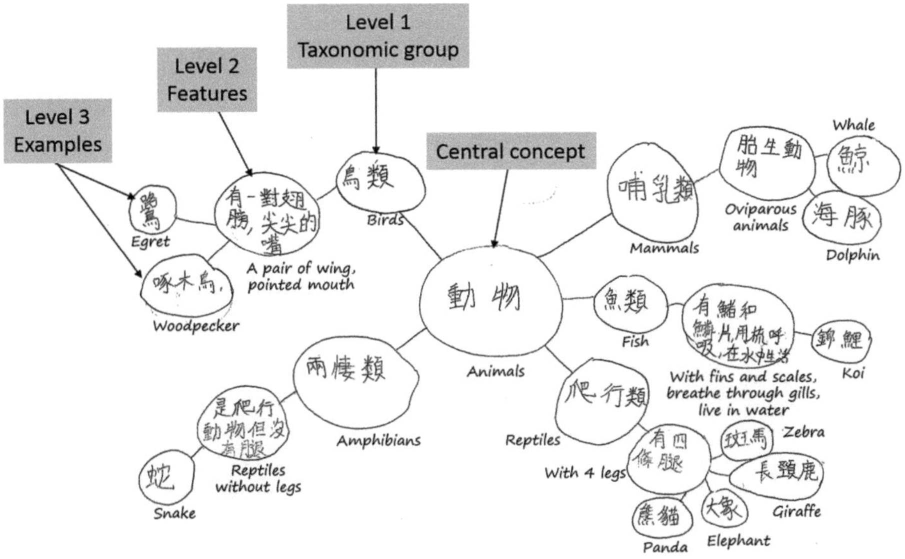

Fig. 12.3 Sample pretest concept map

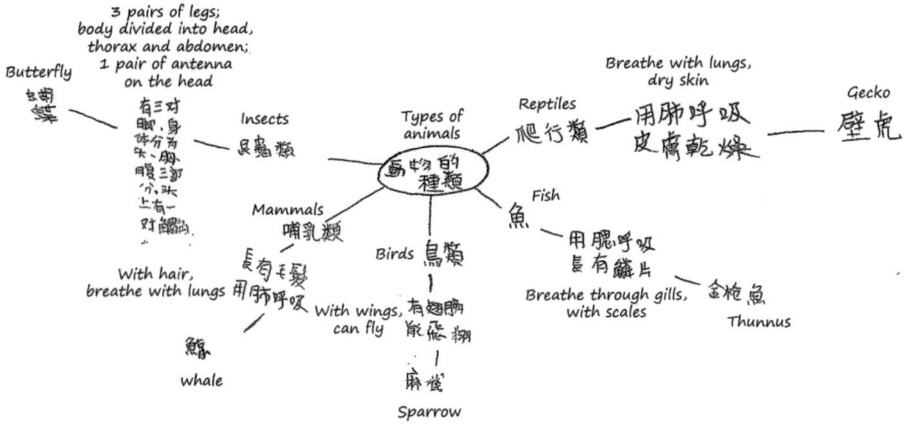

Fig. 12.4 Sample posttest concept map

guide. Since the taxonomic group ‘amphibians’ was not in the reference guide, it was considered incorrect, and the features and examples given were ignored.

Interview with Focus Group Students

Face-to-face in-depth interviews were conducted with 23 students from the two classes (12 from Class H and 11 from Class L) chosen randomly by the teacher.

Table 12.1 Reference guide

Level 1	Level 2	Level 3
Taxonomic group	Features	Examples
Fish	Moist skin	Any appropriate examples
	Most have scaly skin	
	Fins	
	Breathe through gills	
Mammals	Hair	Any appropriate examples
	Feed milk to babies	
	Breathe with lungs	
Reptiles	Dry skin	Any appropriate examples
	Body covered in scales	
	Breathe with lungs	
Birds	Feathers	Any appropriate examples
	Beak	
	One pair of wings	
	One pair of legs	
	Breathe with lungs	
Insects	Three pairs of legs	Any appropriate examples
	Three body regions: head, thorax and abdomen	
	One pair of antenna	
	Some have one or two pairs of wings	

Table 12.2 Scoring of the sample student pretest word web shown in Fig. 12.3

	Level 1	Level 2	Level 3
Level	Taxonomic group	Features	Examples
Item	Fish ✓	With fins ✓	Koi ✓
		Scales ✓	
		Breathe through gills ✓	
		Live in water	
	Mammals ✓	Oviparous animals	Whale ✓
			Dolphin ✓
	Reptiles ✓	With 4 legs	Zebra
			Giraffe
			Elephant
			Panda
	Amphibians	Reptiles without legs	Snake
	Birds ✓	A pair of wings ✓	Egret ✓
		Pointed mouth	Woodpecker ✓
Number of correct items	4	4	5

Students were interview individually taking approximately 20 min. A semi-structured interview protocol was developed to guide the interview that encouraged students to describe what they had learned, how learning occurred and the difficulties they had encountered in the lesson. The main questions were: (1) What did you learn about animal classification? (2) How did learning occur? (3) What difficulties did you encounter in the lesson? All the interviews were audiotaped and later transcribed verbatim for analysis.

Results

Changes in Student Conceptions of Animal Classification

The word webs provided two types of information to addressing the research questions. Quantitative data generated by scoring the word webs documented students' conceptions of animal classification at different levels before and after the lesson to be compared statistically. The qualitative data generated by scrutinising the word webs gave detailed information about students' conceptions of animal classification.

The descriptive statistics showed the percentages of students in the two different ability classes having different number of correct items for different levels of the word webs (Tables 12.3 and 12.4). Table 12.3 illustrated that just about everyone in the two groups improved their performance on taxonomic level with the high-ability class achieve near perfect performance (>95 % at highest level), while the low-ability class somewhat lower level of performance but still very acceptable (>65 % at highest level). Similar improvements were noted for the feature of the fish, mammals, reptiles, birds and insects, but none of these performances reached near perfection. Again, the high-ability class outperformed the less able class.

The paired t-tests compared the mean scores achieved by students in the pretest and posttest word webs (Table 12.5), while independent samples *t*-test was used to identify the difference in student learning gains between students in the higher-

Table 12.3 Percentages of students with different number of correct taxonomic groups in Class H and Class L

Number of correct taxonomic group	Class H Pretest $N=37$	Class H Posttest $N=38$	Class L Pretest $N=32$	Class L Posttest $N=31$
0	0.00	0.00	43.75	3.23
1	2.70	0.00	9.38	0.00
2	18.92	2.63	12.50	0.00
3	21.62	0.00	9.38	0.00
4	32.43	5.26	21.88	29.03
5	24.32	92.11	3.13	67.74

Table 12.4 Percentages of students with different number of correct features for different groups of animals in Class H and Class L

Number of correct features for different groups of animals	Class H		Class L	
	Pretest	Posttest	Pretest	Posttest
	$N = 37$	$N = 38$	$N = 32$	$N = 31$
Fish				
0	59.46	2.63	86.67	25.81
1	29.73	7.89	10.00	38.71
2	8.11	34.21	3.33	29.03
3	2.70	26.32	0.00	6.45
4	0.00	28.95	0.00	0.00
Mammals				
0	64.86	5.26	86.67	29.03
1	21.62	15.79	6.67	41.94
2	8.11	28.95	3.33	22.58
3	5.41	50.00	3.33	6.45
Reptiles				
0	84.21	7.89	90.00	29.03
1	10.53	15.79	3.33	45.16
2	5.26	36.84	3.33	25.81
3	0.00	39.47	3.33	0.00
Birds				
0	35.14	0.00	80.00	9.68
1	56.76	5.26	13.33	38.71
2	5.41	28.95	6.67	32.26
3	2.70	18.42	0.00	16.13
4	0.00	44.74	0.00	3.23
5	0.00	2.63	0.00	0.00
Insects				
0	91.89	5.26	96.67	38.71
1	2.70	10.53	3.33	45.16
2	2.70	36.84	0.00	9.68
3	2.70	23.68	0.00	6.45
4	0.00	23.68	0.00	0.00

ability class and those in the lower-ability class (Table 12.6). The following subsections discuss, level by level, the changes in the conceptions of animal classification of students in the two classes before and after the lesson with Internet resources. The within-class pretest-posttest gains were significant for two of the three dimensions (levels 1 and 2) in the high-ability class and for all three dimensions in the low-ability class. The between-class comparison of pretest-posttest gains was significant for two of the three dimensions (levels 1 and 2), favouring low-ability class (level 1) and the high-ability class (level 2), with a nonsignificant gain for level 3 (examples), favouring the low-ability class.

Table 12.5 Pretest and posttest scores for different levels of the word webs of Classes H and L

		Pretest	Posttest			
	N	Mean (SD)	Mean (SD)	*t*	df	Sig. (2-tailed)
Class H						
Taxonomic group (level 1)	37	3.57 (1.18)	4.86 (.54)	6.71	36	.000***
Features (level 2)	37	2.277 (2.28)	12.89 (3.97)	15.87	36	.000***
Example (level 3)	37	8.007 (4.89)	9.43 (6.00)	1.70	36	.097
Class L						
Taxonomic group (level 1)	31	1.71 (1.75)	4.55 (.96)	8.56	30	.000***
Features (level 2)	31	.90 (2.02)	5.45 (3.45)	7.36	30	.000***
Example (level 3)	31	2.42 (3.41)	5.97 (3.41)	5.40	30	.000***

***$p < .001$

Table 12.6 Comparison of the gain scores for different levels of the word webs of Classes H and L

	Class H	Class L			
	Mean (SD)	Mean (SD)	*t*	df	Sig (2-tailed)
Taxonomic group (level 1)	1.30 (1.18)	2.84 (1.85)	4.02	49.12	.000***
Features (level 2)	10.62 (4.07)	4.55 (3.44)	6.57	66	.000***
Example (level 3)	1.43 (5.12)	3.55 (3.66)	1.93	66	.059

***$p < .001$

Level 1 Taxonomic Group

Before the lesson, it was discovered from the analysis of the pretest word webs that the students' prior conceptions of animal classification were limited. Students' conceptions of animal classification were limited to the less inclusive concepts that they had come across in their daily life, such as fish and birds. None of the students in either class mentioned the more inclusive concepts of 'vertebrates' and 'invertebrates' in their pretest word webs. Only 24.32 % of the students in Class H and 3.31 % of the student in Class L were able to name all the five taxonomic groups.

After the lesson, students' conceptions of animal classification were still limited to the less inclusive concepts but no longer only limited to those that were familiar to them. Most of the students in the Class H (92.11 %) and two thirds of the students in Class L (67.74 %) were able to name the five taxonomic groups introduced to them through the Internet resources correctly. For example, the group name 'fish' was explicitly shown in the descriptions of the two videos. There were significant differences between the mean scores for taxonomic group (level 1) of the pretest and posttest word webs drawn by the students in both classes, indicating students' knowledge of the proper names for the five taxonomic groups of animals had improved over the lesson. However, Class L had significantly ($p < .001$) higher learning gains for taxonomic group (level 1) ($M = 2.84$, $SD = 1.85$) than their counterparts in Class H ($M = 1.30$, $SD = 1.18$).

Level 2 Features

Before the lesson, it is observed from the students' word webs that the criteria they employed to classify the animals were mixed. Students used a combination of criteria, such as living habitats, body features, appearances, eating habits and types of locomotion. More students in the lower-ability class tended to classify the animals by their living habitats or eating habits, for example, 'animal living in the desert', 'sea animals', 'herbivores' and 'carnivores', while more students in the higher-ability class tried to classify the animals by their body features although some of the features that they had put down on the pretest word webs were incorrect.

Altogether there should be 19 features for the five taxonomic groups of animals. Before the lesson, no student in either class was able to put down half of the 19 features correctly. About half of the students in Class H (51.35 %) could only put down one to two out of the 19 features correctly, and 75 % of the students in Class L failed to put down any features correctly.

As for the features of each taxonomic group of animals, it is found that students' knowledge was poorest for insects and reptiles and best for birds among the five groups of animals. In Class H, 81.21 % of the students failed to put down any feature for reptiles, and 91.89 % failed to put down any feature for insects correctly. In Class L, a very large majority of the class could not provide any feature for insects (96.67 %) and reptiles (90.00 %) correctly. In contrast, students' knowledge of the features of birds seemed to be the best because over half of the students in Class H (56.76 %) were able to put down one feature for birds and in most cases that feature was 'have wings'. Despite the fact that 80 % of the students in Class L could not provide any feature for birds correctly, their knowledge of the features of birds was still considered the best among the five groups the animals because the remaining 20 % of the students were able to put down one to two features for birds correctly. Surprisingly, students' knowledge of features of mammals was quite limited. Over half of the students in Class H (64.86 %) and a large majority of the students in Class L (86.67 %) were unable to put down any feature for mammals correctly.

After the lesson, as observed from the posttest word webs, students' conceptions of animal classification had become more systematic. Most of the features that students had put down on their posttest word webs were body features of the five taxonomic groups of animals even though the accuracy had yet to be improved, particularly for Class L. Student learning of the features occurred through watching of the Internet resources, recording the information on the worksheets and discussion with group members. For example, while watching the videos, photos and diagram of fish, students recorded the features of fish on the group worksheet and discussed whether the features were correct.

Results of the paired t-tests showed that there were significant differences between the pretest and posttest scores for features (level 2) in both classes, indicating that students' knowledge of the features of different taxonomic groups of animals had improved over the lesson. The number of correct features put down by students in Class H had increased from 0 to 9 in the pretest to 7 to 18 in the posttest, from only half of the class was able to provide one to two correct features in the pretest to close to three-fourth was able to provide 10 or more features correctly in the posttest. In Class L, there was also noticeable improvements in students' knowledge of the features of different groups of animals after the lesson, with over half of the class being able to put down at least five features correctly compared to less than one-tenth of the class could do so before the lesson (Table 12.6). Even though students in the two classes had an increased understanding of the features of the five taxonomic groups of animals after the lesson, students in the higher-ability class ($M = 10.62$, $SD = 4.07$) appeared to have significantly ($p < .001$) higher learning gains than students in the lower-ability class ($M = 4.55$, $SD = 3.44$).

The pretest word webs illustrated that students in both classes were particularly unfamiliar with the features of reptiles and insects. After the lesson, in Class H, close to 40 % of the class could put down all the three features of reptiles correctly, and close to a quarter of the class could put down all the four features of insects correctly on the posttest word webs, whereas in Class L, around 70 % of the students were able to provide one or two features of reptiles correctly, and around 60 % were able to provide one to three features of insect correctly. These data suggest that students' knowledge of the features of reptiles and insects as well as for other taxonomic groups of animals, including those for mammals, had improved after the lesson. However, the considerable percentage of students in the lower ability who failed to provide any features for different groups of animals warrants our attention.

Level 3 Examples

Before the lesson, students' conceptions of animal classification were reflected in the examples that they provided for the different groups of animals. The correctness of the examples provided depended largely on whether they had named the taxonomic groups and identified critical features correctly in the first place. Moreover, the examples that the students provided were influenced by the students' understanding of the features of different groups of animals. For example, a student named 'reptiles' correctly, but she believed that 'reptiles were animals that crawl', so the examples she provided for 'reptiles' were lions, tigers, dogs and cats. In fact, many students in both classes held the misconception that reptiles were crawling animals, and therefore the examples provided could range from mammals (e.g. horses, lions) to insects (e.g. grasshoppers, ants) and reptiles (e.g. snakes, lizards).

After the lesson, the examples that students provided for different groups of animals were more accurate. Students in the lower-ability class tended to provide examples that they had learned during the lesson, while students in the higher-ability class could provide examples other than those introduced in the lesson. For example, many of the examples for fish provided by students in Class L were actually mentioned in one of the videos (http://www.56.com/u93/v_MTMwNTU2NzQ.html).

Results of the paired t-tests indicate a significant ($p < .001$) increase in the number of the correct examples that students in the lower-ability class provided in the posttest word webs ($M = 5.97$, SD = 3.41) when compared with that in the pretest word webs ($M = 2.42$, SD = 3.41). However, there was no significant difference between the number of correct examples students in the higher-ability class provided before and after the lessons. The difference in learning gains between the two classes for examples (level 3) was not significant.

Students' Perception of Their Own Learning of Animal Classification with Internet Resources

Student learning in the lesson was captured from the 23 individual face-to-face interviews after the lesson. During the interviews, students were asked to rate their learning in the lesson as 'I learned all the content', 'I learned most of the content', 'I did not learn most of the content' or 'I learned nothing'. All 12 students from the higher-ability class rated their learning as 'I learned most of the content', while three of the students from the lower-ability class rated 'I learned all the content' and the remaining eight students 'I learned most of the content'. This showed that students perceived their learning was successful. When asked about how they learned the content, students' answers fell into the following: 'from information found on the Internet', 'from the videos and photos on the Internet' and 'by discussing with group members'. Pseudonyms are used to protect the student's identity. For example, the students said:

> From the videos and information on the Internet that the teacher provided, and from discussing with my group members. The worksheets also helped me learn. (Gary from Class H)
>
> We searched for information from the Internet resources, watched the videos and discussed with each other. (Susan from Class H)
>
> The videos and Internet resources could help us learn. The discussion with other classmates as well. (Linda from Class L)
>
> Searching for information from the resources on the Internet together with my group members. (Jeremy from Class L)

Results from the student interview suggested that students learned the taxonomic groups and features of different groups of animals from the Internet resources, which were carefully selected by the teacher. The videos, drawings and diagrams and photos from the Internet functioned well to enhance students' understanding of the features of different taxonomic groups of animals. The worksheets were tools to help students organise the information found from the Internet, and they were indispensable in student learning.

Students' perceptions about learning with Internet resources rated their level of engagement in the lesson as 'very engaged' or 'kind of engaged', revealing a favourable level of engagement by the students in the lesson. More students in the lower-ability class than the higher-ability class found themselves 'very engaged' in the lesson.

Furthermore, the interview results indicated that most students (19 out of 23) preferred the Internet-based lesson in this study over traditional lessons that involved whole class lecturing by the teacher, mainly relying on the textbook and occasionally supplemented with a few photos or pictures. The remaining four students interviewed believed that both methods had their own advantages. However, students did encounter problems related to the use of computers and the Internet such as the slow speed of the video streaming, the English language barrier in understanding videos and the difficulty of inputting Chinese characters.

Discussion and Conclusion

It is shown in the findings of this study and previous research studies that classifying animals by their habitats and movement patterns is common among primary students (e.g. Kattmann, 2001; Yen, Yao, & Chiu, 2004). The present study presented possible ways and positive findings in introducing Internet resources for students' learning of animal classification based on animals' features for a better understanding of their taxonomic groups.

Moreover, although getting first-hand experience rearing or working with different kinds of animals contributes to students' learning about animals (Prokop et al., 2008; So & Cheng, 2008; Trowbridge & Mintzes, 1988; Yen et al., 2004), it is almost not possible to keep a wide range of animals in the school campus due to the limitations of space. This study also reported encouraging findings of the use of Internet resources to facilitate students' learning of animal classification.

Understanding Students' Alternative Conceptions

The analyses of pretest word webs showed that the students held alternative conceptions about how animals were classified. This finding is line with the findings of previous studies investigating students' conceptions about animal classification using students from elementary school to university (Trowbridge & Mintzes, 1985, 1988; Yen et al., 2004). Unfortunately, the study of Gomez-Zwiep (2008) showed that despite knowing that students have alternative conceptions, teachers do not consider these prior conceptions during lesson planning and often underestimate the impacts that alternative conceptions have on student learning. Teachers can make good use of the pretest results, whether it is concept mapping, word webbing or other forms of assessment for learning to inform instruction and empower learning,

This study demonstrated how assessing students' concepts could influence lesson planning so that appropriate instructional materials can be selected and effective teaching strategies can be implemented to facilitate conceptual change in students.

Learning Animal Classification with Internet Resources

The use of Internet resources and small-group discussion appeared to facilitate improvements in both classes based on the comparison of pretest and posttest word web scores. The selection of Internet resources by the teacher focused on showing the features of different taxonomic groups of animals to the students. These resources helped students to name the different groups of animals, identify their features and determine appropriate examples. The online data collected at the application stage of lesson where student groups were required to apply their learning from the Inquiry stage by deciding which group penguins or dolphins belong with reasons suggested that most of the students in both classes were able to classify penguins as birds and all the students were able to classify dolphins as mammals. This seems to have supported that using the rich information available on the Internet was a practical way to overcome the limitations by virtually extending the physical boundaries of the classroom and allowing students to 'get in touch' with a wide range of animals – how the animals look on the outside and the inside, how they move and other attributes. This claim resonates with the interview data that students learned about animal classification from the Internet resources that illustrate the features, behaviours and habitats of different groups of animals.

Learning in Internet Resource-Rich Environments

Blumenfeld et al., (2006) suggested that online resources acted as 'hooks' that engaged the students to participate and sustain their interest in learning. Other studies have also demonstrated the positive effect of the Internet on science learning – students' interest and participation in some less interesting science topics are promoted with the use of online materials (Ruthven, Hennessy, & Deaney, 2005); students are more motivated and empowered to learn science in Internet-rich programme (Mistler-Jackson & Songer, 2000). The students in this study were found to be engaged in the lesson with rich Internet resources and they preferred it over traditional lecturing. Internet resources should be supplemented with other educational techniques to make information more concrete to the students, including models, mock-ups, etc. (Cardak, 2009). The finding that more students in the lower-ability class found themselves 'very engaged' in the lesson suggests that future research should focus on looking for the differential effects that the Internet brings to students with different academic abilities and how the Internet should be effectively utilised to enhance the learning of students with different academic abilities.

Acknowledgements The project contributing to this paper is funded by a grant from the Research Grants Council of the University Grants Committee of Hong Kong. The author would like to thank the teachers and students who participated in the project.

References

Andrews, K. E., Tressler, K. D., & Mintzes, J. J. (2008). Assessing environmental understanding: An application of the concept mapping strategy. *Environmental Education Research, 14*(5), 519–536.

Bell, B. F. (1981). When is an animal, not an animal? *Journal of Biological Education, 15*(3), 213–218.

Blumenfeld, P. C., Kempler, T. M., & Krajcik, J. S. (2006). Motivation and cognitive engagement in learning environments. In R. K. Sawyer (Ed.), *The Cambridge handbook of the learning sciences* (pp. 475–88). Cambridge, UK: Cambridge University Press.

Braund, M. (1998). Trends in children's concepts of vertebrate and invertebrate. *Journal of Biological Education, 32*(2), 112–118.

Cardak, O. (2009). Science students' misconceptions about birds. *Scientific Research and Essay, 4* (12), 1518–1522.

Census and Statistics Department. (2006). *Thematic household survey report* (Report No. 26). Retrieved from http://www.statistics.gov.hk/pub/B11302262006XXXXB0100.pdf

Curriculum Development Council. (2011). *General studies for primary schools curriculum guide (primary 1 – primary 6)*. Retrieved from http://www.edb.gov.hk/attachment/en/curriculum-development/kla/general-studies-for-primary/gs_p_guide-eng_300dpi-final%20version.pdf

Education and Manpower Bureau. (1998). *Information technology for learning in a new era five-year strategy 1998/1999 to 2002/2003*. Retrieved from http://www.edb.gov.hk/index.aspx?nodeID=423&langno=1

Education Bureau. (2009). *Working group on textbooks and e-learning resources development main report*. Retrieved from http://www.edb.gov.hk/attachment//en/curriculum-development/resource-support/textbook-info/wg%20final%20report.pdf

Gomez-Zwiep, S. (2008). Elementary teachers' understanding of students' science misconceptions: Implications for practice and teacher education. *Journal of Science Teacher Education, 19*(5), 437–454.

Hay, D., & Kinchin, I. (2008). Using concept mapping to measure learning quality. *Education + Training, 5*(2), 167–182.

Hill, J. R., & Hannafin, M. J. (2001). Teaching and learning in digital environments: The resurgence of resource-based learning. *Educational Technology Research and Development, 49*(3), 37–52.

Internet World Stats. (2013). *List of countries classified by Internet penetration rates*. Retrieved from http://www.internetworldstats.com/stats.htm

Jungwirth, E. (1971). A comparison of the acquisition of taxonomic concepts by B.S.C.S. and non-B.S.C.S. pupils. *The Australian Science Teachers Journal, 17*(4), 80–82.

Kattmann, U. (2001). Aquatics, flyers, creepers and terrestrials—Students' conceptions of animal classification. *Journal of Biological Education, 35*(3), 141–147.

Kinchin, I. M. (2000). Concept mapping in biology. *Journal of Biological Education, 34*(2), 61–68.

Krauss, D. A., Salame, I. I., & Goodwyn, L. N. (2008). Using photographs as case studies to promote active learning in biology. *Journal of College Science Teaching, 40*(1), 72–76.

Lee, S. W. Y., Tsai, C. C., Wu, Y. T., Tsai, M. J., Liu, T. C., Hwang, F. K., et al. (2011). Internet-based science learning: A review of journal publications. *International Journal of Science Education, 33*(14), 1893–1925.

Martin, M. O., Mullis, I. V. S., & Foy, P. (2008). *TIMSS 2007 international science report: Findings from IEA's trends in international mathematics and science study at the fourth and eighth grades*. Chestnut Hill, MA: TIMSS & PIRLS International Study Center, Boston College.

McClure, J. R., Sonak, B., & Suen, H. K. (1999). Concept map assessment of classroom learning: Reliability, validity, and logistical practicality. *Journal of Research in Science Teaching, 36* (4), 475–492.

Mistler-Jackson, M., & Songer, N. B. (2000). Student motivation and Internet technology: Are students empowered to learn science? *Journal of Research in Science Teaching, 37*(5), 459–479.

Nielsen. (2011). *Hong Kong's digital landscape is dynamic and evolving*. Retrieved from http://blog.nielsen.com/nielsenwire/online_mobile/hong-kong%E2%80%99s-digital-landscape-is-dynamic-and-evolving/

Novak, J. D. (1998). *Learning, creating and using knowledge: Concept maps as facilitative tools in schools and corporations*. Mahwah, NJ: Lawrence Erlbaum.

Novak, J. D., & Gowin, D. B. (1984). *Learning how to learn*. New York, NY: Cambridge University Press.

Novak, J. D., & Cañas, A. J. (2008). *The theory underlying concept maps and how to construct and use them*. Retrieved from http://cmap.ihmc.us/Publications/ResearchPapers/TheoryUnderlyingConceptMaps.pdf

Pinto, M., Doucet, A.-V., & Fernández-Ramos, A. (2010). Measuring students' information skills through concept mapping. *Journal of Information Science, 36*(4), 464–480.

Prokop, P., Prokop, M., & Tunnicliffe, S. D. (2008). Effects of keeping animals as pets on children's concepts of vertebrates and invertebrates. *International Journal of Science Education, 30*(4), 431–449.

Reiser, R. A., & Dempsey, J. (2007). *Trends and issues in instructional design and technology* (2nd ed.). Upper Saddle River, NJ: Pearson.

Rosas, S. R., & Camphausen, L. C. (2007). The use of concept mapping for scale development and validation in evaluation. *Evaluation and Program Planning, 30*(2), 125–135.

Ruthven, K., Hennessy, S., & Deaney, R. (2005). Incorporating Internet resources into classroom practice: Pedagogical perspectives and strategies of secondary-school subject teachers. *Computers & Education, 44*, 1–34.

Ryman, D. (1974). Children's understanding of the classification of living organisms. *Journal of Biological Education, 8*(3), 140–144.

Shepardson, D. P. (2002). Bugs, butterflies, and spiders: Children's understandings about insects. *International Journal of Science Education, 24*(6), 627–643.

So, W. M. W. (2012). Creating a framework of a resource-based e-learning environment for science learning in primary classrooms. *Technology, Pedagogy and Education, 21*(3), 317–335.

So, W. M. W., & Watkins, D. (2005). From beginning teacher education to professional teaching: A study of the thinking of Hong Kong primary science teachers. *Teaching and Teacher Education, 21*(5), 525–541.

So, W. M. W., & Cheng, M. H. M. (2008, November). *Accelerating science thinking: The infusion of cognitive conflict strategies into primary science education*. Paper presented at the APERA 2008 Conference, Singapore

Task Force to Review Learning and Teaching Materials. (2011). *Report of the task force to review learning and teaching materials*. Retrieved from http://www.edb.gov.hk/attachment/en/about-edb/publications-stat/major-reports/report%20of%20the%20task%20force%20to%20review%20learning%20and%20teaching%20materials_english%20version_20111213.pdf

The Partnership for 21st Century Skills. (2004). *21st century learning environments*. Retrieved from California K–12 High Speed Network website: http://www.k12hsn.org/files/research/Technology/le_white_paper-1.pdf

Trowbridge, J. E., & Mintzes, J. J. (1985). Students' alternative conceptions of animals and animal classification. *School Science and Mathematics, 85*(4), 304–316.

Trowbridge, J. E., & Mintzes, J. J. (1988). Alternative conceptions in animal classification: A cross age study. *Journal of Research in Science Teaching, 25*(7), 547–571.

Williams, B. (2004). Using digital video in teaching materials. Retrieved from http://www.excellencegateway.org.uk/media/AskButler/examples/78/Using%20digital%20video%20in%20teaching%20materials.doc

Yen, C. F., Yao, T. W., & Chiu, Y. C. (2004). Alternative conceptions in animal classification focusing on amphibians and reptiles: A cross-age study. *International Journal of Science and Mathematics Education, 2*(2), 159–174.

Yen, C. F., Yao, T. W., & Mintzes, J. J. (2007). Taiwanese students' alternative conceptions of animal biodiversity. *International Journal of Science Education, 29*(4), 535–553.

Yin, Y., Vanides, J., Ruiz-Primo, M., Ayala, C. C., & Shavelson, R. J. (2005). Comparison of two concept-mapping techniques: Implications for scoring, interpretation, and use. *Journal of Research in Science Teaching, 42*(2), 166–184.

Part V
Inquiry-Based Science Instruction

Chapter 13
Packaging Inquiry-Based Science Learning for Students: A Discursive Analysis of One High School Teacher's Talk from Singapore

Shien Chue and Yew-Jin Lee

Introduction

Many reform documents worldwide promote inquiry as *the* central strategy for the teaching and learning of science in schools (Anderson, 2002, NRC, 1996). While teachers do attempt to teach using scientific inquiry, they also find it complex and difficult to do so based on a variety of reasons (e.g., Chue & Lee, 2013; Lee, 2008, 2014). Specifically, most teachers have themselves learnt science through didactic methods and hence often struggle to make sense of inquiry-based instruction or may only adopt its surface features during teaching. This is not helped when active, hands-on strategies effective for increasing conceptual understanding (Minner, Levy, & Century, 2010) are often jeopardized because of a lack of relevant content knowledge, concerns about classroom management, and erroneous beliefs about inquiry among teachers (Roehrig & Luft, 2004). More broadly, there can be varying or uneven degrees of direction and decision-making from teachers through their talk that are extremely consequential for science learning.

While studies on teacher instructional discourses in (science) education are not rare, the emphasis has remained largely on students' abilities to ask/raise productive questions or to engage in argumentative discourse with their teachers or peers (Louca, Zacharia, & Tzialli, 2012). Little research has focused on the functions of teacher talk other than teacher questioning and giving useful feedback, which form a small component of the more complex processes of classroom discourse. Yet, studies have shown that when teachers offer structured opportunities for students to explore scientific ideas during scientific investigations, students' conceptual learning can be much enhanced (Osborne, 2010). In a nutshell, the subtle discursive features of teacher instructional talk—what is said and how it is said—can be critical in supporting students' construction of scientific understanding

S. Chue (✉) • Y.-J. Lee
National Institute of Education, Nanyang Technological University, Singapore, Singapore
e-mail: shien.chue@nie.edu.sg; yewjin.lee@nie.edu.sg

M.S. Khine (ed.), *Science Education in East Asia*,
DOI 10.1007/978-3-319-16390-1_13

(Collins, Palincsar, & Magnusson, 2005). Early studies in mathematics reforms have already reported how teachers can be thoroughly confused with regard to providing the "right" resources such as new textbooks or using new vocabularies when teaching but still get it wrong with respect to conforming to the actual spirit of the innovation (e.g., Cohen, 1990). Even though evidence for the significance of teacher instructional talk on student learning has been known for some time, it is still largely beyond the reach of policymakers to effect control of what and how subject matter is to be delivered in the classroom (e.g., Lefstein, 2008).

The aforementioned therefore points towards the importance of teachers' orchestration of structures and the functions of language interactions to create opportunities for students to learn complex scientific meanings (Lemke, 1990; Márquez, Izquierdo, & Espinet, 2006). This is especially so when instructional talk is coextensive with helping students model and learn ways of hypothesizing, evaluating, and questioning knowledge—important mechanisms of inquiry (Yore & Treagust, 2006). Knowing that teachers' discourse plays an important role in influencing students' learning (Roth, 2008), we focus on how these concretely engage or hinder students during a school-based innovative science inquiry program. Our research question therefore asks how instructional talk can shape both positively and negatively inquiry activities in a science classroom in Singapore. From discourse analyses of a high school chemistry teacher's talk over a term, we examine how the former influenced the evolving theories and explanations for students' scientific investigations. Alluding to students' abilities and downplaying important processes and content knowledge during instructional time, little did he anticipate that it in fact afforded restricted forms of classroom inquiry activities. Our results locate his repertoire of talk framing students' scientific abilities and personal understandings of investigations as two factors having a major influence over how these activities were to be ideally carried out. The following section reviews key research on inquiry-based teaching to underscore our efforts in adding knowledge in the area of discourse emerging during the implementation of science inquiry curriculum (Keys & Bryan, 2001). We next describe from our case study data the instructional talk of a teacher to highlight how teacher talk constrained and supported students' inquiry along the trajectory of scientific inquiry. As we do all these, we believe that researchers and educators will be better placed to understand what it means to conduct inquiry-based lessons in the real-world classroom.

Literature Review

While inquiry-based science teaching has been an increasingly prominent theme in science education, the transition from theory to practice has been unsatisfactory (Wilson, Taylor, Kowalski, & Carlson, 2010). Teachers are constantly informed by research to engage students in authentic investigative questions (Crawford, 2005) through project-based science instruction (Krajcik, Blumenfeld, Marx, & Soloway, 1994) or design-based inquiry (Chue & Lee, 2013). These instructional techniques while advantageous for students to actively apply scientific methods to answer meaningful questions about everyday experiences are challenging for teachers to implement (Bodzin & Beerer, 2003). Little is known about how teachers conceptualize and translate ideas about inquiry into classroom practices. This is not helped when the ubiquitous term of inquiry is seen both as a characteristics of classroom instruction and as a certain type of activity. Teachers are left to create their own ideas of what constitutes this form of teaching (Anderson, 2002). Critically, there is little research that investigates systematically how teachers use language to introduce, describe, and implement such inquiry programs for students (Hsu & Roth, 2009). In order to achieve a fuller understanding of how inquiry science is taught in the classroom, it is necessary to look at the discourses that emerge during the implementation of such curricula.

Importantly, this push for inquiry-based teaching highlights the role of language in science teaching and learning, which encompasses the three goals for science education: the conceptual, epistemic, and social (Duschl, 2008). By focusing on how teachers and students use language to formulate and assess knowledge claims, these processes of discourse have potential to offer evidences of conceptual understanding and changes in learner beliefs (Sandoval, 2014). Just as Lemke (1990) suggests that to learn science is to talk science, within the classroom community, members of this community need to bring together their respective experiences and local knowledge through the use of language for creation and communication of knowledge (Wells, 2000). Learners, when engaging in the discourses of science, will hence develop new repertoires for interacting with people, texts, and technologies which is part of the socialization process of education, learning to be scientifically literate (Kelly, 2007).

Therefore, teachers need a functional view of language as a prerequisite for learners to develop an understanding of inquiry (Bartholomew, Osborne, & Ratcliffe, 2004). Teachers need to use language to invite students to formulate arguments and evidences as they can direct how learners meet new discourses through interaction with different forms of language from myriad sources (Leach & Scott, 2003). While this requires strong instructional guidance for success in inquiry-based teaching (Kirschner, Sweller, & Clark, 2006), it certainly does not preclude talk that fosters students' capacity to reason in the classrooms (Alexander, 2006). Particularly instruction that emphasizes student-active thinking and drawing conclusion from data, these teaching strategies that engage students in the learning process through scientific investigations are more likely to increase conceptual understanding (Minner et al., 2010).

Yet, discourse-based studies in science education continue to focus on teachers' abilities to ask questions and provide students with feedback (Louca et al., 2012). Data is often gleaned from traditional full-class discussion that follows a triadic pattern in which the teacher initiates discussion by asking students questions for which responses are subsequently evaluated by the teacher (i.e., IRE). Such classroom interactions place teachers in a position of power in which they control the participants, direction of conversation topics, and what contributions count as legitimate without revealing other discursive features at work in crafting the inquiry lessons (Scott, Mortimer, & Aguiar, 2006; Lemke, 1990). While Mortimer and Scott (2003) suggest that teachers should engage in the dimensions of dialogic-interactive communication to attend to the different ideas emerging in a lesson, this cannot be the sole pattern of discourse in an inquiry science classroom. In fact, actual science teaching discourse is a complex event that requires teachers to also identify and interpret the context during science teaching (Louca et al., 2012).

As a result, we know even less how teachers use talk to manage and organize their inquiry-based teaching. This is especially so when teachers are concerned with classroom management rather than on melding theory and practices (Blumenfeld, Krajcik, Marx, & Soloway, 1994). Teachers often circumvent good intentions of inquiry programs in order to implement reform-based science curriculum (Jones & Eick, 2007), and we are not wiser with regard to how to help teachers who lack the experience in providing appropriate scaffolds for students through discourse during such inquiry-based teaching. Needed are more in-depth knowledge about how teachers give instructions/directions during inquiry-based lessons before research-based inquiry programs can be successfully implemented on a larger scale. This study therefore provides a close description of inquiry-based teaching implementation to suggest how a teacher can actually use talk to consolidate/scatter students' inquiry learning experience.

To demonstrate how teachers deploy language to implement an innovative science curriculum, we draw upon analytical procedures from discursive psychology. This requires us to understand language use as the arena of action (Wiggins & Potter, 2008). Moving away from the individual mind to processes that manifest only in and through talk (Potter, 2005), this approach aims to demonstrate how language sustains reality; "personal" views, beliefs, and attitudes are constituted through talking and writing (Gilbert & Mulkay, 1984). Hence, a discursive view provides a convenient route for us to account for the classroom phenomena such as cognition and learner ability as produced through teacher talk rather than assuming these as the end product of some unseen underlying psychological phenomena (Potter & Wetherell, 1987). With this useful tool that takes a rather agnostic viewpoint about the origin of human cognition, we bring analytical focus on a teacher's instructional discourse that accomplishes the social act of implementing science inquiry in the classroom. By doing so, we join other researchers to identify teacher repertoires in their idealized conceptions of learners, curriculum, and teaching (Hökkä, Eteläpelto, & Rasku-Puttonen, 2010; Hsu & Roth, 2009).

Context of Study

The cell energy module (CEM) is a school-based science-enrichment curriculum designed by Hugo, division head of the chemistry department in Karri High School, a high-achieving school in Singapore for the secondary three-chemistry cohort. We had first approached the science department of the school to offer teacher training sessions on how teachers can use a microbial fuel cell (MFC) to engage students in engineering design projects. Providing hands-on sessions lasting about 3 h each time, our research team demonstrated as well as assisted workshop participant teachers in learning about the setup and functions of the microbial cell. Our workshops ended with calls for teachers to design science-enrichment modules for their students to engage in inquiry learning using the microbial cell as an investigative tool. Hugo designed the CEM with a junior colleague, Jason, after a series of workshop and planned to implement the module over 15 weeks as a science-enrichment module for students interested in conducting science experiments.

The program focuses on giving students the authentic experience of tinkering with scientific tools for investigation. The program consists of a 15-week term of a 1.5-h lesson each week. The first six weeks aims to equip students with the necessary skills such as molarity calculation, preparing standard solutions, and knowledge of redox reactions to prepare them to investigate the effects of various factors affecting the voltage produced in a microbial cell. The remaining nine weeks comprises an investigative project where students are first led by the teacher to investigate the effect of a variable selected by the teacher, on the microbial cell. This is followed by students working in groups of four to investigate two other variables and their effects on the voltage of the microbial cell. Students are expected to plan, design, and conduct their science experiment in their own group and seek the help of their teacher when necessary. Students are also expected to present their project findings during the last two weeks of the program.

Method

In this study, we focus on the use of language and generalize from a sample of discourse from Hugo as he interacted with his students. Specifically, we focus on Hugo whose discourse serves as an example for the kind of discourses teachers use while implementing science inquiry programs. We first collected data by video recording the classroom sessions, and this resulted in 22.5 h of video-recorded lessons with Hugo as the main teacher in charge of the program. After he co-taught the first four lessons with Jason, he remained as the main teacher in charge of the lessons, while Jason assisted with logistics such as contacting the technician when the projector was not working. In addition, student artifacts (e.g., worksheets, reports) and students' written reflections were also collected. During the lessons, one video camera was directed at Hugo's teacher workbench in front of the science laboratory to capture his talk. In this way, researchers were able to zoom in on his talk as well as inscriptions on the whiteboard if any.

Hugo drew on multiple practices including lectures, experiments, and demonstrations to begin each lesson. After lecturing or demonstrating a certain phenomenon, Hugo would typically announce to students that they were free to get work on their experiments in their own groups. Students had to complete their investigation within a given time frame and return to their seats to listen to Hugo's summary of the lesson as part of the lesson closure. Thus, Hugo's classroom talk was an essential resource for students' investigation process and action in the laboratory.

Focusing on language in use, i.e., how Hugo used talk to thread the classroom events for the students, we analyzed the instructional talk that was available for making sense of Hugo's conception of curriculum and students. Firstly, we constructed an event map to document the classroom activities as they unfolded in each lesson. This provided a macro view of how the classroom activities led from one to another and provided a view of how the module was conducted over the 15 sessions. Importantly, this map localized the occurrence of instructional talk within time and space as a function of talk against the opportunities and constrains for the actor/teacher (Edley, 2001). From the field notes taken in the classroom, we noticed that these instructional talk episodes usually happened at the beginning and ending segments of each lesson. That is, Hugo introduced science activities or science concepts at the beginning of each lesson and related it to what inquiry science requires out of the student. Similarly, Hugo summarized his observations and opinion of the students' work during the ending segment of each lesson.

We transcribed all the segments and found patterns of variation and consistency in a range of features which allowed us to map the pattern of interpretative repertoires that Hugo was drawing on (Potter & Wetherell, 1995). We formulated tentative repertoires and descriptions and subjected these to discussion with members of our research team. We then tested the interpretative repertoires in the entire dataset. This was a tedious process that necessitates familiarity with the data and repeated readings of the transcripts. By reading and rereading the transcripts of Hugo's talk, certain images, metaphors, and figures of talk start to emerge (Edley, 2001). In this study, the analysis involved close reading of the transcriptions, and consequently all the sections which included curriculum talk were extracted. Thereafter, the data were thematized. This process was in many respects the same as in other types of qualitative analysis (especially thematic analysis) in which the purpose is to recognize the prevailing patterns of the data (Braun & Clarke, 2006). Ultimately, we derived two sets of interpretative repertoires presented here that influenced students' inquiry process.

Results

This section identifies two interpretative repertoires used in the teaching of inquiry science. Each of these repertoires demonstrates how Hugo had used spoken language to construct the performance of inquiry science in class. Through an in-depth analysis of the instructional discourse, we demonstrate the repertoires Hugo engaged to move students along a predictable track of inquiry science learning. However, his good intentions unknowingly subverted students' learning of investigative practices. This was largely due to how Hugo had inadvertently privileged a simplistic approach towards inquiry learning through two repertoires: inquiry science requires low-level content knowledge and inquiry science as a make-do scientific practice. We conclude by discussing how these two confounding factors hindered the implementation of higher-order scientific inquiry learning as possibly afforded by the tools present in the laboratory.

Assertion 1: Inquiry Science Requires Low-Level Content Knowledge

The interpretative repertoire of inquiry science requiring low-level content knowledge was constructed by Hugo in at least three ways. Firstly, Hugo introduced the CEM module by highlighting the two types of students in his class during the first lesson and attempted to level them in terms of their scientific competency. Namely, the triple science group of students taking biology, chemistry, and physics subjects often perceived as the more intelligent group of students, and the double science group of students taking only two science subjects, namely, chemistry and physics. Despite the differences, Hugo assured the double science group of students that they will not be disadvantaged as the content knowledge required for inquiry in the CEM will be based on students' prior knowledge:

> You will not be disadvantaged even if you are not in triple science. The thing I want to mention over here is, even though it is interdisciplinary, we are trying to include the three sciences, it will be based on what you know. (24/01/2013)

While implying that the double science stream students would only require basic components of chemistry, physics, and biology which they have learned in the general science subject during their lower secondary years, Hugo was quick to acknowledge that he would teach if students were lacking in any content knowledge. Subsequently, as Hugo suggested that the triple science group of students can "pick up things faster," he was also quick to defer back to the double science stream students, suggesting they were not any lesser in their intellectual abilities compared to their triple science counterparts. By downplaying the difference in intellectual abilities of the students, Hugo prepared the stage for his students to engage in inquiry science:

> ...and if you do not know, we will teach, so whether you are taking triple science or not, it does not really matter. But of course, if you are in triple science, maybe, you can pick up things faster, but, I will say, you will be at an advantage. And if you think of it this way, some of you are not taking triple science because you chose not to and because you were not able to. Correct? Am I correct to say that? So in your year 1 and year 2, science modules, you have already learnt the three sciences, and of course, certain things that we mention here may have been taught in your year 1 or year 2, so don't worry about that, we will do that in this module. (24/01/2013)

Secondly, Hugo constructed the CEM as requiring only a set of structured information, which Hugo saw as his responsibility to dispense. This set of knowledge included redox chemistry and laboratory skills such as weighing and preparation of standard solutions during the second lesson. Hugo engaged a didactic style of teaching after the introductory session, focusing on cookbook styles of chemistry experiments for students to experience redox reactions. These lectures and hands-on experiments were monitored by Hugo, and students played a passive role during Hugo's lectures and performed the redox experiments following given worksheets. Other topics covered also included fuel cell, differences between voltaic and electrochemical cells, and reactivity between metals, and no attempts were made to link these topics to how they might be related to the main objective of the CEM where students had to investigate factors affecting the voltage produced by a fuel cell. As Hugo had planned the curriculum and lectured according to what he felt was appropriate, it was not surprising that there was very little discussion or lectures from Hugo, who was a chemistry teacher, about the biological aspect of the fuel cell. While the topics were delivered through a lecture-style manner, students were generally quiet, and it was only during week 4 that students had a chance to have a glimpse of how microbes can be energy producers through a teacher's demonstration. During this demonstration, Hugo set up a microbial cell with a Visking tubing and carbon rods attached to a voltage data logger sensor. Students watched the experiment by crowding in front of the teacher's bench. While we may expect students to be given the opportunity to raise questions about microbial fuel cells, the demonstration was restricted to an observatory one with Hugo assuming that students have had forgotten how to use the electronic balance for weighing of dry yeast and "exempting" students from wearing safety goggles as he had deemed the demonstration as a "safe" experiment:

> For voltage, for preparation of slurry, look at instructions, how many of you can still remember how to use the electronic balance ? I assume you remember a bit but not all of it and so just for the benefit for all you, I will demonstrate how to use the electronic balance...This one is quite safe, no need goggles, since we are using yeast, it is still quite safe, I will exempt all of you from wearing goggles, just gather over here. (07/02/2013)

Students were also instructed to produce a one-page write-up after the demonstration to be handed up during week 7 of the module. The one-page write-up by students mostly contained statements about what microbial fuel cells are, how they work, and how they might be applied for daily life. While these one-page write-ups could have served as discussion boards for students to clarify their thoughts, Hugo only returned these write-ups at the end of the module, and no attempts were made to link students' investigations to their initial one-page write-up.

By restricting students' inquiry practices to a predetermined set of chemistry concepts during the first 4 weeks of the module, Hugo successfully portrayed inquiry science as requiring low-level content knowledge of redox, titration skills, preparation of standard solutions, and differences between voltaic and electrochemical cells. Lowering the bar further, Hugo justified for students why they would need to learn how to prepare standard solutions by directly helping them link to the investigative possibility of testing the effect of different concentrations of solution on the voltage produced in the fuel cell.

> The purpose of teaching is that when you are doing your MFC, some of you may want to dilute your solution, and what is important that you can transfer exactly the volume, and the variable you want to investigate?. ..K concentration, so that is the variable you want to set, and if you do not transfer the volume accurately, there is no way you can obtain that concentration. (28/2/2013)

Even as Hugo attempted to touch on concepts related to the fuel cell, he did not move beyond descriptive science as observed below when he was teaching the biology behind the fuel cell:

> This module over here, there is a lot of chemistry, biology and I would say a bit of physics. Like what you did just now, measuring voltage, that was physics. But the reactions that occur, that was chemistry. And we are looking at today, essentially biology. Ok, what do you understand by this words, microbes. Is it something big or something small? Ok, small, something small just like the word micro and over here, it refers to organisms that are rather small. Ok, can anyone give me examples? Examples? Any examples before I show you some examples. (7/2/2013)

Describing the functions of different microbes, drawing upon examples such as yogurt drinks, Hugo offered elementary ideas about biological reactions and did not venture into topics such as microbial reactions or how microbes such as yeast function as catalysts in fuel cells. Similarly, by confining voltage as a phenomenon in the realm of physics, Hugo conveniently binds the concept to the physic subject for which he has not been trained to teach and hence, does not need to further elaborate.

Assertion 2: Inquiry Science Is a Series of Make-Do Scientific Practice

While inquiry practices may provide students with opportunities to learn science at the conceptual, epistemic, and social level (Duschl, 2008), Hugo constructed the work of inquiry as a scientific practice that requires compromises. Firstly, Hugo demonstrated for students how experiments can be conducted and discussed in a laissez-faire manner. After he had demonstrated for students the Visking tubing experiment, he highlighted to students that he was not particular about how the experiment was set up:

> ...so that when you do the investigation of microbial fuel cell, as you notice just now as I tie up the tubing, I was not very particular about how I tie up the tube, just that, as long as it is secured, so for the volume, I am not particular, just about equal, so that is about. (7/2/2013)

After the demonstration, students worked in groups to repeat the Visking tubing experiment. When students had completed their experiment, Hugo questioned students about their voltage recordings and his previously recorded readings. However, it was a monologue where students were bombarded with various possible scenarios to consider without further discussion. Without giving space for students' response, Hugo portrayed students' investigation as the real work of doing science which can result in a variety of results and the way to obtain consistent results between groups would be to conduct experiments over a longer period of time:

> . . .and another reason for the difference could be the design, k, let's say we compare your results with another group results. Why is there a difference? Can anyone tell the difference? Is it the different volumes of the components used, what else? Notice the contents? Did you do that? Some of you may have just put in the components and your voltage may not be high compared to those who mix it well. K compare, is the voltage constant? Or is there a slight increase or slight decrease? Can we say that there is slight decrease? What's the reason for that? Probably need to investigate over a longer period of time. (14/02/2014)

Yet, there was no time for these students. While Hugo had constructed inquiry science as problematic in nature, students were not given any opportunities to discuss about how and why the voltage recordings differ. Hugo went on to conclude the lesson by informing students that they would keep the questions in mind for later weeks while they focus on how to prepare standard solutions for the following two weeks of lessons. This disjoint in classroom activities does not help students engage in inquiry practices. In fact, students will need to compromise on their results/analysis as Hugo delays the repetition of experimentation by spending the next two weeks on standard solution preparation and assessment. One may even suggest that the two-week gap may just eliminate any burning questions about the experiments that students had just observed and performed.

Secondly, when some groups of students faced technical issues in linking their fuel cells to the voltage sensors and were only able to obtain readings for one factor investigated, Hugo adjusted his assessment criteria for the investigative work of students. He stepped in to assure the class that he would be fair and changed the assessment criteria from having to report on the effects of two factors on voltage of cell to only one factor and its effect on voltage. This potentially removed the incentive for students who had successful investigated the change of two variables on the fuel cell to repeat their experiments:

> How many of you? Hands down. K I understand that some groups have some problems with the sensor, either the sensor was faulty, some things happened or you had forgotten to save the data. So in order to be fair to all, so what we will do, is to change the assessment criteria a bit. . ..for those who have investigated two variables, in your report just write one. For those who have problems today, don't worry. . . (28/3/2013)

Thirdly, Hugo presented cookbook style of instructions and comments for his students which inadvertently offered students a mock-up view of inquiry practices. Critically, this series of instructional talk supported the make-do scientific practice of Hugo which students subsequently emulated. For example, when students were preparing to write up a research proposal, Hugo provided a framework for students to follow and added that the purpose of the proposal was to enable teachers to assess the feasibility of students' ideas for investigation:

> . . .you will need to submit the proposal at the start of the lesson so that during the rest of lesson, we can go and tell you what needs to be changed. The teacher can tell you what to change, if we wait till the end, we will not have enough time to tell you what are the things for improvement and what should be included in the proposal. Factors that you are investigating, materials that you will need, for example, beaker, how many, write down

> specifically, because if you don't write down, later when you carry out the experiment, you will still be thinking of all these things. (28/3/2013)

While Hugo acknowledged to students that a research proposal was to help students "plan what you want to do for investigation," he also commented that "it is messy in a way but it is more about a collection of ideas, why you want to do it." These ideas were not often deemed feasible from Hugo's point of view and students had to amend their proposal. This was especially so for a group of students who were adamant about investigating the effect of temperature on the fuel cell. Students had mentioned in their reflection that they "were determined to create the most unique experiment" and were "demoralized" when Hugo "told them that the factor of temperature of environment was not feasible because the plastic material of the fuel cell was a poor conductor and using hot water or ice for temperature control would take a huge amount of time to complete the experiment."

Just as Hugo may have offered students the opportunity to write a research proposal that mapped out their inquiry plan, unknowingly, Hugo limited the depth in which students were able to investigate by simplifying their investigative process as one that does not have any correct answer and simply a way of life as demonstrated by his summative comments at the end of Lesson 7:

> ...hope you are not too lost, at the end of the investigation, the results may be very confusing for you. For example you expect that by changing a factor, you expect higher efficiency, but then, it may go the other way round, or you may not even detect any change. K that is fine, because after that, you need to think of what is the reason for this result, is there something that went wrong? Or is there something that you can do better ? K this is the whole part about the investigation. It is not just about getting result, it is not about getting the correct answer. And like what I have mentioned, there is no correct answer...and honestly, even you ask any of the teachers around here, we do not really know what happen when we change the factor. So it is about knowing what goes on in the investigation and sometimes, things do not go your way. And you need to find out why. You are clear. So do not be too concerned that there must be an answer, a correct answer. Things do not always work that way. Any questions? (7/3/2013)

During the student-led inquiry sessions which lasted for only two sessions, Hugo noticed that students were not getting the results they were hoping for. However, instead of discussing with students why the results were off their expectations, students' inquiry activities were watered down to requiring some luck for accurate results:

> ...For evaluation of results, if you are studying the factor of concentration, and you think that by increasing concentration the voltage will get higher, but your results proof otherwise, k can discuss, why did you results proof otherwise. K, this is what I talk about evaluation of results....I hope that through the experiment that you have done for these two weeks, sometimes you realize things do not work your way and that is actually how scientists actually conduct experiments. They do not expect everything to work well in the first time, if it is does, I would say, it is really luck. Not really skill. And if you can repeat the experiment and get the same results for more than 3 times, then that is actually skill. So sometimes it is really down to luck. (4/4/2013)

And since students did not have opportunities to repeat their experiments for the third time, it was quite easy for them to attribute their lack of accurate results to the lack of luck. This was also reflected in a student's reflection when she mentioned "in all experiments, fair and reliable results are always the ultimate goal. In this experiment, we were introduced to more precise and accurate instruments (fuel cell and voltage sensor), even though there were some factors which could not be controlled, I realized conducting a fair experiment was indeed not easy." Hence, it is also not surprising that another student reflected that his group "does not trust the settings of the computer" and that their experiments show that "not everything can be reliable at times."

Discussion

Understanding the impact of teachers' implementation of innovative science inquiry program is a complex task. This study focused on how instructional talk of an experienced science teacher shaped science inquiry activities in the classroom. Our findings revealed how a teacher had used two interpretative repertoires to shape inquiry science learning for his students. Firstly, by constituting inquiry science as requiring low-level content knowledge, Hugo structured content knowledge as bite-size information lacking in depth and assured students of different intellectual abilities that they possess sufficient knowledge in order to move them along their investigative work. Secondly, Hugo presented inquiry science as a make-do scientific practice, which involved compromising with inaccurate readings, lowering assessment criteria, and attributing results of experiments to luck. These discursive repertoires provided comprehensible and accountable resources for students' inquiry albeit in limited ways. Critically, these repertoires demonstrate how science inquiry is locally produced and exhibited in the classroom.

While our work illuminates the manner in which science is taught in the classroom, we need to acknowledge that it is difficult to make clear boundaries of particular repertoires (Potter, 1996). However, with repeated viewing of video data, readings of transcripts, and debate, we ultimately derived the two repertoires as presented in this study to sufficiently account for the instructional talk analyzed. Our results provide further evidence to what Lynch and Macbeth (1998) observed about scientific activities in the classroom: that science is not a packet of knowledge held by the expert community, but a series of distinctive mock-up situations of inquiry that constitute and animate the science classroom. Just as researchers often bemoan the fact that teachers adapt innovative research-based curriculum to present their own versions in the classroom (Fogleman, McNeill, & Krajcik, 2011), we may do well to recognize such adaptations as practical social accomplishments as much as enactments of curriculum, thematically played out across as scenes of practices in the larger scheme of scientific occasions (Lynch & Macbeth, 1998).

References

Alexander, R. (2006). *Towards dialogic teaching: Rethinking classroom talk*. Cambridge, UK: Dialogos.

Anderson, R. D. (2002). Reforming science teaching: What research says about inquiry. *Journal of Science Teacher Education, 13*(1), 1–12.

Bartholomew, H., Osborne, J., & Ratcliffe, M. (2004). Teaching students "ideas-about-science": Five dimensions of effective practice. *Science Education, 88*(5), 655–682.

Blumenfeld, P. C., Krajcik, J. S., Marx, R. W., & Soloway, E. (1994). Lessons learned: How collaboration helped middle grade science teachers learn project-based instruction. *The Elementary School Journal, 94*(5), 539–551.

Bodzin, A. M., & Beerer, K. M. (2003). Promoting inquiry-based science instruction: The validation of the science teacher inquiry rubric (STIR). *Journal of Elementary Science Education, 15*(2), 39–49.

Braun, V., & Clarke, V. (2006). Using thematic analysis in psychology. *Qualitative Research in Psychology, 3*(2), 77–101.

Chue, S., & Lee, Y. J. (2013). The proof of the pudding?: A case study of an "at-risk" design-based inquiry science curriculum. *Research in Science Education, 43*(6), 2431–2454.

Cohen, D. K. (1990). A revolution in one classroom: The case of Mrs. Oublier. *Educational Evaluation and Policy Analysis, 12*(3), 311–329.

Collins, K. M., Palincsar, A. S., & Magnusson, S. J. (2005). Science for all: A discursive analysis examining teacher support of student thinking in inclusive classrooms. In *Establishing scientific classroom discourse communities: Multiple voices of teaching and learning research* (pp. 199–224). Mahwah, NJ: L. Erlbaum Associates.

Crawford, T. (2005). What counts as knowing: Constructing a communicative repertoire for student demonstration of knowledge in science. *Journal of Research in Science Teaching, 42*(2), 139–165.

Duschl, R. (2008). Science education in three-part harmony: Balancing conceptual, epistemic, and social learning goals. *Review of Research in Education, 32*(1), 268–291.

Edley, N. (2001). Analysing masculinity: Interpretative repertoires, ideological dilemmas and subject positions. In M. Wetherell, S. Taylor, & S. J. Yates (Eds.), *Discourse as data. A guide for analysis* (pp. 189–228). Milton Keynes, UK: Open University.

Fogleman, J., McNeill, K. L., & Krajcik, J. (2011). Examining the effect of teachers' adaptations of a middle school science inquiry-oriented curriculum unit on student learning. *Journal of Research in Science Teaching, 48*(2), 149–169.

Gilbert, N., & Mulkay, M. (1984). *Opening Pandora's box: A sociological analysis of scientists' discourse*. Cambridge, UK: Cambridge University Press.

Hökkä, P., Eteläpelto, A., & Rasku-Puttonen, H. (2010). Recent tensions and challenges in teacher education as manifested in curriculum discourse. *Teaching and Teacher Education, 26*(4), 845–853.

Hsu, P. L., & Roth, W. M. (2009). An analysis of teacher discourse that introduces real science activities to high school students. *Research in Science Education, 39*(4), 553–574.

Jones, M. T., & Eick, C. J. (2007). Implementing inquiry kit curriculum: Obstacles, adaptations, and practical knowledge development in two middle school science teachers. *Science Education, 91*(3), 492–513.

Kelly, G. J. (2007, May). *Scientific literacy, discourse, and knowledge*. In: A paper presented at the Linnaeus Tercentenary 2007 symposium "Promoting scientific literacy". Uppsala: Uppsala University.

Keys, C. W., & Bryan, L. A. (2001). Co-constructing inquiry-based science with teachers: Essential research for lasting reform. *Journal of Research in Science Teaching, 38*(6), 631–645.

Kirschner, P. A., Sweller, J., & Clark, R. E. (2006). Why minimal guidance during instruction does not work: An analysis of the failure of constructivist, discovery, problem-based, experiential, and inquiry-based teaching. *Educational Psychologist, 41*(2), 75–86.

Krajcik, J. S., Blumenfeld, P. C., Marx, R. W., & Soloway, E. (1994). A collaborative model for helping middle grade science teachers learn project-based instruction. *The Elementary School Journal, 94*(5), 483–497.

Leach, J., & Scott, P. (2003). Individual and sociocultural views of learning in science education. *Science & Education, 12*(1), 91–113.
Lee, Y.-J. (2008). Thriving in-between the cracks: Deleuze and guerrilla science teaching in Singapore. *Cultural Studies of Science Education, 3*, 917–935.
Lee, Y.-J. (2014). Science education in a straightjacket: The interplay of people, policies, and place in an East-Asian Developmental State. In A.-L. Tan, C. L. Poon, & S. S. L. Lim (Eds.), *Inquiry into the Singapore science classroom: Research and practices* (pp. 151–171). Dordrecht, The Netherlands: Springer.
Lefstein, A. (2008). Changing classroom practice through the English National Literacy Strategy: A micro-interactional perspective. *American Educational Research Journal, 45*(3), 701–737.
Lemke, J. (1990). *Talking science: Language, learning, and values.* Norwood, NJ: Ablex Publishing.
Louca, L. T., Zacharia, Z. C., & Tzialli, D. (2012). Identification, interpretation—Evaluation, response: An alternative framework for analyzing teacher discourse in science. *International Journal of Science Education, 34*(12), 1823–1856.
Lynch, M., & Macbeth, D. (1998). Demonstrating physics lessons. In J. G. Greeno & S. V. Goldman (Eds.), *Thinking practices in mathematics and science learning* (pp. 269–298). Mahwah, NJ: Lawrence Erlbaum.
Márquez, C., Izquierdo, M., & Espinet, M. (2006). Multimodal science teachers' discourse in modeling the water cycle. *Science Education, 90*(2), 202–226.
Minner, D. D., Levy, A. J., & Century, J. (2010). Inquiry-based science instruction—What is it and does it matter? Results from a research synthesis years 1984 to 2002. *Journal of Research in Science Teaching, 47*(4), 474–496.
Mortimer, E., & Scott, P. (2003). *Meaning making in secondary science classrooms.* Milton Keynes, UK: Open University Press.
National Research Council (Ed.). (1996). *National science education standards.* Washington, DC: National Academy Press.
Osborne, J. (2010). Arguing to learn in science: The role of collaborative, critical discourse. *Science, 328*(5977), 463–466.
Potter, J. (1996). *Representing reality. Discourse, rhetoric and social construction.* London, UK: Sage.
Potter, J. (2005). Making psychology relevant. *Discourse and Society, 16*(5), 739–747.
Potter, J., & Wetherell, M. (1987). *Discourse and social psychology: Beyond attitudes and behaviour.* London, UK: Sage.
Potter, J., & Wetherell, M. (1995). Discourse analysis. In J. Smith, R. Harré, & R. van Langenhove (Eds.), *Rethinking methods in psychology* (pp. 80–92). London, UK: Sage.
Roehrig, G. H., & Luft, J. A. (2004). Constraints experienced by beginning secondary science teachers in implementing scientific inquiry lessons. *International Journal of Science Education, 26*(1), 3–24.
Roth, W. M. (2008). The nature of scientific conceptions: A discursive psychological perspective. *Educational Research Review, 3*(1), 30–50.
Sandoval, W. (2014). Science education's need for a theory of epistemological development. *Science Education, 98*(3), 383–387.
Scott, P. H., Mortimer, E. F., & Aguiar, O. G. (2006). The tension between authoritative and dialogic discourse: A fundamental characteristic of meaning making interactions in high school science lessons. *Science Education, 90*(4), 605–631.
Wells, G. (2000). Dialogic inquiry in education: Building on the legacy of Vygotsky. In C. D. Lee & P. Smagorinsky (Eds.), *Vygotskian perspectives on literacy research: Constructing meaning through collaborative inquiry* (pp. 51–85). Cambridge, UK: Cambridge University Press.
Wiggins, S., & Potter, J. (2008). Discursive psychology. In S. Willig & W. Stainton-Rogers (Eds.), *The Sage handbook of qualitative psychology* (pp. 73–91). London, UK: Sage.
Wilson, C. D., Taylor, J. A., Kowalski, S. M., & Carlson, J. (2010). The relative effects and equity of inquiry-based and commonplace science teaching on students' knowledge, reasoning, and argumentation. *Journal of Research in Science Teaching, 47*(3), 276–301.
Yore, L. D., & Treagust, D. F. (2006). Current realities and future possibilities: Language and science literacy—Empowering research and informing instruction. *International Journal of Science Education, 28*(2–3), 291–314.

Chapter 14
Practice and Effect of Lessons on Inquiry Activities in Senior High School Chemistry: Focusing on Students' Attitudes Toward Chemistry

Takuya Kamon and Hiroki Fujii

Introduction

Inquiry activities in high school science curriculum in Japan were introduced in the 1970s, brought by the influence of a modernization movement of science education in the USA (Ministry of Education, Science and Culture, 1973). The science education in those days was focused on fostering students' skills in science through inquiry and also on enhancing their understanding of basic concepts of science through the process of inquiry. Ever since, inquiry activities have been considered as of great importance in high school science lessons. Recently, promoting students' abilities in scientific thinking, judgment, and expression through inquiry learning is also emphasized. In the first part of this chapter, we present the current situations of lessons on inquiry activities in senior high school science, with particular emphasis on chemistry.

Lessons on inquiry activities may contribute not only to promote students' skills, understanding, and abilities in science but also to improve their attitudes toward science. By highlighting chemistry, previous studies have shown that lessons on inquiry activities improve students' attitudes toward chemistry. However, such studies were not observed in Japan. For this reason, in the second part of this chapter, we present a lesson on inquiry activities with the subject matter content on acid/base and neutralization in senior high school chemistry and examine the validity of such a lesson in terms of changes in students' attitudes toward chemistry.

The guidelines of the lesson based on inductive experiments are as follows: conducting guided inquiry pattern on which teachers give the topics of activities to

---

T. Kamon
Saidaiji Senior High School, 2-1-17, Saidaijiue, Higashi-Ku, Okayama 704-8112, Japan

H. Fujii (✉)
Department of Science Education, Graduate School of Education, Okayama University, 3-1-1, Tsushimanaka, Kita-ku, Okayama 700-8530, Japan
e-mail: fujii-hi@okayama-u.ac.jp

M.S. Khine (ed.), *Science Education in East Asia*,
DOI 10.1007/978-3-319-16390-1_14

students, adapting students' discussion for planning experiments, and selecting of contents relevant to students' daily life experiences. The students are divided into two groups, namely, the experimental group who will take the lessons based on inductive experiments and the control group who will follow the usual lessons on deductive experiments.

After conducting the lessons in a five-unit period for 3-week duration for both groups, the changes in students' attitudes toward chemistry were evaluated based on an 18-item questionnaire. Unfortunately, the results showed that there was only one item that was significantly different for the experimental and control groups. We finally present some points as major causes that led to such results and also suggest that these points must be discussed in future studies in order to improve students' positive attitude toward chemistry through lessons on inquiry activities.

School Chemistry and Inquiry Activity

School Curriculum

According to "Course of Study" designated as the official guidelines for school teaching in Japan (Ministry of Education, Culture, Sports, Science and Technology, 2009), science subjects for senior high school are composed of the following ten subjects: science and our daily life, basic physics, basic chemistry, basic biology, basic earth science, advanced physics, advanced chemistry, advanced biology, advanced earth science, and science project study. The aims of basic chemistry and advanced chemistry are shown in Table 14.1. One of the important aims is to promote students' abilities in inquiry and attitudes toward inquiry. In conformity with this aim, inquiry activities are incorporated in each learning content. Subsequently, the points to note in both subjects are also shown in Table 14.1. Some of the important points to note are to promote students' abilities in scientific thinking, judgment, and expression through inquiry learning and at the same time to foster various skills through inquiry including collecting information. The former is emphasized in Course of Study, as it stands now.

Science project study is a subject that is newly organized to make further promotion of students' abilities in inquiry and attitudes toward inquiry. The aim of the subject is shown in Table 14.1. The distinguishing feature is to develop students' creativity, which is one of the current concerns in the field of science education in Japan. Subsequently, the points to note in science project study areas are as follows: introducing students into setting proper research topics by themselves; collaborating with research institutes like university, science museum, and the like in order to progress students' research; and providing an opportunity for students to present a report based on the results of the research. Therefore, the inquiry activities of science project study mean "open inquiry" (Allen, Barker, & Ramsden, 1986), in which students make topics of the activities by themselves. The

Table 14.1 Aim of and points to note in basic chemistry, advanced chemistry, and science project study

Basic chemistry
The aims of basic chemistry are as follows: to enhance students' interest in substances and the changing in relation to their daily life and society; to promote their abilities and attitudes on inquiry in chemistry, conduct observations, and experiments with a sense of purpose; to promote their understanding of basic concepts and principles/rules of chemistry; and finally to foster their scientific viewpoint and way of thinking
The points to note in basic chemistry are as follows: (1) to promote students' formation of basic concepts of science, considering a relation between junior high school science and senior high school science, and to promote their abilities in scientific thinking, judgment, and expression through learning about a way of inquiry in chemistry; (2) regarding inquiry activities, to set opportunities for students to make and present a report on observations and experiments in relation to each learning content; and to foster various skills on a way of inquiry like collecting information, formulating hypothesis, planning of experiment, verifying by experiment, analyzing/interpreting data of experiment, *etc.*
Advanced chemistry
The aims of advanced chemistry are as follows: to enhance students' inquiring mind on objects and phenomena of chemistry; to promote their abilities and attitudes on inquiry in chemistry, conduct observations, and experiments with a sense of purpose; to promote their deep understanding of basic concepts and principles/rules of chemistry; and finally to foster their scientific view of nature
The points to note in advanced chemistry are follows: (1) to promote students' formation of basic concepts of science, considering a relation between basic chemistry and advanced chemistry, and to promote their abilities in scientific thinking, judgment, and expression through learning about a way of inquiry in chemistry; (2) inquiry activities in this level are the same as the abovementioned "points to note in basic chemistry" portion
Science project study
The aims of science project study are as follows: to promote students' abilities and attitudes on scientific inquiry and to foster their basis of creativity through conducting research on the set of topics of science with observations and experiments

students conduct a long-term (during few months to 1 year) research about the topic.

Regarding the selection of subjects related to chemistry, almost all of science course students complete both basic chemistry and advanced chemistry because they need to study both subjects in taking entrance examination under the Faculty of Sciences in the university. On the other hand, literature course students have a tendency to complete just basic chemistry. Although there is such difference on each course, both students have an opportunity to learn inquiry activities in chemistry.

School Textbook

All school textbooks in Japan are edited by textbook publishers in conformity with the designated Course of Study. Examples of inquiry activities covered in basic and

Table 14.2 Themes of inquiry activities in chemistry textbooks

Basic chemistry
Chemistry and human life
Various plastics and metals
Separation of mixture
Chromatography
Constitution of substance
Writing out periodic table by using computer
Polarity of molecule presumed from model of molecule
Chemical bond and property of substance
Change of substance
Quantitative relations of chemical change
Drawing up titration curve by using spreadsheet program
Distinction of acid, base, and salt
Quantitative relations of oxidation-reduction reaction
Establishment of the atomic theory and the molecular theory
Electrolytic refining of cupper (utilization of oxidation-reduction reaction)
Advanced chemistry
State and equilibrium of substance
Charles' law
Measurement of molecular weight by using the freezing point depression
Change of substance and equilibrium
Verification of Hess's law by using computer
Verification of Faradays' law
Reaction of Daniell cell
Measurement of reaction rate
Quantification of chloride ion in aqueous solution by Mohr's method
Inorganic substance
Halogen and halogen compound
Preparation and property of NO and NO_2
Separation of metal cation
Organic compound
Synthesis and property of soap
Selective reaction of functional group
Separation and identification of aromatic compound
High molecular compound
Melting and remaking of polystyrene foam
Synthesis of cuprammonium rayon
Chemical distinction of plastic goods
Property of lysozyme as enzyme

advanced chemistry textbooks are shown in Table 14.2 (Saito et al., 2011; 2012). The inquiry activities are divided into two categories of the activities based on inductive experiments (italicized themes in Table 14.2) and those based on deductive experiments (non-italicized themes in Table 14.2). In the textbooks, the latter is shown more than the former.

The inquiry activities based on inductive experiments follow guided inquiry pattern in which students plan and do experiments by themselves after teachers give the topics of activities to them. For instance, the topics are as follows.

Basic Chemistry, Separation of Mixture
Let's separate salt (sodium chloride) and silica gel from a mixture of salt, limestone (calcium carbonate), and silica gel powder. Moreover, let's identify whether separated salt and silica are pure, namely, that these substances don't contain other substances.

The inquiry activities based on deductive experiments follow a pattern in which students conduct experiments after teachers have assigned the topics of activities and explained the ways of conducting the experiments. The pattern is called "confirmation inquiry" (Cheung, 2008). For instance, the topics are as follows.

Basic Chemistry, Chromatography
Each substance has an inherent property. Using differences in such property, we can separate a mixture of two and more kinds of substances. Based on an inherent property, we can also make a distinction, detection, and identification of the substance. Here, let's identify by paper chromatography that the ink of water-based pen is a mixture of coloring matters.

Subsequently, the textbooks show how to progress inquiry activities. The inquiry activities that lead to the presentation of the report are composed of the following procedures: selecting topics of experiment, formulating hypothesis, planning experiment, conducting experiment, analyzing results, making recommendations, preparing a report about the topics, and presenting the report.

Moreover, the textbooks introduce the ways of preparing a report and the slides for presentation. Thus, the textbooks are edited with an intention of promoting students' understanding about inquiry activities in chemistry.

School Practice

In the lessons of basic chemistry and advanced chemistry, there is an actual condition that attaches too much importance to teaching a body of science. The teacher cannot spare enough time for inquiry activities. Indeed, they are limited to conduct some themes of inquiry activities that are introduced in the textbooks. This tendency is, in most cases, caused by university entrance examination that is focused on assessing students' understanding and knowledge of science. However, the entrance examination that is focused on assessing their scientific thinking is gradually increasing. Moreover, the admission system in universities has diversified in recent years. Most of the universities' admission systems are based on recommendations, in which the products of inquiry activities in school science are

Table 14.3 Themes of inquiry activities in science project study

Property of superabsorbent polymer
Relations between color development of plant pigment and groups of mordanting metals
Electrolysis of tin and formation of "metal tree"
Relations between molecular structure of azo compounds and absorption spectrum
Synthesis of ester
Chemistry for environmental problems: absorption of CO_2 by lithium silicate
Determining concentration of HNO_3 not to generate NO_2
Conditions to efficiently produce bioethanol

evaluated. Therefore, the awareness of the importance of promoting inquiry activities is increasing, and consequently, the necessity for improving chemistry lessons is increasing as well.

Subsequently, science project study is normally conducted for science course students. Some high schools lay emphasis on science education, and they are called "Super Science High School" (204 schools in the whole of Japan on 1st April 2014). These high schools are developing inquiry activities in science. For instance, themes of the activities in an unnamed school are shown in Table 14.3. The open inquiry in which students make the topics of activities by themselves is adapted in these themes.

Based on the aforementioned current situations of inquiry activities in senior high school chemistry in Japan, we present hereinafter a lesson on inquiry activities that highlights students' attitude toward chemistry. The reason for this focus is that the lesson on inquiry activities may contribute not only to promote students' scientific thinking, judgment, and expression in chemistry but also to improve their attitudes toward chemistry.

Research

Background and Goal

It is a long time since students' alienation from science subjects was advocated as a serious problem in Japan. The result from the ROSE project (Sjøberg & Schreiner, 2010) showed that Japanese students' interest in school science was at a low level in international comparisons. The results from PISA 2006 (National Institute for Educational Policy Research, 2007) and TIMSS 2011 (National Institute for Educational Policy Research, 2013) revealed that the students' self-confidence and positive perceptions toward science learning were also low. Subsequently, the results from a domestic survey in Japan accompanying with PISA 2006 (National Institute for Educational Policy Research, 2008) suggested that senior high school students did not have enough motivation for learning science compared to junior high school students. Moreover, PISA 2006 and TIMSS 2011 reported that Japanese students' positive feelings toward usefulness of science and technology were also low.

Interest in school science, self-confidence, and positive perception toward science learning, motivation for learning science, and feelings toward usefulness of science and technology belong to the fields of emotion and will. Therefore, these are generally considered as attitudes toward science and school science (Simon & Osborne, 2010). Improving students' attitudes toward those is one of the important tasks of science education.

Previous studies have shown that lessons on inquiry activities improve students' attitudes toward science and school science (European Commission, 2004; 2007). Chemistry is no exception to this situation (Colburn, 2000; Roehrig & Luft, 2004). For instance, Palmer (2009) mentioned that experiencing experiments based on inquiry activities aroused junior high school students' interest in chemistry. Lewis, Shaw, Heitz, and Webster (2009) mentioned that the lessons on inquiry activities improved university students' self-concept like self-confidence in learning chemistry. Putti (2011) indicated that lessons that emphasize the making of reports in inquiry activities exerted favorable influence upon senior high school students' attitudes toward chemistry. Moreover, Lin, Hong, Chen, and Chou (2011) mentioned that an acceptance of questions in inquiry activities in chemistry, e.g., "What if..." and "Re-see," enhanced students' motivation to learn science.

Although aforementioned studies are composed of various subjects and types of the lessons, all studies prove that the lessons on inquiry activities are effective in improving students' attitudes toward chemistry. However, such studies were not observed in Japan, while there were many reports concerning to the practices of those lessons (e.g., Murakami, 2010; Okayama Prefectural Education Center, 2001).

For this reason, in this study, a lesson on inquiry activities in senior high school chemistry will be presented, and the validity of such lesson will be examined with respect to changes in students' attitudes toward chemistry.

Development of the Lesson

Previous studies focused on whether students' attitudes toward chemistry varies, depending on type, term, and content of the developed lessons and at the same time in number, age, and academic achievement of students as a subject of the study. It is therefore difficult to identify which kind of the lessons are effective in improving students' attitudes. Nevertheless, most studies (Chattejee, Williamson, McCann, & Peck, 2009; Lewis et al., 2009; Lin et al., 2011; Palmer, 2009; Putti, 2011; Vaino, Holbrook, & Rannikmae, 2012) commonly adopt "guided inquiry" in which teachers give the topics of activities to students, while others use "open inquiry" in which students make the topics by themselves. For instance, Chatterjee et al. (2009) mentioned that comparing the results of chemistry lessons on open inquiry and those on guided inquiry in a liberal arts course in university, the latter was more effective in improving students' attitudes toward chemistry.

Inquiry activities are composed of the following procedures: setting of task, formulating hypothesis, planning experiment, conducting experiment, analyzing results, and making recommendations. Some studies pay attention to the phases before and after the experiment. Palmer (2009) showed that not only a phase of experiment but also a phase of proposal was important to enhance students' interest in chemistry. Putti (2011) mentioned that discussions before and after the experiment contributed to improving students' attitudes toward chemistry.

Moreover, Vaino et al. (2012) introduced the lesson contents highly relevant to students' daily life and society, e.g., "Should we use vegetable oils as a fuel?" and "Can alcohols help human life?". They mentioned that these lessons enhanced students' motivation for learning chemistry.

Consequently, we develop the lessons on inquiry activities using the following guidelines:

- Conducting guided inquiry in which teachers assign the topics of activities to students
- Adapting students' discussion for planning experiments
- Selecting of contents relevant to students' daily life experiences

Trial of the Lesson

Subject and Term of the Lesson

The subject of the study was the first year students of a prefectural senior high school in Western Japan. The students were chosen for the reason that they would have various characteristics regarding attitudes toward chemistry, because they were not divided into science course and literature course yet. Subsequently, an experimental group (105 students of three classes) and a control group (70 students of two classes) were organized. The former group carried out the lessons based on inductive experiments, while the latter group carried out the lessons based on deductive experiments. There was no big difference between both groups in terms of the results of an examination in chemistry just before the trial of the lessons.

The trial of the lessons was conducted for both groups from 5th to 26th of March 2013. The lessons were composed of five periods having 50 min for each period.

Contents of the Lesson

The subject matter content of the lessons was acid/base and neutralization in basic chemistry (Table 14.4). Before the lesson presentation, the students of both groups have already learned the contents of definitions of acid/base, acidity/basicity, and

Table 14.4 Subject matter contents of the lessons

Period	Exp. group		Cont. group	
1st	Practice for improving basic skills on experiment			
2nd	Neutralization	Experiment about quantitative relations	Neutralization	Experiment of acid-base titration
3rd		Experiment of acid-base titration		Experiment about quantitative relations
4th	pH of chemical substances in daily life	Plan and presentation of experimental methods	pH of chemical substances in daily life	Experiment by using red cabbage
5th		Experiment planned by students		Experiment of electrophoresis on ion

neutralization reaction and salt. In addition, the students from the control group have learned a formula on quantitative relations of neutralization reaction.

In the first period of the lessons, the students of both groups prepared the pH test paper by soaking an extraction of perilla leaf in filter paper and made an experiment on changes in the colors of the paper by acid/base reagents. The experiment was conducted with the aim of improving students' basic skills of handling test tube, dropping pipet, operating the pH meter, and *etc.*, and at same time that of fostering their understanding of how to record the results of experiments.

From the second to the fifth periods of the lessons, the students of both groups learned the same contents similar to the first period. However, the students from the experimental group took the lessons based on inductive experiments, while the students from the control group took it based on deductive experiments.

The topic of activities in the second and third periods was on neutralization reaction. In line with conducting the guided inquiry, the teacher gave the topic to the students. In the second period, for the students from the experimental group, the teacher presented diluted hydrochloric acid in a test tube whose concentration and volume were already known. In order to neutralize the acid perfectly, the students presumed the volume of sodium hydroxide with same concentration of the hydrochloric acid, the volume of that with different concentration, and the volume of calcium hydroxide with same concentration. They prepared the presumed bases, mixed the acid with the base, and made a judgment through changes in colors of universal pH indicator whether the acid was perfectly neutralized with the base. On the basis of this result, they deduced a formula on quantitative relations of neutralization reaction that displayed the relations among valence, concentration, and volume of acid/base.

In the third period, the teacher presented a standard solution of oxalic acid for the students to conduct an experiment of acid-base titration with a buret. The students neutralized the acid with sodium hydroxide with unknown concentration and recorded the volume of the dropped base. Applied the results to the formula on quantitative relations of neutralization reaction, they calculated accurate concentration of the sodium hydroxide.

On the other hand, after reviewing the formula on quantitative relations of neutralization reaction, the students from the control group conducted the same experiment of acid-base titration in the second period, the same experiment done by the experimental group in the third period. In the third period, the students conducted the same experiment by the students from the experimental group in the second period and confirmed that the formula holds.

Subsequently, the topic of activities in the fourth and fifth periods was to examine the pH of chemical substances in daily life. Similar to the second and third periods, the teacher gave this topic to the students. The teacher prepared chemical substances like table vinegar, lemon juice, carbonated water, soapy water, detergents for kitchen and for toilet, and lime water as discussed in the lesson guide on selecting contents relevant to students' daily life experiences.

In the fourth period, the students from the experimental group are divided into small groups of four to five persons and have considered how to examine the pH of these substances. Referring to bibliographical sources in chemistry, they discussed and presented the ways of experiments, experimental apparatuses, and presumed results of the experiments. These activities fitted in the guideline of the lessons on adapting students' discussion for planning experiments.

In the fifth period, the students examined acidity/basicity of the substances by the experiments planned by students. They understood that there was a variety from high acidity to high basicity. Thus, the students took the lesson on inquiry activities composed of setting of task, formulating hypothesis, planning of experiment, conducting of experiment, analyzing of results, and making recommendation.

Shown below are the lists of ways planned by the students to conduct the experiment:

- Using an extraction of purplish cabbage: The pH of chemical substances in daily life was presumed through changes in the color after dropping the extraction in the substances. The 13 groups of 23 groups proposed this way. One group chose black soybean and another one group chose skin of eggplant instead of purplish cabbage.
- Using plural pH indicators: Several pH indicators having different color change intervals like thymol blue, methyl orange, methyl red, bromothymol blue, phenolphthalein, *etc.*, and litmus paper were combined. The pH of the substances was presumed through changes in the colors of the indicators. Six groups were proposed this way. Each group differed in the choices of pH indicators.
- Using pH meter and universal pH paper: The pH of the substances was taken an accurate measurement by using known apparatuses and materials. Six groups were proposed this way.
- Using pH test paper by an extraction of perilla leaf: The method was introduced by the teacher in the first period of the lessons. Two groups adapted it.

The students from the control group did not plan experiments in the fourth period. After the teacher proposed the use of purplish cabbage as the only experimental way, the students examined the pH of chemical substances in daily life. In the fifth period, they examined acidity/basicity of these chemical substances by applying a method of electrophoresis on oxonium ion and hydroxide ion. This method was also proposed by the teacher.

The method of electrophoresis, which the students from the control group had conducted in the fifth period, was set so that the students from the experimental group would come up some proposals on how to conduct it during the planning experiments. However, in actuality, those students from the experimental group were not able to do it. Therefore, in the end of the fifth period, the teacher introduced the method to those students with the aim that the subject matter content between the two groups might not differ.

Evaluation of the Lesson

Data Collection

Before and after the trial of the lessons, a survey with a questionnaire was conducted. The aim of the survey was to clarify the changes in the students' attitudes toward chemistry through the lessons on inquiry activities. As previously mentioned, the attitudes include the fields of emotion and will like interest in school science, students' self-confidence and positive perception toward science learning, motivation for learning science, and feelings toward usefulness of science and technology.

The questionnaire with 18 items was a modified version of Attitude Toward Chemistry Lessons Scale (Cheung, 2009) that was developed to measure students' attitudes toward chemistry lessons. The scale was composed of three items each in four dimensions: liking for chemistry theory lessons, liking for chemistry laboratory work, evaluative beliefs about school chemistry, and behavioral tendencies to learn chemistry (non-italicized items in Table 14.5). The scale included items pertaining to chemistry lessons as well as chemistry as a subject and a science. It also covered various items on attitudes. Thus, it was a good fit for the survey in the current study. Six additional items measuring students' positive perceptions toward chemistry learning, self-confidence in chemistry learning, and feelings toward usefulness of chemistry (National Institute for Educational Policy Research, 2007, 2012; Sjøberg & Schreiner, 2010) were also added to the scale (italicized items in Table 14.5).

The items on the questionnaire were rated on a four-point scale; (1) disagree, (2) partly disagree, (3) partly agree, and (4) agree. The survey with the questionnaire was conducted on the 5th of March before the lessons and on the 26th of March 2013 after the lessons, with a duration of ten minutes.

Table 14.5 Items of the questionnaire

Liking for chemistry theory lessons
Q1. I like chemistry more than any other school subjects
Q2. Chemistry lessons are interesting
Q4. Chemistry is one of my favorite subjects
Q5. Chemistry lessons are fun
Q14. I have confidence in studying chemistry
Liking for chemistry laboratory work
Q6. I like to do chemistry experiments
Q7. When I am working in the chemistry lab, I feel I am doing something important
Q10. Doing chemistry experiments in school is fun
Evaluative beliefs about school chemistry
Q3. Chemistry is useful for solving everyday problems
Q8. People must understand chemistry because it affects their lives
Q17. Chemistry is one of the most important subjects for people to study
Q9. If I study chemistry, my daily life becomes meaningful
Q11. Chemistry lessons are useful for thinking something scientifically
Q16. Chemistry lessons are useful for getting a job I desire
Q18. Chemistry contributes to the development of our society
Behavioral tendencies to learn chemistry
Q12. I am willing to spend more time reading chemistry books
Q13. I like trying to solve new problems in chemistry
Q15. If I had a chance, I would do a project in chemistry

Data Analysis and Results

The 83 students from the experimental group and the 62 students from the control group responded to all items on the questionnaire before and after the lessons. The responses of the students were the data for the analysis.

Validity of Student Responses to Items of the Questionnaire

The mean and the standard deviations were first calculated from the pre- and post-questionnaires. Because the mean was in a range from 1.71 to 3.09 and the standard deviation was in a range from 0.73 to 0.99, there were no ceiling effects or floor effects for all items. Therefore, the students' responses to all items were valid.

Subsequently, the factor analysis through the principal factor method/promax rotation was done for the 18 items. Based on range attenuation of eigenvalue (8.87, 1.80, 1.42, 0.80, 0.70 ...) and an interpretability of factors, three factors were extracted (Table 14.6). The factor 1 was composed of nine items. Five of these items (Q1, Q2, Q4, Q5, and Q14) belonged to a dimension of liking for chemistry theory lessons that was set beforehand. The factor 2 composed of seven items. The six of those seven items (Q3, Q8, Q9, Q11, Q17, and Q18) belonged to a

Table 14.6 Results of factor analysis

Question numbers	Factor 1	Factor 2	Factor 3
1	**.84**	−.09	.08
2	**.69**	.04	.24
3	.04	**.67**	−.08
4	**.92**	−.06	.02
5	**.57**	−.02	.35
6	.05	−.02	**.84**
7	.26	**.44**	.16
8	−.13	**.88**	.00
9	.32	**.49**	.01
10	−.08	.05	**.89**
11	.12	**.70**	.04
12	**.59**	.15	.02
13	**.64**	.15	−.08
14	**.88**	−.07	−.11
15	**.68**	.17	−.01
16	**.66**	.25	−.13
17	.21	**.65**	−.03
18	−.05	**.68**	.11

(Factor pattern after promax rotation)

dimension of evaluative beliefs about school chemistry. The factor 3 included two items. Both items (Q6 and Q10) belonged to a dimension of liking for chemistry laboratory work.

Although the Q12, Q13, and Q15 were expected to be in a dimension of behavioral tendencies to learn chemistry, these items belonged to a dimension of liking for chemistry theory lessons in fact. The Q7 and Q16 belonged, respectively, to a dimension of evaluative beliefs about school chemistry and to a dimension of liking for chemistry theory lessons. Therefore, the measured items for students' attitudes toward chemistry belonged to three dimensions.

Effectiveness of the Lessons

The mean of student responses to the items is shown in Table 14.7. The mean of experimental groups' responses increased for nine items (Q3, Q4, Q7, Q8, Q9, Q13, Q14, Q16, and Q17), while that of control groups' responses increased for seven items (Q1, Q3, Q5, Q7, Q9, Q12, and Q14). There was a tendency for experimental group to have more increments than the control group.

The analysis of variance was done for the 18 items, setting groups (experimental group and control group) and times (before and after the lessons) as free variables and at the same time the mean of the items as a bound variable. The results indicated that for Q5 and Q16, there was an interaction on significant level of nearly 5 % between groups and times, although there was not an interaction on significant level for all items (Q5, $F(1, 143) = 3.84, p < .10$; Q16, $F\ (1,143) = 2.95$,

Table 14.7 Average of answers

	Exp. group		Cont. group	
Question numbers	Pretest	Posttest	Pretest	Posttest
1	2.11	2.05	2.03	2.06
2	2.47	2.42	2.42	2.32
3	2.67	2.76	2.50	2.56
4	2.23	2.36	2.13	2.11
5	2.75	2.72	2.34	2.55
6	3.05	2.89	3.18	3.15
7	2.48	2.53	2.45	2.47
8	2.80	2.90	2.85	2.76
9	2.36	2.37	2.11	2.23
10	3.11	2.99	3.35	3.31
11	2.78	2.75	2.77	2.68
12	2.02	2.00	1.98	2.02
13	1.67	1.80	1.71	1.71
14	1.71	1.88	1.60	1.74
15	2.04	2.04	2.10	2.00
16	2.13	2.20	2.21	2.05
17	2.40	2.48	2.40	2.40
18	3.14	3.04	3.16	2.90

$p < .10$). When comparing these results with the mean of the responses, the feeling "Chemistry lessons are fun" (Q5) decreased in experimental group and increased in control group. The feeling toward usefulness of chemistry "Chemistry lessons are useful for getting a job I desire" (Q16) increased in experimental group and decreased in control group. Thus, the students who took the lessons on inquiry activities improved partially feelings toward usefulness of chemistry.

Moreover, the results showed that for Q10 and Q14, there was a significant simple main effect on between times, namely, an effect of times observed in both groups in common (Q10, $F(1, 143) = 8.75, p < .01$; Q14, $F\,(1{,}143) = 7.89, p < .01$). Comparing these results with the mean of the responses, the feeling "Doing chemistry experiments in school is fun" (Q10) was reflected in both groups and the feeling "I have confidence in studying chemistry" (Q14) increased in both groups. These results can be inferred as, whether in inductive experiments or in deductive experiments, experiments in this study were difficult for the students of both groups; and it follows that a successful performance in doing experiments boosted their self-confidence in learning chemistry.

Discussion

In this study, the lessons on inquiry activities in senior high school chemistry were conducted, and the validity of such lessons was examined in terms of changes in students' attitudes toward chemistry. The result of an 18-item questionnaire showed that the experimental group differed significantly from the control group only on one item. We present three assumptions as major causes for the aforementioned result.

Firstly, the number of unit periods spent on inquiry investigation lessons and the duration of the term were short. Previous studies have not shown the relations between the number of unit periods or the duration of the term and the effectiveness in changing students' attitudes toward chemistry. However, these studies had conducted at least the lessons in 15-unit periods or more for the duration of relatively longer term, e.g., lessons of once a week for one semester (15-unit periods) (Chattejee et al., 2009), 24-unit periods (Vaino et al., 2012), 28-unit periods (Palmer, 2009), lessons of twice a week for two months (Lin et al., 2011), and lessons of fifth a week for 1 year (Putti, 2011). The number of unit periods and the duration of the term in this study were short in contrast to these studies. Thus, it is thought that an expected result was not produced. Therefore, in order to promote the changes in the students' attitudes toward chemistry, a certain unit period and duration of term should be set for the lessons on inquiry activities.

Secondly, students' discussion for planning experiments lacked depth. The students from the experimental group conducted a discussion in units of a group of four to five persons during the fourth period of the lessons. However, the discussion did not work well because only one or two students in a group planned the way of the experiment, referring to bibliographical sources in chemistry; and other students just simply confirmed the plan. Therefore, it is thought that the lessons did not improve the positive attitude toward chemistry. After all, students did not make use of the opportunity to conduct a discussion in chemistry lessons. Tsuruoka (2012) indicated a problem in this direction – senior high school textbook of chemistry in Japan hardly uses a space for activities for planning experiments.

Finally, the students' experiences of and exposures to inquiry-related lessons as well as their adequate abilities to conduct guided inquiry activities were limited. Indeed, the students' comments on their worksheets displayed that inductive experiments in the lessons were difficult for them. Palmer (2009) mentioned that students who have little experience of inquiry activities tend to lack the adequate abilities to go through the process of inquiry. The students themselves in this study may also have such a problem.

In order to improve students' positive attitude toward chemistry through lessons on inquiry activities, we expect that future studies must address the undermentioned things:

- Trial of the lessons on inquiry activities for a relatively longer term
- Development of the lessons on inquiry activities focused on students' discussion for planning experiments
- Promoting students' abilities of inquiry through their experiences of inquiry activities

Additional Remark The work described in this chapter was supported by Japan Society for the Promotion of Science, Grant-in-Aid for Scientific Research(C), no. 22500818. The authors gratefully acknowledge the helpful suggestions of Derek Sin-pui Cheung and the assistance of Gerald Tembrevilla in preparing the original manuscript.

References

Allen, J. B., Barker, L. N., & Ramsden, J. H. (1986). Guided inquiry laboratory. *Journal of Chemical Education, 63*(6), 533–534.

Chattejee, S., Williamson, V. M., McCann, K., & Peck, M. L. (2009). Surveying students' attitudes and perceptions toward guided-inquiry and open-inquiry laboratories. *Journal of Chemical Education, 86*(12), 1427–1432.

Cheung, D. (2008). Facilitating chemistry teachers to implement inquiry-based laboratory work. *International Journal of Science and Mathematics Education, 6*(1), 107–130.

Cheung, D. (2009). Developing a scale to measure students' attitudes toward chemistry lessons. *International Journal of Science Education, 31*(16), 2185–2203.

Colburn, A. (2000, March). An inquiry primer, *Science Scope*, pp. 42–44.

European Commission. (2004). *Europe needs more scientists: Report by the High Level Group on Increasing Human Resources for Science and Technology in Europe.*

European Commission. (2007). *Science education now: A renewed pedagogy for the future of Europe.*

Lewis, S. E., Shaw, J. L., Heitz, J. O., & Webster, G. H. (2009). Attitude counts: Self-concept and success in general chemistry. *Journal of Chemical Education, 86*(6), 744–749.

Lin, H. S., Hong, Z. R., Chen, C. C., & Chou, C. H. (2011). The effect of integrating aesthetic understanding in reflective inquiry activities. *International Journal of Science Education, 33* (9), 1199–1217.

Ministry of Education, Culture, Sports, Science and Technology. (2009). *Course of study in upper secondary school* (in Japanese).

Ministry of Education, Science and Culture. (1973). *Instruction of school science focused on process of inquiry* (in Japanese).

Murakami, T. (2010). Development of new scenario on investigation learning in science education -messing about and communication. *Journal of Educational Research. Kyoto University of Education, 10*, 91–100 (in Japanese).

National Institute for Educational Policy Research. (2007). *Knowledge and Skill for Life. Report of International Results on PISA 2006 Survey by OECD*. Tokyo, Japan: Gyousei (in Japanese).

National Institute for Educational Policy Research. (2008). *Prompt report: Totaled result of a survey based on questionnaires of PISA for grade three students of junior high school* (in Japanese).

National Institute for Educational Policy Research. (2012). *Summary: Report of the results of international survey on TIMSS 2011* (in Japanese).

National Institute for Educational Policy Research. (2013). *International comparison of science education. Report of survey on TIMSS 2011*. Tokyo, Japan: Akashisyoten (in Japanese).

Okayama Prefectural Education Center. (2001). *Research on development and instruction of teaching materials for chemical experiment. Part 2 -inquiry activity of senior high school chemistry-* (in Japanese).

Palmer, D. (2009). Student interest generated during an inquiry skills lesson. *Journal of Research in Science Teaching, 46*(2), 147–165.

Putti, A. (2011). High school students' attitudes and beliefs on using the science writing heuristic in an advanced placement chemistry class. *Journal of Chemical Education, 88*(4), 516–521.

Roehrig, G. H., & Luft, J. A. (2004). Inquiry teaching in high school chemistry classrooms: The role of knowledge and beliefs. *Journal of Chemical Education, 81*(10), 1510–1516.

Saito, I., et al. (2011). *Basic chemistry*. Osaka, Japan: Keirinkan (in Japanese).
Saito, I., et al. (2012). *Advanced chemistry*. Osaka, Japan: Keirinkan (in Japanese).
Simon, S., & Osborne, J. (2010). Students' attitudes to science. In J. Osborne & J. Dillon (Eds.), *Good practice in science teaching: What research has to say* (pp. 238–258). Maidenhead, UK: Open University Press.
Sjøberg, S., & Schreiner, C. (2010).*The ROSE project: An overview and key findings.*
Tsuruoka, M. (2012). Enquiry-related analysis of chemistry laboratory tasks in the textbook for upper secondary school. *Journal of Research in Science Education, 52*(3), 113–120 (in Japanese).
Vaino, K., Holbrook, J., & Rannikmae, M. (2012). Stimulating Students' intrinsic motivation for learning chemistry through the use of context-based learning modules. *Chemistry Education Research and Practice, 13*(4), 410–419.

Chapter 15
Developing Technology-Infused Inquiry Learning Modules to Promote Science Learning in Taiwan

Ying-Shao Hsu, Hsin-Yi Chang, Su-Chi Fang, and Hsin-Kai Wu

Background of the Study

To achieve curriculum goals linked to scientific literacy, inquiry-based learning and instruction have been emphasized progressively in science education over the last two decades (Ford & Wargo, 2007; Minner, Levy, & Century, 2010). This trend is reflected worldwide in educational reforms and curriculum standards. Currently, a framework for K-12 Science Education (National Research Council, 2012) stresses that both skills and knowledge are important to scientific practices, such as conducting scientific investigations. "Students cannot comprehend scientific practices, nor fully appreciate the nature of scientific knowledge itself, without directly experiencing those practices for themselves" (National Research Council, p. 30). The Grades 1–9 Curriculum Guidelines in Taiwan (Ministry of Education, 2008) and the current education reform also point out that secondary students are expected to develop both scientific knowledge and inquiry competencies through scientific inquiry activities.

Taiwanese secondary students have demonstrated outstanding performance in many international large-scale science assessments. In the 2003, 2007, and 2011 TIMSS (Trends in International Mathematics and Science Study; Martin, Mullis, Foy, & Stanco, 2012), Taiwanese students were ranked in the top 5. In the 2006 and 2009 PISA (the Programme for International Student Assessment; OECD, 2007), they were ranked number 4 and 12, respectively. Despite their high rankings in these international assessments, Taiwanese students expressed less positive attitudes toward and were less identified with science.

Y.-S. Hsu (✉) • S.-C. Fang • H.-K. Wu
National Taiwan Normal University, Taipei City, Taiwan
e-mail: yshsu@ntnu.edu.tw

H.-Y. Chang
National Kaohsiung Normal University, Kaohsiung, Taiwan
e-mail: hsinyichang@nknucc.nknu.edu.tw

M.S. Khine (ed.), *Science Education in East Asia*,
DOI 10.1007/978-3-319-16390-1_15

In fact, inquiry-based instruction is not new to Taiwanese teachers. There were also many graduate studies devoted to teaching and learning scientific inquiry (Wang et al., 2011). The meta-analysis study of Wang et al. pointed out that in Taiwan, inquiry-based instruction not only enhances students' conceptual development and scientific process skills but also their attainment of positive attitudes. The research findings also revealed that the effects of inquiry-based instruction appeared more obvious for elementary than for secondary students. They speculated that this might be due to less willingness on the part of secondary teachers to try novel instruction and less experience in teaching with inquiry-based approaches.

Indeed, science teachers play a key role in inquiry science learning. Their beliefs and values about teaching and learning, knowledge about science, and experience of teaching inquiry projects all have a significant influence on their teaching practice (Chang, 2013; Gyllenpalm, Wickman, & Holmgren, 2010; Liu, Lee, & Linn, 2010). It is noted that one significant challenge for science teachers to implement inquiry-based instruction successfully is to transform the national curriculum guidelines into operational curricula in terms of inquiry activities (Krajcik, McNeill, & Reiser, 2008; Penuel, Fishman, Gallagher, Korbak, & Lopez-Prado, 2009). Therefore, our research team aims to put forward the notion of inquiry learning in Taiwanese secondary science classrooms by developing inquiry learning modules in a web-based learning environment: the Collaborative Web-based Inquiry Science Environment (CWISE) This online platform was originally developed by the University of Berkeley and named WISE (Linn, Clark, & Slotta, 2003; Linn & Eylon, 2011; Slotta & Linn, 2009). We translated it into Chinese and extended its function to cater for collaborative learning. The learning modules designed in the study possess two significant characteristics. First, they are designed based on the national middle-school Science and Technology curriculum guidelines (Ministry of Education, 2008); therefore, they can be used widely in secondary science classes. On the one hand, teachers and students do not need to spend extra time on teaching/learning scientific inquiry. On the other hand, we hope that adding inquiry elements in science learning can also promote conceptual understanding and raise students' positive attitudes toward science. Second, CWISE is a web-based learning environment which supports the presentation of multimedia. With the support of technology such as animation, simulation, and visualization, it is expected that teachers can effectively engage students in inquiry learning activities (Linn, Clark, & Slotta, 2003).

In the following sections, we first explain the rationale behind the instructional design framework and illustrate how the framework is used for developing the learning modules. Then, we used two examples: plate tectonics in the subject area of earth science and buoyancy in physics, to demonstrate how the inquiry learning modules are constructed in the web-based learning environment.

Instructional Design Framework

In order to put the design of learning modules into operation, we first explicate the underlying rationale by transforming it into four distinct features. These four features will be described in the first part of the section. In the second part, we will illustrate the scientific inquiry abilities involved in the study (including experimental abilities and explaining abilities) and how secondary students are expected to demonstrate these abilities. Based on the rationale described in the first part, the third part introduces the four-step instructional design framework used for designing inquiry learning modules. We will summarize the steps in the framework together with examples.

The Rationale for Designing Learning Modules

1. The modules are designed based on the national curriculum guidelines.

The established national curriculum guidelines state what kind of science content and competency should be taught and developed at a particular level. The guidelines about science content in general are very big ideas that are normally presented as topics such as the periodic table, digestive system, or motion. Therefore, it is very challenging for teachers to design learning activities simply based on the guidelines. To solve this problem, we adopted the method suggested by Krajcik, McNeill, and Reiser (2008), namely, to unpack and articulate the key ideas embedded in the guidelines. This method allows us not only to align the teaching materials with the national curriculum guidelines but also to pave the way for the alignment of the guidelines, teaching practice, and assessments, which are the three fundamental elements of science teaching and learning.

2. The modules are inquiry orientated.

It is accepted worldwide that science is a practice that encompasses more than just established facts and concepts, but also includes scientific ways of thinking and reasoning (McNeill, 2011). Based on this perspective, learning science implies that students are able to construct their scientific knowledge such that they are able to explain the phenomena they observe, as well as developing inquiry abilities for engaging in scientific activities (Bybee, 2006). However, previous studies have shown that to understand science as inquiry is actually challenging and problematic for students, because it involves an extensive range of cognitive and social skills that develop during secondary levels (Bybee; Flick, 2006; McNeill, Lizotte, Krajcik, & Marx, 2006). Students do not naturally develop these cognitive and social skills. In other words, appropriate instructional strategies are needed to support and guide students when they are learning and practicing scientific inquiry (Osborne, Erduran, & Simon, 2004). In the study, we incorporate inquiry elements into the modules by means of the web-based learning environment. For example, we use visualizations or simulations to create a scenario, in which students can manipulate relevant variables, record data, and observe the patterns of the data.

They are also required to provide scientific explanations based on their investigations. Apart from creating inquiry learning environments, we also used embedded prompts to scaffold students in the designed inquiry activities.

3. The learning activities (in the learning module) are structured according to the knowledge structure.

The key concepts unpacked are like "ingredients" for making a dish. They have different textures so that they need different cooking times. Therefore, the ingredients have to be cooked in a certain order. Similarly, the "ingredients" for designing learning modules, i.e., the key concepts, need to be organized in a way that can be learned or taught in a logical sequence. We name this "knowledge structure."

Two aspects need to be carefully considered when structuring learning activities. First, from a longitudinal point of view, it is crucial to take students' prior knowledge into account. What students have learned in earlier levels not only provides an entry point for engaging them in learning activities but also forms an important foundation for building on their knowledge at the present level. In addition, how the learning goals at the present level are going to form a foundation for the next level also need to be examined. Second, regarding the learning goals at the present level, it is vital to take students' common alternative conceptions into account and design instructional activities to clarify these alternative ideas.

4. The modules are revised and refined according to empirical studies.

The study employed a design-based approach (Cobb, Confrey, diSessa, Lehrer, & Schauble, 2003; Hsu, Chuang, & Lin, 2012) to understand how the instructional innovations impact on a *learning ecology* in terms of the changing of norms in teacher-student discourses and interactions and in the ways students deal with tasks or problems (Cobb et al., 2003, p. 9). In other words, we used the design-based research method to evaluate and look into the influence of the instructional innovations (Anderson, Zuiker, Taasoobshirazi, & Hickey, 2007; Hickey, Taasoobshirazi, & Cross, 2012). Apart from that, for the purpose of developing modules specifically, we used design experiments to improve the initial design of the learning modules by testing and analyzing the ongoing student learning status as well as the learning environment. Therefore, in line with the learning modules, we also developed assessments, which are used for refining existing learning modules. Since the core content of the chapter is to describe the design of the learning modules, the content of the assessment framework and how they are used to inform the construction of the inquiry learning environment will be reported and discussed in future publications.

Scientific Inquiry Abilities: Experimental Abilities and Explaining Abilities

Scientific inquiry is a multifaceted activity involving multiple steps: asking questions and defining problems, developing and using models, planning and carrying

Table 15.1 Inquiry abilities involved in the study

Experimental abilities	Identifying and choosing variables (CV)	Be able to select appropriate dependent and independent variables and design a study based on the variable
	Planning an experiment (PE)	Be able to design feasible investigations and use appropriate measuring instruments and methods
	Transforming and Representing data (TR)	Be able to use a different way to represent data and describe the relationships between variables
Explaining abilities	Making a claim (MC)	Be able to summarize data and find out its distribution or trend in order to propose a claim
	Using evidence (UE)	Be able to use knowledge or data as evidence to support hypotheses
	Describing the reasoning process (DR)	Be able to apply scientific principles, concepts, or prior experience to reason and connect evidence to the claim. The process also requires interpreting or inferring the meaning of data
	Evaluating explanations (EE)	Be able to use scientific knowledge or data to evaluate explanations, in order to include or exclude other possibilities

out investigations, analyzing and interpreting data, using mathematics and computational thinking, constructing explanations and designing solutions, engaging in argument from evidence, and obtaining, evaluating, and communicating information (NRC, 2012). Depending on their purposes, researchers may look into students' performance on a complete inquiry cycle (e.g., Cuevas, Lee, Hart, & Deaktor, 2005; Hofstein, Shore, & Kipnis, 2004) or only focus on selected types of inquiry abilities (e.g., Kyza, Constantinou, & Spanoudis, 2011; McNeill et al., 2006; Wu & Hsieh, 2006; Wu, Wu, & Hsu, 2013).

With reference to the national middle-school Science and Technology curriculum guidelines (Ministry of Education, 2008), we found that experimental and explaining abilities are two major inquiry abilities that need to be applied in learning the content. Therefore, the design of the learning modules focused on developing students' experimental and explaining abilities. Each of the two key abilities includes several sub-abilities. Table 15.1 lists the definitions of the inquiry abilities involved in the study.

The Framework for Instructional Design

In light of the four features explicated earlier, we developed a framework for designing inquiry learning modules (see Fig. 15.1). The framework summarizes the design process into four steps.

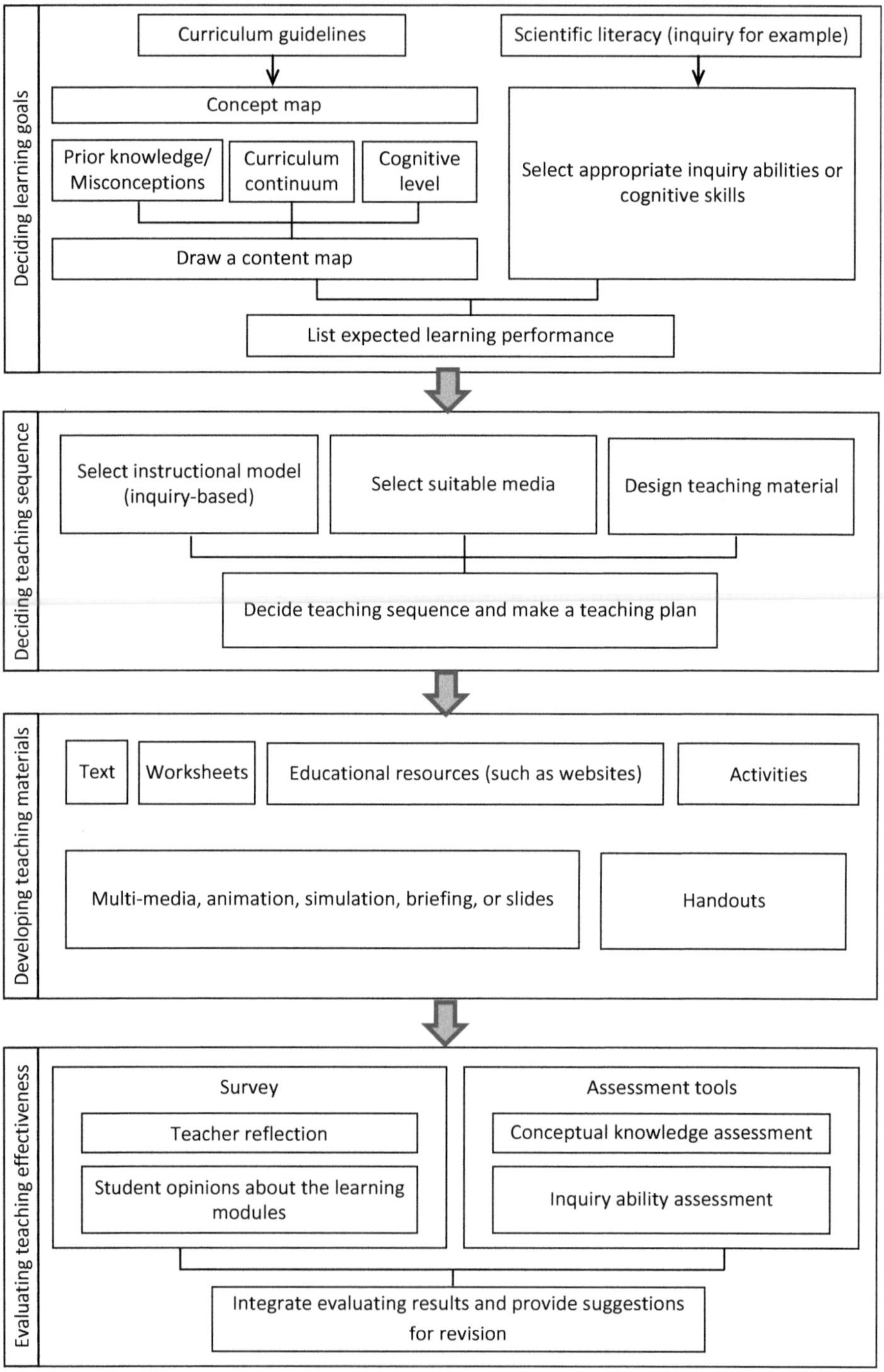

Fig. 15.1 The instructional design framework

Table 15.2 Examples of unpacking curriculum guidelines

Curriculum guidelines	Unpacked content
Subtopic 210 The surface of the earth and diastrophism	
Plate tectonics	
4d. To understand that plate tectonics cause volcanic eruptions, earthquakes, and the formation of mountains	Seafloor: continental margin, ocean basin, and mid-ocean ridge
	Driving force of plate motion
	The definition of tectonic plate
	The relationships between seismic zone, volcanic belt, and plate boundaries
	Three types of plate boundary (convergent, divergent, and transform) and accompanying geological phenomena
4e. To know the continental drift hypothesis and plate tectonics	The origin, the content, and the limitations of the continental drift hypothesis
	Seafloor spreading hypothesis

Step 1: Deciding Learning Goals

1. Unpacking the curriculum guidelines and making connections between the contents/concepts

To start with, the curriculum guidelines are unpacked and broken down into several key concepts. For example, with the topic of plate tectonics, one guideline states "to understand that plate tectonics cause volcanic eruptions, earthquakes and the formation of mountains." To make sense of the guideline, one has to understand several embedded component ideas, for example, the structure of the seafloor, the definition of tectonic plate, and the driving force of plate motion. Therefore, the first step is to clarify what key concepts are included in the guideline. Table 15.2 shows two full examples of the unpacking process.

In the next step, these component ideas are organized and structured in a logical order. By logical, we mean that some ideas need to be introduced before the others in order to make the latter meaningful. For example, the definition of tectonic plate needs to be introduced before the mechanism of plate tectonics, or learners might not be able to grasp the concept of plate tectonics. However, it should also be noted that there are a myriad of ways to structure the component ideas. Depending on the purposes and storyline they have in mind, teachers/teaching material developers may use different ways to organize the ideas. Figure 15.2 shows one way to structure the component ideas unpacked in Table 15.1. Here, we intend to use a chronological approach to introduce relevant hypotheses about the movement of the earth's continents. We also decide to give learners relevant ideas about the structure of the earth before we introduce the hypotheses.

2. Considering student prior knowledge/experience and possible alternative conceptions

For successfully engaging students in learning a topic, it is usually helpful to start with relevant experience they have in everyday life or with some ideas they

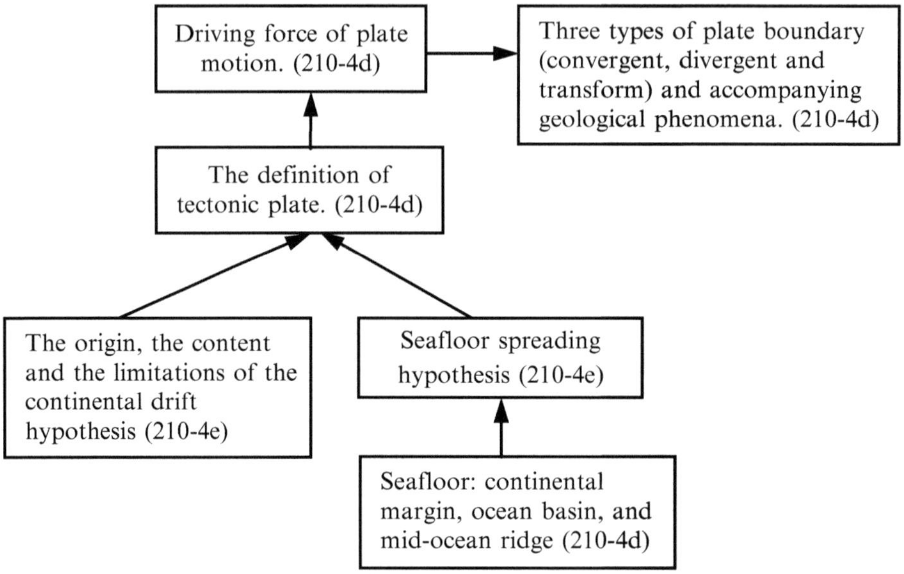

Fig. 15.2 Examples of structuring the component ideas

have learned before. For example, one common entry point to introduce the concept of plate tectonics in Taiwanese secondary classrooms is to talk about the experience of earthquakes. This not only produces some interesting elements in the lessons but also makes the concept closer to learners' everyday life. The other aspect that needs to be considered before designing teaching materials is students' possible alternative conceptions. Some quizzes can be designed in the modules to check whether learners understand the concept correctly. Also, designed activities or visualizations can be used to clarify and address possible alternative conceptions. For instance, two common alternative conceptions for the concept of plate tectonics are "the lithosphere is the same as a tectonic plate" and "plate boundaries are coastlines." Both of them are used in our development of the plate tectonics learning module.

For clearly representing the content included in the learning module, a "content map" is created. The map encompasses three content-related materials: the structured content, students' prior knowledge, and possible alternative conceptions.

3. Creating learning performance

After specifying the content in the learning modules, we need to think about how we expect students to exhibit their understandings, i.e., learning performance. As the module is positioned as inquiry oriented, inquiry elements (or lower-level cognitive skills) are added in this step to form the expected learning performance. In other words, learning performance is composed of a piece of unpacked content and an inquiry ability/cognitive skill. The following represents how learning performance is constructed.

Content × Scientific Inquiry ⇒ Learning Performance

For example, we expect students to "reason" (inquiry ability) the relationship between plate boundaries and seismic/volcanic belts (content) by observing relevant data about the distributions of volcanoes and earthquakes. The learning performance is described as "be[ing] able to infer the relationships between tectonic plates, and volcanoes/earthquakes from the given data." Another example of learning performance given here is "be[ing] able to develop a claim (inquiry ability) about the type of plate boundary (content) based on cross-section figures of earthquake-focus along with depth."

Step 2: Deciding Teaching Sequence

Based on the established content map, the second step further organized the teaching activities and teaching sequences in the learning module. Three aspects are considered in the step: teaching model, suitable media, and teaching material design. The teaching model is chosen according to the purpose of the learning module. For instance, the study aims to develop inquiry abilities, so we adopted an inquiry-based learning module. Moreover, depending on the variation of inquiry, a structured or an ill-structured inquiry model (National Research Council, 2000, p. 29) can be selected. The selection of suitable media and the design of the teaching material are intertwined. The reason for constructing the inquiry learning modules on CWISE is because the platform not only provides learners with an online interactive environment to observe, explore, and record relevant data/phenomena but also supports them in expressing their ideas (Linn & Eylon, 2011).

Step 3: Developing Teaching Materials

The third step involves the development of the teaching materials in detail, such as text, activities, worksheets, animations, simulations, and slides. Since students have different "Zones of Proximal Development, ZPDs," we may consider providing different types of scaffolds or relevant supporting resources, like visualizations, peer interactions, and teacher guidance, as "distributed scaffolds" (Tabak, 2004) in the activities.

Step 4: Evaluating Teaching Effectiveness

After completing the development of the learning modules, the modules will be implemented together with relevant assessments in real science classrooms. The implementation of the learning modules has two purposes. First, in the first or second round of implementation, students' responses in the lessons and their answers in the assessments can help teachers/module developers to revise and refine the modules. Second, with more mature modules, the implementation of the modules and assessment can be used to evaluate the effectiveness of the teaching practice. The step reflects the fundamental idea of a design-based study as we have foreshadowed in the first part of this section.

In the following sections, we use two examples in different subject areas to exemplify a complete process of development of inquiry learning modules.

Example One: Earth Science (Plate Tectonics)

The theory of plate tectonics offers an explanation of "the past, present, and future geographic distribution of the earth's landmasses and oceans" (Gobert, 2000, p. 938). It is proposed that the earth's crusts are broken into pieces, and they are moving continuously on the mantle driven by the slow convective movement of hot magma (in the upper mantle) (Plummer & Carlson, 2008). Plate tectonics is a challenging topic in secondary science learning because learners are required to build models and use causal reasoning to comprehend the hidden, explanatory mechanisms, such as mountain formation, convection, and continental drift. In Taiwan, this topic is first introduced in the ninth grade and then revisited in the tenth grade with deeper content. The learning module developed in the study is for the ninth grade students who are learning the topic of plate tectonics for the first time. In the following sections, we will start with unpacking the curriculum guidelines and describe in more detail how we construct the learning module on plate tectonics in CWISE.

Unpacking National Curriculum Guidelines

To begin with, we selected the curriculum guidelines that related to the topic of tectonics. For beginners who are first learning about plate tectonics, it is necessary to include relevant contents such as the internal structure of the earth, fast earth changes, and common geological phenomena as basic ideas for constructing the ideas of the theory of plate tectonics. Therefore, as Table 15.3 shows, in addition to the guidelines of plate tectonics (subtopic 210, 4d and 4e), we included curriculum guidelines that involved necessary concepts (subtopic 110, 4b and subtopic 210, 4b and 4c). These five guidelines were then unpacked into smaller content areas.

After unpacking the curriculum guidelines, we drew a concept map for organizing these ideas in a logical way. Fundamentally, these ideas were classified into two groups: the structure of the earth and plate tectonics. The former group includes the

Table 15.3 Unpacking the curriculum guidelines related to the topic of plate tectonics

Curriculum guidelines related to plate tectonics	Content unpacking
Subtopic 110 The composition of earth	
Lithosphere, atmosphere, and hydrosphere	
4b. To understand the internal structure of the earth by models	The internal structure of the earth can be determined by seismic waves
	The three main layers of the earth: crust, mantle, and core
	The physical properties of the layers of the earth
	The composition of the layers of the earth
	The differences between continental crust and oceanic crust
Subtopic 210The surface of the earth and diastrophism	
Changing landforms	
4b. To understand fast earth changes, such as volcanic eruption and earthquakes	The causes of earthquakes
	The focus, the epicenter, and the depth of focus
	The differences between earthquake magnitude and earthquake intensity
	Preparedness and emergency management strategies
4c. To understand common geological phenomena, such as folds and faults. Try to explain the causes of the different types of terrains in Taiwan	The causes of folds and faults
	The differences between normal, thrust, and strike-slip faults
	Taiwan tectonics.
	The causes of the different types of terrains in Taiwan.
Plate tectonics	
4d. To understand that plate tectonics cause volcanic eruptions, earthquakes, and the formation of mountains	Seafloor: continental margin, ocean basin, and mid-ocean ridge
	Driving force of plate motion
	The definition of tectonic plate
	The relationships between seismic zone, volcanic belt, and plate boundaries
	Three types of plate boundary (convergent, divergent, and transform) and accompanying geological phenomena
4e. To know the continental drift hypothesis and plate tectonics	The origin, the content, and the limitations of the continental drift hypothesis
	Seafloor spreading hypothesis

concepts of the layered structure of the earth, their composition, and physical properties. The latter group encompasses the ideas of convection, types of plate boundaries and accompanying geological phenomena. The structure of the ideas was also connected to the curriculum guidelines at grade 10 and elementary levels in the content map, as shown in Fig. 15.3. On the left-hand side of the map, we listed common alternative conceptions regarding these concepts.

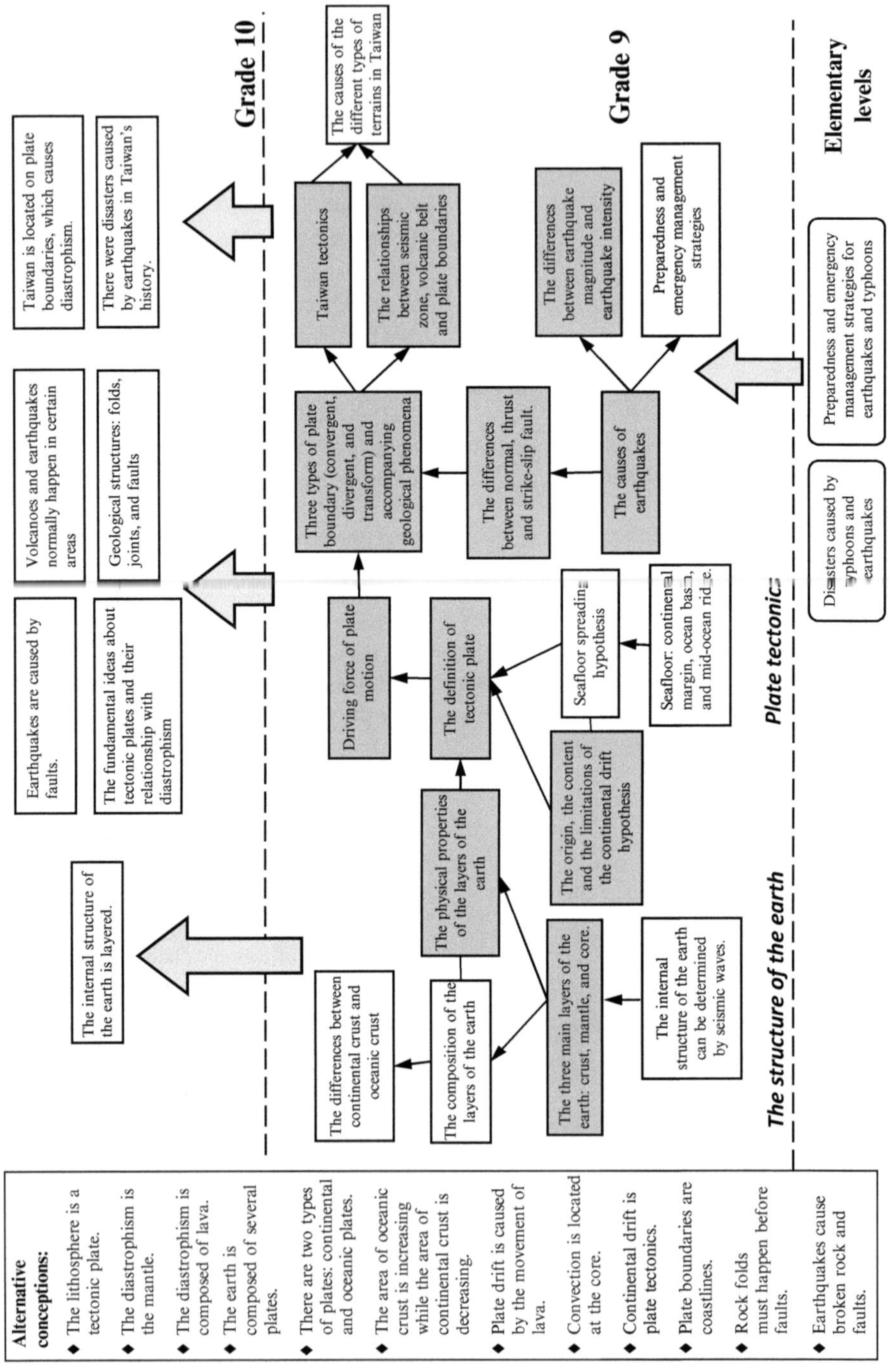

Fig. 15.3 Content map showing structure of the ideas

Creating Learning Performance

Previous research has indicated that learners have difficulties in learning plate tectonics due to several reasons (Ault & Charles, 1984; Gobert, 2000; Gobert & Clement, 1999; Jacobi, Bergeron, & Malvesy, 1996): (1) the size scale is too large to be comprehended; (2) the time scale of geological processes is difficult to grasp because it exceeds our reference of a human life; (3) it involves phenomena that cannot be observed directly such as convection and the earth's internal layers; and (4) to make sense of the concepts requires learners to appreciate and integrate different types of information, i.e., "spatial, causal and dynamic information" (Gobert, p. 939). In order to support students to overcome the difficulties, we considered incorporating visualizations and simulations (Chang, 2013; de Jong & van Joolingen, 1998; White & Frederiksen, 1998) into the design of the learning module. For example, the software package "Seismic Eruption"[1] is used to help students to explore the relationships between seismic zone, volcanic belt, and plate boundaries. On top of that, students are guided to use "Seismic Eruption" to construct cross-section figures of earthquake focus along with depth by which they can explore the plate boundaries underground and meaningfully understand different features of different types of plate boundary (convergent, divergent, and transform).

In this module, students are expected to perform their understandings through manipulating, observing, and summarizing the data in the software and correctly using the data as evidence to explain the causal relationships of the phenomena. As a result, we put emphasis on three inquiry abilities: making a claim, using evidence, and reasoning. Table 15.4 shows two examples of the expected learning performance, and Table 15.5 shows all of the learning performance involved in this learning module.

Table 15.4 Two examples of the developing learning performance in the unit of plate tectonics

Content	Scientific inquiry	Learning performance
The relationships between seismic zone, volcanic belt, and plate boundaries	Making a claim	Be able to deduce the relationships between tectonic plates and volcanoes/earthquakes from the given data
	Descrbing the resoning process	Be able to use relevant evidence to deduce the relationships between tectonic plates and volcanoes/earthquakes
Three types of plate boundary (convergent, divergent, and transform) and accompanying geological phenomena	Making a claim	Be able to summarize relevant information about cross-section figures of earthquake focus along with depth and to make a claim about the type of plate boundary
	Using evidence	Be able to use correct evidence to determine the type of plate boundary

[1] Seismic Eruption was created by Alan L. Jones, State University of New York at Binghamton. It is a program for the visualization of seismicity and volcanic activity in space and time.

Table 15.5 Learning performance in the module of plate tectonics

Curriculum guidelines	Content unpacking	Bloom cognitive dimension	Inquiry ability	Learning performance
Subtopic 110 The composition of earth *Lithosphere, atmosphere, and hydrosphere* 4b. To understand the internal structure of the earth by models	The three main layers of the earth: the crust, mantle, and core	Remember		Understanding that the internal earth can be divided into three layers, from outside in: the crust, mantle, and core
Sub-topic 210 The surface of the earth and diastrophism	The focus, the epicenter, and the depth of focus	Remember		Be able to explain the meaning of the depth of (earthquake) focus
Changing landforms		Understand		Be able to differentiate between the focus and the epicenter
4b. To understand fast earth changes, such as volcanic eruptions and earthquakes			MC	Be able to deduce the depth of focus from relative motion between tectonic plates
4c. To understand common types of geological structures, such as folds and faults. Try to explain the causes of the different types of terrains in Taiwan	The causes of folds and faults	Remember		Understanding the causes of folds and faults
	The differences between normal, thrust, and strike-slip faults	Understand		Understanding how different relative motions cause different types of faults
			MC	Be able to determine the location and the type of the fault from pictures
Plate tectonics 4d. To understand that plate tectonics cause volcanic eruptions, earthquakes, and the formation of mountains	Driving force of plate motion		MC	Be able to apply the concept of convection to deduce water movement
				Be able to deduce how water moves with different heating positions
				Be able to apply the concept of convection to deduce the relationship between

(continued)

Table 15.5 (continued)

Curriculum guidelines	Content unpacking	Bloom cognitive dimension	Inquiry ability	Learning performance
				plate movement and the asthenosphere
	The definition of tectonic plate	Understand		Understand physical properties of plates, and be able to differentiate between the lithosphere and asthenosphere
				Be able to mark the positions of the lithosphere and asthenosphere on a map of the earth's structure
			MC	Be able to apply the knowledge of the properties of the lithosphere and asthenosphere to deduce whether or not the plates can move
	The relationships between seismic zone, volcanic belt, and plate boundaries	Understand		Be able to use related data to determine plate boundaries
			MC	Be able to infer the relationships between tectonic plates and volcanoes/earthquakes from the given data
			DR	Be able to explain why the seismic zones overlap volcanic belts
				Be able to use relevant evidence to deduce the relationships between tectonic plates and volcanoes/earthquakes
	Three types of plate boundary (convergent, divergent, and transform) and	Understand		Be able to apply the knowledge of the formation of divergent plate boundaries to deduce the

(continued)

Table 15.5 (continued)

Curriculum guidelines	Content unpacking	Bloom cognitive dimension	Inquiry ability	Learning performance
	accompanying geological phenomena			accompanying geological features
				Be able to apply the knowledge of the formation of convergent plate boundaries to deduce the accompanying geological features
				Be able to recognize geological features and to mark the directions of plate movement in the pictures of convergent plate boundaries
				Be able to describe how volcanoes are formed from the relative motion of tectonic plates
				Be able to use correct evidence to determine the type of plate boundary
			MC	Be able to apply the knowledge of a transform plate boundary to deduce the relative motion between San Francisco and Los Angeles in the future
				Be able to use the movement of the asthenosphere to deduce the directions of forces on tectonic plates, the types of faults, and the relative motion of plates
				Be able to summarize relevant information about cross-section figures of earthquake focus along with depth and to make a

(continued)

Table 15.5 (continued)

Curriculum guidelines	Content unpacking	Bloom cognitive dimension	Inquiry ability	Learning performance
				claim about the type of plate boundary
				Be able to develop a claim about the type of plate boundary based on cross-section figures of earthquake focus along with depth
			UE	Be able to use cross-section figures of earthquake focus along with depth to determine the type of plate boundary
				Be able to use correct evidence to determine the type of plate boundary
			DR	Be able to explain the relative motion between San Francisco and Los Angeles in the future
				Be able to explain why deep-focus earthquakes usually occur at convergent plate boundaries
				Be able to explain how to use cross-section figures of earthquake focus to deduce the type of plate boundary
4e. To know the continental drift hypothesis and plate tectonics	The origin, the content, and the limitations of the continental drift hypothesis		MC	Be able to apply the knowledge of continental drift to infer that the distributions of continents and seas will continue to change
			DR	Be able to explain the causes of continental drift

MC making a claim, *UE* using evidence, *DR* describing the resoning process

Deciding Learning Sequence

In fact, in the step of creating learning performance, we were also designing the activities that corresponded to the learning performance and structuring them in a logical sequence. The two steps (i.e., creating learning performance and deciding learning sequence) are mutually affecting. For example, in activity 3, we intend to introduce the features of the three types of plate boundaries first, and then we use the data from the Seismic Eruption software to help students understand how scientists determine the types of plate boundaries and why they can do so. As a result, the first four steps in the activity (steps 3-1 to 3-4) do not involve inquiry abilities; they are designed to give students an overview of how plate boundaries are categorized by scientists. In contrast, all of the learning performance created in activity 4 requires students to apply inquiry skills to make sense of the content. Table 15.6 shows the learning sequence of the learning module.

Developing Learning Materials

Next, according to the sequence decided upon, we began to go into the detail of each step, including writing the text, creating animation and interactive visualizations, and designing quizzes inserted in the steps to help monitor students' learning status. We will use four examples, as four different strategies used in the module, to demonstrate how we developed the learning materials.

Interactive Visualization

In the first example, we used interactive visualizations to present earthquake distribution, volcano distribution, and plate boundaries. The step was designed for students to observe and explore the relationships between plate boundaries and the seismic zones/volcanic belts. As Fig. 15.4 shows, students can select the information they need, click the boxes, and observe the information presented. They are required to use the data as evidence to test their hypothesis and to reason the relationships between plate boundaries and the seismic zones/volcanic belts.

Animation

Since most of the concepts in plate tectonics are related to dynamic changing of geological phenomena, our second strategy is to take advantage of animation to

Table 15.6 The learning sequence of the module of plate tectonics and the inquiry abilities involved in each step

Activity	Step	The title of the step	Inquiry ability involved in the step
1. Forces that shape the surface of the earth	1-1	Common geological activities	
	1-2	Folds and faults	
	1-3	The types of faults	
	1-4	Quiz: What is the type of fault in the picture?	MC
	1-5	The focus and the epicenter (of an earthquake)	
	1-6	Earthquakes and volcanoes	MC, DR
	1-7	Earthquakes, volcanoes, and tectonic plates	DR
2. Convection and the theory of plate tectonics	2-1	The internal structure of the earth	
	2-2	The properties of tectonic plates	
	2-3	The properties of tectonic plates (cont.)	MC
	2-4	Tectonic plates and the asthenosphere	
	2-5	What is convection?	MC
	2-6	How do convection currents move?	MC
	2-7	The movement of convection	
	2-8	Convection in the asthenosphere	MC
	2-9	Continental drift and plate tectonics	MC, DR
3. The types and characteristics of plate boundaries	3-1	Three types of plate boundaries	
	3-2	Divergent plate boundaries	
	3-3	Convergent plate boundaries	
	3-4	Convergent plate boundaries (cont.)	
	3-5	Transform plate boundaries	MC, DR
	3-6	The characteristics of the three types of plate boundaries	MC
	3-7	The depth of focus of the three types of plate boundaries	MC, DR
4. Seismic Eruption	4-1	The introduction of "Seismic Eruption"	
	4-2	Introducing the data resource of the depth of focus	
	4-3	Screenshot step by step	
	4-4	Determine the type of plate boundaries from the construct cross-section figures of earthquake focus (1)	MC, UE, DR
	4-5	Upload your file to CWISE	UE
	4-6	Determine the type of plate boundaries from the construct cross-section figures of earthquake focus (2)	MC, UE, DR
5. Review	5-1	Summary	
	5-2	Geological activities in Taiwan	MC
	5-3	End	

Website of the module of plate tectonics: http://cwise.nccu.edu.tw/webapp/preview.html?projectId=55

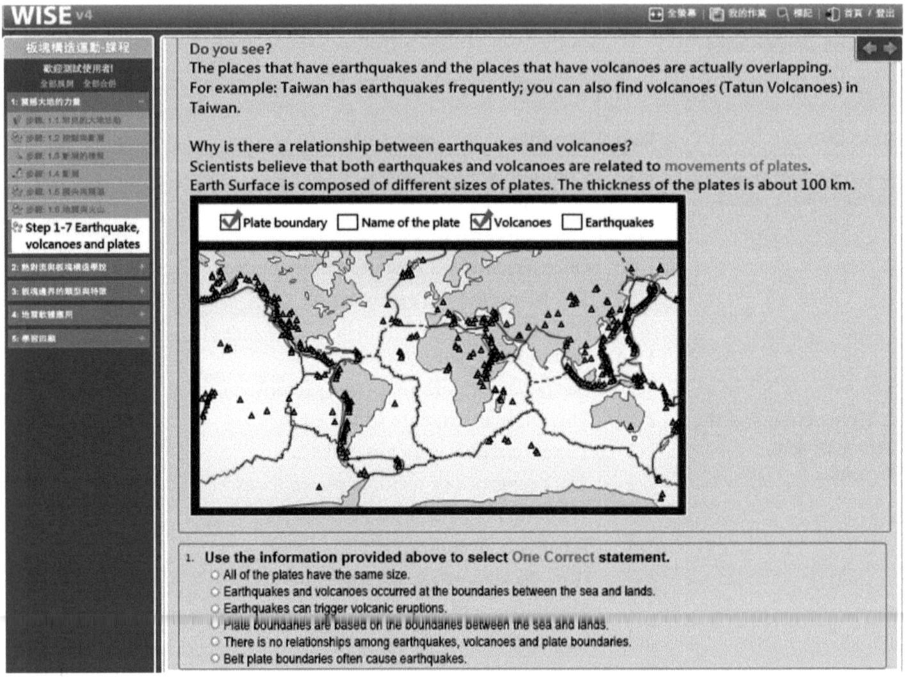

Fig. 15.4 An example of using interactive visualizations to develop learning material

present the processes. Figure 15.5 shows an example. It is a screenshot of steps 1–3: the types of faults. In the step, students can repeat the process and observe the relationship between the force direction and the movement of the blocks. Another example is step 2–9, which displays the process of continental drift.

Seismic Eruption Software[2]

One distinct feature of the plate tectonic module is to include the Seismic Eruption software for activity design. Specifically, rather than just describing the features of plate boundaries, we use the software to help students understand what scientists actually do to determine and classify the types of plate boundaries. The software is used as a data resource and as an analyzing tool. Students can access relevant data about the volcanoes and earthquakes which happened all over the world from 1960 to 2010. In addition, they are guided to use one function of the software to create

[2] Seismic Eruption was created by Alan L. Jones, State University of New York at Binghamton. It is a program for the visualization of seismicity and volcanic activity in space and time. (http://bingweb.binghamton.edu/~ajones/#Computer%20Programs)

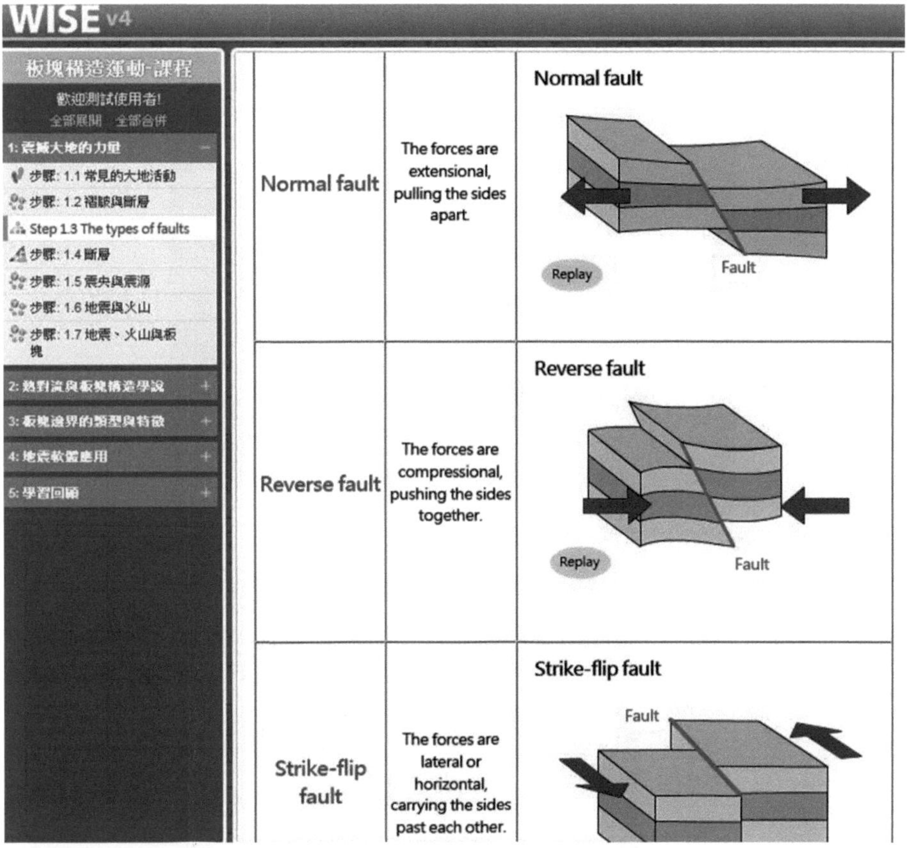

Fig. 15.5 An example of using animations to develop learning material

"cross-section figures of earthquake focus along with depth (Fig. 15.6)." The figures are analyzed and used as evidence to determine the type of plate boundaries.

Embedded Quiz (Immediate Assessment)

In order to monitor student learning status, we designed one or two quizzes to insert into the steps. In general, the purposes of the quizzes are to check whether a student holds alternative conceptions and to summarize the concepts that they have just learned. Therefore, the quizzes appear after introducing a new concept or in the last steps of an activity. Figure 15.7 shows an example of a quiz that helps students to summarize the relationships among force directions, the types of faults, and the types of plate boundaries.

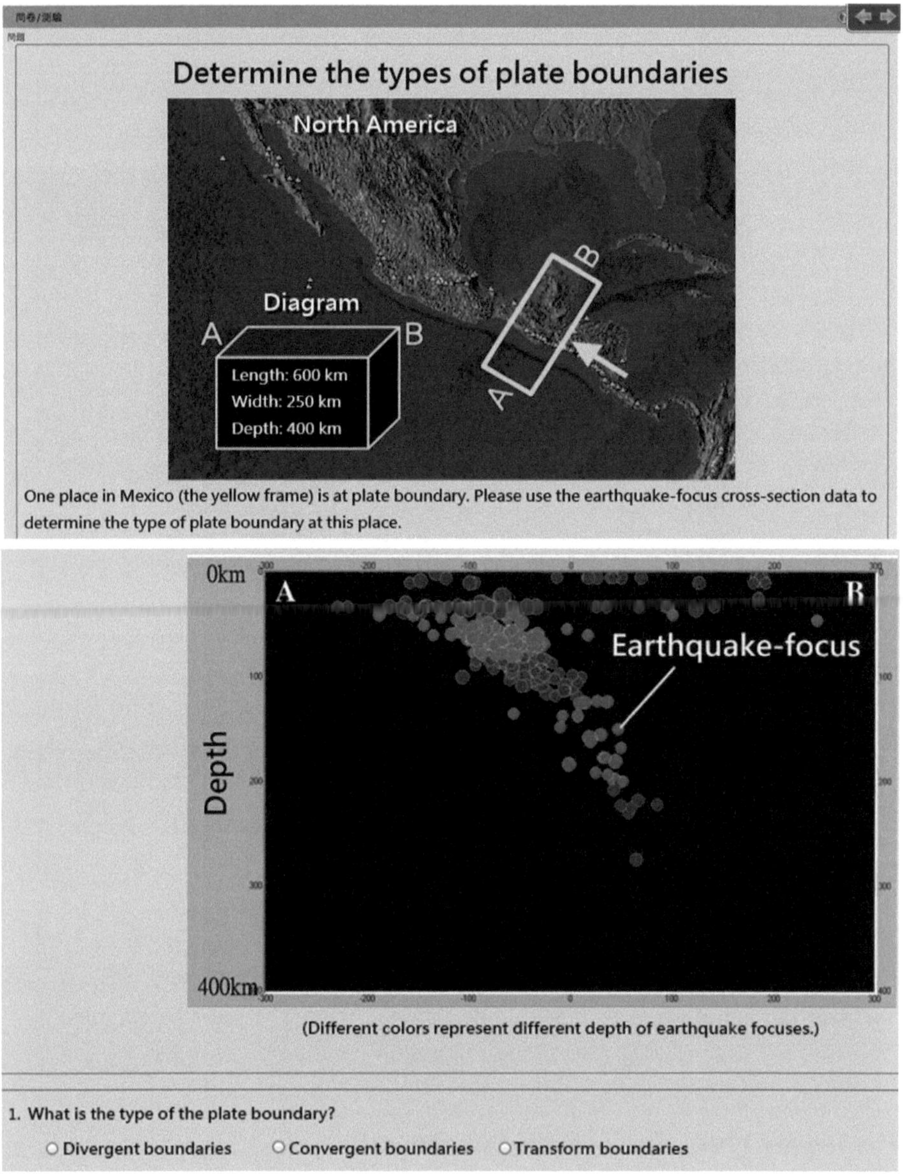

Fig. 15.6 Screenshots of the steps related to the Seismic Eruption software

Summary

In sum, the development of the plate tectonics module is developed according to the instructional design framework that stresses two important features: (1) it is designed based on national curriculum guidelines, and (2) it incorporates inquiry

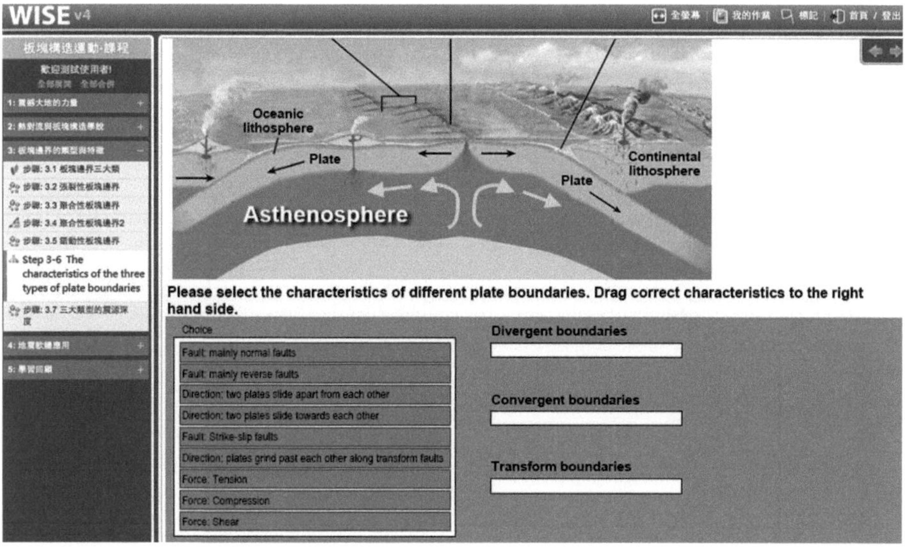

Fig. 15.7 An example of a quiz embedded in activity 3

abilities in learning activities. From common geological features to complex, large-scale geological phenomena, the module intends to help students build on their knowledge from their everyday life and experience. The module takes advantage of the learning platform, CWISE, to incorporate different technologies into the design of inquiry learning activities, such as interactive visualizations, animations, and the Seismic Eruption software. It is hoped that the incorporation of the inquiry elements and the technology can effectively support students' learning of the topic of plate tectonics.

Example Two: Physics (Buoyancy)

Unpacking the National Curriculum Guidelines

Three national curriculum guidelines are related to the topic of buoyancy (Ministry of Education, 2008) for grades 7–9: students understand that (1) the net upward buoyancy force of an object wholly or partially immersed in a fluid equals the weight of the fluid displaced by the object; (2) an object can be motionless even though multiple forces are acting upon it; (3) for a motionless object, the vector sum of all external forces acting upon the object is zero and the sum of the moments of all external forces about any line is zero. The first guideline is central to our buoyancy module, whereas the second and third guidelines are related and applied as students make sense of the phenomena of sinking and floating.

We unpacked the guidelines into six smaller concepts including:

(1) Properties of objects related to the phenomena of sinking or floating.
(2) Definition of the buoyancy of an object in a fluid referring to the net force exerted by its surrounding fluid.
(3) Understanding of buoyancy resulting in the weight changes of an object measured in air and in a fluid.
(4) The buoyancy of an object in a fluid is equal to the weight of the fluid displaced in terms of magnitude (Archimedes' principle).
(5) Reasoning of the Archimedes' principle to understand why the buoyancy of an object in a fluid in terms of magnitude also equals the density of the fluid times the volume of the object immersed in the fluid.
(6) Understanding that for a floating object, the buoyant force is equal to the weight of the object.

Creating Learning Performance

Rather than using rote-learning methods to learn the concepts outlined above, in the module we engage students in inquiry practices to learn the concepts. Considering the combination of concepts and inquiry abilities, we delineate the learning performances that students need to achieve in this module, including (1) students conduct virtual experiments to reason that the density of an object is related to whether the object can sink or float in a fluid; (2) students use computer simulations to visualize the multiple forces including buoyancy and gravity acting upon the object when it is floating or sinking in a fluid; (3) students use their visualization of forces acting upon the object to reason why the weight of the object in a fluid reduces; (4) students analyze the data of their virtual experiments to reason that the buoyancy of an object in a fluid is equal to the weight of the fluid displaced in terms of magnitude (Archimedes' principle); (5) students analyze the data of their virtual experiments to reason that the buoyancy of an object in a fluid in terms of magnitude also equals the density of the fluid times the volume of the object immersed in the fluid; and (6) students analyze the data of their virtual experiments to reason that for a floating object, the buoyant force is equal to the weight of the object.

Developing Learning Materials

After creating learning performances in correspondence with the national curriculum guidelines, we developed learning materials including five activities. We discuss each activity as follows.

Activity 1: The Floating Future The first activity contextualizes students in an environmental issue of sea level rise and its consequences to introduce the idea of

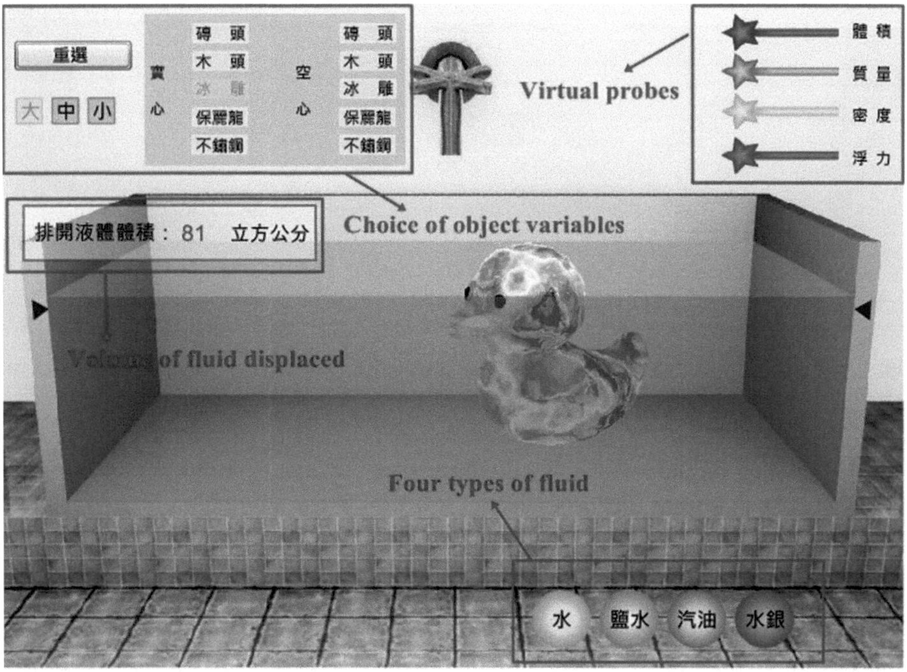

Fig. 15.8 The *bath duck* simulation for students to conduct virtual experiments

building floating houses in some coastal cities or areas such as in the Netherlands. This issue is also significant in Taiwan since Taiwan is an island and the majority of the population on the island lives in coastal cities. An animation is embedded in this activity to show students how a floating house is built. Students are asked to write down their reflection to think about why a house may float on the surface of the sea.

Activity 2: The Phenomenon of Sinking and Floating We developed a simulation (Fig. 15.8) for students to conduct virtual experiments to learn factors related to an object sinking or floating in a given fluid. Students can change the values of four variables in the simulation to conduct their experiments: (1) the material of the object (duck) (brick, wood, ice, Styrofoam, or steel), (2) the size of the object (duck) (large, medium, or small), (3) the composition of the duck (solid or hollow), and (4) the type of fluid (water, saline water, gasoline, or mercury). In addition, on the upper right-hand side, four virtual probes are provided so that students can use them to measure the volume, mass, density, or buoyant force of the object and the density of the liquid. A small pop-up window appears on the left side to indicate the volume of the fluid displaced after the duck is placed in the fluid. Students are asked to synthesize from their experiments to reason variables directly related to the phenomenon of sinking and floating.

Activity 3: Buoyancy In this activity, we introduce the definition of buoyancy. Students are guided to first conduct experiments focusing on only those objects

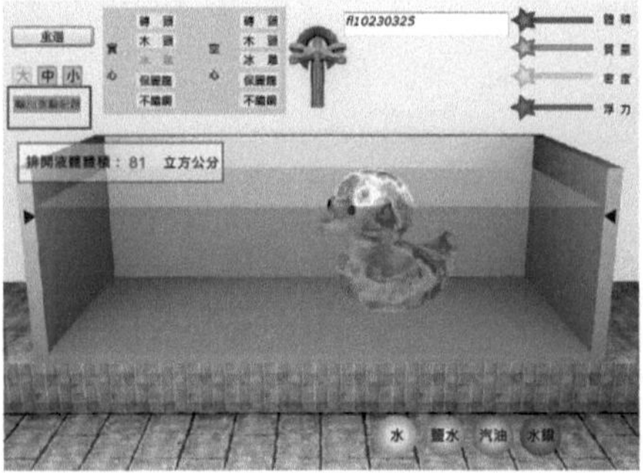

實驗次數	ID	步驟	物體	實心/空心	是否為沉體	物體所受浮力(gw)	物體在液面下體積(cm³)	液體與其密度(g/cm³) 油0.7 水1 鹽水1.1 水銀13.6	物體重量(gw)
1	fl10230325	2.3	小木鴨	實心	浮體	15	15	水1	15
2	fl10230325	2.3	中冰鴨	實心	浮體	54	54	水1	54
3	fl10230325	2.3	大鐵鴨	空心	沉體	135	90	水1	90
4	fl10230325	2.3	小保麗龍鴨	空心	浮體	2.1	2.1	水1	2.1
5	fl10230325	2.3	大銅鴨	實心	沉體	270	90	鹽水1.1	99
6	fl10230325	2.3	中木鴨	空心	浮體	18	25.7	油0.7	18
7	fl10230325	2.3	大冰鴨	實心	浮體	81	6	水銀13.6	81
8	fl10230325	2.3	中鐵鴨	空心	浮體	90	6.6	水銀13.6	90
9	fl10230325	2.3	大銅鴨	空心	浮體	45	45	水1	45
10	fl10230325	2.3	小木鴨	實心	浮體	15	15	水1	15
11	fl10230325	2.3	大冰鴨	空心	浮體	45	45	水1	45
12	fl10230325	2.3	中銅鴨	實心	沉體	180	60	水1	60
13	fl10230325	2.3	大保麗龍鴨	空心	浮體	6.3	6.3	水1	6.3
14	fl10230325	2.3	小鐵鴨	空心	沉體	45	30	水1	30
15	fl10230325	2.3	中鐵鴨	實心	沉體	120	60	水1	60
16	fl10230325	2.3	中保麗龍鴨	實心	浮體	12	12	水1	12
17	fl10230325	2.3	小冰鴨	空心	浮體	15	15	水1	15
18	fl10230325	2.3	大木鴨	實心	浮體	45	40.9	鹽水1.1	45
19	fl10230325	2.3	小鐵鴨	空心	沉體	45	30	鹽水1.1	33
20	fl10230325	2.3	中銅鴨	實心	沉體	180	60	油0.7	42
21	fl10230325	2.3	中冰鴨	空心	浮體	30	2.2	水銀13.6	30
22	fl10230325	2.3	小木鴨	實心	浮體	15	1.1	水銀13.6	15
23	fl10230325	2.3	大保麗龍鴨	實心	浮體	18	1.3	水銀13.6	18
24	fl10230325	2.3	小木鴨	空心	浮體	9	0.700005	水銀13.6	9
25	fl10230325	2.3	中鐵鴨	空心	浮體	90	6.6	水銀13.6	90

Fig. 15.9 *Left*: the button (highlighted in the *red* box) to automatically create worksheets of experiment records; *Right*: a worksheet recording the student's experiments

(ducks) floating on the surface of a fluid and then conduct experiments focusing on those sinking in a fluid. We embedded a function in the simulation for students to create a worksheet of their experiments recording the properties of the ducks and fluids they have experimented with (Fig. 15.9). Using the worksheets, the students are guided to reason that for a floating object, the buoyant force is equal to the weight of the object and that for a sunken object (and a floating object), the buoyant force of an object in a fluid equals the density of the fluid times the volume of the object immersed in the fluid. Moreover, they are guided to draw their visualizations of the buoyant forces acting upon floating and sunken objects.

Activity 4: Building Models of Buoyancy In this activity, students are guided to use textual, symbolic, and visual representations to build models to represent their ideas of buoyancy. For example, they are requested to select from several sentence options from dropdown menus to create a principle (i.e., textual representation of a buoyancy model) to summarize patterns and rules found in relation to the phenomena of sinking and floating (Fig. 15.10). After they create their principle, they are required to post it online and critique their classmates' principles. Embedding a critiquing activity enables students to examine diverse and alternative ideas and to reflect on the most productive ones (Chang & Linn, 2013). The goal of the critiquing activity in this module is to provide chances for students to generate a principle of buoyancy that is adequate in terms of consisting with the scientific principle.

In this activity, students also create a visual and symbolic model to represent their idea of buoyancy (Fig. 15.11). A free online software package, MySystem (Svihla, Ryoo, Linn, & Dorsey, 2011), allows students to use customizable stamps and objects to depict relationships among variables. In this module, students use MySystem to represent their idea of variables and their relationships associated with the phenomena of sinking and floating and the concept of buoyancy.

Activity 5: Simulating a Floating House In this activity, students are guided to revisit the idea of a floating house and apply what they have learned to use a

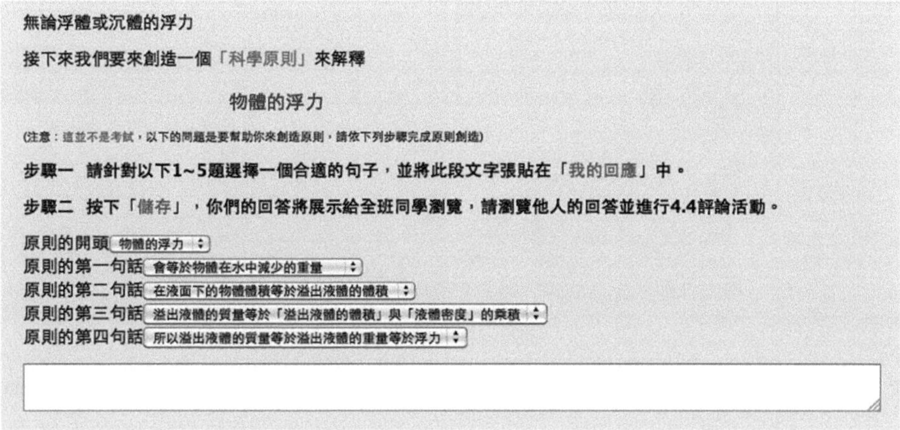

Fig. 15.10 Dropdown menus to support students in creating a principle of buoyancy

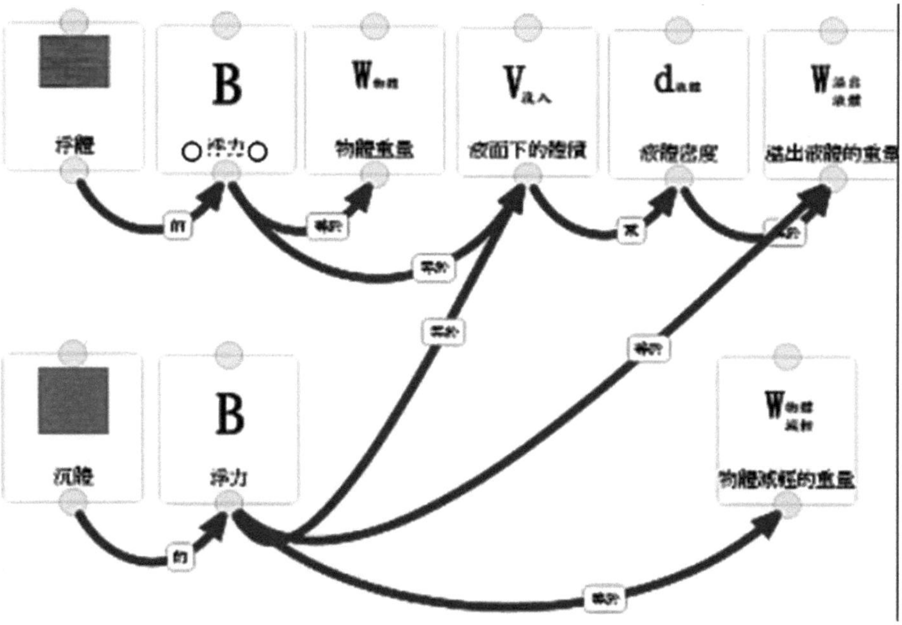

Fig. 15.11 A visual and symbolic model of buoyancy made by a student

computer simulation to design and build their own floating house. We developed the computer simulation to allow students to select from a list of virtual objects such as bases, walls, doors, and windows made of different materials such as wood, stone, or iron to build a virtual house (a maximum of four stories, Fig. 15.12, left). In the *building* mode, students can also have the computer program calculate the

Fig. 15.12 *Left*: the *building* mode of the simulation in which students select different objects from the left panel to build a house. *Right*: the *simulation* mode in which students test their design to see how the house floats on the surface of the sea

average density (including the air in the house), the loading capacity, and the cost of the house by clicking on the "calculate" button (the red button, and the result is shown in the upper right-hand side of the screen in a box as in Fig. 15.12). In the *simulation* mode (when they click the orange button on the screen), students can observe how well the built house floats on the surface of the sea. A design fails when not only the base but also part of the house immerses under the sea. At the end of the activity, students are asked to generate explanations that discuss why their house floats or sinks.

Summary

The buoyancy module has four characteristics: (1) encouraging students to use multiple representations to build their model of buoyancy; (2) incorporating virtual experiments for students to conduct inquiry and test their ideas about sinking and floating; (3) engaging students in critiquing activities to help them generate scientific principles of buoyancy; and (4) providing a design task to contextualize learning of buoyancy in real-life situations. The goal of our design of the module is to address the common difficulty of student understanding of buoyancy through engaging students in the practice of scientific inquiry including experiments, model building, use of multiple representations, and critique and communications. Contextualizing the topic in daily life examples also helps students link between school science and everyday experiences to develop useful knowledge and skills.

Summary and Implications

In summary, this chapter demonstrates a systematic way to develop technology-infused inquiry learning modules that are based on the national curriculum guidelines. A four-step instructional design framework (see Fig. 15.1)—deciding learning goals, deciding learning sequence, developing teaching materials, and evaluating teaching effectiveness—was developed and guided us to develop the inquiry learning modules. The two learning modules on the topics of plate tectonics and buoyancy exemplified the development process and demonstrated how technology such as interactive visualizations, animations, multiple representations, and virtual experiments can be employed in the design of inquiry activities.

The instructional design framework proposed in this study has two levels of implications. At the first level, the framework provides a systematic method for science teachers to develop inquiry learning activities from the national curriculum guidelines. At the second level, the learning modules developed on CWISE can be shared and used as a "science learning module bank." They can be used in general scheduled science lessons since the contents involved are closely related to the national curriculum guidelines. In addition, depending on the needs of different science classrooms, science teachers can modify these modules in terms of the content or the sequence of the steps.

We hope that this study not only bridges the gap between the national curriculum guidelines and inquiry teaching practice but also serves as a base to promote students' science learning by inquiry.

References

Anderson, K. T., Zuiker, S. J., Taasoobshirazi, G., & Hickey, D. T. (2007). Classroom discourse as a tool to enhance formative assessment and practice in science. *International Journal of Science Education, 29*(14), 1721–1744. doi:10.1080/09500690701217295.

Ault, J., & Charles, R. (1984). The everyday perspective and exceedingly unobvious meaning. *Journal of Geological Education, 32*(2), 89–91.

Bybee, R. W. (2006). Scientific inquiry and science teaching. In L. Flick & N. Lederman (Eds.), *Scientific inquiry and nature of science: Implications for teaching, learning, and teacher education* (pp. 1–14). Dordrecht, The Netherlands: Springer.

Chang, H.-Y. (2013). Teacher guidance to mediate student inquiry through interactive dynamic visualizations. *Instructional Science, 41*(5), 895–920. doi:10.1007/s11251-012-9257-y.

Chang, H.-Y., & Linn, M. C. (2013). Scaffolding learning from molecular visualizations. *Journal of Research in Science Teaching, 50*(7), 858–886. doi:10.1002/tea.21089.

Cobb, P., Confrey, J., diSessa, A., Lehrer, R., & Schauble, L. (2003). Design experiments in educational research. *Educational Researcher, 32*(1), 9–13. doi:10.3102/0013189x032001009.

Cuevas, P., Lee, O., Hart, J., & Deaktor, R. (2005). Improving science inquiry with elementary students of diverse backgrounds. *Journal of Research in Science Teaching, 42*(3), 337–357. doi:10.1002/tea.20053.

de Jong, T., & van Joolingen, W. R. (1998). Scientific discovery learning with computer simulations of conceptual domains. *Review of Educational Research, 68*(2), 179–201. doi:10.3102/00346543068002179.

Flick, L. B. (2006). Developing understanding of scientific inquiry in secondary students. In L. B. Flick & N. G. Lederman (Eds.), *Scientific inquiry and nature of science: Implications for teaching, learning, and teacher education* (pp. 157–172). Dordrecht, The Netherlands: Springer.

Ford, M. J., & Wargo, B. M. (2007). Routines, roles, and responsibilities for aligning scientific and classroom practices. *Science Education, 91*(1), 133–157. doi:10.1002/sce.20171.

Gobert, J. D. (2000). A typology of causal models for plate tectonics: Inferential power and barriers to understanding. *International Journal of Science Education, 22*(9), 937–977. doi:10.1080/095006900416857.

Gobert, J. D., & Clement, J. J. (1999). Effects of student-generated diagrams versus student-generated summaries on conceptual understanding of causal and dynamic knowledge in plate tectonics. *Journal of Research in Science Teaching, 36*(1), 39–53.

Gyllenpalm, J., Wickman, P.-O., & Holmgren, S.-O. (2010). Secondary science teachers' selective traditions and examples of inquiry-oriented approaches. *Nordic Studies in Science Education, 6*(1), 44–60.

Hickey, D. T., Taasoobshirazi, G., & Cross, D. (2012). Assessment as learning: Enhancing discourse, understanding, and achievement in innovative science curricula. *Journal of Research in Science Teaching, 49*(10), 1240–1270. doi:10.1002/tea.21056.

Hofstein, A., Shore, R., & Kipnis, M. (2004). Providing high school chemistry students with opportunities to develop learning skills in an inquiry-type laboratory: A case study. *International Journal of Science Education, 26*(1), 47–62. doi:10.1080/0950069032000070342.

Hsu, Y.-S., Chuang, F.-T., & Lin, T.-C. (2012). Exploring the framework and implementation of the design research method in science education. *Journal of Research in Education Sciences, 57*(1), 1–27.

Jacobi, D., Bergeron, A., & Malvesy, T. (1996). The popularization of plate tectonics: Presenting the concepts of dynamics and time. *Public Understanding of Science, 5*(2), 75–100.

Krajcik, J., McNeill, K. L., & Reiser, B. J. (2008). Learning-goals-driven design model: Developing curriculum materials that align with national standards and incorporate project-based pedagogy. *Science Education, 92*(1), 1–32.

Kyza, E. A., Constantinou, C. P., & Spanoudis, G. (2011). Sixth graders' co-construction of explanations of a disturbance in an ecosystem: Exploring relationships between grouping, reflective scaffolding, and evidence-based explanations. *International Journal of Science Education, 33*(18), 2489–2525. doi:10.1080/09500693.2010.550951.

Linn, M. C., Clark, D., & Slotta, J. D. (2003). WISE design for knowledge integration. *Science Education, 87*(4), 517–538.

Linn, M. C., & Eylon, B.-S. (2011). *Science learning and instruction: Taking advantage of technology to promote knowledge integration*. New York: Routledge.

Liu, O. L., Lee, H.-. S., & Linn, M. C. (2010). An investigation of teacher impact on student inquiry science performance using a hierarchical linear model. *Journal of Research in Science Teaching, 47*(7), 807–819.

Martin, M. O., Mullis, I. V. S., Foy, P., & Stanco, G. M. (2012). *TIMSS 2011 international results in science*. Chestnut Hill, MA: Boston College.

McNeill, K. L. (2011). Elementary students' views of explanation, argumentation, and evidence, and their abilities to construct arguments over the school year. *Journal of Research in Science Teaching, 48*(7), 793–823. doi:10.1002/tea.20430.

McNeill, K. L., Lizotte, D. J., Krajcik, J., & Marx, R. W. (2006). Supporting students' construction of scientific explanations by fading scaffolds in instructional materials. *Journal of the Learning Sciences, 15*(2), 153–191.

Ministry of Education. (2008). *The outline for the nine-year integrated curriculum*. Taipei, Taiwan: Ministry of Education, Taiwan.

Minner, D. D., Levy, A. J., & Century, J. (2010). Inquiry-based science instruction—What is it and does it matter? Results from a research synthesis years 1984 to 2002. *Journal of Research in Science Teaching, 47*(4), 474–496. doi:10.1002/tea.20347.

National Research Council. (2000). *Inquiry and the national science education standards: A guide for teaching and learning*. Washington, DC: National Academies Press.

National Research Council. (2012). *A framework for K-12 science education: Practices, cross-cutting concepts, and core ideas*. Washington, DC: National Academies Press.

OECD. (2007). *PISA 2006 science competencies for tomorrow's world volume 1: Analysis*. Paris, FR: Author.

Osborne, J., Erduran, S., & Simon, S. (2004). Enhancing the quality of argumentation in school science. *Journal of Research in Science Teaching, 41*(10), 994–1020.

Penuel, W., Fishman, B. J., Gallagher, L. P., Korbak, C., & Lopez-Prado, B. (2009). Is alignment enough? Investigating the effects of state policies and professional development on science curriculum implementation. *Science Education, 93*(4), 656–677.

Plummer, C. C., & Carlson, D. H. (2008). *Physical geology* (12th ed.). Dubuque, IA: McGraw-Hill.

Slotta, J. D., & Linn, M. C. (2009). *WISE science: Web-based inquiry in the classroom*. New York/ London: Teachers College Press.

Svihla, V., Ryoo, K., Linn, M. C., & Dorsey, C. (2011). *Connecting energy across the curriculum*. Retrieved August 18, 2012, from The Concord Consortium web. http://concord.org/publications/newsletter/2011-spring/connecting-energy-across-curriculum

Tabak, I. (2004). Synergy: A complement to emerging patterns of distributed scaffolding. *Journal of the Learning Sciences, 13*(3), 305–335. doi:10.1207/s15327809jls1303_3.

Wang, J. R., Huang, B. Y., Tsay, R. F., Lee, K. P., Lin, S. W., & Kao, H. L. (2011). A meta-analysis of inquiry-based instruction on student learning outcomes in Taiwan. *Asia-Pacific Education Researcher, 20*(3), 534–542.

White, B., & Frederiksen, J. R. (1998). Inquiry, modeling, and metacognition: Making science accessible to all students. *Cognition and Instruction, 16*(1), 3–118. doi:10.1207/s1532690xci1601_2.

Wu, H.-K., & Hsieh, C.-E. (2006). Developing sixth graders' inquiry skills to construct explanations in inquiry-based learning environments. *International Journal of Science Education, 28* (11), 1289–1313. doi:10.1080/09500690600621035.

Wu, P. -H., Wu, H. -K., & Hsu, Y. -S. (2013). Establishing the criterion-related, construct, and content validities of a simulation-based assessment of inquiry abilities. *International Journal of Science Education*, 1–21. doi:10.1080/09500693.2013.871660.

Part VI
Teachers Professional Development

Chapter 16
Developing Teachers Through Professional Learning Communities in Singapore and Shanghai

Salleh Hairon and Charlene Tan

Introduction

In the recent Programme for International Student Assessment (PISA), Shanghai and Singapore once again emerged among the top performing economies. East Asian societies such as Singapore and Shanghai have been recognised by the international education community for their students' impressive performance in international benchmarking tests such as the PISA (for more information, see Organisation for Economic Co-operation and Development [OECD], 2010, 2013). Among the reasons for their educational success are the contextual factors or frames that influence the success of the overall education. Focusing on the structural frame, this chapter critically discusses the investment placed on developing the teachers professionally in both Singapore and Shanghai. This chapter critically discusses the investment placed on teachers' professional development in both Singapore and Shanghai. In particular, this chapter examines professional learning communities (PLCs) as a key school-based professional development platform. This chapter begins by introducing the theoretical framework of 'structural frame', followed by a discussion of the characteristics of effective professional development and PLCs. The second section explores professional development for teachers and PLCs in Singapore and Shanghai, respectively.

S. Hairon (✉) • C. Tan
National Institute of Education, Nanyang Technological University, Singapore, Singapore
e-mail: hairon.salleh@nie.edu.sg; charlene.tan@nie.edu.sg

M.S. Khine (ed.), *Science Education in East Asia*,
DOI 10.1007/978-3-319-16390-1_16

Theoretical Framework: Structural Frame and Professional Development for Teachers

Bolman and Deal (1997) define a *frame* as a mental map or a set of ideas or assumptions one carries in the head to help the person understand and negotiate a particular 'territory' (p. 12). Louis, Toole, and Hargreaves (1999) point out that multi-frame thinking generates not only multiple perspectives but interrelationships and interdependence between frames; the frames point to 'neglected or hidden questions about school improvement and educational change, as well as helping to put more common questions in a new context' (p. 259).

Among the different types of frames is the *structural* frame that refers to the 'policies and procedures, roles and responsibilities, and uses of time and space that express the politics and address the vision' (Hargreaves, Halasz, & Pont, 2007). Louis et al. (1999) elaborate:

> The central question in the structural frame is: How does the school need to be restructured to support the purposes and processes of improvement? Time and space, roles and responsibilities, policies and resources are the focus of this question.... Structural decisions help determine everyday patterns of school life like class schedules, student groupings, teacher assignments, budgets, building layouts, divisions of subject matter, teaching materials, and even teaching content. (p. 262)

Due to space constraint, this chapter shall focus only on the structural frame in the area of professional development for teachers (see Tan, 2011 for more discussion on frames).

Professional Development Supporting Education Systems

While it has been stated that the quality of an education system cannot exceed the quality of its teachers – 'top performing (education) systems demonstrate that the quality of an education system depends on the quality of its teachers' (McKinsey & Company, 2007, p. 23) – what is interesting is how top performing education systems ensure high quality of its teachers. In comparing education systems from diverse education systems, the McKinsey Report (2007) emphasises that besides the image of the teaching profession being closely linked to the perceived level of education and training, professional development is a priority in top performing education systems. In the subsequent report, professional development was identified as the first area of intervention which 'Good to Great' education systems utilised as one of their strategies to professionalise their teachers with the eventual purpose of positively influencing teaching practice and students' outcomes (McKinsey Report, 2010).

In her analysis of the three countries with high achieving nations consisting of Finland, Singapore and South Korea, Darling-Hammond (2012) highlights professional development in the form of ongoing support for teacher learning as one of the strategies used by strong ‘teaching and learning systems’ to ensure high-quality student learning. It is therefore understandable that as countries around the world seek to reform their education systems, the importance of developing teacher competency through effective professional development provisions becomes ever more salient. The importance placed on professional development is further enhanced when high-quality education for all has become a public good (Darling-Hammond).

Effective Professional Development Provisions

However, in order to make the connection between professional development and improvements in teacher practice and student outcomes, it is necessary to enact professional development practices that are considered effective. Based on their literature review, Darling-Hammond, Wei, Andree, Richardson, and Orphanos (2009) outline several propositions. First, professional development should be intensive, ongoing and connected to practice. They regard episodic professional development provisions such as workshops that focus on discrete topics (e.g. classroom management and student motivation) as being undesirable because they are disconnected from practice. They also stress the importance of knowledge applications in teachers’ planning and instructions, which is said to impact on teaching practices leading to student learning outcomes. The importance of the connection between professional development and teacher practice – or job embeddedness – has been well established (Boyle, While, & Boyle, 2004; Corcoran, 1995; Garet, Porter, Desimone, Birman, & Kwang, 2001; Lester, 2003).

Second, Darling-Hammond et al. (2009) posit that professional development should focus on student learning and address the teaching of specific curriculum content. Professional development becomes effective when it addresses the concrete day-to-day teaching and learning challenges specific to the subject matter, which involves learning from role modelling, doing ‘hands-on’ work, being sensitive to context, defining precisely which concepts and skills students are to acquire, identifying concepts and skills that students’ find difficult, analysing students’ performance data, identifying students’ common errors and misunderstanding and using the data to inform decisions on teaching. At the heart of all these, teachers are motivated by students’ learning (Darling-Hammond, 1998).

Third, professional development should be aligned to school improvement priorities and goals. Darling-Hammond et al. (2009) assert that professional development becomes effective when it is an integral part of the larger school reform effort. This is to prevent or minimise the perceived disconnect between what teachers do in professional development experiences and what they are required to teach according to the school curricula discourse such as guidelines, textbooks and assessment.

Fourth, professional development should build strong working relationships among teachers. When teachers are able to work with and learn from one another within and across grade levels and academic departments, there is greater consistency in instruction, more willingness to share practices and try new ways of teaching and more success in solving problems of practice. One suggestion that was given to promote collaborative working relationships is through peer lesson observation followed by a safe environment for teachers to learn from one another's observations (Darling-Hammond et al., 2009). Collaborative professional development experiences provide opportunities for teachers to discuss concepts, skills and problems that intimately relate to teaching (Garet et al., 2001). They also contribute to a shared professional culture where common understanding of instructional goals, methods, problems and solutions can be developed. In addition, they provide a forum for debate and improving understanding that increases teachers' capacity to grow. It seems almost inevitable that collaboration leads to the formation of communities where teachers share not only common practices, but also common values. In this regard, learning is 'as much a socially shared undertaking as it is an individually constructed enterprise' (Alexander & Murphy, 1998, p. 38).

The fourth proposition raised by Darling-Hammond et al. (2009) also suggests the need for teachers to be active agents in the construction and co-construction of knowledge on teaching which impacts on classroom teaching and learning and not purely just moving beyond merely supporting the acquisition of knowledge and skills (Vescio, Ross, & Adams, 2008). This is partly due to the expansion of student learning outcomes consisting of not only academic, but also nonacademic aspects. The articulation of the twenty-first-century skills or competencies is a manifestation of the expansion in student learning outcomes. This would necessarily require transformations in the curricula and curricular development processes, which requires concomitant enhancements in teacher capacity and the appropriate professional development provisions.

The need to co-construct knowledge on teaching to impact classroom teaching and learning also requires teachers to be reflective and engaged in inquiry. As knowledge has become the foundation for all future learning (Alexander & Murphy, 1998), professional development must now be grounded in knowledge about teaching. Knowledge about teaching is questioned and interrogated through reflection and inquiry activities (e.g. reflective practice and action research). The fundamental point regarding teachers engaging in reflection and inquiry for knowledge creation and interrogation is that professional development approaches must move away from the 'transmission of knowledge and skills to teacher by experts to the study by teachers of the teaching and learning process' (Sparks & Hirsh, 1997, p. 12).

Professional Learning Communities (PLCs)

In their review of professional development, Darling-Hammond et al. (2009) have highlighted several models of professional development that are considered effective in bringing about the development of teacher knowledge, teaching practice and student learning outcomes. Professional learning communities (PLCs) along with their attendant benefits were highlighted. PLCs are consistent with the four propositions proposed above (Darling-Hammond et al.). Centrally, PLCs are school-based professional development platforms that provide the structural support to develop teachers. They create conducive platforms to bring about continual teacher learning practices that deepen teachers' knowledge on teaching (curriculum content, pedagogy, instruction, assessment, student learning) which is then translated to improvements in teaching practices and finally to improvements in student learning and student learning outcomes.

Teacher development in PLCs is situated in relationships among teachers within groups (three or more). However, there is currently no universal definition of a PLC, and that there are shades of interpretation of what it means (Darling-Hammond, 1996; DuFour & Eaker, 1998; Hord, 1997; Stoll et al., 2003). Nevertheless, there is still a broad international consensus that PLC has to do with a group of people sharing and critically interrogating their practice in an ongoing, reflective, collaborative, inclusive, learning-oriented, growth-promoting way (Bolam, McMahon, Stoll, Thomas, & Wallace, 2005; Toole & Louis, 2002). The literature on PLCs supports the claim that there is a positive relationship between PLCs and student achievement (Bolam et al., 2005; Hord & Sommers, 2008; Vescio et al. 2008). Student achievement is attained when teachers make collaborative pedagogical decisions which value-add to student learning (Thompson, Gregg, & Niska, 2004). PLCs have also been reported to have an intermediate positive impact on teachers' professional learning, performance and morale, and an ultimate leading to positive impact on student achievement (Bolam et al.). It has been proposed that professional development that focuses on student learning and helps teachers develop skills to teach specific kinds of content, such as PLCs, has strong positive effects on teaching practice (Blank, De las Alas, & Smith, 2007; Darling-Hammond & Richardson, 2009; Wenglinsky, 2000). PLCs have also been seen as having potential to contribute to positive changes to the school culture which are considered as second-order changes (Marzano, Waters, & McNulty, 2005; Wells, 2008) resulting to the potential to improve school capacity that serves the learning needs of students (Boyer, 1995; Fullan, 1997). On a wider scale, PLCs have the potential to contribute to the school improvement agenda (Bolam et al., 2005; Fullan, 2001; Hord, 1997, 2004; McLaughlin & Talbert, 2001, 2006; Senge et al., 2000) and system-wide capacity building (Bolam et al., 2005; Boyer, 1995; Fullan, 1997).

Professional Development for Teachers in Singapore

Singapore was founded as a British trading post and colony in 1819 and granted self-government in 1959. It was briefly occupied by the Japanese from 1942 to 1945 when the British lost the war to the Japanese. After a short-lived union with Malaysia in 1963, it was separated to become a sovereign state in 1965. Singapore is a plural society with foreigners comprising about 36 % of Singapore's population of 4.9 million people. Despite being multi-ethnic, the majority of the population as well as its political leaders are ethnic Chinese (72.4 %) whose ancestors mainly came from South China. Compulsory education is implemented in Singapore where all children have to complete at least 6 years of primary education. Almost all students proceed to another 4 or 5 years of schooling at the secondary level, before continuing their study at a vocational institute, junior college or other post-secondary institutions. There are altogether about 370 schools, which are mostly government schools. Less than 20 schools are independent and specialist schools – but even they must abide by government regulations concerning curricula and assessment. Schools are divided into four zones – East, West, North and South – within which schools are grouped into clusters of about 12–14 schools, each with a superintendent. Each cluster comprises a mixture of primary and secondary schools, and among its functions is mediation between individual principals and schools and the headquarters of the education ministry. The cluster also has designated functions to promote network professional development and school improvement activities across the schools in its group.

The PLC Initiative in Singapore Schools

The PLC policy initiative in Singapore started in 2009 during a Teachers' Mass Lecture by the newly appointed Director-General of Education (Ministry of Education [MOE], 2009). The PLC policy initiative is considered another milestone for the teaching profession and revolves around three themes: 'We Care', 'Teachers Inspire' and 'We are Professionals'. While the phrase 'We Care' speaks of the commitment to care for every child and 'Teachers Inspire' speaks of making a difference in students' learning outcomes, the phrase 'We are Professionals' involves teachers developing each other within a learning community. The essence is encapsulated in the following articulations:

> Singapore Teachers – Lead, Care, Inspire.
> For the Future of the Nation Passes through Our Hands.

The notion of 'Lead' within the vision statement has to do with 'Leading the Learner' and 'Leading Learning'. What is implied here is that teachers are expected to take greater lead or initiative to not only bring about the best of the student through appropriate learning experiences, but also develop their professional colleagues too. It is the latter that PLCs find its resonance. The assumption is that PLCs have the potential to 'level up teacher professionalism in a quick and effective way' (MOE, 2009). PLCs are perceived to bring about the culture of teacher-led professional development which will then raise the quality of instructional expertise across all schools and classrooms in Singapore (MOE, 2011a, 2011b). After piloting 51 PLC schools in 2009, the MOE has strongly encouraged all government schools to come onboard the PLC journey. It has been reported that 305 or 85 % of Singapore schools have embarked on the PLC journey (MOE, 2012).

Along with the introduction of PLCs is also the formation of the Academy of Singapore Teachers (AST) in 2010. The academy provides the platform for teachers to develop their professional knowledge and skills through teacher-led professional development within contexts of professional communities so as to raise the teachers' competence and confidence in the context of increasingly demanding curricula in the twenty-first century (MOE, 2009, 2010). The idea of teacher-led professional dialogue through professional communities to improve professional knowledge and practice is, however, not new. This idea was first introduced in the establishment of Teachers Network (TN) in 1998 – a unit within the then Training and Development Division of the MOE (TDD) (Tang 2000; Tripp 2004). Both the TN and TDD have collapsed to form the Academy of Singapore Teachers (AST). The TN's aim was to develop teacher professional development and build a fraternity of reflective teachers dedicated to excellent practice through a network of support, professional exchange and learning and to serve as a catalyst and support for teacher-initiated development through sharing, collaboration and reflection leading to self-mastery, excellent practice and fulfilment. Its distinctiveness lies in the bottom-up approach to change, as evidenced in its slogan 'For Teachers, By Teachers' (MOE, 2005). The value of teacher-initiated professional development through sharing, collaboration, reflection and inquiry is embodied in its signature platform for teacher learning – the Teachers' Network Learning Circles (LCs). TNLCs consist of self-directed teachers using action research to solve problems that have been collectively identified in relation to the curriculum and classroom pedagogy.

The establishment of TN and AST, along with its aims and objectives, sits within a wider educational policy context. At the international level, there is a strong belief that improving teacher learning brings about enhancement of the quality of teachers, teaching, student learning and educational outcomes. This belief has been reinforced by the report published by McKinsey & Company (2007), which compares a selected range of successful education systems around the world, stating that (1) the quality of an education system cannot exceed the quality of its teachers, (2) the only way to improve outcomes is to improve instruction and (3) high performance requires every child to succeed. Teachers are now seen as pivotal in their role in enhancing the skills of the future workforce in the context of a globally competitive world.

At the local level, the need to improve teacher learning stems from the need of teachers to respond to new demands of the education ministry for greater contextualisation of the curriculum to the needs of individual schools and greater diversification of curricular outcomes. For the former, the education ministry has encouraged all schools to find their own school niches. Since the turn of the twenty-first century, educational policymakers realised that the school curricula must not only continue to produce strong academic results, but also enhance other soft skills such as creative thinking, critical thinking, citizenship, self-directed learning, etc. The education ministry has now identified the core skills required for the twenty-first century, termed as C2015, which include four domains and outcomes: confident person, self-directed learner, concerned citizen and active contributor (MOE, 2008). The new curricular demands require teachers and school leaders to develop new repertoire of pedagogies that are contextualised to the diverse range of school communities. This is where teacher learning, especially the kind that has characteristics of PLCs, finds its greatest relevance.

The Singapore PLC Model

The PLC model introduced by the education ministry conceptualises PLCs as a school instead of a group approach. A PLC integrates and functions within the school strategic framework insofar as it takes into consideration the specific school's vision, mission and values in order to achieve the school goals. Singapore is the first country in the world to have come up with a nationwide PLC model, with an explicit use of the term and concepts relating to PLCs, and formally integrating PLC in the school strategic framework. The ability of the education ministry to standardise a model across all public schools in Singapore is partially due to the smallness of the country and the hierarchical structure of the education system. The Singapore PLC model has specific key features: shared and supportive leadership; school vision, mission, values and goals; action orientation and experimentation; a strong learning and inquiring culture; a community based on trust and commitment; and supportive conditions (TDD, 2010). These features are consistent with the key features of PLCs used in the literature including shared values and vision; collective responsibility for pupils' learning; collaboration focused on learning; individual and collective professional learning; reflective professional enquiry, openness, networks and partnerships; inclusive membership; and mutual trust, respect and support (Bolam et al., 2005).

The Singapore education ministry conceptualises each school as a PLC consisting of several PLTs (professional learning teams) (TDD, 2010). Each PLT may consist of teachers belonging to the same grade level (which is more common in primary schools) or the same teaching subject (which is more common in secondary schools or junior colleges). Although each of these PLTs is essentially a PLC, the originators may see the value of PLTs interacting and collaborating with one another to form a school-wide PLC. In other words, PLTs are sub-PLCs within the school-wide PLC. Small groups of PLCs within a school are to integrate together forming a wider school-wide PLC. The Singapore PLC model focuses on '3 Big Ideas' or aims – enabling students learning, building a culture of collaboration and focusing on student outcomes. It also focuses on '4 Critical Questions' – What is it we expect students to learn? How will we know when they have learned? How will we respond when they don't learn? How will we respond when they already know it? The '3 Big Ideas' and '4 Critical Questions', which were modifications from DuFour's (2004) writing, serve to positively impact on classroom teaching and school learning culture, which would then have a positive impact on student outcomes.

The Singapore PLC model also places importance on school leadership to support the processes of PLC in the form of the Coalition Team comprising the principal, school staff developer (SSDs) and a few heads of department. While the PLTs are to make positive impact on improvements in instructional practice (content and pedagogy), the Coalition Team is to make positive impact on building the school learning culture (structures, processes and resources). Borrowing Fullan's 'Triangle of Success' model, improvements in student outcomes are achieved when the PLTs impact on 'Deep Pedagogy' and the Coalition Team impacts on 'System-ness' (TDD, 2010). PLTs have the choice of adopting a range of collaborative methods/tools, such as 'Learning Circles', 'Lesson Study' and 'Action Research'. 'Learning Circles' is a framework that teachers can use to engage in an action research cycle (Reflect, Plan, Act, Observe, Reflect) to change pedagogy with the emphasis on reflection. 'Lesson Study' emphasises on lesson planning, peer observation and group discussion to improve pedagogy. The 'Action Research' version which has been understood by educational practitioners is one that makes use of experimental designs on inquiry for pedagogical improvements. PLTs also adopt a 'cyclical process of continuous improvement' which includes the following: use of data to identify student needs, study and analysis of teachers' own and peer research, engagement in rigorous reflection, use of research and professional wisdom to make good choices, collaborative experiment with new teaching practices, monitoring and assessment of implementation and communication of information to other stakeholders.

Recently, the importance of PLCs has been felt by teacher educators at the National Institute of Education (NIE), the sole teacher education institute in Singapore, whereby a PLC framework has been constructed for student teachers to engage in reflective practice through learning communities termed as 'Focused Conversations' during their 10-week teaching practicum in schools (Hairon, Goh, & Teng, 2014). All in all, student teachers will be engaged in three focused conversations together with their school mentors collectively with all other student teachers belonging to the same school. Prior to and after these focused conversations, student teachers are to engage in focused conversations with their university supervisors with all other student teachers belonging to the same supervisor. The main objective of these focused conversations is to not only strengthen the theory-practice nexus – that is, encourage and enable student teachers to deepen their theoretical knowledge based on their critical analyses of their practical experiences in schools – but also develop the dispositions of twenty-first-century educators such as reflective practice, inquiry, collaborative learning and networked learning. The exposure to these focused conversations will also prepare student teachers to the new form of teacher learning in schools, specifically PLCs.

Professional Development for Teachers in Shanghai

With a population of 20.7 million, Shanghai is the largest city in China as well as the city with the most developed basic education system in the country (Tan, 2012). It is a cosmopolitan city with a relatively high percentage of non-locals compared to other Chinese cities: 13.8 million are permanent residents and 5.4 million are temporary residents (OECD, 2010). A pioneer in education reforms in China, Shanghai was the first city to implement the 9-year compulsory education in 1978. All children are required to complete at least 9 years of schooling: 5 years of primary education and 4 years of junior secondary education. Almost all students proceed to the senior secondary (high school) level for another 3 years of study where they will either sit for the national college entrance examination to qualify them for tertiary education or opt for vocational institutions. There are currently about 88,600 full-time teachers teaching 701,600 students in 640 154 primary schools and 594,400 students in 979 junior and senior secondary schools (Shanghai Municipal Education Commission, 2011).

There are four formal hierarchical titles for teachers in Shanghai. They are as follows (Tan, 2013):

1. Third-grade (novice) teachers.
2. Second-grade teachers. They are intermediate teachers who are promoted from third-grade teachers after 3–5 years of service. They need to undergo internal evaluation in the school.
3. First-grade teachers. They are advanced teachers who are promoted from second-grade teachers after at least 5 years of service. They need to undergo internal evaluation in the school and external evaluation at the district level.
4. Senior-grade teachers. They are teachers who are promoted from first-grade teachers after at least 5 years of service. They need to undergo internal evaluation in the school and external evaluation at the district level.

Besides the above official titles, there are two honorary titles given to excellent teachers: *gugan* [backbone] teachers and *teji* [special-grade] teachers. Backbone teachers are experienced teachers who have obtained at least a second grade. They comprise about 30 % of the teacher workforce and are usually at least 30 years old. Special-grade teachers are those who are outstanding in their teaching and leadership. They have usually taught for many years and have distinguished themselves in pioneering new and successful practices and initiatives that are acknowledged by the authorities and teaching community.

Professional development for teachers is much emphasised in China, including Shanghai. Its importance stems from the Chinese belief that a teacher is a content expert; in fact, the Chinese word for 'teacher' (*jiaoshi*) is composed of two words that mean 'teach' and 'expert'. To be a 'teaching expert', Chinese teachers are expected to constantly upgrade their knowledge and skills through ongoing professional development. Teacher professional development in Shanghai can be divided into three phases: the beginning period in the 1980s, the exploratory period from

1990 to 1998 and the present period from 1999 onwards (Tan, 2013). Teacher professional development started in the 1980s due to a need to raise the academic and professional qualifications of close to 60 % of the 63,000 teachers at that time. The goal in the exploratory phase was to raise the teachers' standard, and the goal for the current phase is to prepare teachers for the twenty-first century. Teacher professional development in Shanghai generally is carried out at two levels: municipality and district.

A key initiative launched by the Shanghai Municipal Education Commission is the '240, 540' education plan (Shanghai Municipal Education Commission, 2011). This plan stipulates that all teachers except special-grade teachers need to fulfil 240 h of training over 5 years; for senior-grade teachers, it is 540 h. As this is a policy regulation from the top, it is to be enforced strictly and is linked to the teachers' continual employment, salary increase and promotion. This shows the seriousness in which the Shanghai authority takes teacher training. The 240 h of training is converted into credits; ten learning hours is equivalent to one credit. This means a beginning and middle-ranked teacher needs to chalk up 24 credits over 5 years while a senior-grade teacher needs 54 credits over 5 years.

Schools in the district are given a certain level of autonomy to design and implement the school-based training. There are generally four types of school-based training: curriculum training, special topics on teaching, teaching-research training and forum on a special topic. Teachers (generally senior teachers) who wish to offer a school-based training course need to apply formally and receive approval from the district teacher training college. It is noteworthy that teacher professional development in Shanghai covers many essential aspects of teaching such as subject knowledge and research, lesson planning and delivery, teacher mentoring and collaboration and even self-improvement through reading. The rest of the section discusses how Shanghai teachers develop themselves through the structural support of mentoring and professional learning communities.

Professional Learning Communities

Although the term 'professional learning communities' (PLCs) is not used in Shanghai, we do find a similar structure through what are known locally as 'teaching-research group' [*jiaoyan zhu*] and 'lesson preparation group' [*beike zhu*]. The organisation of teachers into teaching-research groups and lesson preparation groups originated from Russia. They were first introduced through a 1957 document issued by the education ministry. Known as 'Work regulation (draft) on research group for secondary teaching' [*guanyu zhongxue jiaoxue yanjiuzhu gongzuo tiaoli* (*caoan*)], the document stated that teachers should be organised into teaching-research group. The objectives are to exchange ideas on teaching experiences and raise the teachers' thinking, professional standard and educational quality. The regulation also covers details of the organisation such as the teaching-research work content, organisational leadership, etc. For example, the document states that three or more teachers will form a subject teaching-research group. If there are insufficient numbers, the group can be combined based on similar subjects.

Since the 1950s, secondary and primary schools progressively and comprehensively formed teaching-research groups which group teachers based on the same subject (e.g. mathematics teaching-research group) and lesson preparation groups which group the same subject teachers based on the same grade (mathematics grade 1). While all schools have teaching-research groups for all subjects, not all schools have lesson preparation groups for all the grades. The size of a teaching-research group varies from three to around ten.

There are three levels of teaching-research: municipal, district and school. At the municipal level is the teaching-research office, followed by the teaching-research offices in the teacher training colleges of all the districts and finally the teaching-research departments in all schools. Each school's teaching-research department is in charge of the teaching-research groups, and below the teaching-research group is the lesson preparation groups. Most teaching-research groups meet once a week for about two or three periods (each period is 40 min). The meeting is planned into the timetable to ensure that structural support is given to the teachers to take the teaching-research activities seriously.

Within the hierarchy and structure, however, is a high level of flexibility for the teaching-research groups. The teachers usually come together to discuss their teaching experiences, share about a new theory-practice, discuss exam questions and problems encountered in teaching and suggestions and conduct research that is linked to their teaching. A key activity for the teaching-research group is to observe and critique each other's lessons. The group may also invite educational experts such as teaching-research officers from the district and university professors and senior teachers from another school to speak to and guide them. It is also not uncommon for teaching-research groups of various schools in the same district to come together to be trained, plan and then exchange ideas.

One of the authors observed a teaching-research group and a lesson preparation group in action in a Shanghai school in May 2011. The Physics teaching-research group comprises five teachers teaching Physics at the senior secondary level. The meeting objectives were to share the problems the teachers faced in the course of teaching and to discuss a new research project. The leader focused on a difficult Physics question, explained how she taught it and asked the rest for their views. The leader reminded the group to teach Physics in a more lively way and to reduce the schoolwork burden of students under the new curriculum reform. She also announced that the group would embark on an action research project focusing on a 'learning case study' for 2 years. One teacher was assigned to prepare the report for the research project, and the leader said that she would conduct the open lesson. What was observed was that the teachers were forthcoming and enthusiastic in sharing their views. The discussion was marked by warmth and politeness with the teacher addressing each other as 'Teacher so-and-so'. The teaching-research group ensures teacher compliance for new initiatives and research projects; at the same time, it lightens the teachers' workload by distributing it among the members and giving them support in sharing their problems and solutions. Through 'teacher research groups', Shanghai teachers are given ample opportunities to actively engage in reflective and inquiry practices to continually develop not only their curriculum content knowledge, but also their pedagogical knowledge (Tan, 2013).

Conclusions

This chapter has argued that PLCs nurture a community of educators who are actively engaged in reflective and inquiry practices to continually develop not only their curriculum content knowledge, but also their pedagogical knowledge. The chapter has also illustrated that PLCs provided the platform for teachers to develop school-based curricula that demands increasing diversity in student learning outcomes. Drawing from Singapore and Shanghai cases, this chapter has highlighted the key approaches and practices in teachers' professional development through PLCs that are consistent with the four propositions of effective professional development espoused by Darling-Hammond et al. (2009). These propositions highlight the need to make professional development (1) intensive, ongoing and connected to teachers' daily teaching practices; (2) focus on student learning and addressing specific curriculum content; (3) align to school improvement priorities and goals; and (4) enhance strong working relationships among teachers. PLCs are therefore highly appropriate and effective at supporting teacher professional development in the current international reform efforts at evolving curricula for the twenty-first century.

It is noteworthy that another reason for PLCs to be effective means of developing teachers is that they support *teacher mentoring* – another effective platform for developing teachers. Mentoring is a 'nurturing process in which a more skilled or more experienced person, serving as a role model, teaches, sponsors, encourages, counsels and befriends a less skilled or less experienced person for the purpose of promoting the latter's professional and/or personal development' (Anderson & Shannon, 1988, p. 40). Mentors are usually teachers with 'authority of experience' (Munby & Russell, 1994, p. 86). Their work usually pertains to the development of beginning teachers including roles such as acting as educational companions, providing emotional support, establishing collegial relationships with other colleagues and helping teachers appreciate the schools' values, expected behaviours and social knowledge (Fletcher & Barrett, 2004; Harrison, Dymoke, & Pell, 2006; Lai, 2005). Although the above activities evidently define the role of mentors, mentoring in reality is an experience that is jointly defined by mentors and mentees. In other words, mentoring is not just a one-sided enactment of a set of action by mentors, rather a dyadic relationship between a mentor and mentee where both parties are committed to engage in a set of actions that deepens and develop the capacities of both parties. Mentoring has also been identified by Darling-Hammond et al. (2009) as the other teacher learning platform to PLCs in relation to the four espoused effective characteristics of teacher professional development.

With its emphasis on mutual learning, mentoring can be said to complement and enrich PLCs. In the case of Singapore, mentoring is more prominent at the initial period of teaching. At the start of their teaching career, beginning teachers are to go through the Structured Mentoring Programme (SMP) that was introduced in 2006. Each beginning teacher is assigned to a mentor in their respective schools that they are posted to by the education ministry. These mentors are usually senior teachers or lead teachers who are both considered to be in the Teaching Track in terms of career development. In the Singapore education system, teachers can aspire to grow in three career tracks: Teaching, Leadership and Senior Specialist. In the Teaching career track, teachers are promoted to the Senior Teacher position, followed by the Lead Teacher, Master Teacher and Principal Master Teacher positions. Out of these four positions, only senior and lead teachers are resident in schools, while master and principal master teachers are resident in the education ministry. The primary role of these teachers is to lead teachers in the development of specific content and pedagogical knowledge. In mentoring, senior and lead teachers thus play an indispensable role in the development of subject content and pedagogical knowledge of beginning teachers, in addition to other roles such as getting them assimilated to the different culture of schools. While the SMP develops teachers in their initial years of their career, PLCs exist as professional development platforms for all existing teachers to further develop their knowledge in teaching regardless of teaching experience especially in regard to the development of school-based curricula. Although they are currently unintegrated, unlike the Shanghai education system, they are conceptualised as at being different stages of a teacher developmental growth.

In the case of Shanghai, teacher mentoring exists in all Shanghai schools and is implemented in two main ways: one-to-one mentoring and group mentoring (Hairon & Tan, 2013). For one-to-one-mentoring, all schools have a structured teacher mentoring system [*shitu daijiao*], whereby novice teachers (new third-grade teachers) are assigned a mentor – an experienced colleague who is usually a *gugan* teacher – for about 3 years. The mentoring process covers all aspects of teaching, such as the discussion of teaching materials, lesson observation and critiquing, teaching methods and setting and marking assignments. Mentoring may take place not only on a one-to-one basis but also as a group whereby novice teachers learn from their more experienced colleagues through daily collaboration and lesson observations. Mentoring occurs within the teaching-research group when the senior teachers guide the junior teachers in various teaching-research activities. The teachers may come together to discuss their teaching experiences, share ideas about a new theory or practice, discuss examination questions and problems encountered in teaching, share suggestions and conduct research related to teaching. Here we see how mentoring and PLCs go hand in hand to nurture teachers to become effective educators.

As governments around the world consider ways to raise the quality of educational outcomes, they will inadvertently be confronted with the formidable task of committing to making comprehensive changes. Increasingly, governments are realising the need to play a more active role in enacting these changes regardless of the 'centralised-decentralised' spectrum of educational policy change and reform dispositions. The comparison between Singapore and Shanghai is a good example to illustrate. While both countries are considered East Asian countries and are influenced by Asian values such as hierarchy (Hofstede, 2001), the degree of hierarchy has slight nuances. For the case of Shanghai, educational policies carried out in a highly bureaucratic way are generally accepted by educators. However, in Singapore, educational policies are carried out with a mixture of 'top-down' and 'bottom-up' approaches even though, on the whole, the education system in Singapore is generally considered a centralised 'command and control' system (Hairon & Dimmock, 2012). For example, while schools are generally and strongly encouraged to implement PLCs in their respective schools, school leaders and teachers are given a degree of latitude to embrace it. Even if school leaders and teachers are compelled to implement PLCs, they are given freedom in choosing the processes and deliverables suitable to their school needs. The implementation of PLCs in Shanghai is, however, more regimented in its application in the sense that PLCs are part of a formal structure under the teaching-research groups. PLC activities are factored into the official timetables of the teachers, and teachers are being appraised on their contributions in their PLCs which is not so in Singapore.

The two cases of teacher professional development in Singapore and Shanghai do not, however, suggest that more centralised education systems are more successful than non-centralised ones. As a further illustration, PLCs in Taiwan are carried out in a more ground-up or community approach, whereby different actors play a more or less equal role in influencing the aims, processes and supporting structures for the enactment of PLCs. These include the regional educational administration, university academics and school leaders and teachers. The two cases of teacher professional development do, however, suggest the importance of investing in professional development as a system-wide approach for countries beyond the East Asian shores to heighten their chances in achieving the desired student learning outcomes through the development of teacher knowledge.

References

Alexander, P. A., & Murphy, P. K. (1998). The research base for APA's Learner-Centered Psychological Principles. In N. M. Lambert & B. L. Coombs (Eds.), *How students learn: Reforming schools through learner-centred education* (pp. 25–60). Washington, DC: The American Psychological Association.

Anderson, E. M., & Shannon, A. L. (1988). Towards a conceptualization of mentoring. *Journal of Teacher Education, 39*(1), 38–42.

Blank, R. K., de las Alas, N., & Smith, C. (2007). *Analysis of the quality of professional development for mathematics and science teachers*. Washington, DC: Council of Chief State School Officers.

Bolam, R., McMahon, A., Stoll, L., Thomas, S., & Wallace, M. (2005). *Creating and sustaining effective professional learning communities* (Research report no. 637). Nottingham, UK: Department for Education and Skills.

Bolman, L. G., & Deal, T. E. (1997). *Reframing organisations: Artistry, choice and leadership* (2nd ed.). San Francisco: Jossey-Bass.
Boyer, E. (1995). *The basic school: A community for learning*. Princeton, NJ: The Carnegie Foundation for the Advancement of Teaching.
Boyle, B., While, D., & Boyle, T. (2004). A longitudinal study of teacher change: What makes professional development effective? *The Curriculum Journal, 15*(1), 45–68.
Corcoran, T. B. (1995). *Helping teachers teach well: Transforming professional development* (Policy Briefs). New Brunswick, NJ: Consortium for Policy Research in Education.
Darling-Hammond, L. (1996). The quiet revolution: Rethinking teacher development. *Educational Leadership, 56*(6), 4–10.
Darling-Hammond, L. (1998). Teacher learning that supports student learning. *Educational Leadership, 55*(5), 6–11.
Darling-Hammond, L. (2012). Soaring systems. *Education Review, 24*(1), 24–33.
Darling-Hammond, L., & Richardson, N. (2009). Teacher learning: What matters? *Educational Leadership, 66*(5), 46–53.
Darling-Hammond, L., Wei, R. C., Andree, A., Richardson, N., & Orphanos, S. (2009). *Professional learning in the learning profession: A status report on teacher development in the United States and abroad*. Dallas, TX: National Staff Development Council.
DuFour, R. (2004). What is a "professional learning community"? *Educational Leadership, 61*(8), 6–11.
DuFour, R., & Eaker, R. (1998). *Professional learning communities at work*. Bloomington, IN: National Education Service.
Fletcher, S. H., & Barrett, A. (2004). Developing effective beginning teachers through mentor-based induction. *Mentoring and Tutoring, 12*(3), 321–333.
Fullan, M. (1997). Broadening the concept of teacher leadership. In S. Caldwell (Ed.), *Professional development in learning-centered schools* (pp. 34–48). Oxford, OH: National Staff Development Council.
Fullan, M. (2001). *Leading in a culture of change*. San Francisco: Jossey-Bass.
Garet, M. S., Porter, A. C., Desimone, L., Birman, B. F., & Kwang, S. Y. (2001). What makes professional development effective? Results from a national sample of teacher. *American Educational Research Journal, 38*(4), 915–945.
Hairon, S., & Dimmock, C. (2012). Singapore schools and professional learning communities: Teacher professional development and school leadership in an Asian hierarchical system. *Educational Review, 64*(4), 405–424.
Hairon, S., Goh, J. W. P., & Teng, A. K. W. (2014). Professional learning communities in teacher internship programme in Singapore. In J. Calvo de Mora & K. Wood (Eds.), *Practical knowledge in teacher education – Approaches to teacher internship programs* (pp. 195–209). New York, NY: Routledge.
Hairon, S., & Tan, C. (2013). Novice teachers learning from others: Mentoring in Shanghai schools. *Australian Journal of Teacher Education, 38*(3), 152–165.
Hargreaves, A., Halasz, G., & Pont, B. (2007). School improvement for systematic improvement in Finland. A case study for the OECD activity improving school leadership. Retrieved April 23, 2008, from http://www.oecd.org/dataoecd/43/17/39928629.pdf
Harrison, J., Dymoke, S., & Pell, T. (2006). Mentoring beginning teachers in secondary schools: An analysis of practice. *Teaching and Teacher Education, 22*, 1055–1067.
Hofstede, G. (2001). *Culture's consequences: Comparing values, behaviors, institutions, and organizations across nations* (2nd ed.). Thousand Oaks, CA: Sage.
Hord, S. M. (1997). *Professional learning communities: Communities of continuous inquiry and improvement*. Austin, TX: Southwest Educational Development Laboratory.
Hord, S. M. (2004). Professional learning communities: An overview. In S. M. Hord (Ed.), *Learning together: Leading together. Changing schools through PLCs* (pp. 5–19). New York: Teachers College Press.

Hord, S. M., & Sommers, W. A. (2008). *Leading professional learning communities: Voices from research and practice*. Thousand Oaks, CA: Corwin Press.

Lai, E. (2005). In-service teachers' perceptions of teaching practice mentoring. *International Journal of Learning, 12*(6), 107–113.

Lester, J. H. (2003). Planning effective secondary professional development programs. *American Secondary Education, 32*(1), 49–61.

Louis, K. S., Toole, J., & Hargreaves, A. (1999). Rethinking school improvement. In K. S. Louis & J. Murphy (Eds.), *Handbook of research in educational administration: A project of the American Educational Research Association* (pp. 251–276). New York: Longman.

Marzano, R., Waters, T., & McNulty, B. (2005). *School leadership that works: From research to results*. Alexandria, VA: Association for Supervision and Curriculum Development.

McKinsey & Co. (2007). *How the world's best-performing school systems come out on top*. Retrieved February 23, 2008, from http://www.mckinsey.com/App_Media/Reports/SSO/Worlds_School_Systems_Final.pdf

McKinsey & Co. (2010). *How the world's most improved school systems keep getting better*. Retrieved November 23, 2012, from Http://mckinseyonsociety.com/downloads/reports/Education/How-the-Worlds-Most-Improved-School-Systems-Keep-Getting-Better_Download-version_Final.pdf

McLaughlin, M., & Talbert, J. (2001). *Professional communities and the work of high school teaching*. Chicago: The University of Chicago Press.

McLaughlin, M., & Talbert, J. (2006). *Building school-based learning communities: Professional strategies to improve student achievement*. New York: Teachers College Press.

MOE [Ministry of Education]. (2005). *Speech by Mr Tharman Shanmugaratnam, Minister for Education, at the MOE Work Plan Seminar 2004, Ngee Ann Polytechnic Convention Centre, Thursday 22 September, 10:00 a.m.* Retrieved February 23, 2006, from http://www.moe.gov.sg/media/speeches/2005/sp20050922.htm

MOE [Ministry of Education]. (2008). *C2015*. Singapore: MOE.

MOE [Ministry of Education]. (2009). *Address by Ms Ho Peng, Director-General of Education, at the teachers' Mass Lecture,* Singapore Expo Hall 2, Wednesday, 26 August 2009, 2.30 pm. Retrieved November 15, 2010, from Http://www.moe.gov.sg/media/speeches/2009/08/26/address-by-ms-ho-peng-at-the-t.php

MOE [Ministry of Education]. (2010). *Speech by Dr Ng Eng Hen, Minister for Education and Second Minister for Defence, at the MOE Work Plan Seminar 2010, on Thursday, 23 September 2010 at 9.30 am at the Ngee Ann Polytechnic Convention Centre*. Retrieved May 20, 2011, from http://www.moe.gov.sg/media/speeches/2010/09/23/work-plan-seminar-2010.php

MOE [Ministry of Education]. (2011a). *Speech by Mr S Iswaran, Senior Minister of State, Ministry of Trade and Industry and Ministry of Education, at the 2011 MOE Promotion Ceremony on Monday, 18 April 2011, 4.00 pm, at Suntec City Ballroom*. Retrieved June 23, 2012, from http://www.moe.gov.sg/media/speeches/2011/04/18/speech-by-mr-s-iswaran-at-the-16.php

MOE [Ministry of Education]. (2011b). *FY 2011 Committee of Supply Debate: 1st Reply by Dr Ng Eng Hen, Minister for Education and Second Minister for Defence on Social Mobility—The Singapore Story: Past, Present and Future*. Retrieved December 12, 2012, from http://www.moe.gov.sg/media/speeches/2011/03/07/fy-2011-committee-of-supply-de-1.php

MOE [Ministry of Education]. (2012). *Speech by Mr Heng Swee Keat, Minister for Education at the 2012 Promotion Ceremony at 4.00 pm on Thursday, 5 April 2012, at the Suntec Convention Ballroom 1–2*. Retrieved February 17, 2013, from http://www.moe.gov.sg/media/speeches/2012/04/05/speech-by-mr-heng-swee-keat-at-16.php

Munby, H., & Russell, T. (1994). The authority of experience in learning to teach: Messages from a physics methods class. *Journal of Teacher Education, 45*(2), 86–95.

OECD [Organisation for Economic Co-operation and Development]. (2010). *PISA 2009 results: What makes a school successful? – Resources, policies and practices* (Vol. IV). Retrieved April 12, 2011, from http://dx.doi.org/10.1787/9789264091559-en

OECD [Organisation for Economic Co-operation and Development]. (2013). *PISA 2012 results in focus: What 15-year-olds know and what they can do with what they know*. Retrieved February 3, 2014, from http://www.oecd.org/pisa/keyfindings/pisa-2012-results-overview.pdf

Senge, P., Cambron-McCabe, N., Lucas, T., Smith, B., Dutton, J., & Kleiner, A. (2000). *Schools that learn: A fifth discipline fieldbook for educators, parents and everyone who cares about education*. New York: Doubleday.

Shanghai Municipal Education Commission. (2011). *Shijiaowei yingfa Shanghaishi jichu jiaoyu gaige he fazhan shierwu guihua* [Municipal Education Commission Releases the 'Twelfth Five' Plan for Shanghai Basic Education Reform and Development]. Retrieved February 10, 2012, from http://www.shanghai.gov.cn/shanghai/node2314/node2319/node12344/u26ai30352.hml

Sparks, D., & Hirsh, S. (1997). *A new vision for staff development*. Alexandria, VA: Association for Supervision and Curriculum Development and National Staff Development Council.

Stoll, L., Bolam, R., McMahon, A., Wallace, M., Thomas, S., Hawkey, K., et al. (2003). *Creating and sustaining effective professional learning communities*. Paper presented at the 16th International Congress for School Effectiveness and Improvement, Sydney, Australia.

Tan, C. (2011). Framing educational success: A comparative study of Shanghai and Singapore. *Education, Knowledge and Economy, 5*(3), 155–166.

Tan, C. (2012). The culture of education policy making: Curriculum reform in Shanghai. *Critical Studies in Education, 53*(2), 153–167.

Tan, C. (2013). *Learning from Shanghai: Lessons on achieving educational success*. Dordrecht, The Netherlands: Springer.

Tang, N. (2000). Teachers' network: A new approach in the professional development of teachers. *ASCD Review, 9*(3), 48–55.

Thompson, S. C., Gregg, L., & Niska, J. M. (2004). Professional learning communities, leadership, and student learning. *Research in Middle Level Education Online, 28*(1), 1–15.

Toole, J. C., & Louis, K. S. (2002). The role of professional learning communities in international education. In K. Leithwood & P. Hallinger (Eds.), *Second international handbook of educational leadership and administration* (pp. 245–280). Dordrecht, The Netherlands: Kluwer.

Training and Development Division. (2010). *Schools as professional learning communities*. Singapore: Training and Development Division, Ministry of Education, Singapore.

Tripp, D. (2004). Teachers' network: A new approach to the professional development of teachers in Singapore. In C. Day & J. Sachs (Eds.), *International handbook on the continuing professional development of teachers* (pp. 191–214). Maidenhead, UK: Open University Press.

Vescio, V., Ross, D., & Adams, A. (2008). A review of research on the impact of professional learning communities on teaching practice and student learning. *Teaching and Teacher Education, 24*(1), 80–91.

Wells, C. (2008). A conceptual design for understanding professional learning community implementation. *Catalyst for Change, 35*(2), 25–37.

Wenglinsky, H. (2000). *How teaching matters: Bringing the classroom back into discussions of teacher quality*. Princeton, NY: Milken Family Foundation and Educational Testing Service.

Chapter 17
Japanese Elementary Teachers' Abilities to Learn How to Teach Science from Curriculum Materials: Preparation for Future Learning Perspectives

Etsuji Yamaguchi

Theoretical Background

Two Perspectives of Science Teacher Education

One of the major trends in current studies on science teacher education is emphasizing the notion of science teacher as learner (e.g., Loughran, 2007). These studies regard in-service and preservice teachers as learners of how to teach science and aim to identify the types of knowledge and beliefs these learners have acquired and how the acquisition of such knowledge and beliefs progresses their competence as teacher. At the same time, based on these findings, these studies are intended to provide intervention and development to support the learning of science teachers.

Many of these previous studies focused on the *learning outcome*. Studies of pedagogical content knowledge (PCK) (Shulman, 1986) for science teaching analyzed PCK learned through teaching experience by comparing PCK between beginning and expert teachers or have discussed whether or not PCK could be obtained through education programs provided by these studies. However, almost no studies have been made focusing on *preparation for future learning*. Such studies are supposed to address issues such as whether or not science teachers have obtained abilities to learn PCK through curriculum materials and what science teachers should learn in order to continue learning how to teach science in the future.

E. Yamaguchi (✉)
Graduate School of Human Development and Environment, Kobe University,
3-11 Tsurukabuto, Nada-ku, Kobe 657-8501, Japan
e-mail: etuji@opal.kobe-u.ac.jp

M.S. Khine (ed.), *Science Education in East Asia*,
DOI 10.1007/978-3-319-16390-1_17

In learning sciences, interdisciplinary field of teaching and learning (Bransford, Brown, & Cocking, 2000; Sawyer, 2006), studies have been pursued focusing on *preparation for future learning* (Bransford & Schwartz, 1999; Schwartz, Bransford, & Sears, 2005; Schwartz & Martin, 2004). These studies are also applied in teacher education research (e.g., Hammerness, Darling-Hammond, & Bransford, 2005). To become not routine expertise but adaptive expertise (Hatano & Inagaki, 1986) in teaching, it is important for teachers to continue updating their knowledge of teaching, even after they have finished teacher education programs offered at public or private sectors. While studies of learning sciences have pointed out the importance of continuously updating knowledge of teaching, this importance has not been fully recognized in the field of science education research. It should also be noted that no research has yet been conducted on the *abilities to learn how to teach science*, which is the theme of this present research.

Teachers' Abilities to Learn How to Teach Science

Abilities to learn how to teach science would become even more important in research on science teacher education. To plan, implement, and evaluate science lessons, both in-service teachers and preservice teachers need to learn how to teach science. This learning will be provided, for example, through teacher education programs offered by public or private sectors. However, science teachers cannot learn everything in these programs. This is because science teachers face changes in teaching content and methods resulting from curriculum reforms, as well as the equity and diversity of their students. They should also continue updating their knowledge by learning themselves.

To explain this issue more concretely, I present the case of elementary teachers in Japan. Elementary science in Japan begins in third grade (8–9-year-olds) and ends in sixth grade (11–12-year-olds). The curriculum covers basic material in the fields of physics, chemistry, biology, and earth sciences. Science lessons are supposed to be provided mainly by a homeroom teacher. Therefore, a teacher in charge of a fourth-year class must on his/her own teach various topics in physics, chemistry, biology, and earth sciences for fourth graders. The teacher might take a science teacher education program that instructs teaching methods incorporating inquiry-based learning, designed to teach a topic of physics to sixth-grade students. Even if the teacher has learned such a teaching method, though, he/she needs to newly learn another method in order to conduct an inquiry-based learning on topics of biology for his/her fourth-grade students.

Currently, curriculum reforms are under way across the entire field of science education around the world. The United States is undergoing reforms toward integration scientific/engineering practices, crosscutting concepts, and disciplinary core ideas (NGSS Lead States, 2013). Curriculum reforms are being implemented not just in the United States but also in Europe, Australia, and various Asian countries (Ministry of Education and Human Resources, Korea, 2007; Osborne &

Dillon, 2008; Tytler, 2007). For example, Japan is experiencing reforms focusing on encouraging students to realize the usefulness of science and the ability to apply scientific concepts (MEXT Japan, 2008).

The speed of these reforms has been accelerating year by year. To actually conduct science lessons in response to such reforms, it is absolutely critical that teachers continue learning. In addition to previous studies on the abilities needed for teaching science, it is also necessary to conduct new studies to clarify current state of teachers' abilities to learn how to teach science, as well as studies to develop and evaluate training programs that will improve teachers' abilities. It is believed that conducting such studies will result in worldwide sharing and more effective use of academic findings that will better facilitate learning about what and how to teach science.

Teacher Education for Elementary Science in Japan

High level of elementary teacher quality and performance in Japan has attracted attention worldwide. For example, Linn, Lewis, Tsuchida, and Songer (2000) have identified eight science activity structures based on observation of ten science lessons taught in five elementary schools. They have also found that each of the eight structures has been implemented by various instructional techniques of teachers. For example, in order to "connect lesson to student interest and prior knowledge" as one of activity structures, "teacher starts lesson with questions or activity designed to intrigue students and build on their ideas" (p. 5). They have presented the conclusion that these structures are the key factor supporting top-level scores of Trends in International Mathematics and Science Study (TIMSS).

However, despite being highly appreciated around the world, the recent decline in elementary teacher quality and performance has become a matter of serious concern in Japan. The Committee to Discuss Teachers' Science Literacy and Teacher Education established by the Science Council of Japan (2007) has been deeply concerned that elementary teachers are required to teach science even though they have got course credits of only one elementary science methods course in preservice education under the current Educational Personnel Certificate Law. Additionally, the Elementary School Subcommittee of the Taskforce to Discuss Science Education Support set up by the Japan Science and Technology Agency (2008) suggests three problem areas regarding elementary science education in Japan: children, teachers, and educational facilities. With teachers, the Subcommittee presented the questionnaire survey results showing more than 60 % of elementary teachers responding that they are not good at teaching science. The survey results also revealed that to the question "Which subject do you emphasize?" only 6 % of teachers answered "Science." The Subcommittee also announced that approximately 60 % of preservice teachers did not take a course in high school physics. To overcome the problem of decline in elementary teacher quality and performance, efforts have been made not only to improve in-service teacher education but also to carry out systematic reforms including introduction of the teacher certification renewal system.

What factors are affecting decline in elementary teacher quality and performance? Most of Japanese science teacher educators consider that the shortage of teachers' subject matter knowledge and skills and their PCK are primary factors. So preservice and in-service teacher education programs for improvement in teachers' knowledge, skills, and PCK have been developed and implemented. However, no suggestions have been made concerning the ongoing deterioration of teachers' abilities to learn how to teach science, which form the backbone of their voluntary day-to-day learning. It is probably true that the reason why Japanese elementary teachers once had high levels of quality and performance in science teaching was not that those who had originally had high levels of quality and performance became teachers but that these teachers continuously learned how to teach science after becoming teachers. Various studies focusing on lesson study in Japan have found that teachers learned much about their lessons by delivering and reflecting on everyday lessons and then discussing their lessons with colleagues, as well as through lesson planning, recording, and observation (Lewis, 2002; Lewis, Perry, & Murata, 2006; Stigler & Hiebert, 1999). As we interpreted the research findings of studies focusing on lesson study from the perspective of the abilities to learn how to teach science, I can make a conjecture that Japanese elementary teachers have cultivated their abilities of learn through lesson study. If the conjecture is true, teacher education for elementary science in Japan could be faced with a challenge that teachers lack abilities of learn how to teach science.

A Case Study of Japanese Preservice Elementary Teachers' Abilities to Learn How to Teach Science from Curriculum Materials

Overview

As the first step in addressing the issues described above, in order to develop a method for capturing teachers' abilities of learn how to teach science, I conducted a case study focusing on learning from curriculum materials and preservice elementary teachers before the science methods course.

In Japan, curriculum materials are easily available to every teacher. For example, all six publishers provide Japanese elementary science textbooks. All public schools and almost all private schools adopt the textbook of one publishing company out of the six based on neighboring regional units. All elementary schools purchase one of teacher edition textbooks and make them available to their teachers. For another example, many excellent commercial materials, especially for beginning elementary teachers, have been sold in the form of books or distributed via the Internet for free. When the national curriculum is reformed, new commercial materials have been developed at the same time. Therefore, it can be said that learning from curriculum materials is one of the useful learning resources for elementary teachers.

We decided to conduct this case study on preservice elementary teachers before the science methods course for the following reason. The Educational Personnel Certificate Law rigidly regulates the teacher education system and its certificate requirements in Japan (MEXT, 2008a; see also Ogawa 2001). Therefore, almost all national, public, and private universities have similar program structures. Preservice elementary teachers should complete a 4-year program. In the first and second year of the program, they take courses in educational foundations such as educational psychology and curriculum and instruction. Then, in the third and fourth years, in addition to the science methods course that is the focus of this study, the preservice teachers take multiple subject methods course such as language arts, mathematics, and social studies and go to elementary schools for student teaching. Given these circumstances in Japan, we can take clue to the reform of preservice science education programs from the case study of understanding the actual conditions of the abilities to learn how to teach science that preservice teachers have already acquired in the first phase of the teacher education program, i.e., before the science methods course and student teaching. Moreover, we hope that this case study facilitates critical review of science education programs not only in Japan but also in other countries.

Method

Participants

A total of 84 Japanese preservice elementary teachers of a national university were the participants of this study. Most are traditional fourth-year seniors (about 21 years old) in their third year of the university. They were 61 female and 23 male preservice teachers. Most of them come from middle or upper middle class families. According to the latest statistics, the ratios of male and female elementary school teachers are 38.2 and 61.8 % in Japan (Ministry of Education, Culture, Sports, Science and Technology, 2008b). The ratio in this study is similar to that of the nationwide statistics. Of 84 participants, I obtained answers from 77.

Data Sources

I set the test environment that teachers can interact with curriculum materials because teachers' abilities to learn from curriculum materials are realized in the interaction with curriculum materials. In the test, preservice teachers were asked to read the materials assuming that they had to draw up lesson plans for their own lessons. They were then asked to freely describe the points they found informative in preparing their lesson plans and the reasons why. They took about an hour to finish the assessments on an individual basis.

The curriculum materials were part of a commercial book designed for beginning teachers and generally available and on the curricular unit of the period of pendulum for fifth-grade students (10–11-year-olds) (Hioki, 2005). The materials described outline of the curricular unit and lesson plans with learning activities for each school hour. The materials also indicated the goals for each lesson and the focal points of the instruction, what should be prepared for experimentation, the flow of overall lesson, examples of writing on the blackboard, and examples of students' note-taking. The commentary provided teaching rationales on why teachers should ask such questions and instruct their students to do such activities.

Analysis

To identify what content and how deep preservice teachers learned from curriculum materials, the following two types of analyses were performed. One type of analysis was conducted using the categories shown in Table 17.1 to identify *what content.* These categories were based on Davis (2006). Davis (2006) developed the categories to explore how US preservice elementary teachers critiqued curriculum materials. These categories cover important aspects of inquiry-based science teaching at elementary schools. Japanese elementary teachers are traditionally encouraged to use inquiry-based approach in their science lessons. Additionally, curriculum materials used in the case study contain important aspects of inquiry-based science teaching. So I have applied the categories of Davis (2006) to analysis for identifying *what content.* In the analysis, at first, I coded all of the data according to categories of Davis. For the data not falling under any of the categories, I added new categories to classify these data. These newly added categories are also included in Table 17.1. Next, a second coder coded all of the data according to categories listed in Table 17.1. No more new categories were added. I and a second coder discussed any discrepancies to agree upon a final characterization of the data, in terms of the identification of instances and quality ratings.

Another type of analysis was conducted using the following three categories to identify *how deep*: *pedagogical knowledge for elementary teaching*, *PCK for elementary science teaching*, and *topic-specific PCK for elementary science teaching.* I used *pedagogical knowledge for elementary teaching* to describe the pedagogical knowledge not related to any specific subject and that might be applied to all subjects of elementary school. *PCK for elementary science teaching* was used to indicate the pedagogical content knowledge for science but was not focused on any specific topics. A final category—*topic-specific PCK for elementary science teaching*—was used to capture pedagogical content knowledge unique to specific topics, in case of this study, period of pendulum. I illustrate the difference between three categories with some examples. For example, the answer "You (teacher) need to design driving questions to be asked while attaching importance to students' ideas"

Table 17.1 Categories to identify what content preservice teachers learned from the curriculum materials

Category	Example
Lesson-related categories	
Student ownership and engagement	The lesson engages students in science learning that is meaningful and engaging to them
Questioning	The lesson expects students to ask or answer challenging questions
Instructional goals	The lesson sets a limited number of worthwhile science learning goals. The goals include goals for learning science concepts and scientific inquiry
Real-world applications and connections	The lesson makes connections to real-world examples of the scientific ideas or other connections such as to other lessons
Instructional representations	The lesson represents the science content in scientifically accurate ways and will not promote alternative ideas
Communication	The lesson supports students in productively discussing ideas with each other and with the teacher
Finding out and/or discovery[a]	The lesson expects students to find out and/or discovery
Introducing scientists[a]	The lesson introduces scientists who related learning content to students
Criteria of evaluation[a]	The lesson represents criteria of evaluation
Unit planning[a]	The lesson represents unit planning
Flow of lesson[a]	The lesson represents the flow of each lesson
Writing on blackboard and note-taking[a]	The lesson represents examples of writing on blackboard and students' note-taking
Inquiry-related categories	
Inquiry and investigations	The lesson supports students in engaging in scientific inquiry
Specific aspects of inquiry	Combines messing about, finding information, making predictions, making observations, making sense of data, and planning and designing
Messing about	The lesson allows for students to explore, make initial observations, manipulate objects, and play with materials (Krajcik et al., 1999, p. 93)
Finding information	The lesson allows for students to ask others, read purposefully, and evaluate information (Krajcik et al., 1999, p. 93)
Making predictions	The lesson allows for students to ask and answer questions like What would happen if...? (building from Krajcik et al., 1999, p. 93)
Making observations	The lesson allows for students to assemble experimental apparatus and gather data (Krajcik et al., 1999, p. 93)
Making sense of data	The lesson allows for students to analyze, transform data, and make inferences (Krajcik et al., 1999, p. 93)
Planning and designing	The lesson allows for students to consider how they might answer their questions (Krajcik et al., 1999, p. 93)

Note: Almost all categories are based on Davis (2006).
[a]These categories are unique to this study.

was coded as *pedagogical knowledge for elementary teaching*. The answer "You need to make students think about factors affecting the time period of a pendulum for themselves in a group" was coded as *PCK for elementary science teaching*. This was because the description in the answer related to common features of elementary science teaching, which try to make students think about factors that would change natural phenomena, although the term "pendulum" was used. One example of the answers coded as *topic-specific PCK for elementary science teaching* was as follows: "You and students should not determine in advance that which are factors affecting the time period of a pendulum, the weight of the pendulum, length of string, or angle from which the pendulum is dropped. If you and students are not determined in advance, the result of first experiments varies by students group. It will allow for students to ask questions." This answer refers to factors specific to the period of pendulum and includes the instructional guidance that suggests how teachers deal with these factors. Furthermore, this answer also explains the rationale why instructional guidance is important.

The above three types of knowledge were used because there is evidence that in order for teachers to design and implement better science lessons, having general pedagogical knowledge alone is not enough; teachers must acquire PCK and topic-specific PCK for elementary science teaching (Abell, 2007; Veal & MaKinster, 1999). Moreover, the curriculum materials used in the case study have included PCK for elementary science teaching and topic-specific PCK related to the period of pendulum. In the analysis, I and the second coder coded all of the data according to categories. All data were coded into any of the three categories. We discussed any discrepancies to agree upon a final characterization of the data, in terms of the identification of instances and quality ratings.

Results

What Content Preservice Teachers Learned from Curriculum Materials?

Table 17.2 shows that many preservice teachers can learn about *student ownership and engagement* (53.2 %), *real-world applications and connections* (46.8 %), *planning and designing* (40.3 %), *making sense of data* (37.7 %), and *making predictions* (33.8 %). In regard to *student ownership and engagement*, the curriculum materials propose initial meaningful activities such as riding on a swing or playing a game in which students get past big swinging pendulum. Some preservice teachers were able to learn it for their science lesson planning. In regard to *making sense of data*, the curriculum materials suggest that results of students' experiment are better represented in a table or a graph. Some preservice teachers were able to learn that using a graph was effective in making sense of data.

In contrast to these results, only a few teachers can learn *finding information* (0.0 %), *inquiry and investigations* (2.6 %), *specific aspects of inquiry* (3.9 %), *introduction to what scientists do* (3.9 %), *instructional goals* (7.8 %), *making*

Table 17.2 Results of what content preservice teachers learned from curriculum materials

Category	Number of teachers	(%)
Lesson-related categories		
Student ownership and engagement	41	53.2
Questioning	9	11.7
Instructional goals	6	7.8
Real-world applications and connections	36	46.8
Instructional representations	18	23.4
Communication	3	3.9
Finding out and/or discovery[a]	7	9.1
Introducing scientists[a]	3	3.9
Criteria of evaluation[a]	16	20.8
Unit planning[a]	16	20.8
Flow of lesson[a]	14	18.2
Writing on blackboard and note-taking[a]	15	19.5
Inquiry-related categories		0.0
Inquiry and investigations	2	2.6
Specific aspects of inquiry	3	3.9
Messing about	25	32.5
Finding information	0	0.0
Making predictions	26	33.8
Making observations	8	10.4
Making sense of data	29	37.7
Planning and designing	31	40.3

$N = 77$

[a]These categories are unique to this study.

observations (10.4 %), and *questioning* (11.7 %). In regard to *finding information*, such activities of students are not included in the curriculum materials. In regard to *inquiry and investigations* and *specific aspects of inquiry*, because these categories include all aspects of scientific inquiry, many data are coded to other categories of inquiry and investigations. So results of *finding information*, *inquiry and investigations*, and *specific aspects of inquiry* are not problematic. However, results of *instructional goals*, *making observations*, *and questioning* are problematic because only a few teachers can learn them although these activities of students are included in the curriculum materials. In regard to *questioning*, the curriculum materials propose the driving question and its appropriate implementation. For example, the driving question is "What are the factors that change the time period of the pendulum?" and its appropriate implementation is that teachers should ask the driving question after letting students think by themselves about reasons why the time period of the pendulum is changed and then having all class members share their ideas. In regard to *making observations*, the curriculum materials introduced laboratory equipment to help students make measurements easily and accurately and then explained the focal points of instructions on measurements.

Table 17.3 Results of how deep preservice teachers learned from curriculum materials

Category	Number of teachers	(%)
Pedagogical knowledge for elementary teaching	34	44.2
PCK for elementary science teaching	68	88.3
Topic-specific PCK for elementary science teaching	31	40.3

$N = 77$

How Deep Preservice Teachers Learned from Curriculum Materials?

Table 17.3 shows that while almost preservice teachers can learn *PCK for elementary science teaching* (88.3 %), fewer than half of the teachers can learn *topic-specific PCK for elementary science teaching* (40.3 %). For example, in case of *student ownership and engagement*, many preservice teachers learn that playing games or other activities, as an introduction to the unit, would enhance students' motivation for learning. However, only a few teachers can learn the implementation guidance that it would be better to use large-size pendulum for playing games or other activities as an introduction and the rationale why it was better to have students ride on a swing or play a game in order to encounter the pendulum.

In case of *making predictions*, the curriculum materials informed that teacher should encourage students to ask questions on why the experimental results were different from group to group, despite all the groups conducting the same experiment. Such a question would motivate students to consider on their own factors that would affect the time period of the pendulum, such as the length of string, weight of the pendulum, and angle from which the pendulum is dropped. Many preservice teachers learn that it is important for students to share their predictions and reasons with the class before the experiment. However, only a few teachers can learn what kind of predictions students make and what timing would be better for students to make predictions.

Concluding Remarks

In this chapter, I discussed the importance of teachers' abilities to learn how to teach science from preparation for future learning perspectives. And in order to develop a method for capturing teachers' abilities to learn how to teach science, I conducted a case study focusing on learning from curriculum materials and preservice elementary teachers before the science methods course.

My case study clarified what content preservice teachers learned from curriculum materials. Japanese preservice elementary teachers do not crucially lack abilities to learn how to teach science from curriculum materials before the science methods course. When they start the science methods course, they can learn some elements of basic structure of Japanese elementary science lessons such as real-world applications and connections, making predictions, and so on. However, preservice teachers cannot learn all the elements of basic structure of Japanese elementary science lessons. As already discussed, high level of elementary teacher quality and performance in Japan has attracted attention worldwide. Basic structure of Japanese elementary science lessons has been evaluated as the key factor supporting top-level scores of TIMSS. Whatever curriculum materials informed strong points of Japanese elementary science lessons, it would be difficult for preservice elementary teachers to learn adequately how to teach science from the materials.

My case study also clarified how deep preservice teachers learned from curriculum materials. While Japanese preservice elementary teachers can learn PCK for elementary science teaching from the curriculum materials, they tend to have difficulties in learning topic-specific PCK for elementary science teaching. As discussed above, in order for teachers to design and implement better science lessons, having general pedagogical knowledge alone is not enough; teachers must acquire PCK and topic-specific PCK for elementary science teaching. However, if curriculum materials informed topic-specific PCK for elementary science teaching, it would be difficult for preservice elementary teachers to learn adequately how to teach science from the materials.

Consequently, preservice elementary teachers need to develop their abilities to learn how to teach science from curriculum materials. This chapter indicates that we need to continue to support preservice teachers in developing their abilities to learn from curriculum materials. For example, in regard to "how deep," only bottom-up learning, i.e., learning topic-specific PCK from curriculum materials, may not be sufficient. Because PCK "goes beyond knowledge of subject matter per se to the dimension of subject matter knowledge for teaching" (Shulman, 1986, p. 9), top-down learning, i.e., transforming PCK into topic-specific PCK by applying their PCK for designing, implementing, and reflecting on elementary science lessons of particular topics, may be necessary. Since these ideas are only conjecture, further studies are needed in the future.

This chapter also indicates research questions for future studies of teachers' abilities to learn how to teach science. For empirical studies, these include the following: What abilities of learning how to teach science teachers can be developed through teaching experience? How do we design teacher education program for cultivating their abilities to learn how to teach science? For theoretical studies, these include the following: How do teachers' abilities to learn how to teach science develop from beginning to experienced? What resources (curriculum materials, mentoring, coteaching, and so on) can teachers learn from, and how different are those learnings? Through these studies, we can deepen our understanding of science teachers as "adaptive experts" (Hatano & Inagaki, 1986) and how to support science teachers to become adaptive experts.

Author Note An earlier version of the work was presented at NARST 2009 conference. This research was supported by JSPS KAKENHI Grant Number 25282038.

References

Abell, S. K. (2007). Research on science teacher knowledge. In S. K. Abell & N. G. Lederman (Eds.), *Handbook of research on science education* (pp. 1105–1149). Mahwah, NJ: Lawrence Erlbaum.

Bransford, J. D., & Schwartz, D. L. (1999). Rethinking transfer: A simple proposal with multiple implications. *Review of Research in Education, 24*, 61–100.

Bransford, J., Brown, A. L., & Cocking, R. R. (2000). *How people learn: Brain, mind, experience, and school (expanded edition)*. Washington, DC: National Research Council.

Davis, E. A. (2006). Preservice elementary teachers' critique of instructional materials for science. *Science Education, 90*(2), 348–375.

Hammerness, K., Darling-Hammond, L., & Bransford, J. (2005). How teachers learn and develop. In L. Darling-Hammond & J. Bransford (Eds.), *Preparing teachers for a changing world* (pp. 358–389). San Francisco, CA: Jossey-Bass.

Hatano, G., & Inagaki, K. (1986). Two courses of expertise. In H. Stevenson, J. Azuma, & K. Hakuta (Eds.), *Child development and education in Japan* (pp. 262–272). New York, NY: W. H. Freeman & Co.

Hioki, M. (Ed.). (2005). *Fifth grade science lessons: All of lessons of all units with samples of and teachers' writing on blackboard and students' writing on notes (in Japanese)*. Tokyo, Japan: Toyokan Publishing Co., Ltd.

Krajcik, J., Czerniak, C., & Berger, C. (1999). *Teaching children science: A project-based approach*. Boston, MA: McGraw-Hill College.

Linn, M. C., Lewis, C., Tsuchida, I., & Songer, N. B. (2000). Beyond fourth-grade science: Why do U. S. and Japanese students diverge? *Educational Researcher, 29*(3), 4–14.

Lewis, C. (2002). *Lesson study: A handbook of teacher-led instructional improvement*. Philadelphia, PA: Research for Better Schools.

Lewis, C., Perry, R., & Murata, A. (2006). How should research contribute to instructional improvement? The case of lesson study. *Educational Researcher, 35*(3), 3–14.

Loughran, J. (2007). Science teacher as learner. In S. Abell & N. Lederman (Eds.), *Handbook of research on science education* (pp. 1043–1065). Mahwah, NJ: Lawrence Erlbaum Associates.

Ministry of Education and Human Resources, Korea. (2007). *Science education curriculum*. Seoul, Korea: Ministry of Education and Human Resource.

Ministry of Education, Culture, Sports, Science and Technology (MEXT) Japan. (2008). *Expository of elementary science curriculum guidelines* [in Japanese]. Retrieved August 30, 2014, from http://www.mext.go.jp/a_menu/shotou/new-cs/youryou/syokaisetsu/index.htm

Ministry of Education, Culture, Sports, Science and Technology (MEXT). (2008b). *At what university can the teacher's license qualification be acquired now?* (in Japanese). Retrieved August 30, 2014, from http://www.mext.go.jp/a_menu/shotou/kyoin/daigaku/index.htm

Ministry of Education, Culture, Sports, Science and Technology (MEXT). (2008c). *Statistical research on school teachers* (in Japanese). Retrieved August 30, 2014, from http://www.mext.go.jp/b_menu/toukei/chousa01/kyouin/kekka/1268581.htm

Lead States, N. G. S. S. (2013). *Next generation science standards: For states, by states*. Washington, DC: The National Academies Press.

Ogawa, M. (2001). Reform Japanese style: Voyage into an unknown chaotic future. *Science Education, 85*(5), 586–606.

Osborne, J., & Dillon, J. (2008). *Science education in Europe: Critical reflections*. London, UK: The Nuffield Foundation.

Sawyer, R. K. (Ed.). (2006). *Cambridge handbook of the learning sciences*. New York, NY: Cambridge University Press.

Shulman, L. S. (1986). Those who understand: Knowledge growth in teaching. *Educational Researcher, 15*(2), 4–14.

Schwartz, D. L., & Martin, T. (2004). Inventing to prepare for future learning: The hidden efficiency of encouraging original student production in statistics instruction. *Cognition and Instruction, 22*(2), 129–184.

Schwartz, D. L., Bransford, J. D., & Sears, D. (2005). Efficiency and innovation in transfer. In J. Mestre (Ed.), *Transfer of learning from a modern multidisciplinary perspective* (pp. 1–51). Greenwich, CT: Information Age Publishing.

Stigler, J. W., & Hiebert, J. (1999). *The teaching gap: Best ideas from the world's teachers for improving education in the classroom.* New York, NY: Free Press.

The Committee to Discuss Teachers' Science Literacy and Teacher Education established by the Science Council of Japan. (2007). *The ideal future for teachers' science literacy and teacher training* (in Japanese). Retrieved August 30, 2014, from http://www.scj.go.jp/ja/info/kohyo/pdf/kohyo-20-y1.pdf

The Elementary School Subcommittee of the Taskforce to Discuss Science Education Support set up by the Japan Science and Technology Agency. (2008). *New development of elementary science education through collaboration school and society* (in Japanese). Retrieved August 30, 2014, from http://rikai.jst.go.jp/center/jstcpse_report_001.pdf

Tytler, R. (2007). *Re-imagining science education: Engaging students in science for Australia's future.* Camberwell, VIC: Australian Council for Educational Research.

Veal, W. R., & MaKinster, J. G. (1999). Pedagogical content knowledge taxonomies. *Electronic Journal of Science Education*, 3(4). Retrieved August 30, 2014, from http://unr.edu/homepage/crowther/ejse/ejsev3n4.html

Etsuji Yamaguchi is an associate professor of science education at Kobe University, Japan. He received his Ph. D. from Kobe University. His research interests include examining the ways that students and teachers develop more sophisticated understandings and abilities. In addition, he is interested in investigating and exploring the impact of designing innovative science learning environments in formal and informal contexts. He was awarded the Young Researchers Award (the Japan Society for Science Education (JSSE), 1999; the Society of Japan Science Teaching (SJST), 2003), the Outstanding Paper Award (JSSE, 2007, 2008; the Japanese Association of Educational Psychology, 2011), the Best Presentation Award (JSSE, 2002, 2004, 2007), and the Outstanding Science Education Practice Award (JSSE, 2005).

Chapter 18
Chinese Secondary School Science Teachers' Perceptions of the Nature of Science and Chinese Native Knowledge

Hongming Ma

Introduction

China's science education has been undergoing a new round of curriculum and assessment reform from the beginning of this century. Realizing that the conventional curriculum attaches too much importance to knowledge teaching and book learning, the new curriculum standards have paid particular attention to quality education, which includes intellectual, affective, ethical, and physical development of all students. All science curriculum standards are developed under the framework of three dimensions – knowledge and skills, process and methods, and attitudes and values. The new curriculum emphasizes more on teaching the nature of science and its importance to enhancing scientific literacy. The curriculum standards of junior secondary school science (Ministry of Education, 2011, pp. 5–6) explicitly state:

> Understanding the nature of science will enhance students' scientific literacy and promote the development of inquiry skills and scientific attitude among students. Firstly, for students as life-long learners, understanding the nature of science helps the students to learn the formation of scientific knowledge and the connections between knowledge in different disciplines. As a result, the students will better understand scientific achievements, methods and limitations, appropriately evaluate the practical value and social influence of science and pay close attention to the new development of science. Secondly, as future citizens, the students need to understand the science and technological processes in everyday life, to understand science-related social issues, to be able to participate in the decision-making, to understand that science is an important part of contemporary culture, and, understand the rules of scientific community.

This document emphasizes that developing an understanding of the nature of science is a prerequisite to teaching scientific literacy. This requires a better

H. Ma (✉)
Faculty of Education and Arts, Federation University Australia, Churchill, VIC 3842, Australia
e-mail: hongming.ma@federation.edu.au

M.S. Khine (ed.), *Science Education in East Asia*,
DOI 10.1007/978-3-319-16390-1_18

understanding of the nature of science by both science teachers and students. However, compared with research in English-speaking countries (e.g., Deng, Tsai, & Chai, 2011; Lederman, Lederman, & Antink, 2013) research into the nature of science in Mainland China is rare. As part of the effort to promote the study on the nature of science in Mainland China, this paper examines a group of Chinese secondary school science teachers' perceptions of the nature of science and discusses the implications for science education in China.

Views of the Nature of Science

The "nature of science" is different from the "scientific knowledge." Scientific knowledge refers to knowledge about the natural world, such as scientific laws, theories, and hypotheses introduced in school science textbooks, whereas the "nature of science" covers various topics such as the status of scientific knowledge (levels of uncertainty), relationship between scientific theories and reality, what science is, and the social aspects of science. Lederman (1992) defines the term as "the values and assumptions inherent to the development of scientific knowledge" (p. 331). According to McComas, Clough, & Almazroa (1998):

> The nature of science is a fertile hybrid arena which blends aspects of various social studies of science including the history, sociology, and philosophy of science combined with research from the cognitive sciences such as psychology into a rich description of what science is, how it works, how scientists operate as a social group and how society itself both directs and reacts to scientific endeavors. (p. 4)

Given the broad issues this concept involves, it is almost impossible to cover every aspect of the nature of science in one study. Considering the context of contemporary China, this section pays particular attention to three dimensions of the nature of science which consist of realism/constructivism, universalism/multiculturalism, and scientism. The purpose of reviewing these particular aspects of the nature of science is to provide background information for building the theoretical framework of the empirical study described later in this chapter.

Realism/Constructivism

This dimension of the nature of science considers the relationship between scientific theories and reality. The issue of scientific realism has been one of the central concerns in many debates within the philosophy of science. The core issue is this: Do theoretical constructs – unobservable entities postulated by scientific theories and hypotheses (e.g., genes, dark matter, magnetic field) – exist? Put simply, scientific realists believe in the existence of theoretical entities. A positive argument for scientific realism is that certain theories in science have been very successful in explaining and predicting phenomena (empirical evidence). However, in the history of science, there have been many theories that we now regard as false and many of these theories were to various degrees successful in explaining and predicting observed phenomena. Responses to considerations like this have led to a more cautious version of scientific realism. A statement formulated by Leplin (2006) is representative:

> Theoretical entities that are needed to explain or predict empirical results, and that are posited by well-supported theories free of empirical or conceptual difficulties, exist and have those of the properties these theories attribute to them that enable them to fulfil their explanatory and predictive roles. (p. 688)

A more detailed explanation of this statement can be found in Leplin (1984, pp. 1–2), which consists of the following basic ideas:

1. The best current scientific theories are at least approximately true.
2. The central terms of the best current theories are genuinely referential.
3. The approximate truth of a scientific theory is sufficient explanation of its predictive success.
4. The (approximate) truth of a scientific theory is the only possible explanation of its predictive success.
5. A scientific theory may be approximately true even inferentially unsuccessful.
6. The history of at least the mature sciences shows progressive approximation to a true account of the physical world.
7. The theoretical claims of scientific theories are to be read literally, and so read are definitively true or false.
8. Scientific theories make genuine, existential claims.
9. The predictive success of a theory is evidence for the referential success of its central terms.
10. Science aims at a literally true account of the physical world, and its success is to be reckoned by its progress toward achieving this aim.

Although there have always been difficulties, even with a more careful scientific realism, there is a strong argument (e.g., Franklin, 1999) that no scientists in laboratories are antirealists. Also because it seems to be most compatible with an intuitive position, scientific realism has a strong representation in public arenas (e.g., mass media).

Another view that is inconsistent with various aspects of scientific realism is constructivism. During the last two decades, the term "constructivism" became popular and has been widely used in different academic fields including philosophy, sociology, cognitive science, art, and architecture. This has caused confusion about the meaning of the term. It is less controversial if it is considered as a theory of learning that views learning as a process in which the learner actively constructs or builds new ideas or concepts. However, as a view of the nature of scientific knowledge, some of the constructivist perspectives are provocative.

One controversial position that the constructivists hold is that scientists do not "discover" a given reality and scientific knowledge is experience-dependent.

Glasersfeld (2001) argues "like other rational knowledge, scientific theories are derived from human experience and formulated in terms of human concepts, it seems no more than a pious hope to expect that these theories reflect anything that lies beyond the experiential interface" (p. 41). Glasersfeld (2001) therefore suggests that the role of scientific knowledge should be modified from true representation of reality to "functional fit" as there is no justification for an exclusive claim of objectivity. This position is sometimes interpreted as "constructivism denies an externally exist reality." As a response to this, Glasersfeld argues that constructivism should be seen as a theory of knowing, rather than a theory of being, which means that constructivism does not make ontological claims about the nature of reality.

Hung (1997) further explains statements made by constructivists that knowledge is socially and intellectually constructed:

> What these statements assert is that the "shape" and nature of the reality that scientists claim to have discovered are social and intellectual constructs. The reason is this. It is the form adopted that determines the "look" of the contents. Since the form adopted is socially caused and mind-dependent, the resulting contents are social and intellectual products as well. (p. 449)

Even though constructivism does not necessarily deny an external reality, a fundamental controversy remains in relation to the role of reality in choosing between rival theories and settling disagreements. Constructivism, or in a more precise term, social constructivism, argues that what counts as supportive evidence is substantially determined by social factors in any knowledge-generating process. Such social factors may include collaboration and competition between scientists, their research students, and institutions; power struggle within scientific communities and between different social groups; and social, economic, and political contexts of scientific research. Social constructivism is criticized for merely reducing justification of scientific knowledge to social consensus as a result of power struggle. To many realists, asserting that production of scientific knowledge is subject to social factors may lead to uneasy relativism, which claims that there are multiple realities, none having precedence over the other in terms of claims to represent the truth (Gross & Levitt, 1994). Since realists still have difficulties in solving problems such as the theory-ladenness of observation and experiment, questions on the relationship between theory and reality remain unsolved.

Universalism/Multiculturalism

This dimension is closely related to the debate between realism and constructivism. It has been paid more attention within and has more immediate influence on the field of science education. Opinions are divided among science educators and science education researchers on whether and for what purpose native knowledge (or "folk thought," "ethnosciences") about nature should be included in school science together with Western modern science (WMS).

The universalists insist that WMS (or Eurocentric science) provides a much more superior form of knowledge than any other ways of knowing nature. Among them, Matthews's view is representative:

> The core universalist idea is that the material world ultimately judges the adequacy of our accounts of it. Scientists propose, but ultimately, after debate, negotiation and all the rest, it is the world that disposes. The character of the natural world is unrelated to human interests, culture, race or sex. Ultimately, the concept is judged by the object, not the other way around. Just as volcanic eruptions are indifferent to the race of those in the vicinity, . . . so also the science of lava flows will be the same for all. For the universalist, our science of volcanoes is assuredly a human construction with negotiated rules of evidence and justification, but it is the behavior of volcanoes that finally judges the adequacy of our volcanology, not the reverse. (1994, p. 182)

Matthews's view shows that the underpinning assumptions of universalists are, in many ways, consistent with a realist standpoint. First, reality is mind independent, so as the representation of reality, science is free from human subjectivity; second, because the law of nature applies universally, the law of nature "discovered" by science is culture-free and value-free; third, the justification of the output of scientific research is ultimately determined by reality itself irrespective of sociocultural contexts within which the scientific research is located. For universalists, WMS is the best form of knowledge to represent "objectivity" and to account for particular natural phenomena. As Matthews (1994) states:

> Modern challenges, as distinct from traditional life-style challenges, are best met by modern science. How to teach this science effectively in multicultural settings is the challenge: this challenge is not met by diversions into multi-science doctrines, or adoption of questionable relativisms. (p. 198)

Universalists acknowledge that students' cultural backgrounds should be catered for in science teaching. However, they argue against treating native knowledge as correct as WMS simply because educators have moral obligation of treating other cultures with respect (Siegel, 1997). Given this position, native knowledge can only be used for the purpose of enhancing understanding of WMS.

Also acknowledging the effectiveness and power of WMS, the multiculturalists disagree with the universalists mainly on the purposes of adopting native knowledge in school science. Stanley and Brickhouse (2001) argue:

> Rather than merely showing students that IK [Indigenous Knowledge] and TEK [Traditional Ecological Knowledge] are different from WMS [Western Modern Science] (and possibly suggesting they were inferior), our approach would be to show students how these different views of science are firmly rooted in certain cultural assumptions that influence how they go about formulating and solving problems of significance. Furthermore, rather than teach a particular view of science or pretend as though the nature of what we were teaching was uncontroversial, we would then teach students about the controversy over what is to be included in the science curriculum and how such decisions might be made. (p.47)

For multiculturalists, native knowledge has more value to be included in school science curriculum rather than enhancing the superiority of WMS. Aikenhead and Michell (2011) give several reasons for indigenous knowledge to be placed in school science curriculum. These reasons include to promote equity and social justice, to strengthen a nation's economy, to improve Eurocentric science, to prepare science-oriented students for science careers, to support indigenous sovereignty and cultural survival, and to enhance human resiliency.

Stanley and Brickhouse (2001) propose a more "local" and "multiple" view of science and base their proposal on several assumptions including:

> 1) Our ability to understand nature is constrained by the limits of human cognitive abilities; 2) The observer is part of the reality that is observed, thus social construction plays a role in the scientific account of physical reality; 3) We cannot determine if reality is either uniform or invariant; reality may best be described as a flux. (p. 39)

These assumptions are in line with a constructivist view of scientific knowledge. In essence, debates between universalism and multiculturalism can be seen as the representation of debates between realism and constructivism in the field of science education. Because of the political and ethical nature of education, the issue has been made more complex than merely ontological and epistemological arguments.

In contemporary Chinese society, universalism dominates school science curriculum and mass media. Like in many developed countries, WMS (often taken as equivalent to science and technology) is seen as the main power (if not the sole power) for the development of national economy. Along with the development of WMS in China, much Chinese native knowledge has been labeled either as "superstitious" or "outdated." Science textbooks are formulated under the framework of WMS with occasional mentioning of examples of Chinese contributions. The Chinese contributions mentioned in textbooks are those that can be understood within the paradigm of WMS.

One form of Chinese native knowledge that is difficult to be fully understood by WMS is traditional Chinese medicine (TCM). Like much long-standing native knowledge, TCM has shown its effectiveness in practice in many ways. Unlike much other native knowledge that lacks sound theoretical foundation, TCM has its own theoretical framework; however, it is incompatible with WMS in many ways. There has always been tension between TCM and Western medicine (endorsed by WMS) ever since the introduction of WMS to China. Like WMS, Western medicine dominates the contemporary medical system in China. Although marginalized, TCM also has its place in many hospitals and medical service centers. In public arenas, forums on the Internet, for instance, debates about whether TCM is science, are intensive. The tension, however, is invisible in general school science curricula as there is no mention of TCM at all – students have to go to specialized TCM school (from postsecondary to postgraduate) to study TCM. It is doubtful to what extent science education has contributed to people's understanding of the nature of TCM and further to the understanding of the nature of science. It is in this sense that discussion of universalism and multiculturalism in the Chinese context is made realistic.

Scientism

Scientism is an ideology that concerns the relationship between natural science and other ways of human inquiry such as social, ethical, political, cultural, and religious inquiries. Scientism is closely associated with a logical positivist position that insists that true knowledge of the world can only be obtained through empirical methods and reasoning of natural science. Any explanation about phenomena (natural or social) that cannot be directly experienced and empirically proven is ungrounded speculation. The all-inclusive view of scientism believes that the methods of science are the only appropriate and authoritative approaches to explore not only the natural world but also human social world. According to Poole (1998, p.195) scientism "denies that anything other than the natural world exists," "denies that there are first causes or final causes," and "denies that there could ever be behavior other than law like (anti-miraculous)." Poole (1998) distinguishes science from scientism and argues that views of scientism are not necessary presuppositions for science.

Scientism became a topic of debates in public arena in China in the early 1900s after WMS had been introduced systematically to Chinese academia. During this period, Chinese society was undergoing cultural upheaval. Traditional beliefs and values were challenged by scholars who had overseas study experience. Science (WMS) was promoted to be an essential agent of social change. However, even among those who had been greatly influenced by Western culture, there were different understandings of the social function of science. Radical social reformers emphasized that the epistemology of natural science was the most effective means to social reform and should be the only source of an individual's worldview and of solutions to social problems (Yang, 2001). Others insisted that science could not replace humanities such as ethics, aesthetics, philosophy, and religion in seeking value of human life and the power of science should not be abused.

Discussions about the role of science in human inquiry led to the famous debate of "science versus metaphysics" in 1923. Lasting for more than a year, this event profoundly promoted the formation of scientism in Chinese modern thought. Kwok (1965) identified two varieties of scientism that emerged during the debate: "materialistic scientism" which reduces all aspects of human activities to matters of material reality and scientific laws and "empirical scientism" which believes "the scientific method" is a panacea for all social problems. As Kwok observes, "scientism exhibits first, a particular understanding of the power of science, second, a critique of tradition, and, third, a form of substitute religion" (p. 30). The empirical scientism identified by Kwok was still part of Chinese thought and had profound political influence at least until the late 1980s (Hua, 1995). Hua adds a third type of scientism, "technological determinism," that became influential around the 1980s. Hua describes "technological determinism" as "a distortion of orthodox Marxism plus some modifications of the indigenous Chinese science tradition" (p. 19). Further, Hua explains:

> It borrows from Marxist scientism crucial notions such as objective law and the idea that productive forces are the most important in the development of human society. It differs from Marxism in that it downplays the role of relations of production. . . . China's scientific development was characterized by: 1) politicisation (*zhengzhihua*); 2) technologization (*jishuhua*); 3) downplay of comprehensive theories. (p. 19, italics in original)

In contemporary China, because the ruling party endorses Marxist materialism that is, in many ways, consistent with materialistic scientism, materialistic scientism is still playing an important role in political arena. Technological determinism also has many advocates due to the increasing reliance on technology for economic development. The followers of empirical scientism, in contrast, have declined as a result of the rise of nationalism, which promotes traditional philosophical ideas in solving contemporary social problems.

As stated earlier in this section, reviewing these three dimensions of the nature of science serves the purpose of building the theoretical framework for an empirical study. Findings of the study are presented in relation to these three dimensions in the next section, after a description of research purpose and method.

The Study

Purpose

The study reported in this chapter was conducted with a group of Chinese secondary school science teachers. The aim of the research was to find out this group of teachers' understandings of the nature of science. Particular attention was given to teachers' views in relation to the above reviewed dimensions of the nature of science: realism/constructivism, universalism/multiculturalism, and scientism.

The Method

This study adopted a qualitative approach and used semi-structured interview. The design of the interview protocol consisted of three parts. The first part was topics about nature. Questions were about human ability to explore nature and the relationship between humans and nature. It was hoped that teachers' responses to questions like these would help to show their understanding of scientific theory and reality, as well as the role of science in investigating nature. The second part was questions about science and culture. Teachers were asked whether and to what extent they believed that culture influences science. The third parts involved Chinese native knowledge – TCM. TCM was treated as everyday phenomena rather than professional medical practice in the interview. This was mainly because most of the participant teachers did not study TCM formally – they might use TCM services and learn something about TCM from family and friends or from media.

Because they were not TCM practitioners or TCM students, they may not know much about the theoretical knowledge of TCM. Teachers were asked whether they trust TCM, how they view the differences between TCM and Western medicine, and how their understanding of TCM may influence their science teaching.

Twenty-five participant teachers were selected from two secondary public schools on a voluntary basis. Among them, eight were Biology teachers, eight Chemistry teachers, and nine Physics teachers. The interviews were conducted face to face with each teacher in Chinese in their school precinct. Chinese transcripts were used for initial analysis and only reported teachers' responses were translated into English.

Interview data were analyzed from two dimensions. The first dimension looked at themes and trends across cases (each teacher was seen as a case). Teachers' understanding of the nature of science inferred from their worldview discussions and their views of teaching culturally related knowledge have been reported elsewhere (Ma, 2009, 2012). This chapter reports findings from the second dimension of analysis: By examining teachers' responses to interview questions within each case, the structure of individual teacher's philosophical perspectives was examined. Related findings are reported in the next section.

Findings

The review of three dimensions of the nature of science shows that there is a certain degree of consistency between realism and universalism, as well as between constructivism and multiculturalism. Teachers' views, however, showed hybrid perspectives with combinations of different philosophical positions. That is, a teacher may have a realist view of scientific theory and reality while holding a multiculturalist position on the relationship between WMS and Chinese native knowledge about nature. Three categories of philosophical inclination among participant teachers were identified and presented below as case studies.

Universalist Scientism: Meng

Meng was a junior high school Biology teacher. At the time of interview, she was in her 40s and had more than 15 years of teaching experience. Meng had a strong inclination to scientism in terms of science (WMS) versus other worldviews. As for the status of Chinese native knowledge, like almost all other participant teachers, she believed TCM could be seen as a science. Although she did not argue that TCM was a less worthy form of knowledge, she intended to believe that WMS was more powerful in providing explanation to natural phenomena (such as health and disease). Compared with other participant teachers, her view is closer to a universalist position. Meng represents those participant teachers who showed both inclinations to universalism and scientism; therefore, her position is categorized as "universalist scientism."

As a Biology teacher, it is not a surprise that Meng described nature as an ecosystem defined in biological science:

> Nature consists of animals, plants, and humans. Living things, such as bacteria, fungi and viruses. Also non-living things, such as sunlight, air, soil, water, and so forth. The universe.

Meng believed that nature is material and did not believe in nonmaterial spiritual existence such as a soul. She asked ironically "how many grams does a soul weigh?" For one thing, she did not think supernatural phenomena, such as channelling, transpersonal experience, or faith healing, exist. For another, she labeled the spiritual explanations of these phenomena as "fallacy, heresy, and superstition."

Meng firmly believed that the best way of understanding nature is "guided by scientific theories and through scientific methods." She explained:

> An ideal situation between humans and Nature is living together in harmony. This is based on humans' thorough understanding of the law of Nature. Study the intrinsic law, respect and follow the law, and then benefit from it.

She was also very confident that given enough time, humans would eventually understand nature thoroughly. This confidence is based on her view of how science progresses:

> [Scientific] methods and theories should stand up to scrutiny. No matter being tested by which authority, the results should be the same. Then it can be accepted as scientific knowledge. It might not be perfect, but it approaches truth gradually.

Although not expressed in exact terms, this view is close to a positivist view of science, which holds that scientific knowledge verified by empirical evidence is the only authoritative knowledge. Consistent with this view, Meng held that science is a more superior form of knowledge than other ways of knowing. She made a comparison between evolutionism and creationism:

> It is very convincing teaching evolutionism with direct fossil evidence. I definitely don't believe that life was created by any spirit or God. That's just a kind of culture, Buddhist culture. It can't be taken seriously. School Biology should teach evolutionism rather than creationism.

It is worth noting that she thought that creationism was Buddhist rather than a Christian belief, which shows that she had some misunderstandings of religion. She probably did not take religious doctrine seriously. Although she thought that different cultural beliefs could exist pluralistically and should be treated with tolerance, deep in her heart, she hoped that "more people believe in science."

More in line with a universalist view, Meng believed that epistemologically, WMS is superior to Chinese native ways of knowing.

> People think metaphysics, such as Zhouyi, is essence of traditional Chinese culture. I feel that using Zhouyi in fortune telling is like a wild guess, I don't really know it though.... Zhouyi is too mysterious. Are there really those intrinsic associations? ...It's just a guess, superstitious, better not to promote it. ...People tend to believe superstition because they don't know how to investigate things using scientific methods. It is the right way to go to try to find a scientific explanation.... Metaphysics has misled China.

Zhouyi is one of the most ancient Chinese classic belief systems. It uses a symbol system to describe change and relationships between things and events in the universe. Zhouyi was originally used in divination. However, its underpinning philosophical ideas have profoundly influenced many aspects of traditional Chinese thinking such as in philosophy, religion, medicine, literature, arts, and military study. It is obvious that Meng did not realize the philosophical distinction between traditional Chinese metaphysics and WMS. She mainly criticized the use of Zhouyi for fortune telling.

The representative of Chinese native knowledge, TCM, has also been influenced by the dialectical thinking embodied in Zhouyi. However, like most of the participant teachers, Meng did not realize this. It is understandable because participant teachers were not TCM practitioners or learners. They paid more attention to the healing effect of TCM rather than its theoretical assumptions. Because teachers trusted the practical effect of TCM, they believed that TCM is scientific. Meng was not an exception. It seems that by acknowledging TCM as science, Meng showed a multicultural position. However, a closer examination of her view has revealed that she still thought, unconsciously, epistemology of WMS is the universal way of developing both Western medicine and TCM.

> Science can't explain TCM now. There might be, for example, another field, not electric or magnetic fields, that can be used to explain TCM. It just hasn't been discovered yet. ... Western medicine is more convincing and more easily accepted because of its methods. It can be seen through autonomy and tested through experiments. It's a good idea to study TCM by adopting methods of Western medicine. ... Western medicine and TCM can finally be unified. There is only one truth. As long as appropriate methods are adopted, truth can be approached gradually.

It is obvious that Meng did not think that TCM could represent a different paradigm that is incompatible with WMS. She believed that TCM has to prove itself through methods of WMS. Meng's opinions on science teaching also reflect her position of universalist scientism. She strongly argued that science education should play a more effective role in promoting scientific worldview.

> Superstition will find its way around if science isn't promoted enough. Education could do more to promote evolutionism,... to promote science, then more people will accept scientific worldview.

In terms of teaching, Meng intended not to mention TCM too much, though she thought students would benefit from being introduced to TCM.

> I tend to explain things from Western medicine's perspective since it is more familiar to me. TCM is too mysterious and obscure. ...Better just mention those [TCM] that have evident effect and recognized by all. For those that you can't explain, better not to mention. ... TCM lacks complete scientific theoretical framework, if you introduce it in your classroom, it may cause confusion among students.

Meng's cautious attitude toward teaching Chinese native knowledge like TCM is representative among the participant teachers. However, in terms of the relationship between WMS and TCM, not all teachers held the universalist view.

Realist Multiculturalism: Yang

Yang was a senior high school Chemistry teacher. He was in his late twenties and had taught for more than 6 years. The reason that his position is labeled as "realist multiculturalism" is that on the one hand, he showed a strong inclination to view scientific theories and reality as equivalent, and on the other hand, he did not think that WMS is more superior in explaining certain natural phenomena.

Yang's view of nature was similar to that of Meng's. However, he placed particular importance on the "materiality" of nature:

> Nature is the material world, the Universe, the Solar system, the Galaxy, matter on the Earth. Everything that is material, including living things, including 'fields'. ... I simply think that Nature is material, while social phenomena, political phenomena, relationships between people, between nations, are not Nature.

In the discussion about nature, he repeatedly emphasized that nature is material and pointed out that human social activities are not nature. Particularly, he gave the example of "fields," electronic or magnetic fields, and equated them to reality instead of theoretical entities. This is a view shared by some realists. Further, he talked about unobservable phenomena:

> Looking at the world from a perspective of chemistry is studying the structure of matter. If you had a pair of fire eyes like the Monkey King, you would be able to see through matter as if you were using X ray. ... Matter consists of molecules, atoms, and ions. Inside an atom, there is a big space around the nucleus. Electrons, with a very small mass, move around the nucleus. Surely this can also be seen from photos.

Monkey King was a famous character in Chinese literature who had sharp eyes that could identify a creature's (usually a demon's) original form from its changed appearance. The metaphor of the Monkey King seems to suggest that if human eyesight was good enough (or more likely, with the help of extended probes), they would be able to directly "see" the structure of matter through its appearance. Here again, he viewed theoretical entities such as "atoms" as real existence and was very confident that we can have photos that show the "real" image of these entities.

This realist position explains Yang's view of justification of scientific knowledge. He firmly believed that no matter which cultural background the scientists have, as long as the phenomenon being investigated is the same, the output of scientific research should be the same:

> Science investigates material things. Take the 'bird flu' virus, H5N1, for example. No matter who discovered it, Western or Eastern, the structure of the virus is the same. ... The development of science is both accidental and inevitable. Kekule's dream is accidental. However, maybe decades later, the structure of benzene would be discovered without having a dream.

It is obvious that Yang's view is in line with the universalist's idea that "the material world ultimately judges the adequacy of our accounts of it" (Matthews, 1994, p. 182). However, the reason that his position is categorized as "realist multiculturalism" instead of "realist universalism" is mainly because of his tolerant attitudes toward other ways of knowing and toward Chinese native knowledge.

Like Meng, Yang tended to believe in evolutionism and hoped that science would finally answer the question of "how life began." However, unlike Meng, Yang was doubtful about the power of science in solving problems in every aspect of human life. As a result, he thought that religious belief plays an important role in human life:

> Religious belief is good, because life is happy only when people have faith.... Spiritual stuff influences human inner life, I doubt that science can achieve it. This is the limitation of science. No matter how developed science is, ... spiritual stuff can't be influenced by scientific studies. The human nature is too complicated.

In terms of the relationship between Western science and Chinese native knowledge (TCM), Yang believed that both are science and equally good. His view of the scientificity of TCM was consistent with his realist position:

> Science is material. Herbal medicine and acupuncture are material. . . . As long as they have healing effect, they are scientific laws. Laws of matter are scientific laws. . . . I believe in "Qi Gong". Qi is matter, similar to "field". It can't be seen, but it cures disease.

Qi Gong is an exercise developed in ancient China thousands of years ago. Qi Gong aims to improve body function by controlling breath and mind through meditation. Qi is seen as some kind of inner energy. It is said that Qi Gong has the power to heal many diseases by touch healing or distant healing. However, it has not been supported by modern medical research. Obviously, Yang did not pay much attention to the spiritual base of Qi Gong and he believed that Qi, instead of a concept, really exists like the concept of "field" in WMS.

Different from Western universalists, he did not believe that WMS is superior, at least, in the field of medical research. He also argued that Western medicine and TCM have different theoretical frameworks and will not be completely unified:

> Just like all roads lead to Rome. Although TCM and Western medicine have different theoretical frameworks, they all succeed in healing disease. . . . Both TCM and Western medicine are good science. I feel each has advantages and disadvantages. . .. Along with the development of science, both theories will develop respectively, each keeping their own essence.

Yang thought that TCM could be mentioned in secondary science teaching for the purpose of broadening students' knowledge base. Like most other participant teachers, he suggested that more professional knowledge should be taught in postsecondary stages. When being asked whether it will cause confusion among the students because TCM is different from WMS introduced in the textbooks, he insisted that feeling confused would make the students to pay attention to different aspects of knowledge in their learning:

> Being flexible is important in learning, whether it is science, or anything else, including TCM and Western medicine. Maybe for someone, there is only one truth. I don't think so. You must be flexible. I mean, knowing the world from different perspectives. Learn from all perspectives, learn critically.

When discussing science in general, Yang was most persistent that the output of scientific research should be the same if studying the same phenomenon. However, when TCM was involved in the discussion, he was more inclined to a multicultural or plural view. The inconsistency can be found in many participant teachers' responses, which makes Yang's "realist multiculturalism" representative.

Compromised Constructivism: Lei

Lei was a senior high school Physics teacher. She was in her late twenties and had taught for 4 years. Like in Yang's case, there was some inconsistency in her view. In fact, her views, especially in terms of the relationship between WMS and Chinese native knowledge, were more ambiguous. This is not unusual among participant teachers. Different from many other participant teachers, she believed human social factors play an important role in knowledge production – a view close to a constructivist position.

Unlike Meng and Yang, who separated human spiritual and social activities from nature without much hesitation, Lei struggled to distinguish human social factors from nature:

> The originally existent, such as animals and plants, belongs to Nature. We can't call houses and buildings Nature, can we? But if animals, such as bees and ants, build their houses, we call it Nature. . . . If humans build a house, there will be lots of social activities involved. . . . Humans are both natural and social. . .. Humans are social production as well. . .. Human social activities [Think], I feel, should belong to Nature. In my mind, Nature and society can't be separated definitely.

In thinking about nature, Lei first said that human-made products such as buildings were not nature. However, this led her to think about human social activities and she finally concluded, not very convinced, that nature also consists of human social factors.

Because human social activities are seen as part of nature, understanding nature is seen as including both investigating the physical world and exploring human social activities.

> Different disciplines have different ways of investigation. Archaeologists, for instance, may study Nature by examining an artifact's age, decorative patterns and symbols on it. However, for High-energy Particle Physicists, they study the microstructure of atoms to see if they can be further divided. It's hard to find a unified method.

Realizing the complicity of nature, Lei expressed a view that acknowledges multiple methods of exploration rather than claiming that the scientific method is the only way of knowing nature.

The involvement of social factors also influences her ideas about scientific activities. She emphasized more on the collective nature of scientific research and portrayed an image of scientists that cannot transcend their social environment.

> An individual's idea is influenced by his prior knowledge and what other people round him think. An individual is definitely not independent. Humans are products of society and influenced by social environment. . . . If being isolated, an individual wouldn't be able to make great contribution. . . . He would have no chance to exchange ideas with others.

In line with this view, instead of emphasizing the role of the material world in justifying scientific theories, Lei viewed the process of knowledge justification more as consensus to be made among scientists. This is also the basis on which she argued for the scientificity of TCM:

> You can't say that TCM is not scientific.... Once somebody has established a theory and the theory is then gradually carried forward by his successors. ... Once more and more people accept the theory, it will naturally be accepted by most of the people. ... Other people may think it is not very scientific. But scientific or not, it is up to how people define it.

She even argued that for people with Chinese cultural background, TCM's explanations of the cause of disease might sound more reasonable. Her argument reflects her view of the importance of sociocultural environment on people's judgments:

> Western medicine and TCM explain the cause of illness in different ways. I feel explanations of experienced doctors of TCM sound more reasonable than those of doctors of Western medicine. . . .Because we have been raised in this environment – when a doctor of TCM says something like 'suffering from excessive internal heat', although you don't know exactly where the heat is, you understand at once what is the cause of the illness. Isn't it also very scientific?

Lei's attitude toward the relationship between Western medicine and TCM is ambiguous. In her own words, "Western medicine and TCM are different, but not much." She mistakenly thought that Western medicine also use concept such as "internal heat" and TCM teaches the same anatomy as in Western medicine. Therefore, she tended to focus on their similarities rather than differences:

> Speaking of Western medicine, it comes to mind the word 'synthetic'. Many Western medicines have Chinese herbal elements. You can't distinguish them. ... They have similarities. Either decocted in TCM, or made into pills or tablets in Western medicine, the effects are the same. Maybe one's effect is better than the other's. I'm not sure. I'm not a medical practitioner. You can't say that Western medicines have nothing to do with TCM, and vice versa.

Reflected in her ideas of teaching science, she said that she would try to find the similarities and tell the students that it is not that the knowledge is different, it is just that they are expressed in a different way. In terms of involving TCM in science teaching, she thought that the students should be given the option to choose whether they would like to learn something about TCM. This view is slightly different from many other participant teachers' views in that it is the students who initiate the learning of TCM. As for other Chinese native knowledge, Lei tended not to mention too much of ancient Chinese science and technological achievements; instead, she argued that teachers should focus more on current Chinese science achievements, which are, according to her, more inspiring. Current Chinese science achievements are mainly those achieved with the framework of WMS.

Implications and Conclusions

The findings present three categories of philosophical positions found in the participant teachers: universalist scientism, realist multiculturalism, and compromised constructivism. In terms of the realism/constructivism dimension, more teachers held a realist inclination. As for scientism, almost all teachers believed that science is the most powerful way of knowing nature. However, their views on the value of other ways of knowing nature (e.g., religion) differed. Meng represents an extreme view of promoting a scientific worldview only, while at the other end of the spectrum, Yang held a more tolerant view. Teachers' views in the dimension of universalism/multiculturalism are very complex. Most of the teachers thought that TCM should be included in school science teaching. This seems to suggest that they held a multiculturalist view. However, a closer examination of teachers' responses shows that quite a few teachers believed that TCM would be finally explained by WMS, such as Meng, which is a universalist view. Because teachers did not know much about TCM, to different extents, they expressed a compromised view toward Western medicine and TCM. As curriculum planners, teachers play a crucial role in assisting students to achieve the goals set by science curriculum standards. It is of particular relevance to examine teachers' own understanding of the nature of science for teaching scientific literacy, as it is the focus of the current science education reform.

First of all, is teacher's understanding of the nature of science robust enough? Findings from this study show that teachers' perceptions of the nature of science were not informed. Whether teachers held a realist or constructivist inclination, they did not realize the tension between these two philosophical positions. As a result, their arguments were more out of intuition rather than theoretical thoughts, and inconsistency can be identified in the same teacher's views. This is not surprising, given that the nature of science was not an explicit topic in teacher education courses taken by this group of teachers. In fact, the nature of science has become an issue in discussion among science educators and science education researchers in China since the early twentieth century. This does not mean that there has not been implicit involvement of the nature of science in these teachers' education programs. Historically, science was taught in modern China as value-laden. Science was seen by Marxists as an ideological weapon against idealism (Xu & Fan, 1957/1982). Marxist "dialectical materialism" holds a pro-science, pro-positivism philosophy and has been promoted as the main source for philosophical thinking in almost every sociocultural arena including science education. Consequently, science teachers may have developed their own understanding of the nature of science without being more sophistically informed.

The junior science curriculum standards place a particular importance on teaching the nature of science to secondary school students. However, there are no further suggestions or resources to support the teaching in this document. Although some teaching resources may start to include materials for discussing the nature of science, without teachers having a more sophisticated view, these materials may be used in a superficial way or just ignored. This raises the need for teaching the nature of science explicitly in preservice teacher education courses and in in-service teacher professional development programs.

A subsequent issue is how to teach and what to be included in the course. A study by Wan, Wong, & Yung (2011) shows that science teacher educators in China valued introducing the nature of science to preservice science teachers for diverse reasons. Also, many science teacher educators are still influenced by Marxism in thinking about the nature of science (Wan, Wong, & Zhan, 2013). Given the complexity of the nature of science, teachers will benefit more if diverse views or positions are made available to them. By presenting competing theories and challenging teachers' own naïve views, teacher educators can help preservice or in-service teachers to critically think about diverse ideas and develop more informed views that make sense to themselves.

Another related issue is how to view the status of TCM in science. It seems to be an issue involving the debate between universalism and multiculturalism. However, as TCM remains a popular everyday phenomenon in contemporary China, the ignorance of TCM in mainstream science teaching also urges a reconsideration of how to teach scientific literacy in China. Enhancing students' scientific literacy is the focus of the current science education reform. According to Rennie, Goodrum, and Hackling (2001, p.466), scientifically literate people are "skeptical and questioning of claims made by others about scientific matter" and are able to "make informed decisions about the environment and their own health and well-being." TCM is an everyday science-related phenomenon and closely related to people's health and well-being. Findings from the present study show that no matter which philosophical position about the nature of science that the teachers held, few teachers cast doubt on the scientificity of TCM and questioned the relationship between practice and theory of TCM. As teachers' justification was mainly based on personal experiences and lacked input from the understanding of the nature of science, it can hardly be said that their decisions were informed.

Universalism is a dominant feature of the current science education curriculum in China. TCM promoters have long lost their battle against WMS in the field of science education since TCM was removed from the basic education (primary and secondary) system more than a hundred years ago. Nowadays, TCM practitioners are still struggling to find their legitimate place in the modern medical system, as many holistic claims made by TCM cannot be supported by reductionist analytical methods of WMS. In folk culture, TCM is still very popular and a certain number of people have blind faith in both its healing effect and theoretical explanations. Because almost no people in contemporary China have had formal TCM education in their primary or secondary schooling, again, it can hardly be said that science education has contributed anything in helping people better understand TCM as a science-related issue.

Given the complexity of TCM, it may be too early to suggest that TCM be taught as science in primary or secondary schools. However, as an initiative effort, it could be included in the teacher education course discussing the nature of science. From a multiculturalist point of view, Stanley and Brickhouse (2001) propose a few questions that might help students better understand the nature of both WMS and native knowledge:

> What questions is WMS good at answering? What kinds of problems do TEK solve? In addition to the ability of these sciences to solve particular problems, is one of them simply more likely to be true than others? If so, how would you know? What historical conditions may explain why WMS arose primarily out of Western European culture rather than elsewhere in the world? What are the similarities between TEK and WMS? Would it have been possible for TEK to have become a part of our culture at some point in history? If it had, how different might our culture now be? How do other belief systems (e.g., religion) interact with TEK, Chinese science, and WMS? (p.47)

If replacing "TEK" with "TCM," the first few questions will be very relevant to the Chinese context. Since TCM has already become a part of Chinese culture, the last three questions could be modified to better suit the Chinese cultural discourse: What Chinese cultural factors made TCM readily acceptable to Chinese people? What epistemological tools provided by TCM can help us better understand the world? What research initiatives can be proposed to ensure the continuous development of TCM? Thinking about these questions will raise the awareness of the cultural embeddedness of both WMS and Chinese native knowledge.

The third implication is related to the influence of scientism on teaching the social aspects of science. Unlike Lei, many teachers saw nature and human social activities as opposite and believed that science does not deal with sociocultural phenomena. Further, many believed that science is superior to other ways of knowing, though to different extent, teachers acknowledge the value of humanities. This attitude can be an obstacle to fulfilling the affective dimension of the new science curriculum standards. Teachers may only focus on teaching the "cold hard facts" of science and avoid teaching moral, ethical, and political issues that are associated with science-related problems. However, the real-world scientific problems are often entangled with sociopolitical issues and without a better developed understanding of social factors involved, it will be difficult to make informed decisions.

A possible solution would involve socio-scientific issues in the teacher education courses on the nature of science. Socio-scientific issues (SSI), as defined by Zeidler and Nichols (2009), involve

> the deliberate use of scientific topics that require students to engage in dialogue, discussion, and debate. They are usually controversial in nature but have the added element of requiring a degree of moral reasoning or the evaluation of ethical concerns in the process of arriving at decisions regarding possible resolution of those issues. (p.49)

SSI pays particular attention to the psychological, social, and emotive growth of students and contributes to the increase of scientific literacy in general. Involving SSI in discussions on the nature of science will provide chances for teachers to critically think the role of science in society and consciously incorporate moral and ethical reasoning as part of their science teaching pedagogy.

In conclusion, the empirical study presented in this chapter implies that teachers (both in-service and preservice) will benefit from courses that explicitly include topics on the nature of science. Given the Chinese cultural background, purposefully involving Chinese native knowledge and current socio-scientific issues in the discussion will help teachers develop more sophisticated understanding of the nature of both WMS and Chinese native knowledge, as well as the role of both knowledge systems in society.

References

Aikenhead, G., & Michell, H. (2011). *Bridging culture – Indigenous and scientific ways of knowing nature*. Toronto, ON: Pearson Canada.
Deng, F., Tsai, C. C., & Chai, C. S. (2011). Students' views of the nature of science: A critical review of research. *Science Education, 95*, 961–999.
Franklin, A. (1999). *Can that be right?* Dordrecht, The Netherlands/Boston, MA/London, UK: Kluwer Academic Publishers.
Glasersfeld, E. V. (2001). The radical constructivist view of science. *Foundations of Science, 6*, 31–43.
Gross, P., & Levitt, N. (1994). *Higher superstition: The academic left and its quarrels with science*. Baltimore, MD: John Hopkins University Press.
Hua, S. (1995). *Scientism and humanism: Two cultures in post-Mao China (1978–1989)*. Albany, NY: State University of New York Press.
Hung, E. (1997). *The nature of science: Problems and perspectives*. Belmont, CA: Wadsworth Publishing Company.
Kwok, D. W. Y. (1965). *Scientism in Chinese thought 1900–1950*. New Haven, CT/London, UK: Yale University Press.
Lederman, N. G. (1992). Students' and teachers' conceptions of the nature of science: A review of the research. *Journal of Research in Science Teaching, 29*(4), 331–359.
Lederman, N. G., Lederman, J. S., & Antink, A. (2013). Nature of science and scientific inquiry as contexts for the learning of science and achievement of scientific literacy. *International Journal of Education in Mathematics, Science and Technology, 1*(3), 138–147.
Leplin, J. (1984). Introduction. In J. Leplin (Ed.), *Scientific realism* (pp. 1–7). Berkeley, CA/Los Angeles, CA/London, UK: University of California Press.
Leplin, J. (2006). Realism. In S. Sarkar & J. Pfeifer (Eds.), *The philosophy of science – An encyclopedia* (Vol. 2, pp. 686–696). New York, NY/London, UK: Routledge.
Ma, H. (2009). Chinese secondary school science teachers' understanding of the nature of science – Emerging from their views of Nature. *Research in Science Education, 39*(5), 701–724.
Ma, H. (2012). *The images of science through cultural lenses – A Chinese study on the nature of science*. Rotterdam, The Netherlands/Boston, MA/Taipei, Taiwan: Sense Publishers.
Matthews, M. R. (1994). *Science teaching: The role of history and philosophy of science*. New York, NY/London, UK: Routledge.
McComas, W. F., Clough, M. P., & Almazroa, H. (1998). The role and character of the nature of science in science education. In W. F. McComas (Ed.), *The nature of science in science education: Rationales and strategies* (pp. 3–39). Dordrecht, The Netherlands/Boston, MA/-London, UK: Kluwer Academic Publishers.
Ministry of Education. (2011). *ChuzhongKexueKechengBiaozhun* [Junior high school science curriculum standards]. Beijing: Beijing Normal University Publishing Group
Poole, M. W. (1998). Science and science education: A Judeo-Christian perspective. In W. W. Cobern (Ed.), *Socio-cultural perspectives on science education. An international dialogue* (pp. 181–201). Dordrecht, The Netherlands: Kluwer Academic Publishers.
Rennie, J., Goodrum, D., & Hackling, M. (2001). Science teaching and learning in Australian schools: Results of a national study. *Research in Science Education, 31*, 455–498.
Siegel, H. (1997). Science education: Multicultural and universal. *Interchange, 28*, 97–108.

Stanley, W., & Brickhouse, N. (2001). Teaching sciences: The multicultural question revisited. *Science Education, 85*, 35–49.

Wan, Z. H., Wong, S. L., & Yung, B. (2011). Common interest, common visions? – Chinese science teacher educators' views about the values of teaching nature of science to prospective science teachers. *Science Education, 95*(6), 1101–1123.

Wan, Z. H., Wong, S. L., & Zhan, Y. (2013). When nature of science meets Marxism: Aspects of nature of science taught by Chinese science teacher educators to prospective science teachers. *Science & Education, 22*, 1115–1140.

Xu, L. Y., & Fan, D. N. (1982).*Science and socialist construction in China* (J. C. S. Hsu, Trans.). Armonk, NY/New York: M. E. Sharpe, Inc. (Original work published 1957).

Yang, G. R. (2001). Lunwusishiqi de kexuezhuyi [Scientism during the May 4th period]. In B. Hao & Z. S. Ouyang (Eds.), *WusiYundong Yu ErshiShiji De Zhongguo* [The May 4th Movement and China in the 20th Century] (pp. 183–197). Beijing: ShehuiKexueWenxianChubanshe.

Zeidler, D. L., & Nichols, B. H. (2009). Socio-scientific issues: Theory and practice. *Journal of Elementary Science Education, 21*(2), 49–58.

Chapter 19
Developing Science Teachers' Pedagogical Reasoning Through Reflective Practices in Overseas Professional Development Program: A Case Study

Do Yong Park and Young Hak Kim

Introduction

One of the key components in fulfilling the educational initiatives of improving in-service teachers' teaching in many countries is professional development (OCED, 2009). In 1985, an overseas science teacher professional development (OSTPD) program was developed by the Korean Ministry of Education. Its aim was to improve science teacher training. Done as part of the ministry's educational reform, the program has continually sent several hundred science teachers and supervisors to Western countries including the UK and United States. The ministry put the program on hiatus from 1988 to 1994 due to the country's unstable political and societal situation.

When the OSTPD program was resumed in 1994, around 25–33 secondary Korean biology teachers were sent to a couple of universities in the United States. A group went each year during the summer, so it has often been called "a summer intensive professional development program." From 2007 to 2010, Illinois State University provided the Korean Science Teachers Professional Development (KSTPD) program for secondary Korean biology teachers. The objective was to make them current with (1) biological content knowledge and (2) biology education (pedagogical knowledge) and also (3) to observe secondary school classrooms (Park, 2011; Park & Han, 2012). The program design focuses on content knowledge and is based on the research outcomes of a couple of professional development programs.

Research findings have reported that teacher professional development based on academic subject matter knowledge is positively correlated to instructional practices and student outcomes (Cohen & Hill, 2000). Birman, Desimone, Porter, and Garet (2000) also found that "the degree to which professional development focuses

D.Y. Park (✉) • Y.H. Kim
School of Teaching and Learning, Illinois State University, Normal, IL, USA
e-mail: dpark@ilstu.edu; yhkim22@gmail.com

M.S. Khine (ed.), *Science Education in East Asia*,
DOI 10.1007/978-3-319-16390-1_19

on content knowledge is directly related to teachers' reported increases in knowledge and skills" (p. 30). This connection of content knowledge with instructional practices in professional development is well documented (Jeanpierre, Oberhauser, & Freeman, 2005; Loucks-Horsley, Hewson, Love, & Stiles, 1998). Specifically, Jeanpierre et al. (2005) reported that teachers' increased content knowledge of monarch butterfly ecology successfully translated into instructional practices. This occurred during a 2-week-long resident institute and through numerous practices of inquiry. In addition, a significant increase in secondary science teachers' content knowledge was demonstrated in an intensive field-science professional development that combined scientific research and pedagogy in a contextual setting (Nelson & Hanegan, 2003). The authors went on to report that the intensive professional development model positively affected the secondary teachers' teaching practices, engendering more student-centered instruction.

Likewise, evaluation of the KSTPD program indicated a high degree of teacher satisfaction with most of the academic programs provided in the summer intensive program (Park, 2011). Korean teachers evidently have positive experiences with the program in terms of the above objective of the program. Nevertheless, when it comes to their biology classes in Korea, they tend to struggle to apply their acquired knowledge and skills. One of the key reasons behind such a struggle may be the fact that they had too few reflective opportunities to develop and apply pedagogical reasoning during the program (Boud, 2001; Demet, 2009). Therefore, in this paper, we report the findings of our case study in which participants used pedagogical reasoning through multiple reflective practices both during and after the professional development program.

Theoretical Framework

Our professional development program is based on what Putnam and Borko (2000) suggested for teacher learning. Teacher learning and cognition is "(a) situated in particular physical and social contexts; (b) social in nature; and distributed across the individual, other persons, and tools" (Putnam & Borko, p. 4). The nature of knowledge, thinking, and learning are interactive and situational in meaningful contexts (Lave & Wenger, 1991). Hence, the learning tasks in teacher professional development programs should engage teachers in rich discourse, such that it fosters teacher thinking, interactions, and sharing experiences. The situated nature of cognition focuses on interactive systems that include teachers, interacting with each other as well as materials (Putnam & Borko). This is a way that determines what is learned and how the teacher learning takes place.

Our KTPD program was developed to help Korean teachers learn biological content knowledge and biology education in the real contexts of American university research labs and secondary school settings. Being situated in real contexts can help them gain authentic and meaningful understanding. Putnam and Borko (2000) recommended this as a way to determine essential elements that help change teacher thinking and practices. Our professional development model provides multiple authentic contexts including various up-to-date biology laboratories. Here, teachers are able to witness and participate in the latest research project, experiencing a top-notch biological study. After the lab experience, teachers have a chance to discuss how to apply what they learned to their teaching. In addition, teachers are able to see, through school visits, how the research-produced knowledge is taught in classrooms using biology pedagogy. When time is offered to discuss, evaluate, reflect, and form new comprehension, this professional development experience is transformed into practice in teachers' classrooms (Shulman, 1987).

Korean Science Teachers Professional Development

From the 1960s through the 1980s, professional development programs for Korean science teachers focused on general education theories and strategies. Since that time science teachers have wanted to participate in programs that focus on helping deepen their subject content knowledge and skills for teaching (Kwak, 2008). Science teachers especially hope to have opportunities to participate in overseas professional development programs. The overseas science teacher professional development program, a 4-week summer residential institute, has been implemented on a large scale thanks to the support of the Korean government. Since the 1980s, hundreds of Korean science teachers have visited Western countries to participate in overseas science teacher professional development programs. The programs provide an opportunity for the participants to experience the latest scientific content knowledge and science teaching pedagogies (Cho, Yager, Park, & Seo, 1997).

Illinois State University's Korean Science Teachers Professional Development (KSTPD) program has provided secondary Korean biology teachers with advanced biological science content knowledge, pedagogical knowledge of biological science, and academic and cultural experiences in the United States (Park, 2011). Figure 19.1 depicts a model of the Korean Science Teachers Professional Development (KSTPD) offered in 2010, the program used for this study.

The KSTPD program of 2010 consisted of (1) biological sciences including microbiology, cell/molecular biology, genetic biology, neuroscience, ecology, entomology, and immunology; (2) biology education and science education, i.e., pedagogical content knowledge (PCK), inquiry-based science teaching and learning, nature of science, assessment, standards, reforms; and, finally, (3) cultural experiences including visits to schools, community, science museums, cultural artifacts with symbolic cultural statues, music, food, and visiting homes. Most importantly, every session of the workshop program was formed with two critical components: (a) updated latest theories of each topic followed by (b) laboratory experiments or activities that the teacher participants could utilize in their classrooms right away.

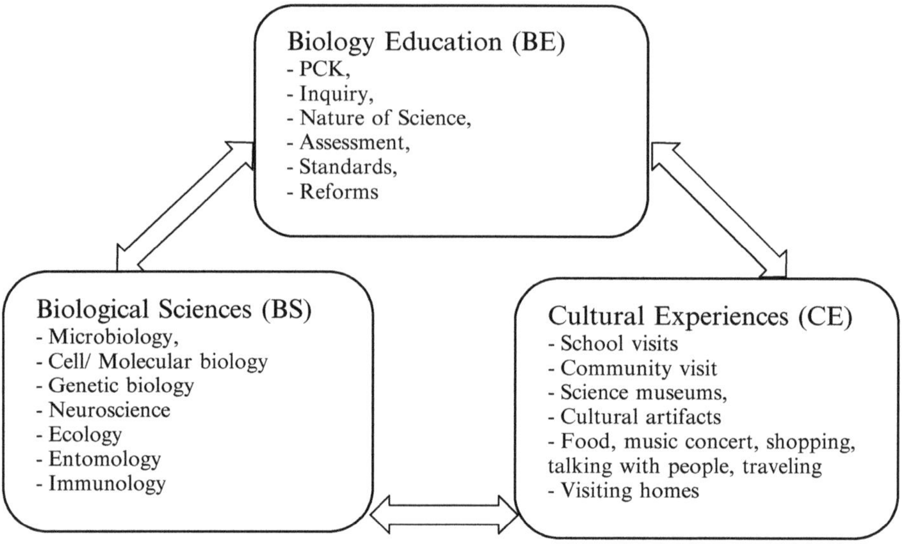

Fig. 19.1 A model of overseas professional development program for secondary biology teachers

The overseas professional development programs were effective when science teachers applied what they learned to their classes. For example, some Korean teachers who participated in the teacher-training programs of the United States applied in their classrooms a small-scale chemistry (SSC) program, which they learned from the KSTPD. The participants' students responded positively to the SSC lab program (Yoo, 2007). In addition, Cho et al. (1997) reported findings of a study in which teachers taught 5 days in their own classes and continued to refine their modules of constructivist teaching and learning through a bimonthly meeting with peers after they returned to Korea. They found that "the effect of the workshop was significant and that the learning was retained" (Cho et al., 1997, p. 400). However, few studies have looked at how the teachers develop, in terms of pedagogical reasoning and reflection, what Shulman called "good teaching."

Pedagogical Reasoning

Pedagogical reasoning is an attempt to illustrate reflective practice during the teaching process. With pedagogical reasoning, a teacher must act in order to transform his/her comprehension of subject matter into "forms that are pedagogically powerful and yet adaptive to the variations in ability and background presented by the students" (Shulman, 1987, p. 15). Shulman's model of pedagogical reasoning and action suggests that a teacher go through a cycle of pedagogical activities for good teaching: comprehension, transformation, instruction,

Table 19.1 A model of pedagogical reasoning and action (Shulman, 1987, p. 15)

Comprehension
Of purpose, subject matter structures, ideas within and outside the principle
Transformation
Preparation: critical interpretation and analysis of texts, structuring and segmenting, development of curricular repertoire, and clarification of purposes
Representation: use of a representational repertoire which includes analogies, metaphors, examples, demonstrations, explanations, and so forth
Selection: choice from among an instructional repertoire which includes models of teaching, organizing, managing, and arranging
Adaptation and tailoring to student characteristics: consideration of conceptions, preconceptions, misconceptions, and difficulties, language, culture, and motivations, social class, gender, age, ability, aptitude, interests, self-concepts, and attention
Instruction
Management, presentations, interactions, group work, discipline, humor, questioning, and other aspects of active teaching, discovery or inquiry instruction, and observable forms of classroom teaching
Evaluation
Checking for student understanding during interactive teaching
Testing student understanding at the end of lessons or units
Evaluating one's own performance and adjusting for experience
Reflection
Reviewing, reconstructing, reenacting, and critically analyzing one's own and the class's performance and grounding explanations in evidence
New comprehension
Of purpose, subject matter, students, teaching, and self
Consolidation of new understandings and learning from experience

evaluation, reflection, and new comprehension (Shulman, p. 15). Table 19.1 indicates a cycle of pedagogical reasoning and action.

In this study, we adopted Shulman's pedagogical reasoning model and only focused on "comprehension and transformation." This was because, during this study, teachers had limited opportunities to fully teach students by using what they learned from the program. This is, in part, because some of the contents and approaches of the KSTPD programs did not fit exactly the situations and conditions of Korean classrooms (Cho et al., 1997). Pedagogical reasoning provides an opportunity, through context-based teaching, for teachers to improve their practices as well as to develop their pedagogical content knowledge. During the professional development program, teachers can experience comprehension and transformation in their pedagogical reasoning process (Ward & McCotter, 2004). In Shulman's (1987) model, the essence of the act of pedagogical reasoning is the process of transformation "wherein one moves from personal comprehension to preparing for the comprehension of others" (p. 16). For this study, we developed an operation definition of each based on Shulman's model. The following operational descriptions of comprehension and transformation guided us in conducting research with this study.

- Comprehension: *Teachers need to comprehend the purpose and focus of the subject matter. They also need to develop the subject matter knowledge in as many ways as possible to accommodate a variety of student understandings.*
- Transformation: *Teachers' knowledge of subject matter needs to be transformed in ways that are pedagogically powerful and yet adaptive to the variations in students' abilities and backgrounds.*

Reflective Journal

An effective way of providing meaningful reflection is the writing of a reflective journal. It helps teachers to critically reflect on their teaching practices and pedagogical reasoning (Boud, 2001; Ward & McCotter, 2004). In Ward et al.'s (2004) study, preservice teachers were asked to write their reflections on student learning through two research projects. Based on the analysis of participants' reflections, a rubric was developed to demonstrate the dimensions and quality of reflection in four different levels including Routine, Dialogic, Technical, and Transformative. These were adopted as a framework for our data analysis. These levels are likely to be revealed in teachers' teaching practices as they reflect on and evaluate what is presented with their knowledge, conceptions, and experiences in class. Boud also suggested that journal writing was an effective way to enhance pedagogical reasoning.

In fact, a number of studies have illustrated the benefits of a reflective journal. Harland and Myhill (1997) claimed that the use of a reflective journal allows students to take ownership of the learning process. The act of reflective journal writing helps to make explicit the interconnections among school experience, university seminars, background reading, and students' own wider experiences. Through reflective journal writing, one is able to clarify confusing or difficult experiences and ideas and begin to develop one's own philosophy for the teaching of a subject (Harland & Myhill). Demet (2009) explored types of reflection in the reflective journals written by 62 preservice teachers. The findings of the study indicated that the preservice teachers developed a positive attitude toward journal writing and planned to use it in their future teaching. Black and William (1998) suggested that it be used as an instructional tool to assess student learning as a form of self-assessment.

Several studies have claimed that a reflective journal could be used as an effective way to develop reflective practice (Morrison, 1996) and improve knowledge (Cantrell, Fusaro, & Dougherty, 2000) and learning (Dart, Boulton-Lewis, Brownlee, & McCrindle, 1998). Dart et al. (1998) reported that participants' insights presented in reflective journals became more profound over time along with an improved quality of thinking and reflection as well as their influence on practice. Sparks-Langer and Colton (1991) synthesized research on teachers' reflective thinking and concluded that "critical reflection may be promoted through close examination of cases that illustrate particular aspects of context, pedagogy, content, ... of teaching and learning that will help teachers develop a rich, flexible, repertoire of ideas, attitudes, and skills" (p. 43). As teachers critically reflect the context and pedagogy of their teaching, they are moved to reframe an image of teaching that helps them better understand the complexities of teaching.

Research Question

The purpose of this study is to enhance the pedagogical reasoning of teachers going through the professional development program. Specifically, it aims to develop a strategy of reflective journal writing and to explore the relationship between journal writing and pedagogical reasoning. The following research question guided this study:

> *How does the reflective practice of reflective journal writing affect the development of Korean secondary biology teachers' pedagogical reasoning in professional development program?*

Methodology

Research Design and Sampling: This study is a case study that investigates how reflective writing helps teachers develop their pedagogical reasoning. A case study is particularly useful when the research answers how and why questions (Yin, 2014). The study utilized a qualitative methodology with selective and purposeful sampling design. Patton (2002) stated that "qualitative inquiry typically focuses in depth on relatively small samples, even single cases ($N = 1$), selected purposefully" (p. 230). Of the 20 secondary biology teachers (16 females and 4 males with 8–20 years of teaching experience) who went through the program, two female volunteer teachers ($N = 2$; Eunmi Chang and Sumi Lee—both pseudonyms) were purposefully selected for this case study. Eunmi had 15 years of high school teaching experience and Sumi had 10 years of middle school teaching experience. The two volunteer teachers were highly interested in applying the contents and approaches of the program to their classrooms though they felt it to be a burden because some of the contents and approaches of the program did not readily fit the conditions of Korean science classes.

Data Collection and Analysis: Data were collected from the following three sources: a pedagogical reasoning practice worksheet (see Appendix A), a post-survey (see Appendix B), and a lesson plan. The worksheet and the post-survey were completed during the program in the United States. The lesson plan was written in Korea after participants returned to their classrooms. Per the pedagogical reasoning practice worksheet (reflective journal), at the first meeting of the Korean Science Teachers Professional Development (KSTPD) program in the United States, we explained the purpose of the worksheet. The worksheet was completed during the 4 weeks of the KSTPD program. Two participating teachers wrote a pedagogical reasoning practice worksheet four times (once a week for 1 month) altogether, thinking of what and how they planned to apply what they had learned. As participants completed the worksheets, we uploaded them to a web board (removing all identifiers so that everyone involved in this study could read for interactive comments). The post-survey questionnaire responses were administered and uploaded at the end of the program. In a case study, what are particularly useful are analytic techniques of pattern-matching, explanation-building, and time-series analysis (Yin, 2014). The journal worksheet was analyzed by two authors of this paper to see how the participants developed their pedagogical reasoning reflection. One author read teachers' reflective journals each week using an open code looking it over to find/match patterns and thematic ideas from the first reading (Merriam, 1998). The second author interpreted and explained the patterns and then categorized them into either Focus or Inquiry or Change. As the first author read the journal over, up to fourth one, each journal was analyzed in a time series. Once the person finished reading each journal, the next person repeated the same process. When the readers found thematic ideas through multiple times of readings of a journal, they then analyzed them as being one of four levels including Routine, Technical, Dialogic, and Transformative by using the reflection rubric (see Appendix C). We as the readers sat down and discussed each reflective journal one by one to justify and finally determine the level and category (Lincoln & Guba, 2000). When we were in disagreement on the rubric-level decision, we discussed it until we closed the gap in between. Regarding the post-survey questionnaire responses, they were collected at the end of the program as conclusive data. They were analyzed by coding them on Excel to see if there was any insight and thought in the participants' beliefs and behaviors, which helped us thoroughly interpret teachers' development of pedagogical reasoning. Concerning the lesson plans, participants were asked to write a lesson plan during 4 weeks of study in Korea. As participants wrote the lesson plans, they sent them to us via email and we then uploaded them to a web board (see Fig. 19.2) after having removed all identifiers for interactive comments. The lesson plans were read by two authors in this study to see if there were any thoughts and ideas that reinforced or contradicted what teachers had written in the reflective journals and survey.

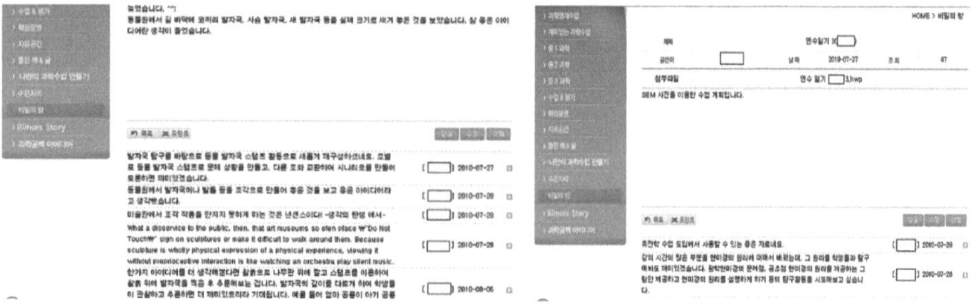

Fig. 19.2 Sample web board of participants

In qualitative research, validity and reliability are often a concern mainly because of the researcher's interpretation of the data. However, we as researchers inevitably seek to construct how teachers develop their pedagogical reasoning through reconstructions and interpretations of their experiences. In other words, we seek to construct "what these objects, events, and behaviors *mean* to the people engaged in" (Maxwell, 1992, p. 288). We used member checking to secure the two parameters of research credibility. Lincoln and Guba (1985) argued that "member checking is directed at a judgment of overall credibility" (p. 315). Member checking is "the most crucial technique for establishing credibility" (Lincoln & Guba, p. 314). In our study, one science teacher selected among the participants and one external science educator served as members for checking. The two members checked our findings to ensure that we captured the true meaning of the teachers' experiences. In addition, they read the responses and the thematic ideas to see if they matched. When two members disagreed on an item of our findings, they had an extended discussion and justified a final decision based on the big theme and time series of the participants' development of pedagogical reasoning demonstrated in three data sets. They then "may be able to agree that reconstructions are fair even if they are not in total agreement with them" (Lincoln & Guba, p. 315). This process helps us reassure the accuracy of the participants' constructions of pedagogical reasoning.

The Context of this Study: For over 4 years, Illinois State University has provided secondary biology teachers the Korean Science Teachers Professional Development (KSTPD) program. Each year, the KSTPD program offered the program with a slightly different focus, trying to meet the teachers' needs and reform goals of Korean science education. Such education was typically claimed to be lacking inquiry-based science and the nature of science due to an overemphasis on the national college entrance exam in secondary schools. This particular study was conducted in 2010; the program theme was "Inquiry-based Teaching and Learning in Biology Education" which emphasized the increase of teachers' subject matter knowledge and pedagogical skills. Secondary Korean biology teachers came to the United States to learn up-to-date biological science knowledge and how it is taught in the real context of US schools. Their learning is situated in particular US physical and social contexts and is social in nature as they engage in rich discourse, thinking, interacting with local school teachers and students, and sharing experiences during the program activities (Putnam & Borko, 2000). The KSTPD program invited prominent scholars of biological sciences and science education to the program. During the program, the Korean staff helped translate English presentations and discussions. However, Korean teachers still struggled to understand unfamiliar learning tasks. For four full weeks, participants attended each session from 9 am to 3 pm weekdays. After 3 pm each day, the program staff offered, as fulfillment of the program components, afternoon programs that usually consisted of cultural/educational visits, e.g., museum, farmers' market, music concert, talking with community people, shopping, having a chance to try indigenous food, community festival, and the like. These cultural experiences helped enhance their understanding of the community where the schools are situated. A total of 13 full-day lectures

(9 content workshops and 4 biology educations) were offered for the entire period of the workshop (see Fig. 19.1 for major topics). Each lecture was structured with lectures in the morning, followed by laboratory activities in the afternoon so that the participants understood effectively what the theories were and how they were applied.

Pedagogical Reasoning Practice Worksheet and a Reflection Rubric: This study was interested in knowing how reflective journal writing was effective at developing teachers' pedagogical reasoning. However, participants were familiar with pedagogical reasoning. Thus, we developed a pedagogical reasoning practice worksheet. In fact, the worksheet was modified from what was suggested in Shulman's work (1987), using the key process of pedagogical reasoning and action. Participants had a chance to exercise, over the entire 4 weeks, practicing the pedagogical reasoning by filling out the worksheet. This training helped teachers to increase their confidence in using the worksheet before they actually implemented it into their own classrooms. This process of self-reflection journal writing was metacognitive in nature. Implementing the worksheet was a necessary step because it showed that teachers were using reflective practices for teacher professional development.

The participants wrote the pedagogical reasoning practice worksheet and uploaded it to a web board after they finished instructions using what they had learned from the program (see Fig. 19.1). The web board, which enabled participants to exchange ideas, was created. When the participants' worksheets were uploaded, everyone in this study was able to see and share their opinions on the worksheet, making comments and asking questions. We took the comments and responses to the questions and organized them into a spreadsheet (without identifiers) for the purpose of analysis. We adapted a reflection rubric from Ward and McCotter's work (2004) to analyze the worksheet (see Appendix C). Using the rubric, we modified it to fit the purpose of our study. Table 19.2 shows the extent to which teachers developed their pedagogical reasoning reflection in the areas of comprehension and transformation as a result of reflective writing. As shown in Table 19.2, we grouped Focus and Purpose under Comprehension based on similar concepts that asked, "What have you learned?" We also grouped Inquiry and Change under Transformation. Transformation was divided into five steps: preparation, representation, selection, adaptation, and tailoring. Inquiry included preparation, representation, and selection.

Table 19.2 Analysis scheme of teachers' reflective journal

	Comprehension	Transformation	
		Inquiry	Change
Topic	Focus, Purpose	(Does the teacher consider the various classroom environments?)	(Is the teacher able to actively use what they have learned and tailor it to their class?)
Journal	(What have you learned?)	Preparation, representation, selection	Adaptation, tailoring
Reflection level of journal			

Change included adaptation and tailoring. Table 19.2 presents the evaluation scheme of the reflective journal used in this case study.

This rubric includes four levels for evaluating the journal: Routine, Technical, Dialogic, and Transformative. Routine reflections tend to contain very definitive statements that revealed either a lack of curiosity or a lack of attention to complexity. Technical reflection can be instrumental, in that the reflection is used as a means to solve specific problems but does not question the nature of the problem itself. Dialogic reflection is an ongoing process. The term itself reveals discussion and consideration of the views of others. Transformative reflection considers a scope beyond that of the study, into the real world, and grasps important implications (Ward & McCotter, 2004).

Survey Questionnaire

The post-survey questionnaire was designed to probe the thoughts of participants at a deeper level. This activity helped up to gather more in-depth data to be able to better answer the research question. The listed questions focused on how the teachers felt about the journal-writing activity. The questions could elicit teachers' opinions on the effectiveness of pedagogical reasoning writing practice as a technique to improve professionalism. As conclusive data collection, we administered a post-survey questionnaire. This was done after the final practice worksheet had been administered. We transcribed the data and removed all identifiers and posted the results on our web board so that everyone in this study could have access and make comments and pose questions.

Lesson Plan

When participants went back to Korea, they made lesson plans based on what they had learned in the ISU's Korean Science Teachers Professional Development (KSTPD) program. The lesson plans could be good evidence to show adaptation and tailoring in the transformation stage of pedagogical reasoning. To explore that possibility, we constantly compared the participants' lesson plans with their reflective journals in terms of any consistency, changes, insights, and thoughts that helped us better answer the research question.

Results

To find common themes that answered the research question, we read, multiple times, the participants' reflective journals, lesson plans, and questionnaire. Some of the terms that best fit the common themes and patterns that emerged were *Routine*,

Technical, and *Dialogic*; these were adopted from a reflection rubric (Ward & McCotter, 2004). The results are presented in three sections. The reflective journal analysis of two participants is first presented, followed by the lesson plan analysis, and lastly presented with the questionnaire analysis. Combining the analysis of the three data sets helped triangulate, as validation, the qualitative research data of this study.

Pedagogical Reasoning Practice Worksheet (Reflective Journals)

We had two volunteer participants (Eunmi Chang and Sumi Lee) write reflective journals. Each participant kept four journals, and we analyzed them by using a reflective journal rubric.

Eunmi Chang: A High School Teacher's Reflection Journal Analysis

With 15 years of teaching experience in high school, she was interested in plant taxonomy and identification. Table 19.3 presents the evaluation results of the reflective journals.

Reflective journals in Focus and Purpose Stage: In this stage, two levels of reflection emerged including *Routine level and Technical level* (see Table 19.3). Routine reflections tended to contain definitive statements (Ward & McCotter,

Table 19.3 Analysis of Eunmi's reflective journal

	Comprehension	Transformation	
Topic Journal	Focus, purpose (What have you learned?)	Inquiry (Does the teacher consider the various classroom environments?), preparation, representation, selection	Change (Is the teacher able to actively use what they have learned and tailor it to their class?), adaptation, tailoring
Reflection level of Journal 1 (class observation)	Routine	Dialogic	Technical
Reflection level of Journal 2 (inquiry)	Technical	Dialogic	Technical
Reflection level of Journal 3 (genetics)	Technical	Dialogic	Dialogic/Transformative
Reflection level of Journal 4 (ecology)	Technical	Transformative	Dialogic/Transformative

2004). This was indicated in Journal 1, "I observed the class about the theory of continental drift." Technical reflection can best be thought of as instrumental in the reflection (Ward & McCotter). Technical reflection was continuously presented in Journals 2, 3, and 4, "From the workshop, I learned that strawberries and bananas are also good materials for extracting DNA in a classroom experiment" (Journal 2); "I learned how to better teach genetics through student centered learning activities" (Journal 3); and "One instructor introduced how to teach genetics through student centered learning activities" (Journal 4). Therefore, Eunmi's reflection demonstrated a stage of Focus and Purpose in her development of pedagogical reasoning by demonstrating definitive statements (Routine) and instrumental reflection (Technical) about what she learned from the professional development.

Reflective Journals in Inquiry Stage: This stage showed two levels of reflection including *Dialogic level and Transformative level* (see Table 19.3). Dialogic reflection is "best thought of as ongoing process. The term itself connotes discussion and consideration of the views of others" (Ward & McCotter, 2004, p. 252). Dialogic reflection was shown in Journals 1, 2, and 3, "I am going to try to develop my own peer evaluation strategies in large classes, then I will try to find examples of strategies used by other teachers" (Journal 1); "I need to find a new way to extract DNA without other materials. After the workshop, I found that using methylene blue dye will show the DNA" (Journal 2); "In my lesson planning, I am considering students who either don't need much science knowledge or need a lot of experience based on the career they want to pursue" (Journal 3); and "It is very difficult for students to understand genotypes and Mendel's Law of Dominance because they only solve the problem on paper not doing activity. I want to find inquiry activities in genetics and I would like to make student centered lesson plan from them" (Journal 3).

On the other hand, transformative reflection questions more deeply fundamental assumptions and purpose (Ward & McCotter, 2004). For example, she wrote, "I taught the quality of water with BOD, COD and DO, and I didn't explain it in the perspective of diversity of living things. I already knew how to measure the quality of water by using the equation, but national curriculum doesn't include the equation, so I didn't teach it. It is very difficult for students to understand the importance of variation in nature with only one definition or method. I would like to use the method with my class to emphasize the importance of variation of living things. When I have a field trip with students, I will give them activities that use water insects and ask them to measure the level of variation. From these activities, students will be able to understand the importance of variation" (Journal 4). Overall, in the Inquiry stage, Eunmi showed her inquiry about what she had learned in the program in terms of how to utilize it in her own teaching.

Reflective Journals in Change Stage: In Change stage, there were three levels of reflection demonstrated in Eunmi's reflective writings: *Technical level*, *Dialogic level*, and *Transformative level* (see Table 19.3). For the Technical level of reflection, what is characteristic of the technical qualities of the Change dimension is narrow change or change without new insight. The following journal quotes demonstrated this point: "From the workshop, I learned peer evaluation and I will apply the idea to my wrap-up class" (Journal 1). Dialogic reflection considers and synthesizes new ideas. For example, the participant wrote, "I will assign a group project that requires the students to find a method to extract DNA from substances" (Journal 2); "When I cannot find a fruit fly, I will use an electro-microscopic diagram that shows the eye color and shape" (Journal 3); "I will give a set of data to each group, and they will analyze it. Then, they can discuss their findings with other groups" (Journal 3); "I would like to make lesson plan for the advance students in science. In this activity, students will create their own new species of fruit fly offspring using Mendel's law" (Journal 4); and "It is difficult for Korean high school students to participate in outside research. Instead, I assign worksheets that give different scenarios like clean water, slightly polluted water and heavily polluted water" (Journal 4).

Transformative reflection "questions fundamental assumptions and purpose more deeply" (Ward & McCotter, 2004, p. 253). Eunmi became more conscientious about how to change her own teaching toward the end of the 4-week-long program evident as follows, "I will give my students the opportunity to think about meiosis. In the daughter cell, you have unique DNA information, therefore you are special and unique. You have to be appreciated and you and your classmates have to respect each other. From the meiosis study, students can apply the knowledge to the real world and they understand nature and how to see each person as unique. I think this aspect of uniqueness in the meiosis study is essentially important to my students who need always to be competitive in class and Korean society. They need to learn how to help and respect each other to make a better society" (Journal 3); "When students catch water insects and identify them, it promotes students' interest in nature. When I give a lesson about the diversity with pollution topic, students will be able to think about the relationship between human beings and nature" (Journal 4); and "I learned how to measure the quality of water with water insects. I would like to know the way to measure the quality of water with plants" (Journal 4). In sum, Eunmi's reflection presented a rich information about how to transform what she learned in the program into her own teaching as she reflected through narrow changes, synthesizing of new ideas and considering of Korean contexts in teaching, and proposing deeply thought ideas of teaching that fits in the Korean classroom contexts.

Sumi Lee: A Middle School Teacher's Reflection Journal Analysis

She had taught for 10 years in middle schools. She was interested in teaching inquiry-based science and gifted education. Table 19.4 shows the evaluation results of Sumi's reflective journals.

Reflective Journals in Focus and Purpose Stage: In this stage, four levels of reflection were identified in Sumi' reflective writings: *Routine*, *Technical*, *Dialogic*, and *Transformative* (see Table 19.4). Routine reflection is demonstrated in the following quotes, "When I visited ISU high school labs, I saw the lab setup and facilities. I also had a meeting with the science teachers and principle where we

Table 19.4 Analysis of Sumi's reflective journal

	Comprehension	Transformation	
Topic Journal	Focus, Purpose (What have you learned?)	Inquiry (Does the teacher consider the various classroom environments?), preparation, representation, selection	Change (Is the teacher able to actively use what they have learned and tailor it to their class?), adaptation, tailoring
Reflection level of Journal 1 (visiting a high school)	Routine/ Transformative	Dialogic	Technical
Reflection level of Journal 2 (field trip)	Routine	Technical	Technical
Reflection level of Journal 3 (microbiology)	Technical	Dialogic	Dialogic
Reflection level of Journal 4 (genetics)	Dialogic	Transformative	Transformative

discussed the curriculum and school life" (Journal 1) and "I saw 'Bean' at the Millennium Park" (Journal 2).

According to Ward et al. (2004), technical reflection can best be thought of as instrumental. It was evident in the following reflection, "The microscope can catch the clear images and collect clear images of the whole slide. Then we can put the images together to recreate the slide. The reason we can't see the whole slide clearly is because light illuminates only part of the slide at a time" (Journal 3). Dialogic reflection is best thought of as an ongoing process, well demonstrated in the following quote, "This program uses student-centered learning strategies." She promoted students' interest with the questions which are interesting to students. For example, "Why can some people drink more alcohol and others?" (Journal 4). As Ward et al. (2004) explained, transformative reflection questions fundamental assumptions and purpose more deeply. Sumi showed in Journal 1 that she reflected on how to transform her lab experience into her teaching, "When I visited the labs in a high school, it was very impressive to see the teacher's ideas and philosophy. For example, there were some posters about lab safety directions which students drew on their own. Those posters were very simple but they were able to present powerful messages to students" (Journal 1). Through four of Sumi's reflective writings, she demonstrated four levels of reflections using statements of her classroom observations, instrumental reflection, ongoing process of student learning, and change of ideas and philosophies into her own teaching.

Reflective Journals in Inquiry Stage: Three levels of reflections emerged here: *Technical*, *Dialogic*, and *Transformative*. Technical reflection can be seen in Sumi's passage, "It is very interesting that when I move to the left, the mirror shows the movement to the right side. That means the mirror reverses the sides. I was wondering about it. I stopped thinking because I was too tired" (Journal 2). As noted above, dialogic reflection is best thought of as an ongoing process. This level is evident in the following, "There are plastic bags with numbers on them. It is very useful for the teacher to organize and check the classroom materials. I would like to use this idea in my class" (Journal 1) and "When we use this electron microscope, we can see the image without coating the surface of the materials. But the materials are dried and contracted with this process, and it is not easy to see the correct images. I am wondering how to see the images perfectly when they are distorted" (Journal 3).

Transformative reflection more deeply questions fundamental assumptions and purpose, "He always smiles and makes difficult concepts clear and easy for the students. His classroom behavior encourages positive classroom participation. To be a good teacher, what are the prerequisites? I think the good teacher has two things. One is the teacher's positive attitude in the classroom; the other is the teacher's ability to teach difficult concepts to students easily" (Journal 4). Sumi seems to have put a lot of thought into what she observed through reflective writings over time. As seen in Journal 4, she acquired not only the strategies and approaches and new subject knowledge but also tried to transform them into a knowledge base and philosophy through transformative pedagogical reasoning.

Reflective Journals in the Change Stage: This stage of Sumi's pedagogical reasoning showed three levels of reflection: *Technical*, *Dialogic*, and *Transformative*. First, technical reflection showed how Sumi stated what she observed about the lab activity, "In fact, the height of the water is very shallow. I had thought the reason was because the ceiling of the cave was so high but it wasn't really high. Why does the water look shallower than it is? I am considering another reason why and I would like to make a new lesson plan based on the question and my problem solving process" (Journal 2). Second, dialogic reflection described how Sumi considered and synthesized new ideas about an electron microscope: "In the past, I taught the mutation of color in fruit flies. From the workshop, I have learned that using an electron microscope to examine the fly's eye shape is a good way to teach mutation. I will distribute one electron microscope picture to each group showing the fruit fly's eye shape" (Journal 3). Lastly, transformative reflection more deeply questions fundamental assumptions and purpose: "Today, we had a workshop on meiosis and daughter cells. Although I knew what meiosis is and taught it many years, I think it was very special because it updated my previous knowledge about meiosis and learned a new meaning of its process. Why is each individual different and unique? The fundamental answer is meiosis. The instructor suggested that we should give students the opportunity to think about meiosis as the creation of individual daughter cells. In the daughter cell, every single person has unique DNA information; therefore he/she is unique. I will teach meiosis in terms of why each individual becomes so unique in the entire universe – you are the being that no one resembles. Therefore, everyone in classrooms has to respect each other. In daily life, students judge their classmates based on money or social status. Based on what I learned about meiosis today, I will give my students an opportunity to think about meiosis and apply to the real world. I hope they then understand the diverse natural world and learn how to respect each person as unique being" (Journal 4).

As Eunmi and Sumi were writing their reflective journals over time, they noticeably advanced the level of their pedagogical reasoning from routine and technical to dialogic or transformative reasoning (see Tables 19.3 and 19.4). When teaching meiosis, neither teacher had ever thought about why each individual was so unique. After the workshop, Eunmi and Sumi realized that they could connect meiosis to real life, not confining the knowledge within the classroom but taking it out of the classroom boundary and using it to expand students' understanding about the real world. This is the purpose of science education that helps students become scientifically literate (Bybee, 1997; OECD, 2006). Therefore, it stands to reason that reflective journal writing can help participants increase their pedagogical reasoning ability.

Lesson Plans

When participants returned to the Republic of Korea, they wrote four lesson plans based on what they had learned in the Korean Science Teachers Professional Development (KSTPD) program. After they sent them to us, we uploaded them to a web board. The lesson plans can provide good evidence of how participants adapted to the reality of Korean classrooms in the transformation stage of pedagogical reasoning. To explore that possibility, as an example, we compared the participants' lesson plans with their reflective journals.

Eunmi—Meiosis Lesson Plan: Eunmi explained, in Journal 3, the relationship between the formation of a daughter cell and how this related to the dignity of life, "I will give my students the opportunity to think about meiosis. In the daughter cell, you have unique DNA information, therefore you are special and unique. You have to be appreciated and you and your classmates have to respect each other. From the meiosis study, students can apply the knowledge to the real world and they understand nature and how to see each person as unique. I think this aspect of uniqueness in the meiosis study is essentially important to my students who need always to be competitive in class and Korean society. They need to learn how to help and respect each other to make a better society" (Journal 3). From Eunmi's reflections on meiosis, she transformed what she learned about DNA and meiosis into her own lesson plan. The goals of the lesson were:

> Students are able to explain that daughter cells obtain different DNA information through meiosis.
>
> Students are able to explain how they are special and unique through the perspective of meiosis.

These goals of the lesson were appreciated and pedagogically reflected in the way that Eunmi understood the content of the workshop, "During meiosis, daughter cells have various combinations from the parental chromosomes. There are $2^{23} = 8.4 \times 10^6$ possible combinations. Every person has unique DNA information, therefore you are all special and unique. Everyone should be appreciated because of this diversity, and you and your classmates must respect each other." Eunmi next elaborated on how to teach meaningfully, "In meiosis, variation occurs because each gamete (either sperm or egg) contains a mixture of genes from two different parent chromosomes in sexual reproduction. In other words, the genetic coupling of non-identical DNA takes place in meiosis. It results in an offspring, which has the genetic material of two different individuals. Therefore, in meiosis, the characteristics of parent chromosomes are combined with the characteristics of offspring chromosomes, which ultimately results in a new and unique set of chromosomes." She then reflected on how to approach the students with questions to be asked, "Through the perspective of meiosis, are you special and unique? Do you and your classmates have to be appreciated and do we have to respect each other?" The reflection in this lesson demonstrated how Eunmi's reflective journal writing helped her to adapt and apply the subject knowledge in her class.

Sumi—Mutation Activity Lesson Plan: Sumi adapted a fruit fly mutation activity having performed one during the workshop of the KSTPD program. In fact, she applied the electron microscope pictures to her genetics class. The following reflection piece depicted how she applied it, "In the past, I taught the mutation of color in fruit flies. From the workshop, I have learned that using an electron microscope to examine the fly's eye shape is a good way to teach mutation. I will distribute one electron microscope picture to each group showing the fruit fly's eye shape" (Journal 3). An electron microscope, an example of the newest, sophisticated technology, was introduced for participants during the workshop to produce a three-dimensional image of a fruit fly's eye that helped thoroughly understand its shape. She was excited to use the technology because it was something that she had never tried in her teaching. So she was excited and thought deeply about how to pedagogically approach the students when teaching the topic of genetics. Many Korean biology teachers believed it was too difficult to use an electron microscope in their class. When she went back to Korea, however, she visited a research institute that had an electron microscope. She learned how to take a picture of fruit fly's eyes using the microscope and made use of them in her classroom activities. In Sumi's reflections about this lesson, we noticed that she meaningfully adapted a new technology and applied it to her class. In other words, she was aided in organizing her ideas and searching for an appropriate way to teach through pedagogical reasoning by writing a reflective journal.

Frequency of Thematic Languages: We read, multiple times, three data sets including reflective journals, post-surveys, and lesson plans. We were looking for thematic languages using a nonjudgmental, open code (Merriam, 1998). We then invited two experts to member-check our findings to confirm perceived accuracy and reactions (Lincoln & Guba, 1985). Thematic languages were chosen when the frequency (the number of appearance in two teachers' three data sets) had more than 5, thus meeting one of the fundamental assumptions of chi-square analysis.

Table 19.5 Frequency of thematic languages appeared in two teachers' reflective journals, post-survey, and lesson plans

Thematic languages (Eunmi + Sumi)	Observed frequency[a]
Application	21
Reinforcement	5
Change	15
Inquiry	12
Appropriateness	6
Student-centered instruction	5
Problem-solving/critical thinking	5
Subject matter knowledge	5
Pedagogical knowledge	5

[a]($p < 0.001$)

As shown in Table 19.5, nine thematic languages represent their perception and reflected outcome of pedagogical reasoning. Application, change, and inquiry are the most used languages as a theme in their pedagogical reasoning, which reflected significantly the objectives of the 2010 KSTPD program. Specifically, the program focused on inquiry-based teaching and learning in secondary biology in 2010. Many of the program sessions provided teachers with the scope of inquiry-based teaching with the understanding of the nature of science, subject matter knowledge, and classroom observations that included curriculum materials and lesson plans, teaching in class and lab, and assessment. The chi-square test demonstrated that there was a significant difference between the observed frequencies and the expected frequency for each word (χ^2 (8) = 32.77, $p = .000$). We set up the expected frequency with no differences in the expected number of times each language might appear in our three different data sets because our assumption was that it appears equally in any data during pedagogical reasoning.

Post-survey Questionnaire

The post-survey questionnaire was designed to probe teachers' thoughts and ideas at a deeper level. The questions focused on how the teachers felt about the activity. The questionnaire elicited participants' opinions on the effectiveness of pedagogical reasoning writing practice. To achieve this purpose, a couple of questions were selectively analyzed below.

Why did you participate in the reflection journal writing? Teachers' responses indicated that the reason they participated in the reflective writing strategy was to reinforce what they learned, to make their instruction more meaningful, and ultimately to apply what they learned to the Korean students. Therefore, two key themes were found: *reinforcing* and *applying*. Teachers wrote, "I would like to reinforce what I have learned"; "I want to write and reorganize the instruction so it is more applicable to my classroom and meaningful and memorable to myself"; and

"I think reflective journal writing is an innovative way to reflect on my class and my teaching skill." Basically the two volunteer teachers continually reflected how they could justify and apply what they had learned into their own classes after reinforcing their knowledge clearly and meaningfully.

Was "the reflective writing" enjoyable and helpful for you as a teacher? (describe three things specifically): By keeping a reflective journal, they were able to reflect more upon adapting ISU's program to Korean classrooms. Through a number of responses, key themes emerged: *remembering, applying, organizing,* and *thinking critically*. For example, two teachers responded, "By writing reflections, I was able to remember what I had learned, then come back to it later and analyze it"; "I developed the good habit of writing down ideas, so when I have participated in another program I will write down my reflection"; "I had a good opportunity to think about the lesson in various ways"; "Because I have written and analyzed the reflection journal, I would like to apply this information to my class"; "Writing a reflection journal helps to organize my ideas and makes them easier to apply"; "The journal writing is very valuable for me to think critically about the program"; and "Because I was always thinking about what I would write in my journal, I concentrated more in my classes." These responses may imply that reflective writing can serve as a tool to be utilized by teachers in remembering, applying, organizing, and thinking critically what they learned in the professional development program. The teachers also showed an in-depth level of thinking regarding pedagogical reasoning:

> Even though I had a positive experience in other professional development programs, I often struggled to apply the programs to my class when I came back to my class. But now I am considering the applications of what I am learning while writing in the reflection journal after the training sessions.
>
> I can consider specifically about how to deliver the instruction to a specific audience, the Korean students.

As shown in the above survey responses, reflective journal writing helped organize participants' ideas and make those ideas relevant to students and readily to apply. Reflective journal writing also gave a specific direction to what teachers needed to do to improve their class by comprehending or *reinforcing* (teachers' term) the subject matter knowledge and then transforming it into a way that students could understand better. As Shulman (1987) noted, a teacher must act to transform his/her comprehension of subject matter into "forms that are pedagogically powerful and yet adaptive to the variations in ability and background presented by the students" (p. 15). Two cases of teachers in this study showed that they came to think about what it meant to learn different kinds of materials for different purposes and how to decide which kinds of learning were most necessary in different learning contexts. Therefore, these reflective journal-writing activities helped to develop the comprehension and transformation that are an integral part to Shulman's theory of pedagogical reasoning.

Would you want to use the pedagogical reasoning practice worksheet in the future? (check which one applies: excellent, good, mediocre, poor, very poor):

Every participant responded "good," suggesting that the journal writing helped them to reflect, analyze, and apply what they had learned.

Discussion

This case study investigated how reflective writing, in professional development, enhances secondary teachers' pedagogical reasoning, a key aspect of "good teaching" (Shulman, 1987, p. 15). In addition to deep understanding of content knowledge, good teaching has a lot to do with teachers' knowledge about and skills at connecting ideas across subject areas to everyday life. This kind of understanding becomes a foundation of "pedagogical content knowledge" (Shulman). Teachers need to comprehend subject matter knowledge deeply as well as understand how to transform it into ideas understandable to students (Darling-Hammond, 1998). In the process of comprehension and transformation of teaching, pedagogical reasoning helps teachers enhance and develop their deep understanding of subject matter and their skills to teach. A number of studies on the positive effects of professional development reported an increased subject matter knowledge and a more fully developed teaching pedagogy (Birman et al., 2000; Cohen & Hill, 2000; Loucks-Horsley et al., 1998; Nelson & Hanegan, 2003). Following the professional development, however, teachers often struggled to apply what they had learned there to their own classroom. This struggle was due, in part, to the different contexts in which these teachers were trying to apply their newfound skills, including various learning environments, a wide demographic spectrum, and the cultural variations of learners.

In our study, the findings imply some possible explanations for the relationship between reflective journal writing and pedagogical reasoning development. The findings indicated that pedagogical reasoning using reflective journal writing during a professional development program was a good way to help them make their knowledge and approaches relevant to students' understanding. As the participants wrote their reflective journals, their level of pedagogical reasoning in the journals grew more sophisticated, from dialogic to transformative reasoning.

Many teachers do not automatically know the meaning of reflection. Reflection is a deliberate thinking about action with the intention to improve through the professional development program (Hatton & Smith, 1995). Teachers deliberately practiced pedagogical reasoning to improve their teaching through writing a reflective journal worksheet. The activity involved a reflective thinking process of mulling what they learned and why and how to teach it to students both relevantly and meaningfully (NRC, 1999). The worksheet included deliberate comprehension and transformation questions as a means for teachers to develop their pedagogical reasoning. The comprehension stage questions used in this study were, for example, "What have you learned from today's lectures?" "How did you teach prior to this lecture content?" "Why do you think the lecture was valuable?" In addition, transformation stage question included, "What do you want to change about your previous lesson?"

By answering these questions in the reflective journal, teachers were able to understand and practice pedagogical reasoning over time. According to our study, writing a reflective journal can be a useful tool in helping teachers to increase their pedagogical reasoning ability. For example, in the change stage of her reflective journal, Eunmi noted that her reflective thinking process helped clarify how to teach meiosis to her students, "I will give my students the opportunity to think about meiosis... From the meiosis study, students can apply the knowledge to the real world and they understand nature and how to see each person as unique. I think this aspect of uniqueness in the meiosis study is essentially important to my students who need always to be competitive in class and Korean society. They need to learn how to help and respect each other to make a better society" (Journal 3). Eunmi came to see that meiosis was a rather boring topic to teach. Through the KSTPD program, however, she not only learned its meaning more deeply but also reasoned pedagogically how to make the concept more accessible to students. She intended to lead the students to realize that "you would not be unique without meiosis and the probability of your uniqueness is $2^{23} = 8.4 \times 10^6$." Eunmi took it to the next level to discuss not only respecting each other but also contributing to making a better society by relevantly putting the concept in Korean contexts of students' competitive lives. The Journal 3 reflection piece above shows that pedagogical reasoning over time helped deepen the subject content knowledge and find a sophisticated and extended way of connecting it to students' lives.

Similarly, Sumi used the same topic in one of her reflective journals, "Why is each individual different and unique? ... In the daughter cell, every single person has unique DNA information; therefore he/she is unique. I will teach meiosis in terms of why each individual becomes unique in the entire universe – you are the being that no one resembles. Therefore, everyone in classrooms has to respect each other. In daily life, students judge their classmates based on money or social status. Based on what I learned about meiosis today, I will give my students an opportunity to think about meiosis and apply to the real world. I hope they then understand the diverse natural world and learn how to respect each person as unique being" (Journal 4). Sumi knew the concept well since she had taught it for many years. She had never, though, connected meiosis to the uniqueness of each person in the real world. Through the workshop, Sumi was able to integrate the meaning of meiosis with each individual's social life. She then hoped for students' deeper understanding of such a diverse natural world of life. Sumi has now integrated content knowledge with the real world and realized that such integration was critical.

Integration becomes a key to a successful reflection, Grossman (2008) argued, because an integration of content-based reflection, metacognitive reflection, and transformative reflection facilitates meaningful reflection. This type of teaching meets the purpose of science education that helps students become scientifically literate (Bybee, 1997; Koballa, Kemp, & Evans, 1997; Millar, 2006; OECD, 2006). In other words, teachers can transform their knowledge such that they apply it to daily life. They do this by questioning their own fundamental assumptions through multiple practices of pedagogical reasoning. This finding implies that, on one hand, pedagogical reasoning can be seen as a developmental process leading to more in-depth and elaborated reflection over time. This is similar to what Pultorak (1996) claimed—that reflection is a developing process. On the other hand, this finding implies that pedagogical reasoning can be seen as a form of reflection that is the analytical interaction between experiences and beliefs about teaching practices (Newell, 1996). By being situated physically and cognitively in the real world, two teachers in this study were able to reflect on their teaching practices and plan actions in teaching over time based on what they learned in the professional development. When teachers have the framework of reflection and action in teaching practice, it helps increase their reflective practice (Reiman, 1999).

The survey questionnaire focused on how the teachers felt about the reflective journal writing, eliciting their opinions on its effectiveness as a technique to improve their professional development. As the questionnaire responses showed, the participants designated the journal writing as "good" and wanted to use the reflective journal worksheet in the future. Participants affirmed a desire to use the reflective journal worksheet in any future program. Teachers were generally positive about the use of journaling in teacher education programs (Bain, Ballantyne, Packer, & Mills, 1999). Apparently, journal writing helps them reflect, analyze, and apply what they learn. Especially, reflection journal writing gives them specific direction on what they have to improve in their own class and their own teaching. They think about what it means to learn different kinds of material for different purposes and how to decide which kinds of learning, in different contexts, are most needed.

The participants' lesson plans offer evidence of their level within the transformation stage of pedagogical reasoning. Transformation is the culmination of pedagogical reasoning, teaching as deliberate reflection, and planning the act of teaching (Shulman, 1987). The lesson plans can reflect teachers' adaptation and tailoring in the transformation stage of pedagogical reasoning. The lesson plans demonstrate how reflection journal writing strongly helps, through the associated pedagogical reasoning, the teachers do their jobs. With the journal writing, the participants recalled what they had learned from the program and figured out what and how to apply that to their classroom. The participants transformed the knowledge into lesson plans. As they wrote, they considered the potential and limitations of it in their classroom context. This demonstrates how the reflective journal can be used at the Transformative level of pedagogical reasoning. In the reflective journal, as they plan lessons, teachers analyze classroom situations to make appropriate pedagogical instructional decisions. These decisions are key to effective instructional planning and teaching (Blackwell, 2008). In other words, writing a reflective journal helps organize a teacher's ideas and makes them easier to adapt and tailor to a certain unit of the lesson plan.

Following the overseas professional development program, participants struggled to apply what they had learned to their biology classes back home; the new context was too different. As Cho et al. (1997) addressed, one of the difficulties encountered in this type of overseas professional development program was unmatched contexts. They reported that "many times the content and strategies of the programs did not fit exactly the reality of Korean classrooms" (p. 404). However, learning is socioculturally context based. Lave and Wenger (1991) premised that "meaning, understanding, and learning are all defined relative to actional contexts" (p. 15). They also argued that "the mastery of knowledge and skill require newcomers to move forward full participation in the socio-cultural practices of a community" (p. 29). Korean teachers participated in the workshop to learn new approaches and contents of biological sciences in a socioculturally different context—the United States. They were motivated by a curiosity about such a new learning experience. Through the project of this study, teachers were asked to reflect and modify their learning into pedagogically reasonable contexts. When the teachers were able to see the potential applicability of what they had learned through the journal writing, our findings show that they could utilize it more effectively in their classrooms.

This study has limitations that include the limited sampling size and varied teaching experience of the participants. The small sample size threatens the external validity (generalizability) of the results of the study. This study only investigated the processes of pedagogical reasoning action that the participants made in comprehension and transformation. To understand the effect of the journal writing for pedagogical reasoning action, the further research should examine the whole process of the pedagogical reasoning, from comprehension to reflection to new comprehension (see Table 19.1). The pedagogical reasoning and action involves "a cycle" and "sequence" through the activities of comprehension, transformation, instruction, evaluation, reflection, and new comprehension (Shulman, 1987). In this study, we initially premised that the overseas professional development model is equally effective as a domestic one. Yet, we admit that this type of professional development model may be different from the domestic professional development model since it occurred in overseas countries in which a cultural factor plays a role in the effect of professional development program. A culture influences a teacher's teaching in one way or another. We assumed that 1 month of professional development in another country may fall short from acquiring full or comprehensive understanding of complexities of teaching. However, studies from the Organization for Economic Co-operation and Development's (OECD's) Trends in International Mathematics and Science Study (TIMSS) and Programme for International Student Assessment (PISA) promote learning from one culture to another or vice versa (Gonzales et al., 2004; OECD, 2012). It is thus hardly avoidable to learn about one another when comparing each country's education.

There are many teacher professional development programs that target the improvement of teachers' teaching and learning. Our findings show that one way that may help hit this target is to utilize the reflective journal writing for developing pedagogical reasoning along with lesson planning in the context of participants' classrooms. This implies that the reflective writing and pedagogical reasoning activity would improve teachers' teaching and learning in a professional development program especially when it occurs in another country's cultural setting.

Conclusion

Pedagogical reasoning is an important part of teachers' professional development programs. Through the practice of reflective journal writing, teachers come up with a repertoire of teaching ideas socially and culturally relevant to their students. At the same time, participants come to focus on their concerns and problems in teaching new approaches and contents in their class. These behaviors of teaching help teachers grow as a professional teacher.

As the literature showed, effective teaching can be achieved through appropriate professional development programs. As evidenced in this research, participants are situated in different sociocultural contexts as they participated in an overseas professional development program. The effect of professional development is achieved when meeting the needs of individual teachers who are in socioculturally different contexts. This study finds a niche for the processional development of secondary biology teachers. Indeed, the teachers with the journal-writing practice during and after the program significantly improved their pedagogical reasoning skills. It is thus suggested that teachers use reflective journal-writing practice during and after a program when the program's goal pertains to pedagogical reasoning. At the same time, as Taylor, Fraser, and Fisher (1997) pointed out the importance of a sociocultural context for implementing teachers' learning, the teachers' teaching is constrained by a sociocultural context. As one way of making a sustainable effort, we implemented a flow-up task through this study so that the participants were provided a chance to reflect what they had learned and to plan a lesson socioculturally relevant to their students. Even if the teachers were confident and favored the reflective journal writing, they may not use it consistently when they teach the newly learned approaches and contents in their class. Therefore, sustainable efforts need to be invested to improve classroom conditions with need-based and consistent instructional programs so that the teachers are able to teach new approaches and contents that are socioculturally relevant and meaningful to their students.

Appendix A: Pedagogical Reasoning Practice Worksheet (Reflective Journal)

Name: ____________________________ Date: ______________________

- Today's lecture topic (or field trip topic):
- Instructor:

1. What have you learned from today's lectures?

   (Summarize the lecture and analyze/apply some of the ideas)

2. How did you teach prior to this lecture content?
3. What do you want to change about your previous lesson?
4. Why do you think the lecture was valuable?
5. My own lesson plan (How can you incorporate today's lecture ideas into your class?)

    (1) Grade Level: ( ), Topic: ( )
    (2) Activities and Instructions
    (3) Expectations of Students

6. Questions or Comments

Appendix B. A Post-survey

Your School Level (Middle School, High School), Years of Teaching: ( ) Years

1. Why did you participate in this reflective journal writing?
2. Was "the reflective journal writing" enjoyable and helpful to you?

    (Describe three things specifically.)

3. Which journal entry would you use in your class and why?
4. Would you want to use the pedagogical reasoning practice worksheet in the future and why?

    Also, check one that best represents your perception as to the effectiveness of it.
    Excellent – Good – Mediocre – Poor – Very Poor

5. What was most difficult when you wrote the reflective journal?
6. Suggestions that you could improve the reflection journal writing.

Appendix C. Reflection Rubric

Reflection rubric[a]

	Level			
	Routine	Technical	Dialogic	Transformative
	Self-disengaged from change	Instrumental response to specific situations without changing perspective	Inquiry part of a process involving cycles of situated questions and action, consideration for others' perspectives, new insights	Fundamental questions and change
Focus (what is the focus of concerns about practice?)	Focus is on self-centered concerns (how does this affect me?) or on issues that do not involve a personal stake. Primary concerns may include control of students, time and workload, gaining recognition for personal success (including grades), and avoiding blame for failure	Focus is on specific teaching tasks such as planning and management but does not consider connections between teaching issues. It uses assessment and observations to mark success or failure without evaluating specific qualities of student learning for formative purposes	Focus is on students Uses assessment and interactions with students to interpret how or in what ways students are learning in order to help them Especially concerned with struggling students	Focus is on personal involvement with fundamental pedagogical, ethical, moral, cultural, or historical concerns and how these impact students and others
Inquiry (what is the process of inquiry?)	Questions about needed personal change are not asked or implied, often not acknowledging problems or blaming problems on others or limited time and resources. Critical questions and analysis are limited to critique of others. Analysis tends to be definitive and generalized	Questions are asked by oneself about specific situations or are implied by frustration, unexpected results, exciting results, or analysis that indicates the issue is complex Stops asking questions after initial problem is addressed	Situated questions lead to new questions Questions are asked with others, with open consideration of new ideas. It seeks the perspectives of students, peers, and others	Long-term ongoing inquiry including engagement with model mentors, critical friends, critical texts, students, careful examination of critical incidents, and student learning. It asks hard questions that challenge personally held assumptions
Change (how does inquiry change practice and perspective?)	Analysis of practice without personal response— as if analysis is done for its own sake or as if there is a distance between self and the situation	Personally responds to a situation but does not use the situation to change perspective	Synthesizes situated inquiry to develop new insights about teaching or learners or about personal teaching strengths and weaknesses leading to improvement of practice	A transformative reframing of perspective leading to fundamental change of practice

[a]Adopted from the work of Ward and McCotter (2004)

References

Bain, J. D., Ballantyne, R., Packer, J., & Mills, C. (1999). Using journal writing to enhance student teachers' reflectivity during field experience placements. *Teachers and Teaching: Theory and Practice, 5*(1), 51–73.

Birman, B. F., Desimone, L., Porter, A. C., & Garet, M. S. (2000). Designing professional development that works. *Educational Leadership, 57*, 28–33.

Black, P., & William, D. (1998). Assessment and classroom learning. *Assessment in Education, 5*(1), 7–74.

Blackwell, S. (2008). The effect of concept mapping on preservice teachers' reflective practices when making pedagogical decisions. *The Journal of Effective Teaching, 8*(2), 77–93.

Boud, D. (2001). Using journal writing to enhance reflective practice. In L. M. English & M. A. Gillen (Eds.), *Promoting journal writing in adult education* (New directions in adult and continuing education no. 90, pp. 9–18). San Francisco: Jossey-Bass.

Bybee, R. (1997). *Achieving scientific literacy: From purposes to practices*. Portsmouth, NH: Heinemann.

Cantrell, R. J., Fusaro, J. A., & Dougherty, E. A. (2000). Exploring the effectiveness of journal writing on learning social studies: A comparative study. *Reading Psychology, 21*(1), 1–11.

Cho, J., Yager, R. E., Park, D., & Seo, H. (1997). Changes in high schools teachers' constructivist philosophies. *School Science and Mathematics, 97*, 400–405.

Cohen, D. K., & Hill, H. C. (2000). Instructional policy and classroom performance: The mathematics reform in California. *Teachers College Record, 102*, 294–343.

Dart, B. C., Boulton-Lewis, G. M., Brownlee, J. M., & McCrindle, A. R. (1998). Change in knowledge of learning and teaching through journal writing. *Research Papers in Education, 13*, 291–318.

Darling-Hammond, L. (1998). Teacher learning that supports student learning. *Educational Leadership, 55*(5), 6–11.

Demet, Y. (2009). Reflective practices of preservice teachers in a listening skill course in an ELT department. *Procedia – Social and Behavioral Sciences, 1*, 1820–1824.

Gonzales, P., Guzman, J., Partelow, L., Pahlke, E., Jocelyn, L., Mak, K., Kastberg, D., & Williams, T. (2004). *Highlights from the Trends in International Mathematics and Science Study (TIMSS) 2003 (NCES 2005–005)*. Washington, DC: U.S. Department of Education/ National Center for Education Statistics.

Grossman, R. (2008). Structures for facilitating student reflection. *College Teaching, 57*, 15–22.

Harland, F., & Myhill, D. (1997). The use of reflective journals in initial teacher training. *English in Education, 31*(1), 4–11.

Hatton, N., & Smith, D. (1995). Reflection in teacher education: Towards definition. *Teaching and Teacher Education, 11*(1), 33–49.

Jeanpierre, B., Oberhauser, K., & Freeman, C. (2005). Characteristics of professional development that effect change in secondary science teachers' classroom practices. *Journal of Research in Science Teaching, 42*(6), 668–690.

Koballa, T., Kemp, A., & Evans, R. (1997). The spectrum of scientific literacy. *The Science Teacher, 64*(7), 27–31.

Kwak, Y. (2008). Research on characteristics of teachers' professionalism by the type of science pedagogical content knowledge. *Journal of the Korean Association for Research in Science Education, 28*(6), 592–602.

Lave, J., & Wenger, E. (1991). *Situated learning: Legitimate peripheral participation*. Cambridge, UK: Cambridge University Press.

Lincoln, Y. S., & Guba, E. G. (1985). *Naturalistic inquiry*. Beverly Hill, CA: Sage.

Lincoln, Y. S., & Guba, E. G. (2000). Paradigmatic controversies, contradictions, and emerging confluences. In N. Denzin & Y. Lincoln (Eds.), *Handbook of qualitative research* (2nd ed., pp. 163–188). Thousand Oaks, CA: Sage.

Loucks-Horsley, S., Hewson, P. W., Love, N., & Stiles, K. E. (1998). *Designing professional development for teachers of science and mathematics*. Thousand Oaks, CA: Corwin Press.

Maxwell, J. (1992). Understanding and validity in qualitative research. *Harvard Educational Review, 62*(3), 279–300.
Merriam, S. B. (1998). *Qualitative research and case study applications in education.* San Francisco: Jossey-Bass Publishers.
Millar, R. (2006). Twenty first century science: Insights from the design and implementation of a scientific literacy approach in school science. *International Journal of Science Education, 28* (13), 1499–1521.
Morrison, K. (1996). Developing reflective practice in higher degree students through a learning journal. *Studies in Higher Education, 21*(3), 317–332.
National Research Council. (1999). *How people learn: Brain, mind, experience, and school.* In J. D. Bransford, A. L. Brown, & R. R. Cocking and Committee on Developments in the Science of Learning (Eds.). Washington, DC: National Academy Press.
Nelson, C. R., & Hanegan, N. L. (2003). *Professional development for secondary science teachers in a contextual setting.* Paper presented at the annual meeting of the National Association for Research in Science Teaching, Philadelphia, PA.
Newell, S. T. (1996). Practical inquiry: Collaboration and reflection in teacher education reform. *Teaching and Teacher Education, 12*(6), 567–576.
OECD. (2006). *Assessing scientific, reading and mathematical literacy: A framework for PISA 2006.* Paris: OECD.
OECD. (2009). *Creating effective teaching and learning environments: First results from TALIS* (pp. 48–86). Paris: OECD.
OECD. (2012). *PISA results in focus: What 15-year-olds know and what they can do with what they know.* Paris: OECD.
Park, D. (2011). *An overseas professional development of inservice secondary biology teachers using a cognitive apprenticeship learning model.* Paper presented at the 55th annual conference of the Comparative and International Education Society (CIES), Montreal, QC.
Park, D., & Han, J. (2012). *Professional development of secondary biology teachers in an overseas country.* Paper presented at the annual meeting of National Association of Research in Science Teaching (NARST), Indianapolis, IN.
Patton, M. (2002). *Qualitative research and evaluation methods* (3rd ed.). Thousand Oaks, CA: Sage.
Pultorak, E. (1996). Following the developmental process of reflection in novice teachers: Three years of investigation. *Journal of Teacher Education, 44*(4), 288–295.
Putnam, R. T., & Borko, H. (2000). What do new views of knowledge and thinking have to say about research on teacher learning? *Educational Researcher, 29*(1), 4–15.
Reiman, A. J. (1999). The evolution of the social role taking and guided reflection framework in teacher education: Recent theory and quantitative synthesis of research. *Teaching and Teacher Education, 15*, 597–612.
Shulman, L. S. (1987). Knowledge and teaching: Foundations of the new reform. *Harvard Educational Review, 57*, 1–22.
Sparks-Langer, G. M., & Colton, A. B. (1991). Synthesis of research on teachers' reflective thinking. *Educational Leadership, 48*(6), 37–44.
Taylor, P. C., Fraser, B. J., & Fisher, D. L. (1997). Monitoring constructivist classroom learning environments. *International Journal of Educational Research, 27*, 293–302.
Ward, J. R., & McCotter, S. S. (2004). Reflection as a visible outcome for preservice teachers. *Teaching and Teacher Education, 20*(3), 243–257.
Yin, R. K. (2014). *Case study research: Design and methods* (5th ed.). Thousand Oaks, CA: SAGE Publications.
Yoo, M. (2007). The effect of small-scale chemistry (SSC) lab program on science-majored student's academic self-efficacy and science-related affective domain in high school chemistry I classes. *Journal of the Korean Chemical Society, 51*(5), 433–435.

Part VII
Emerging Research in Science Education

Chapter 20
Designing Computer-Supported Knowledge Building to Promote Conceptual Change Among High-School Students in Hong Kong

Ivan C.K. Lam and Carol K.K. Chan

Introduction

There has been substantial research on students' scientific concepts and conceptual change since the 1970s (Duit, 2009; Duit & Treagust, 2012; Vosniadou, 2008). Learning science involves complex activity for the gradual transformation of learners' intuitive ideas, commonsense beliefs, or misconceptions into scientifically accepted conceptions. Contemporary science instruction emphasizes enculturation of practice and social construction of scientific knowledge through discursive interactions (Sadler, 2009; Scott, Asoko, & Leach, 2007; Vosniadou, 2012). Inquiry has also been identified as a promising instructional approach to facilitate conceptual understanding and support conceptual change in science that involves formulating and evaluating explanations of natural and scientific phenomena (Chinn, Duncan, Dianovsky, & Rinehart, 2013; Minner, Levy, & Century, 2010; Vosniadou & Mason, 2012).

Despite the fact that inquiry has attracted increasing attention in science classrooms, inquiry is often critiqued as simple inquiry tasks involving highly structured scientific investigations, predetermined research procedures of experiments, and classroom activities focusing on scientific skills rather than creating knowledge and scientific explanations, which is the goal of authentic scientific inquiry (Anderson, 2007; Chinn & Malhotra, 2002). In recent years, many technology-enhanced learning environments involving model building, simulations, and online discussion platforms have developed to support collaborative inquiry learning that in turn facilitates conceptual change in science (Bell et al., 2007; Hennessy, 2006; Lee &

I.C.K. Lam (✉)
Maryknoll Convent School, Hong Kong, China
e-mail: ivanlmhk@gmail.com

C.K.K. Chan
Faculty of Education, The University of Hong Kong, Hong Kong, China
e-mail: ckkchan@hku.hk

M.S. Khine (ed.), *Science Education in East Asia*,
DOI 10.1007/978-3-319-16390-1_20

Jonassen, 2013). "Knowledge building" is an epistemological and educational model, first developed in the 1980s (Scardamalia & Bereiter, 1994), spanning over two decades (Scardamalia & Bereiter, 2006), and a specific emphasis has been placed on students engaging in authentic inquiry and knowledge creation similar to how community of scientists work together to advance knowledge (Chan, 2013; Scardamalia & Bereiter, 2006). Knowledge-building environment is mediated with the technology of Knowledge Forum, a web-based collaborative learning database that supports students to generate own research questions and co-construct explanations on scientific authentic problems through shared discourse (Lee, Chan, & van Aalst, 2006; Zhang, Scardamalia, Lamon, Messina, & Reeve, 2007).

Though knowledge-building inquiry has been much developed for science learning in the last two decades (see review Chan, 2013; Scardamalia & Bereiter, 2006), there are fewer studies involving conceptual change learning. Moreover, most research has been done with the elementary and middle-school students in Western countries. Influenced by the Confucian-heritage cultures, students in Asian classrooms such as Hong Kong Chinese seem in general more willing to accept knowledge from the authority of teachers, less accustomed to engagement in classroom discourse, and they tend to be competitive focusing on examination. On the other hand, research on Chinese learners has shown that when provided with opportunities, Asian students often perceived to be passive can respond actively to constructivist design and instructional innovation (Chan & Rao, 2009; Watkins & Biggs, 2001). More research is needed to explore how the design of innovative learning environments supports knowledge-building discourse in Asian contexts.

This chapter aims to investigate the design and effects of collective knowledge-building environment supported with Knowledge Forum for conceptual change among high-school students in Hong Kong context. Specifically, we investigated how the design can facilitate inquiry-based and reflective thinking among Chinese learners via scientific discourse. We first review the conceptual change theory, knowledge-building model, and learning difficulties about electrochemistry. We then present the design of collaborative knowledge-building environment appropriated for Asian classrooms and report on the evaluation of the designed knowledge-building environment in fostering students' collaborative inquiry and conceptual change learning. We discuss how knowledge building emphasizing metacognition and socio-cognitive dynamics mediated by technology can facilitate students to develop conceptual change, and, finally, several key design principles for conceptual change via knowledge building are examined.

Conceptual Change

The development of students' ideas and explanations in science can be understood from the theories of conceptual change. Over the decades, there is still no consensus on how to explain the nature and process of conceptual change, and there lacks common agreement on well-defined principles or instructional pedagogies to employ in promoting conceptual change to classroom practices (Treagust & Duit, 2008; Vosniadou & Mason, 2012). Science education approaches to conceptual change have focused on various perspectives in their models or theories including epistemological/cognitive (Posner, Strike, Hewson, & Gertzog, 1982), ontological (Chi, Slotta, & De Leeuw, 1994), and social/affective (Pintrich, Marx, & Boyle, 1993) dimensions of learning. Following the conceptual change model by Posner et al. (1982), cognitive conflict is the common conceptual change approach in classroom intervention (Guzzetti, Snyder, Glass, & Gamas, 1993). Related pedagogy such as predict-observe-explain strategy has been developed to elicit cognitive conflict and for inducing conceptual change (White & Gunstone, 1992).

The role of classroom context and learner characteristics has been highlighted for conceptual change research (Sinatra & Mason, 2008). In particular, Dole and Sinatra (1998) describe how the likelihood of conceptual change is determined by the interactions among instructional content, dissatisfaction with existing ideas, motivation, social context, and prior knowledge. Low cognitive engagement describes "superficial, surface-level processing with little reflection," whereas high metacognitive engagement involves "deep processing, elaborative strategy use, and significant metacognitive reflection" leading to strong conceptual change (Dole & Sinatra, 1998, p. 121). Learner's metacognition has an important role in facilitating conceptual change (Georghiades, 2000; review by Zohar & Barzilai, 2013). Georghiades (2000) showed that the experimental group which received metacognitive instruction performed better than the control group in the conceptual change learning on electricity, writing self-reflection led to deeper understanding, and the new conceptions were more durable. As such, conceptual change is viewed as slow and gradual transformation of learners' initial ideas, commonsense beliefs, or misconceptions into scientifically accepted conceptions.

Sinatra and Pintrich (2003) proposed *intentional conceptual change* postulating that conceptual change is a process of intentional learning incorporating "goal-directed" and "learner-control" strategies. Vosniadou (2003) also argued that intentional learning may facilitate conceptual change through the process of monitoring learning, providing metaconceptual awareness – awareness of naïve beliefs and presupposition and understanding of changes in these beliefs (p. 402) – affording multiple representations of a problem, and acquiring more sophisticated epistemologies in science. Classroom social interaction such as peer discussions and teamwork to support self-regulated learning processes can promote conceptual change (Whitebread & Cardenas, 2012).

In the past decades, innovative instructional approaches in science have emphasized inquiry in the classrooms. Considerable research attention has been given to achieve conceptual change and understanding through collaborative inquiry learning (Bell, Urhahne, Schanze, & Ploetzner, 2010; Chinn et al., 2013). In a meta-analysis of research by Schroeder and coworkers (2007), the use of collaborative learning strategies and instructional technology strategies can facilitate student achievement in science. Integrating technology innovations into science classrooms has strong potential to support collaborative inquiry learning and conceptual change, but do not always guarantee success (Bell et al., 2007; Hennessy, 2006; Lee & Jonassen, 2013). Positive instructional effects result with well-designed pedagogical model and scaffolding strategies to support various phases of inquiry learning (Quintana et al., 2004). For example, students who set goals and reflected on their inquiry process highlighting metacognition in the Thinker Tools Inquiry Curriculum improved more in their reasoning and conceptual understanding (White & Frederiksen, 1998). The Web-based Inquiry Science Environment (WISE) supports students to generate and distinguish own ideas using evidence, reflect, and create explanations for conceptual change (Linn & Eylon, 2011).

In sum, research is still limited to explore the effectiveness of the inquiry-based metacognitive instruction supported with technology and scientific discourse in promoting conceptual change within the sociocultural classroom context particularly in Asian classrooms. Moreover, more research has been called for the development of design principles for inquiry instruction to foster conceptual change (Chinn et al., 2013).

Knowledge-Building Environment

Two decades of research on knowledge building, mediated by Knowledge Forum (KF), a computer-supported collaborative learning environment, emphasize knowledge building as a collective work of the community mediated by progressive problem-solving discourse (Scardamalia & Bereiter, 2006). Knowledge building is not merely a pedagogical model but an epistemological theory of intentional learning that explains goals of dynamics of knowledge creation in communities of scientists (Chan, 2013; Zhang, Scardamalia, Reeve, & Messina, 2009). Within the knowledge-building communities, students contribute problems and co-construct explanations for idea improvement on Knowledge Forum and make use of the emerging understandings derived from collaborative discourse to formulate new problems for sustained inquiry. This involves a high level of metacognitive engagement and epistemic agency for intentional conceptual change. Specifically, students generate their own questions, ideas, and theories on inquiry problems, generate different explanations, conduct experiments arising from discourse, synthesize and put ideas together, and make conclusions based on informed evidence. The discourse is supported by affordances of Knowledge Forum such as scaffolds (thinking prompts) of *I need to understand, my theory, new information, a better theory,* and *putting our knowledge together*. Students also take their agency to integrate the discourse from Knowledge Forum and offline classroom for collective understanding and monitor their own progress on the problem-oriented inquiry.

Various empirical studies have shown how knowledge building can promote scientific understanding and support metacognition through shared discourse (Chan, Lam, & Leung, 2012; van Aalst & Truong, 2011; Zhang et al., 2009), and most studies have been conducted with elementary and middle-school science students. To date, fewer studies have examined how students develop complex science knowledge in conceptual change learning (e.g., Chan, Burtis, & Bereiter, 1997; Turcotte, 2012; van Aalst, 2013b). Prior research has shown mixed findings, for example, Vosniadou and Kollias (2003) found that the sixth graders understood better how a hot water system works through the knowledge-building inquiry, but they did not show gains in knowledge about heat and temperature. As highlighted by Vosniadou and her colleagues (2001, 2003), the design of collaborative learning environments and instructional tools needs to incorporate student's prior knowledge, to provide meaningful experiences, to encourage metaconceptual awareness, and to develop intentional learning that increases students' motivation and metacognitive and self-regulation skills for promoting conceptual change. More recently, van Aalst (2013b) employed the epistemological resources perspective focusing on students' questions, prior knowledge, theories, and information to analyze the knowledge-building discourse among elementary Canadian students about heat and matter for conceptual change and has called for further knowledge-building research in the East Asian contexts.

Although much research has been done on knowledge building and scientific understanding, fewer reports have examined the role of knowledge-building processes on conceptual change. This chapter continues the line of inquiry designing knowledge building for conceptual change in chemistry among high-school students in Hong Kong. We enhanced the model employed in Western contexts using the design appropriated for Asian students including principles and augmented with reflective assessment (Chan, 2011; Lee et al., 2006). Despite the fact that assessment is widely recognized as part of the instructional process to scaffold student learning, relatively less research has been given to the alignment of learning and the assessment of individual and collective knowledge advances in the knowledge-building inquiry (van Aalst, 2013a; van Aalst & Chan, 2007; Zhao & Chan, 2014). Following this line of research, this study proposes to use an innovative design that highlights student's *explanatory-based inquiry* and *knowledge-building reflection journals* to scaffold the knowledge-building inquiry and assess the individual and collective scientific understanding from the Knowledge Forum discourse.

Learning and Teaching About Electrochemistry

Electrochemistry has often been reported to be one of the difficult topics to comprehend by students and teachers in high-school chemistry curriculum because it involves abstract concepts to explain the observable phenomena in two areas –

redox reactions and electrochemical cells (Barke, Hazari, & Yitbarek, 2009; de Jong & Treagust, 2002). The misconceptions of electrochemistry have been intensively studied in the literature (Barke et al., 2009; Garnett & Treagust, 1992a, 1992b; Sanger & Greenbowe, 1997a, 1997b). Common misconceptions in electrochemistry encountered by students include mutual dependence of oxidation and reduction, assignment of oxidation numbers, identifying anode and cathode in galvanic and electrolytic cells, flow of current in the electrolyte and the salt bridge, and formation of products at anode and cathode in the electrolysis.

Chemistry should be taught at three representation levels – macroscopic, microscopic, and symbolic – for deep understanding (de Jong & Taber, 2007). Students' misconceptions on electrochemistry are related to the poor interconnection among the three levels of knowledge. Macroscopically, students make observations on chemical changes of galvanic cells and electrolysis. But they need to explain the chemical processes in terms of electrons, ions, and molecules using models microscopically and represent the abstract processes by chemical formulae and equations symbolically. Instructional intervention programs have been focused to visualize the electrochemical processes in order to remedy the misconceptions. Researchers have suggested the inquiry activities using model approach (Cullen & Pentecost, 2011), computer animations, and simulations of electrochemical processes (Osman & Lee, 2014; Sanger & Greenbowe, 2000) as conceptual change strategies to improve students' conceptions on microscopic and symbolic levels. To date, limited studies have been investigated on the effectiveness of collaborative learning instruction and students' reflection on facilitating conceptual change in electrochemistry (Acar & Tarhan, 2007; Cullen & Pentecost, 2011). In particular, little is known about whether and how collaborative knowledge-building inquiry can address students' misconceptions in electrochemistry.

Purpose of the Study

This study aims to design and investigate the collaborative knowledge-building inquiry augmented with reflective assessment mediated by computer-supported Knowledge Forum in fostering conceptual change learning among high-school Chinese students in Hong Kong.

Specifically, three research questions are drawn for this study.

1. Do students who engage in knowledge-building inquiry perform significantly better than the conventional group on conceptual change and academic achievement in chemistry?
2. To what extent do student engage in the Knowledge Forum, and what characterize inquiry and reflection processes on Knowledge Forum?
3. What are the relationships among the knowledge-building measures, conceptual change learning, and academic achievement in chemistry?

Methodology

Participants

Participants included 79 tenth graders from two intact chemistry classes of a girl's high school in Hong Kong, instructed by the same teacher. They were all computer literate, but they had no prior experience of using Knowledge Forum database. The teacher had taught chemistry for more than 10 years in this school at which English is the medium of instruction. He employed knowledge-building pedagogy and Knowledge Forum in teaching various topics of chemistry for 4 years at this school.

Instructional Design

This study employed a quasi-experimental pretest-posttest research design. The experimental class (KB class, $n = 40$) experienced the knowledge-building inquiry, and the comparison class (RI class, $n = 39$) followed the regular instruction in the domain of electrochemistry. There were five chemistry lessons each week and each lesson was of 35-min duration. Prior to the instructional intervention, the teacher took a regular instructional approach focusing on teacher explanation and questioning in both classes during the first semester, and classroom instruction mainly involved PowerPoint presentation, lectures, teacher-guided discussions, and textbook experiments. The curriculum topics included "Reactivity of Metals," "Mole," and "Acids, Bases, and Neutralization" which were the prerequisite to the domain of "Electrochemistry." The other variables were controlled to be very similar for two classes such as the same learning syllabus, the same textbook, and the same experiments conducted over the 18-week research period in the second semester.

Designing Knowledge-Building Environment

Premised on the principles of knowledge building (Scardamalia, 2002) and conceptual change (Vosniadou, Ioannides, Dimitrakopoulou, & Papademetriou, 2001), the design of knowledge-building environment mediated by Knowledge Forum in the present study included four components to promote epistemic agency, inquiry, and metacognition: (a) developing collaborative classroom culture and scientific discourse, (b) engaging in collaborative problem-centered inquiry, (c) deepening the knowledge-building discourse, and (d) aligning assessment for collective learning. This four-phase knowledge-building inquiry pedagogical model was designed in way to appropriate to the sociocultural context of Hong Kong classrooms (Chan, 2011; Lee et al., 2006). Different knowledge-building principles including "epistemic agency" with students formulating problems and reflecting on their progress, "improvable ideas" with students pursuing inquiry through questions and explanations, and "community knowledge" with students working for both individual and collective understanding were emphasized throughout. The highlights of the knowledge-building environment of the present study are summarized below.

(a) *Developing collaborative classroom culture and scientific discourse*

The first phase was important to develop collaborative culture of classroom practices for shifting teacher-centered instruction to constructivist knowledge building, and that is important for Chinese students. Primarily students in Hong Kong are used to following closely teacher instruction and working competitively to obtain good examination results. The initial phase helps them develop epistemic agency, taking more control over their learning and to engage in scientific discourse within collaborative settings.

Various instructional activities were designed so that students were encouraged to activate and reflect on their prior knowledge and articulate their complex ideas for science learning. Activation of prior knowledge is important in conceptual change (Vosniadou et al., 2001), and students made their ideas public for inquiry (Scardamalia & Bereiter, 2006). Students worked in dyads or small groups to discuss the socio-scientific issues and real-world problems in science (e.g., why is it difficult for people to sink in the Dead Sea?). They gave short oral reports on their discussions and questioned the ideas from the other groups. After that, they shared their consensus ideas using concept maps or flow diagrams in the classroom. Second, the predict-observe-explain (POE) tasks (White & Gunstone, 1992) were incorporated to promote cognitive conflict for conceptual change learning. Students first made predictions about an unfamiliar discrepant event or situation (e.g., what happens if a bunch of Mentos candies drops into a 1.5-L bottle of Diet Coke?). Then the students conduct experiments or make observations from the teacher's demonstration, and they finally were guided to articulate the scientific explanations based on their observations and predictions. Third, guided inquiry-based experiments were designed that allowed students to discover the theories and address the authentic science problems for themselves using own experimental data (e.g., which antacids is the most effective?).

Within 4 weeks working on the topic of acid–base neutralization, students began to acculturate the practice of asking productive questions, generating hypotheses, and making informed judgments from multiple sources of information (textbook, Internet, and experiments) in the knowledge-building classroom environment.

(b) *Engaging students in metacognitive and problem-centered inquiry*

Students work on their inquiry problems on the Knowledge Forum 4.5 web version that the asynchronous online platform supports the processes of scientific inquiry through shared discourse. Figure 20.1 illustrates the design of a *view* (inquiry area) which is the shared workspace for different themes of inquiry. Students contribute ideas, theories, and resources using text, graphics, animations, and videos by authoring the *notes* to a *view*. Students can *build-on notes* on their peers to question, reflect, and link own ideas with others' ideas, thus making the

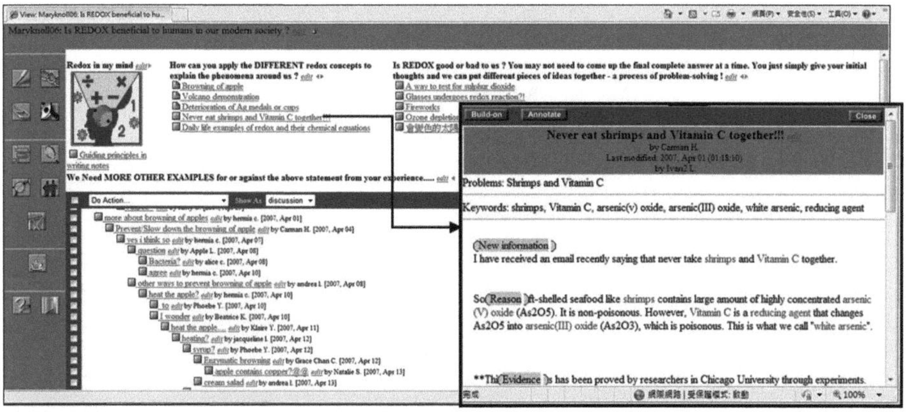

Fig. 20.1 The design of a *view* of Knowledge Forum 4.5 web version. *Note*: The large window shows the feature icons on the left-hand column, the key inquiry questions of the view of redox reactions with high-quality students' notes selected on the top, and the threaded structure of students' build-on notes. The small window illustrates a student' note showing the title, author, date modified, problem worked on, keywords use, and note content with scaffold uses such as (Reason)

knowledge-creation process overt and traceable. The build-on notes spread inward to the original note as shown in Fig. 20.1. The threaded structure of icon of notes represents how the communal conceptual ideas on particular inquiry problem develop over time; idea improvement is important in knowledge building.

The metacognitive prompts or *scaffolds* on Knowledge Forum (e.g., (my theory), (I need to understand), (new information), (a better theory), (putting our knowledge together)) allows students to make thinking explicit for developing metaconceptual awareness of what they know and do not know. Students are encouraged to make their ideas open so that alternative conceptions can be inquired among community members. Aside from these default build-in scaffolds of theory building, additional scaffolds were created to facilitate students' co-construction of chemistry knowledge such as (experiment) and (chemical equation). More supportive features enable students to *coauthor* notes, *annotate* notes, create *rise-above* notes, and add *keywords* and *references* with citations to others' notes that facilitate students to summarize, synthesize, and advance their knowledge-building discourse.

The teacher constructed *views* (discussion areas) in Knowledge Forum aligning with the classroom inquiry of the curriculum topics. During the lessons, students worked in pairs or small groups to pose questions and problems and to generate possible explanations on the Knowledge Forum. Students continued the inquiry discussions, revised their theories and ideas, and addressed each other's questions at home as part of the course assignment. As such, students can model from each other as they engage in knowledge-building inquiry to refine their ideas and questions.

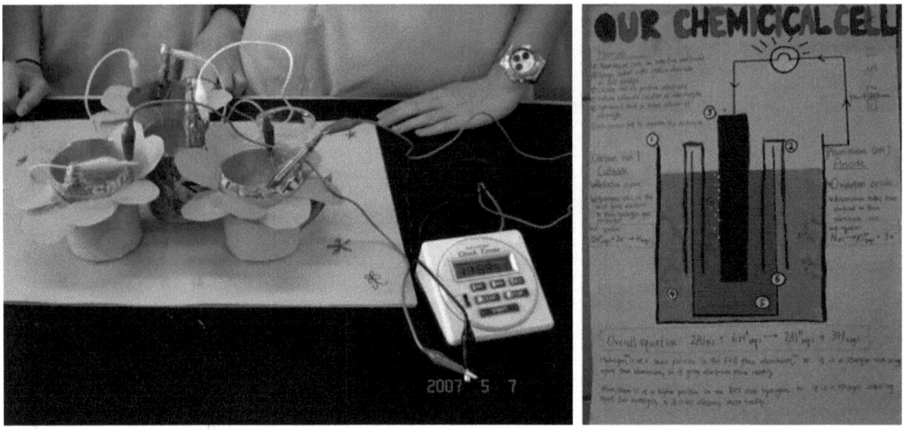

Fig. 20.2 The design of homemade chemical cell model and poster

(c) *Deepening knowledge-building and scientific discourse*

Rather than the teacher setting the problems or questions for inquiry, student-generated questions emerging from the discourse on Knowledge Forum play a prominent role in knowledge-building inquiry. Questions involving cutting-edge ideas and conflicting ideas of science were examined further in classroom inquiry (e.g., what causes the explosion of mobile phones in the charging process?). Specifically, students in groups conducted an inquiry project on designing workable chemical cells using household materials to power simple electrical devices as a *goal* of the curriculum unit. Students constructed their models of chemical cells and explained the designs and working principles of chemical cells using large posters during the class in the eleventh week (Fig. 20.2). The presentation of inquiry project facilitated students to connect the *macroscopic* models of chemical cells and posters with the *microscopic* abstract and complex chemistry processes as well as *symbolic* chemical formula and equations.

Above all, students are motivated to integrate the online and offline discourse in the process of knowledge building and to bridge the old beliefs and current conceptions as rise-above notes in the Knowledge Forum. As such, the offline classroom discourse can result in higher collective cognitive responsibilities among students to engage in sustained asynchronous networked discussions at home and to advance their knowledge-building discourse. Primarily, students worked together to improve both individual and collective understanding, emulating the practice of scientific communities, inquiring, questioning, explaining, testing ideas with experiments and models, and revising their understanding.

(d) *Aligning assessment with learning through reflection*

Reflective assessment is a key component in this knowledge-building design. As students contribute notes on Knowledge Forum, it goes beyond posting a computer note for inquiry. They are encouraged to reflect on their understanding and work together with improving own and others' ideas. Specifically, students constructed three explicit reflection summaries to assess their own and community knowledge and the changes of domain understanding in the knowledge-building process at the fourth, seventh, and eleventh week. The three reflection notes of each student were built on each other to document the student's assessment work in the reflection *view* of the Knowledge Forum. Additional scaffolds of conceptual change such as (*what I think earlier*), (*what I think now*), and (*my new learning*) were specifically created to help students develop metaconceptual awareness of knowledge advancement. As such, these reflection journals provide the learning evidence on the individual and collective knowledge advances and characterize the transformative and embedded assessment of community knowledge on Knowledge Forum. Examples of these knowledge-building reflection journals will be presented in the Results section.

Regular Instruction

Following the conventional approach in the first semester, students of regular instruction (RI) class were engaged with lectures, classroom questioning, teacher-led discussions, and textbook-structured experiments to learn the curriculum units in the second semester. RI students did not use computers during the lessons. Instead, the teacher introduced the key concepts about electrochemistry by playing the computer animations and making PowerPoint presentations in the lessons. While KB students worked on Knowledge Forum after school, the RI students had to finish the textbook problems and experiment report sheets at home, and they took regular quizzes as course assessment which is the common practice of examination-oriented classrooms in Hong Kong.

Instruments

Conceptual Written Tests

Conceptual written tests were designed to assess students' understanding and explanations on the electrochemistry concepts and to probe into their alternative conceptions. The instrument was composed of four structured questions to involve open-ended items, multiple choice, and true/false items followed by giving reasons for the choices which were developed from the common misconceptions identified in the textbook and literature (e.g., de Jong & Treagust, 2002). The questions tested knowledge on oxidation-reduction, galvanic cell, and electrolytic cell concepts at the macroscopic, microscopic, and symbolic levels. The content and face validity of the instrument was provided by a university chemistry professor and an experienced high-school chemistry teacher for the suitability of the questions. Here is an example of test item to assess students' understanding of oxidation-reduction concepts.

(a) What do you understand by the term "reduction"? Give example(s) to illustrate your answer.
(b) Consider the balanced equation of a reaction below:

$$(NH_4)_2Cr_2O_7(s) \rightarrow Cr_2O_3(s) + 4H_2O(l) + N_2(g)$$

The reaction is an example of:

A. Oxidation only
B. Reduction only
C. Oxidation and reduction
D. Neither oxidation nor reduction
I choose the above alternative because______________________________.

For data analysis, the multiple-choice and true/false items were classified as correct (one point) or incorrect/blank (zero point). The written responses of open-ended items and explanations were scored in four ways: no response (zero point), response containing misconceptions and inaccurate information (one point), response showing partially correct ideas or concepts (two points), and clear response demonstrating scientifically accepted ideas or concepts (three points). 30 % of the sample scripts of pretest and posttest were scored independently by the first author and another chemistry teacher. The inter-rater reliability was .82 for pretest and .91 for posttest based on Pearson correlation.

School and Public Examinations

Students took the midyear and final examinations for school summative assessment at Grade 10. These examination papers consisted of multiple-choice and structured questions that followed the standards of public examination at Grade 11. The questions of midyear and final examinations were based on the teaching topics in the first semester (as pretest) and the whole year (as posttest), respectively. All students sat the Hong Kong Certificate of Education Examination at the end of Grade 11 as an overall measure of attainment of subject knowledge. The public examination results of chemistry were used as a delayed posttest to compare the academic achievement in chemistry of two classes.

Knowledge-Building Processes and Dynamics

To investigate students' knowledge-building processes, several measures were included to characterize how students were involved in knowledge-building inquiry. These measures included (a) students' participation and interaction on Knowledge Forum: quantitative indices on student participation and collaboration patterns using built-in assessment tools of KF called Analytic Toolkit (ATK, Burtis, 2002) and Social Network Analyses (SNA, Borgatti, Everett, & Freeman, 2002), (b) student-generated questions examined for depth of inquiry ranging from fragmented factual questions to knowledge-based questions to explanatory questions (Hakkarainen, 2003), and (c) three reflection summaries examined for their individual and collective knowledge advances in the community. Examples of depth of inquiry and reflective journals illustrating knowledge-building processes will be presented in the Results section.

Results

Effects on Conceptual Change and Chemistry Achievement

Conceptual Written Test Scores

The first research question examined whether there were effects of knowledge-building inquiry on students' conceptual change in electrochemistry. Table 20.1 presents the results of pre-/post-conceptual test scores between knowledge-building (KB) and regular instruction (RI) classes.

Analyses were first conducted using the gain scores between pre- and posttests as conceptual change scores for students. Results showed that students of KB class ($M = 37.08$, $SD = 10.04$) outperformed those of the RI class ($M = 31.44$, $SD = 13.78$) on conceptual change scores in electrochemistry, $t(77) = 2.08$, $p = .04$, with a medium effect size (Cohen's $d = .49$).

Additional analyses were conducted to examine the instructional effects on both classes. A 2 (time) × 2 (treatment) repeated measure analysis of variance (ANOVA) was conducted to assess the impact of two instructional approaches (knowledge building and regular instruction) on students' scores of conceptual tests over time (pretest and posttest). Results indicate that there was a significant main effect for time, $F(1, 77) = 640.42$, $p = .000$, a very large effect with partial $\eta^2 = .89$, indicating that students of both classes developed better conceptual understanding on electrochemistry after instruction. A significant interaction effect (time × treatment) was obtained, $F(1, 77) = 4.33$, $p < .05$, a moderate effect with partial $\eta^2 = .05$, indicating that knowledge-building students performed better than comparison students on conceptual change test scores after instruction.

Table 20.1 Means and standard deviations (*SD*) for pre-/post-conceptual test scores of KB and RI classes

	KB class ($n = 40$)		RI class ($n = 39$)	
	Mean	SD	Mean	SD
Pre-conceptual test	18.25	6.92	20.90	7.22
Post-conceptual test	55.33	11.33	52.34	12.12

Fig. 20.3 Misconceptions on the flow of electrons in the electrolytes, salt bridge, and external circuit

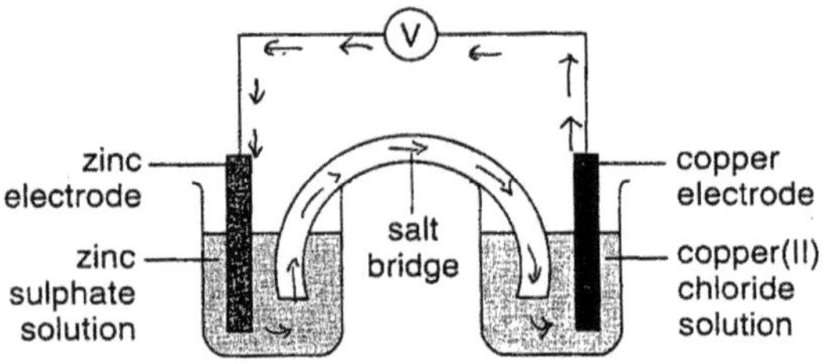

Misconceptions Analysis

The conceptual written tests provided overall performance scores, and further analyses were conducted to investigate students' misconception changes in specific areas of electrochemistry. A total of 9 and 13 misconception statements on electrochemistry were revealed from content analysis of the written responses in pre- and post-conceptual tests, respectively. The students' misconception statements of electrochemistry and their percentages of two classes on the pre-posttests are documented in the Appendix. These problematic conceptions about electrochemistry were consistent with those reported in previous studies (e.g., Barke et al., 2009; Garnett & Treagust, 1992a, 1992b).

As a general evaluation of the data, both collaborative knowledge-building inquiry and regular instruction did have positive effects upon students' conceptual understanding on electrochemistry. Students in both classes showed decreases in misconceptions (negative conceptual changes) over time, but the KB class showed more decreases in several misconceptions than the RI class. The greatest conceptual change occurred with the misconception, *oxidation and reduction reactions occur independently* (item 1) in both KB class (−85 %) and RI class (−77 %). Question 2b tested students on the direction and path of electron flow in the galvanic cell at the microscopic level. Some students wrongly drew the flow of electrons in the electrolytes and salt bridge to complete the circuit, and they reversed the direction of electron flow in the connecting wire as shown in Fig. 20.3. The students in KB class (−55 %) demonstrated greater conceptual change on the misconception that *electrons are drawn through the electrolyte to complete the circuit* (item 2) than those students in RI class (−36 %). Similarly, students in KB class (−33 %) exhibited greater conceptual change on the misconception that *electrons flow from a less reactive metal (copper) to a more reactive metal (zinc)* (item 4) than those students in RI class (−21 %). Moreover, 60 % students in KB class held the misconception that *the salt bridge assists the flow of electrons to complete the circuit* (item 3) in the pretest, and the percentage was reduced to 10 % in the posttest. By comparison, the percentage of students in RI class holding this misconception dropped from 64 to 5 %.

Four additional misconception statements (items 7, 9, 10, and 13) were elicited in the written responses after the instruction. It was noted that students in KB class exhibited lower percentages of all these misconceptions than those students in RI class. Using the misconception item 7 as an example, 5 % of students of KB class and 13 % of students of RI class were confused to assume *the use of graphite as salt bridge* in the posttest, and they did not realize that graphite has no mobile ion to balance the ionic charges in the electrolyte of chemical cell. In sum, the results show that students in KB class exhibited fewer misconceptions than that of the RI students after the instruction, and KB students improved their conceptions more than the RI counterparts especially at the microscopic level.

Chemistry School and Public Examinations

The results of midyear examination (prior to instruction) show that students of two classes exhibited similar level of chemistry knowledge by the end of first semester: KB class ($M = 74.3$, $SD = 11.3$) and regular instruction ($M = 73.2$, $SD = 10.3$). Analysis of the final examination scores, following instruction, showed more difference in mean scores between KB class ($M = 78.2$, $SD = 8.9$) and RI class ($M = 74.9$, $SD = 10.9$), but the difference did not reach statistically significant level.

The attainment of territory-wide public examination on chemistry 1 year later was graded from A to F scored from 5 to 0, respectively. Results showed that the KB class attained an average score of 3.93 ($SD = .86$) and RI class attained an average score of 3.77 ($SD = .96$). Although there were no statistical differences, there were some trend effects, and knowledge-building students seemed to be obtaining higher scores than comparison students in public examination in chemistry 1 year after the knowledge-building instruction.

Processes and Dynamics of Knowledge Building

The second research question focused on the student engagement and collaboration in the socio-cognitive processes of knowledge building to explain the pre-posttest gains on conceptual test scores of the KB class.

Quantitative Indices on Student Participation on Knowledge Forum

Student's Participation on Knowledge Forum: Analytic Toolkit Indices

To examine the role of knowledge-building design, it was first useful to see how students were participating on Knowledge Forum. Quantitative data from the Analytic Toolkit (ATK) provided a rich overview of the degree of participation and the pattern of collaboration among the KB students in the Knowledge Forum over 12 weeks. There were 1,092 written notes created in 11 views, and the size of build-on trees included 97 small clusters (2–5 notes), 41 medium clusters (6–20 notes), 6 large clusters (21–40 notes), and 2 very large clusters (over 40 notes).

Students on average contributed 27.3 notes and read 50.9 % of all the notes written by their classmates (read half of the notes written). The percentage of notes linked was high (73.5 %). The average number of keywords, note revision, and scaffold support uses was 42.8, 8.7, and 15.7 respectively. These ATK quantitative indices suggest substantial participation and interaction among students as compared with the mature knowledge-building communities (e.g., Lee et al., 2006).

Students' Connectivity on Knowledge Forum: Social Network Analysis

The case-by-case matrix of (a) who linked to whose notes (i.e., build-on, rise-above, or references) and (b) who read whose notes was generated by the Analytic Toolkit, and the analyses on network density measures were conducted with UCINET 6.0 program (Version 6.198; Borgatti et al., 2002). The density of note-linking and note-reading interactions was 0.30 and 0.97, respectively. The results indicate that there was a high degree of linking and reading each other's notes as compared with the mature knowledge-building communities (e.g., Zhang et al., 2009). As such, students read many notes written by other students and then built onto some of those notes and used references in constructing their theories which illustrate the collaborative effort in knowledge building.

Depth of Inquiry

Characterization of Depth of Inquiry

The quality of questions posed by students from the database notes was analyzed according to scheme modified from the previous studies (Hakkarainen, 2003; Lee et al., 2006). Student-generated questions were scored using a 5-point scale ranging from factual questions to knowledge-based questions to explanatory questions that are illustrated by the following excerpts.

Level 1 Definition questions
A rating of 1 was given to those questions asking about simple definitions of chemical terms, for example:

What does electrochemical series mean? [#43]

Level 2 Factual and general questions without personal ideas
A rating of 2 was scored to those questions seeking for facts and general information but without personal idea or elaboration, for example:

What determines the memory effect of a cell? [#104]

Level 3 Questions with personal ideas and elaboration
A rating of 3 was scored to those questions seeking for information with personal thoughts, ideas, and elaboration, for example:

According to the video and Shirley's note, ... When acid rain comes into contact with copper, it would most likely dissolve and .... As copper is used in many structures (e.g., statues or buildings), would this reaction have a huge impact on infrastructure? [#356]

Level 4 Puzzlement questions

A rating of 4 was scored to those questions identifying some specific gaps or inconsistencies among different ideas and/or source of information, and asking for different viewpoints for filling the gaps, for example:

(My theory) From my book borrowed outside, it said that an exothermic reaction is a reaction that liberates heat energy.... An endothermic reaction is a reaction that absorbs heat energy....

(I need to understand) As in BBQ burning the charcoal ... the temperature will increase, so I think it is an exothermic reaction. Is my way of saying this correct?

(Putting our knowledge together).... some of you say that there is endothermic reaction when we ignite charcoal using burning paper/charcoal starter. So does it mean a process can have both endothermic and exothermic reaction? Or we should look at this case as two stages? [#464]

Level 5 Explanation-based questions

A rating of 5 was scored to those questions on solving a problem by *metacognitively* recognizing the sources of gaps or inconsistencies among the ideas, moving beyond just asking questions to generate hypotheses, and constructing possible explanations to extend understanding, for example:

(Experiment) From today's experiment, we learnt that a high concentration solution can have a high concentration of negative ions which can be preferentially discharged at the anode .... For example, ... CONCENTRATED NaCl solution,... chlorine gas would be produced at the anode (Cl^- is preferentially discharged but not the OH^-).

(I need to understand)

1. *How can we define "high concentration"? Although dilute solutions are more concentrated than the very dilute ones, very dilute solutions are more concentrated than very very dilute ones. Then in this case, should we say that the very dilute solution is of a "high concentration"?*
2. *From what we have learnt, we know that if a solution has a more concentrated ion than OH^-, it would be preferentially discharged AT THE ANODE. But how about at the cathode? I want to ask, is it that the gas produced (the ion that is preferentially discharged) ALWAYS follows the ECS? Is there any exceptional case?* [#776]

This excerpt illustrates how the student interconnects the microscopic chemical processes and symbolic formula with the macroscopic experimental observations. The student demonstrated her new learning experience and collective knowledge

Table 20.2 Number of questions in each level of inquiry

No. of questions	Level 1	Level 2	Level 3	Level 4	Level 5	*Total*
Period 1 (6 weeks)	40	44	13	4	1	*102*
Period 2 (6 weeks)	49	97	18	17	8	*189*
Whole period	89	141	31	21	9	*291*

Table 20.3 Rating scheme for the knowledge-building reflection

Rating	Description
1	Repeat or restate the isolated facts from the discourse
2	Identify some relevant ideas with minimal elaborations
3	Identify misconceptions or conflicting ideas with elaborations
4	Identify misconceptions or knowledge gaps with collaborative explanations
5	Assess misconceptions or knowledge gaps by comparing different views (models); reflect metacognitively with questions; seek to synthesize and rise above with collaborative explanations

advances in the explanatory statement. Overall, students also made good use of Knowledge Forum scaffolds in posing high-level questions.

All questions were rated by the first author, and a second rater who was a master graduate specialized in learning sciences independently scored 30 % of the questions. The inter-rater reliability of depth of inquiry was .88 based on Pearson correlation.

Analysis on Change Over Time

The levels of questions were compared between two time periods of the study, and Table 20.2 shows the distribution of number of questions in each level of inquiry over two periods. The mean inquiry scores of period 1 and period 2 were 1.84 (SD = 0.86) and 2.14 (SD = 1.04), respectively. Chi-square test shows that students made deeper inquiry over time, $\chi^2(\text{df} = 4, N = 291) = 10.02, p < .05$.

Knowledge-Building Reflection Journals

Students were asked to submit three reflection journals as they reflected on what they have learned in light of their discussions on Knowledge Forum and classroom work. Analyses of students' reflection journals helped to illustrate their metacognition in knowledge building for both personal and community progress and possibly illuminating the conceptual change processes. The three reflection journals differed in emphasis, and they are generally coded from simplistic to complex responses as illustrated in Table 20.3.

A total of 120 reflections were rated by the first author and a second rater independently who was a master graduate in learning sciences. The inter-rater reliability of the first, second, and third reflection statements were .79, .84, and .82, respectively based on Pearson correlation. Overall, the scores for the first, second, and third reflection had an average of 1.98 (SD = 1.03), 3.15 (SD = 1.10), and 2.58 (SD = 1.13), respectively. The total scores of three reflection journals of each student were taken as the reflection scores for post analysis.

We now present an example of a reflection journal to illustrate knowledge-building processes. Specifically, the excerpt below is part of a student's third reflection rated as level 5 to illustrate how knowledge-building reflection and inquiry might scaffold conceptual change via students' increased metaconceptual awareness. As shown in the excerpt, the student was first puzzled from the diverse ideas of her three classmates on the authentic problem of the energy changes of volcano explosion. The community members identified a simple reaction of charcoal burning for collaborative inquiry. Later on, the student began to articulate her initial and current beliefs on the theory of "stages" of a chemical reaction and constructed her own explanation for this inquiry problem.

> *(Argument)… Carman suggested that volcano explosion is an endothermic reaction since it requires heat to start the explosion. Hermia* [5] *Exothermic Reaction and Endothermic Reaction argued that it actually involved both endothermic and exothermic reaction since….. Natalie* [6] *agree searched in web that it should be an exothermic reaction…. **Three different opinions appeared** and that we couldn't come up with the most appropriate answer, therefore simpler examples for explaining the reaction were then discussed on the burning of charcoal.* [7] *Is the ammonium dichromate volcano really an endothermic reaction?*
>
> *(Conflicting ideas) We all agreed that a reaction cannot be both exothermic and endothermic at a time. However different opinions appeared where some people said that a net change in energy should be considered and some said that it is exothermic reaction, and I think that reactions occur in stages….*
>
> *(Different opinion) Linda, Jennifer and Hermia* [8] *is it exothermic reaction? suggested that for the overall reaction, it is an exothermic reaction since there is a net change of temperature gained and loss…. **I was really confused when different opinions popped out*** [9] *Yes it is exothermic(i think)…*
>
> *(What I think now…)… .**Now I agree** that burning of charcoal is overall an exothermic reaction… .**But what I still don't understand** is that whether we can separate the reaction into 2 different stages for consideration.*
>
> *(My new learning)…. I've learnt basic idea of exothermic and endothermic reactions* [10] *Exothermic…. in order to determine the type of reaction, we should consider the net change between energy given out and gained….* [#92]
>
> Note: [1]denotes the title of reference note.

This excerpt from this reflection journal demonstrates how the student reflected on her thinking process for inconsistent ideas in the discursive discourse and examined the diverse ideas and arguments posed by other community members with own interpretation. She explained how the inconsistencies of ideas were progressively resolved based on the constructive use of scientific information from peers' discourse. This excerpt also illustrates the gradual process of

Table 20.4 Correlations among conceptual change, chemistry achievement, and knowledge-building (KB) measures

		1	2	3	4	5
1.	Conceptual change	–				
2.	Chemistry achievement	.57***				
3.	Forum participation	.12	.19			
4.	Forum interactivity	.18	.10	.00		
5.	Knowledge-building inquiry	.47**	.65***	.29	.26	
6.	Knowledge-building reflection	.31*	.31*	.59***	.09	.49**

* $p < .05$; ** $p < .01$; *** $p < .001$, two tailed

conceptual change how the student reorganized her naïve concepts into a more coherent conceptual structure with emerging explanations as the product of collective understanding.

Relationships Among Knowledge-Building Measures, Conceptual Change, and Chemistry Achievement

The third research question examined the relations among knowledge-building measures (Knowledge Forum participation, depth of inquiry, and reflection), conceptual change, and chemistry achievement of public examination, and the correlation results are presented in Table 20.4.

To increase coherence for the analyses, the Knowledge Forum participation scores obtained from ATK were examined using factor analysis; this procedure is commonly practiced in other research on knowledge building (e.g., Lee et al., 2006; Zhao & Chan, 2014). Four of the ATK indices (notes created, keywords use, scaffold uses, and note revision) loaded on Factor 1 called *participation* that explained 46.0 % of the variance, and the other two indices (notes read and notes linked) loaded on Factor 2 called *interactivity* that explained 28.1 % of the variance. Cronbach's alpha reliability coefficients of participation and interactivity scales were 0.78 and 0.61, respectively. For knowledge-building inquiry scores, different number and levels of questions were posed by students on Knowledge Forum. The overall inquiry score for each student was computed based on the frequency and quality of questions (i.e., no. of questions and ratings). For example, a student writing five questions with three questions at level 2 and two questions at level 3 would have an inquiry score of 2.4. The reflection scores from the three reflection journals were combined into an overall mean score.

Correlation results indicate that conceptual change scores were significantly correlated with knowledge-building inquiry ($r = .47$, $p < .01$) and reflection ($r = .31$, $p < .05$). Results suggest that students who engaged more in knowledge-building inquiry and reflection were more likely to achieve greater conceptual change in electrochemistry. Besides, students who achieved greater conceptual change were more likely to attain better chemistry achievement ($r = .57, p < .001$). Next, the chemistry achievement was associated with knowledge-building inquiry ($r = .65, p < .001$) and reflection ($r = .31, p < .05$). Students who asked more high-level inquiry questions and engaged in deeper reflections were more likely to attain better chemistry achievement. However, both KF participation and interactivity did not show any significant correlations with conceptual change or chemistry achievement.

Discussion and Implications

This study addresses the question of how the designs and socio-cognitive dynamics of knowledge building mediated by Knowledge Forum can foster conceptual change in chemistry. We design the collaborative knowledge-building environment characterized with four interrelated pedagogical components to facilitate inquiry, epistemic agency, and metacognitive collaboration for conceptual change. This study provides evidence demonstrating the positive effects of collective knowledge building on conceptual change in electrochemistry and illustrates the roles of explanatory-based inquiry and reflective assessment. In the following, we first examine theoretical aspects of how knowledge building facilitates conceptual change and then consider pedagogical implications of the knowledge-building model for science education in Asian contexts.

Knowledge-Building Inquiry for Conceptual Change and Achievement

The findings of the study show that students who experienced the knowledge-building design had significantly greater conceptual change and exhibited fewer misconceptions on electrochemistry than those students instructed by regular teaching approach, and specifically the knowledge-building students seemed to have sustained some positive effects reflected in their achievement of chemistry in public examination 1 year later. These findings support previous studies that the design of knowledge-building inquiry has the positive instructional effects on science concept learning (e.g., Chan et al., 2012; Vosniadou & Kollias, 2003). Besides, this is one of the few studies in knowledge-building literature that demonstrates the positive effects on conceptual change learning of complex chemistry concepts among high-school Chinese students.

The process of conceptual knowledge gains in the experimental class can be understood through characterizing the socio-cognitive dynamics of knowledge building. Quantitative results show that students have substantial participation, collaboration, and interactivity in the Knowledge Forum, which support the social dynamics and community connectedness in knowledge-building communities (Zhang et al., 2009). Two key themes emerge suggesting the roles of explanatory inquiry and reflective assessment.

First, students engaged in deeper inquiry shifting from fact-seeking questions to explanatory questions over time; they posed questions and constructed explanations to address their own and others' knowledge gaps with sustained questions. Our findings show that students' inquiry on Knowledge Forum is significantly correlated with their conceptual change and chemistry achievement. Apparently, the role of student-generated questions in scientific understanding is well recognized. This study focuses further on questions developed in the context of computer-supported collaborative learning and scientific discourse. As the excerpts illustrate, while some questions were merely paraphrasing questions, higher-level questions illustrate how students were bringing together different views and comparing different models grappling for deeper understanding. Through such collaborative knowledge-building processes, students engaged in explanation and developed higher level of agency that makes it more possible for knowledge restructuring; the high degree of learners' commitment with message (discourse) increases the likelihood of conceptual change to occur (Dole & Sinatra, 1998). Primarily the findings support the importance of depth of inquiry and explanatory-based questions to engage Chinese students in knowledge-building inquiry (Chan et al., 2012, Lee et al., 2006). Moreover, inquiry questions, prior knowledge, and new information provide important epistemological resources for conceptual change (van Aalst, 2013b).

Second, another theme that emerges is the role of reflective assessment in knowledge-building inquiry and conceptual change. Quantitative analyses show that reflection is correlated with conceptual change scores and chemistry achievement. Students who reflected at a deeper level were more likely to achieve better in conceptual change and academic achievement. Qualitative analyses, as the example shows, suggest how students can document the growth of ideas personally and collectively and, through this, developed deeper scientific understanding and conceptual change. These reflective journals using scaffolds and reference notes on Knowledge Forum were more than personal learning journals: Students were asked to reflect on their learning trajectory in light of how the class discourse helped them change their understanding (Lee et al., 2006; van Aalst & Chan, 2007).

The reflective journals demonstrate what we called "meta-discourse," that is, students were reflecting and synthesizing what they discussed in their Knowledge Forum discourse in the past weeks; their assessment involves processing of evidence, explanations, and tracking ideas for knowledge advancement (Goldman & Scardamalia, 2013). More specifically, these reflective journals helped students to assess their own prior and current beliefs including alternative conceptions, conflicting viewpoints, and changes of conceptual understanding over time. Such metaconceptual reflections, which helped students to articulate their understanding by referencing to prior beliefs, alternative conceptions, peer's contributions, and new information from different sources, would help them to restructure their understanding that might bring about conceptual change. Our findings are consistent with the current views on prior knowledge activation and restructuring (Vosniadou, 2008) and collaborative reflection as social mechanism for conceptual change (Miyake, 2008). Our findings further support the notion of embedded and transformative assessment principle in knowledge building (van Aalst, 2013a).

Finally, it is of interest to note Knowledge Forum participation does not correlate with conceptual change or achievement. It is possible that students generally participated at a high level so there is not much variation. As well, such findings could also suggest that reading and writing on computer forum are important to provide the foundation for quality work, yet quantitative counts or mere engagement in technology or computer platform is not adequate for deep change. It would be important to consider the socio-cognitive dynamics and pedagogical principles when designing for computer-supported knowledge building.

Classroom Implications of Designing Knowledge-Building Inquiry for Conceptual Change

This study sheds light on the Asian classrooms of Confucian-heritage cultures that the collaborative knowledge-building inquiry has potential to facilitate conceptual change and student achievement in the public examination. The knowledge-building design and practices of asking productive questions, co-constructing explanations, sustained pursuit of inquiry, reflective assessment, and working as community of scientists constitute key components of an innovative pedagogical model for conceptual change learning for different cultural contexts, including those classrooms emphasizing standard curricula and examinations in the Chinese contexts (Chan, 2008; Zhao & Chan, 2014). The pedagogical design principles for collective knowledge-building inquiry arising from this study are highlighted below that can be useful to those researchers or teachers who are interested in designing technology-enhanced learning environments of similar nature in science classrooms.

1. *Developing authentic problems for collaborative inquiry*

The first pedagogical design principle is important to engage students in various authentic learning tasks for investigating socio-scientific issues or solving real-world problems in the collaborative inquiry (Quintana et al., 2004; Sadler, 2009). Consistent with the knowledge-building principle of *real ideas, authentic problems* (Scardamalia, 2002), students are encouraged to start their inquiry discussions with real-world problems such as mobile phone explosion, apple browning, and antacid effectiveness that make their learning relevant for intrinsic motivation and deep inquiry. They should be provoked to consider something that addresses real ideas and cognitive conflict if conceptual change is to take place. The processes of inquiry involve strategic uses of questions to develop testable ideas and explanations of real-world problems.

2. *Activating prior knowledge for restructuring*

The scientific discourse on Knowledge Forum makes students' thinking and ideas visible in the community so that students articulate their naïve or prior conceptions explicitly in the shared discourse. Building-on and articulating ideas and addressing peers' questions enable students to doubt their prior knowledge, beliefs, and presuppositions, to create socio-cognitive conflicts for conceptual change in the community, and to integrate new information from authoritative sources for idea improvement on the problem of inquiry. This question-driven scientific inquiry facilitates students to activate their prior knowledge and restructure their conceptual knowledge into more coherent frameworks (Vosniadou & Kollias, 2003).

3. *Providing access to multiple visual representations of knowledge*

Students find learning science or chemistry difficult because they have to grasp the abstract scientific concepts in explaining observable phenomena across three levels of representations (macroscopic, microscopic, and symbolic). The third design principle stresses the use of multiple visual representations both offline and online such as concept maps, flow diagrams, flash animations and simulations, real models of chemical cells, posters, and Knowledge Forum. Specifically, student ideas can be represented in multiple ways on Knowledge Forum in the form of written notes, videos, and graphics distributed in different views (discussion area). Knowledge and ideas that are represented in various ways for understanding can bridge students' conceptual understanding among three levels of representations, and particularly important for chemistry learning that needs to interconnect the domains of macroscopic, microscopic, and symbolic knowledge (Chinn et al., 2013; de Jong & Taber, 2007).

4. *Using scaffolds and metacognitive prompts to promote change*

For deep understanding and conceptual change to take place, students need to play active and collaborative roles working with ideas. Scaffolds are prompts and tools that can be developed to help students to engage in deeper thinking. Different scaffolds have been designed in this study that can be adapted based on different classroom contexts. In science classroom, the scaffolding instructional strategies such as predict-observe-explain (POE) tasks and guided inquiry-based experiments can promote collaboration, interchanges of ideas, and constructive feedback among community members. Teachers can use explicit scaffolds to structure the collaborative inquiry, for example, the thinking prompts in the POE tasks such as "*What would happen if……?*" and "*My evidence supports/does not support my prediction because…*" for helping students to articulate prior and current conceptions. The explicit design of scaffolds in Knowledge Forum (i.e., I need to understand, my theory, new information, a better theory, putting our knowledge together) actually reflect scientific inquiry processes, and they are designed to help students engage in theory building, explanations, and idea development similar to communities of scientists. In this study, we also developed conceptual change scaffolds including "what I think earlier," "misconception," "what I think now," and "my new learning." These scaffolds are general metacognitive prompts to support student thinking; with or without computers, science teachers can design scaffolds to help students develop more sophisticated patterns of thinking in scientific inquiry for conceptual change.

5. *Supporting sustained inquiry and intentional learning*

Conceptual change involves not a sudden process of change but sustained efforts to develop understanding. Helping students to engage in intentional learning in science classroom includes setting learning goals, using strategies, and reflecting on understanding. As well, intentional learning also involves students to invest learning efforts going beyond mere task completion to sustained inquiry and pursuit of learning. Consistent with the design principles of conceptual change (Vosniadou et al., 2001), students are encouraged to develop goal-directed learning strategies through the collaborative inquiry and to pursue understanding through designing workable chemical cells. They articulate the theories and ideas and regulate the progress of learning in the knowledge-building discourse that promotes socially shared cognition and collective goal in the community (Hennessy, 2006). Students improve their conceptual understanding through metacognitive engagement and reflective discourse that promote conceptual change. Science teachers need to develop a classroom culture of intentional learning and knowledge building for authentic scientific inquiry beyond task completion.

6. *Develop metaconceptual awareness through reflecting on progress*

The last pedagogical design principle involves reflective assessment in the collaborative inquiry that entails the knowledge-building principle of *embedded and transformative assessment* (Scardamalia, 2002). Assessment is embedded in instruction and it serves to transform understanding. In this study, students were asked to explicitly help each other identify misconceptions in the Knowledge Forum discourse, so they could seek to improve their collaborative understanding. Such assessment is aligned with learning goals; it also helps students to chart their learning progress, shape, and justify their explanations in the scientific discourse that facilitates the process of knowledge building. As students are encouraged to assess and as they become aware of their naïve ideas, they can better integrate new information with alternative conceptions. As such, reflective assessment might help students to engage in the process of metaconceptual awareness, monitoring, and evaluation for reorganizing their knowledge structures and changing their conceptions (Vosniadou, 2003).

We propose that science teachers need to provide opportunities for students to assess their own and others' understanding in light of standards of scientific inquiry. Alternative conceptions are not to be avoided, but students can work together examining and working on areas of difficulties collaboratively. Reflective assessment has been shown to be more effective for low-achieving students (see White & Frederiksen, 1998).

Limitations of this study are noted that point to future research. First, the comparison class is not a strong control. Due to the implementation of complex classroom processes, it would be difficult to make the design totally equivalent. It is also necessary to recognize the halo effects of the use of computer forum, and the novelty might have brought about more interest. Nevertheless, in addition to intergroup comparison, we have also included intra-class comparison suggesting that inquiry and reflection are correlated with conceptual change. Further research on the metaconceptual knowledge and processes would be useful to understand the instructional effects.

A second limitation concerns the predictive role of knowledge building on academic achievement. The results suggest that knowledge-building students performed better in both school and public examinations, and results did not reach significant level of differences. Certainly this study focused on electrochemistry so it is unclear if students could indeed transfer to other areas of chemistry; however, there is still some evidence of relatively better results. This area is generally not so clear, but for instructional innovation to sustain in Asian science classrooms, it would be helpful to explore the instructional effects on both conceptual and achievement results. Further analyses could be developed to see how knowledge-building students differed from comparison students in different areas of performance.

Third, this study has focused on the design and effects of knowledge building on conceptual change, and we obtained quantitative results, but much more analyses are needed for conceptual change processes. Ongoing analyses are undertaken to unravel the process of knowledge building that may bring about conceptual change. Finally, the knowledge-building environment involves very complex components, and it might not be clear, and further investigations need to be developed focusing on the proposed design principles to delineate more clearly which components would be of particular importance.

In conclusion, experiencing conceptual change in science involves a complex and multifaceted process to integrate new information into the existing cognitive structures for the development of a more coherent framework of scientific knowledge. This chapter provides some evidence about how the pedagogical design of collective knowledge-building environment focusing on inquiry and reflection mediated by Knowledge Forum can foster students' deep cognitive, metacognitive, and epistemic processes, which in turn have potentials to empower conceptual change and achievement in science among Chinese students. Prior to integrating the theories of instruction into science classroom practice in Asian context, further research is still needed to understand how the different pedagogical designs and scaffolding strategies influence students' metaconceptual and epistemic processes for knowledge creation and intentional conceptual change in the innovative knowledge-building environment.

Appendix

Students' misconceptions on electrochemistry identified at the pre- and posttests in KB and RI classes

	KB class ($n=40$)			RI class ($n=39$)		
Misconception statements	Pretest %	Posttest %	Conceptual changes (%)	Pretest %	Posttest %	Conceptual changes (%)
1. Oxidation and reduction reactions occur independently	90.0	5.0	**−85**	87.2	10.3	**−76.9**
2. Electrons are drawn through the electrolyte to complete the circuit in an electrochemical cell	62.5	7.5	**−55**	41.0	5.1	**−35.9**
3. Salt bridge assists the flow of electrons to complete the circuit in an electrochemical cell	60.0	10.0	**−50**	64.1	5.1	**−59.0**
4. Electrons flow from cathode (a less reactive metal) to anode (a more reactive metal) of an electrochemical cell in external circuit	37.5	5.0	**−32.5**	20.5	0.0	**−20.5**
5. The more reactive metal is the positive electrode in the metal couple chemical cell	62.5	2.5	**−60**	53.8	0.0	**−53.8**
6. The greater the separation of position of 2 metal electrodes in electrochemical series will give a smaller voltage	62.5	5.0	**−57.5**	59.0	2.6	**−56.4**
7. Graphite is used to replace the salt bridge and there still has remarkable voltage and/or current reading in a chemical cell	–	5.0	**–**	–	12.8	**–**
8. The polarity of the terminals of the applied voltage has no effect on the site of anode and cathode in the electrolytic cell	55.0	2.5	**−52.5**	43.6	7.7	**−35.9**

(continued)

	KB class ($n=40$)			RI class ($n=39$)		
Misconception statements	Pretest %	Posttest %	Conceptual changes (%)	Pretest %	Posttest %	Conceptual changes (%)
9. The products of electrolysis of aqueous salt solution are not predicted by the relative position of ions in electrochemical series	–	17.5	–	–	25.6	–
10. The concentration of anion has no effect of discharge of anions in the electrolysis of dilute salt solution	–	37.5	–	–	51.3	–
11. A zinc-carbon cell goes "dead" when all of the electrons are used up from inside the cell	45.0	30.0	**−15.0**	51.3	30.8	**−20.5**
12. The "+" and "-" ends of the battery are charged, so the electrons are attracted from one end to the other	80.0	20.0	**−60.0**	71.8	25.6	**−46.2**
13. The sum of oxidation number of elements in a polyatomic ion is not equal to the charge of ion	–	25.0	–	–	30.8	–

References

Acar, B., & Tarhan, L. (2007). Effect of cooperative learning strategies on students' understanding of concepts in electrochemistry. *International Journal of Science and Mathematics Education, 5*(2), 349–373.

Anderson, R. D. (2007). Inquiry as an organizing theme for science curricula. In S. K. Abell & N. G. Lederman (Eds.), *Handbook of research on science education* (pp. 807–830). Mahwah, NJ: Lawrence Erlbaum Associates.

Barke, H.-D., Hazari, A., & Yitbarek, S. (2009). *Misconceptions in chemistry: Addressing perceptions in chemical education* (pp. 207–233). London, UK: Springer.

Bell, T., Schanze, S., Graber, W., Slotta, J. D., Jorde, D., Berg, H. B., . . . Evans, R. H. (2007). Technology-enhanced collaborative inquiry learning: Four approaches under common aspects. In R. Pinto & D. Couso (Eds.), *Contributions from science education research* (pp. 451–463). Dordrecht, The Netherlands: Springer.

Bell, T., Urhahne, D., Schanze, S., & Ploetzner, R. (2010). Collaborative inquiry learning: Models, tools, and challenges. *International Journal of Science Education, 32*(3), 349–377. doi:10.1080/09500690802582241.

Borgatti, S. P., Everett, M. G., & Freeman, L. C. (2002). *Ucinet for Windows: Software for social network analysis*. Harvard, MA: Analytic Technologies.

Burtis, J. (2002). *Analytic toolkit for knowledge forum (Version 4.0)*. Toronto, ON: Institute for Knowledge Innovation and Technology, Ontario Institute for Studies in Education, University of Toronto.

Chan, C. K. K. (2008). Pedagogical transformation and knowledge-building for the Chinese learner. *Evaluation & Research in Education, 21*(3), 235–251. doi:10.1080/09500790802485245.

Chan, C. K. K. (2011). Bridging research and practice: Implementing and sustaining knowledge building in Hong Kong classrooms. *International Journal of Computer-Supported Collaborative Learning, 6*(2), 147–186. doi:10.1007/s11412-011-9121-0.

Chan, C. K. K. (2013). Collaborative knowledge building: Towards a knowledge-creation perspective. In C. E. Hmelo-Silver, C. A. Chinn, C. K. K. Chan, & A. M. O'Donnell (Eds.), *The international handbook of collaborative learning* (pp. 437–461). New York, NY: Routledge.

Chan, C. K. K., Burtis, J., & Bereiter, C. (1997). Knowledge building as a mediator of conflict in conceptual change. *Cognition and Instruction, 15*(1), 1–40. doi:10.1207/s1532690xci1501_1.

Chan, C. K. K., Lam, I. C. K., & Leung, R. W. H. (2012). Can collaborative knowledge building promote both scientific processes and science achievement? *International Journal of Educational Psychology, 1*(3), 199–227. doi:10.4471/ijep.2012/12.

Chan, C. K. K., & Rao, N. (Eds.). (2009). *Revisiting the Chinese learner: Changing contexts, changing education*. Hong Kong/Dordrecht, NL: The Comparative Education Research Centre, University of Hong Kong/Springer.

Chi, M. T. H., Slotta, J. D., & De Leeuw, N. (1994). From things to processes: A theory of conceptual change for learning science concepts. *Learning and Instruction, 4*(1), 27–43. doi: http://dx.doi.org/10.1016/0959-4752(94)90017-5

Chinn, C. A., Duncan, R. G., Dianovsky, M., & Rinehart, R. (2013). Promoting conceptual change through inquiry. In S. Vosniadou (Ed.), *International handbook of research on conceptual change* (2nd ed., pp. 539–559). New York, NY: Routledge.

Chinn, C. A., & Malhotra, B. A. (2002). Epistemologically authentic inquiry in schools: A theoretical framework for evaluating inquiry tasks. *Science Education, 86*(2), 175–218. doi:10.1002/sce.10001.

Cullen, D. M., & Pentecost, T. C. (2011). A model approach to the electrochemical cell: An inquiry activity. *Journal of Chemical Education, 88*(11), 1562–1564. doi:10.1021/ed101146u.

de Jong, O., & Treagust, D. F. (2002). The teaching and learning of electrochemistry. In J. K. Gilbert, O. de Jong, R. Justi, D. F. Treagust, & J. H. van Driel (Eds.), *Chemical education: Towards research-based practice* (pp. 317–337). Dordrecht, The Netherlands: Kluwer Academic Publishers.

de Jong, O., & Taber, K. S. (2007). Teaching and learning the many faces of chemistry. In S. K. Abell & N. G. Lederman (Eds.), *Handbook of research on science education* (pp. 631–652). Mahwah, NJ: Lawrence Erlbaum Associates.

Dole, J. A., & Sinatra, G. M. (1998). Reconceptualizing change in the cognitive construction of knowledge. *Educational Psychologist, 33*(2–3), 109–128. doi:10.1207/s15326985ep3302&3_5.

Duit, R. (2009). *Bibliography – Students' and teachers' conceptions and science education*. from http://www.ipn.uni-kiel.de/aktuell/stcse/stcse.html

Duit, R., & Treagust, D. F. (2012). How can conceptual change contribute to theory and practice in science education? In B. J. Fraser, K. Tobin, & C. J. McRobbie (Eds.), *Second international handbook of science education* (Vol. 24, pp. 107–118). Dordrecht, The Netherlands: Springer.

Garnett, P. J., & Treagust, D. F. (1992a). Conceptual difficulties experienced by senior high school students of electrochemistry: Electric circuits and oxidation-reduction equations. *Journal of Research in Science Teaching, 29*(2), 121–142. doi:10.1002/tea.3660290204.

Garnett, P. J., & Treagust, D. F. (1992b). Conceptual difficulties experienced by senior high school students of electrochemistry: Electrochemical (galvanic) and electrolytic cells. *Journal of Research in Science Teaching, 29*(10), 1079–1099. doi:10.1002/tea.3660291006.

Georghiades, P. (2000). Beyond conceptual change learning in science education: Focusing on transfer, durability and metacognition. *Educational Research, 42*(2), 119–139. doi:10.1080/001318800363773.

Goldman, S. R., & Scardamalia, M. (2013). Managing, understanding, applying, and creating knowledge in the information age: Next-generation challenges and opportunities. *Cognition and Instruction, 31*(2), 255–269. doi:10.1080/10824669.2013.773217.

Guzzetti, B. J., Snyder, T. E., Glass, G. V., & Gamas, W. S. (1993). Promoting conceptual change in science: A comparative meta-analysis of instructional interventions from reading education and science education. *Reading Research Quarterly, 28*(2), 116–159. doi:10.2307/747886.

Hakkarainen, K. (2003). Progressive inquiry in a computer-supported biology class. *Journal of Research in Science Teaching, 40*(10), 1072–1088. doi:10.1002/tea.10121.

Hennessy, S. (2006). Integrating technology into teaching and learning of school science: A situated perspective on pedagogical issues in research. *Studies in Science Education, 42*(1), 1–48. doi:10.1080/03057260608560219.

Lee, C. B., & Jonassen, D. (2013). *Fostering conceptual change with technology Asian perspectives*. Singapore: Cengage Learning.

Lee, E. Y. C., Chan, C. K. K., & van Aalst, J. (2006). Students assessing their own collaborative knowledge building. *International Journal of Computer-Supported Collaborative Learning, 1*(1), 57–87. doi:10.1007/s11412-006-6844-4.

Linn, M. C., & Eylon, B.-S. (2011). *Science learning and instruction: Taking advantage of technology to promote knowledge integration*. New York, NY: Routledge.

Minner, D. D., Levy, A. J., & Century, J. (2010). Inquiry-based science instruction – What is it and does it matter? Results from a research synthesis years 1984 to 2002. *Journal of Research in Science Teaching, 47*(4), 474–496. doi:10.1002/tea.20347.

Miyake, N. (2008). Conceptual change through collaboration. In S. Vosniadou (Ed.), *International handbook of research on conceptual change* (pp. 453–478). New York, NY: Routledge.

Osman, K., & Lee, T. T. (2014). Impact of interactive multimedia module with pedagogical agents on students' understanding and motivation in the learning of electrochemistry. *International Journal of Science and Mathematics Education, 12*(2), 395–421. doi:10.1007/s10763-013-9407-y.

Pintrich, P. R., Marx, R. W., & Boyle, R. A. (1993). Beyond cold conceptual change: The role of motivational beliefs and classroom contextual factors in the process of conceptual change. *Review of Educational Research, 63*(2), 167–199. doi: http://dx.doi.org/10.2307/1170472

Posner, G. J., Strike, K. A., Hewson, P. W., & Gertzog, W. A. (1982). Accommodation of a scientific conception: Toward a theory of conceptual change. *Science Education, 66*(2), 211–227. doi:10.1002/sce.3730660207.

Quintana, C., Reiser, B. J., Davis, E. A., Krajcik, J., Fretz, E., Duncan, R. G., . . . Soloway, E. (2004). A scaffolding design framework for software to support science inquiry. *The Journal of the Learning Sciences, 13*(3), 337–386. doi: 10.2307/1466941

Sadler, T. D. (2009). Situated learning in science education: Socio-scientific issues as contexts for practice. *Studies in Science Education, 45*(1), 1–42. doi:10.1080/03057260802681839.

Sanger, M. J., & Greenbowe, T. J. (1997a). Common student misconceptions in electrochemistry: Galvanic, electrolytic, and concentration cells. *Journal of Research in Science Teaching, 34*(4), 377–398. doi:10.1002/(SICI)1098-2736(199704)34:4<377::AID-TEA7>3.0.CO;2-O.

Sanger, M. J., & Greenbowe, T. J. (1997b). Students' misconceptions in electrochemistry: Current flow in electrolyte solutions and the salt bridge. *Journal of Chemical Education, 74*(7), 819–823. doi:10.1021/ed074p819.

Sanger, M. J., & Greenbowe, T. J. (2000). Addressing student misconceptions concerning electron flow in aqueous solutions with instruction including computer animations and conceptual change strategies. *International Journal of Science Education, 22*(5), 521–537. doi:10.1080/095006900289769.

Scardamalia, M. (2002). Collective cognitive responsibility for the advancement of knowledge. In B. Smith (Ed.), *Liberal education in a knowledge society* (pp. 67–98). Chicago, IL: Open Court.

Scardamalia, M., & Bereiter, C. (1994). Computer support for knowledge-building communities. *The Journal of the Learning Sciences, 3*(3), 265-283. doi: http://dx.doi.org/10.1207/s15327809jls0303_3.

Scardamalia, M., & Bereiter, C. (2006). Knowledge building: Theory, pedagogy and technology. In R. K. Sawyer (Ed.), *The Cambridge handbook of the learning sciences* (pp. 97–115). Cambridge, MA: Cambridge University Press.

Schroeder, C. M., Scott, T. P., Tolson, H., Huang, T.-Y., & Lee, Y.-H. (2007). A meta-analysis of national research: Effects of teaching strategies on student achievement in science in the United States. *Journal of Research in Science Teaching, 44*(10), 1436–1460. doi:10.1002/tea.20212.

Scott, P., Asoko, H., & Leach, J. (2007). Student conceptions and conceptual learning in science. In S. K. Abell & N. G. Lederman (Eds.), *Handbook of research on science education* (pp. 31–56). Mahwah, NJ: Lawrence Erlbaum Associates.

Sinatra, G. M., & Mason, L. (2008). Beyond knowledge: Learner characteristics influencing conceptual change. In S. Vosniadou (Ed.), *International handbook of research on conceptual change* (pp. 560–582). Mahwah, NJ: Lawrence Erlbaum Associates.

Sinatra, G. M., & Pintrich, P. R. (2003). *Intentional conceptual change*. Mahwah, NJ: Lawrence Erlbaum Associates.

Treagust, D. F., & Duit, R. (2008). Conceptual change: A discussion of theoretical, methodological and practical challenges for science education. *Cultural Studies of Science Education, 3*(2), 297–328. doi:10.1007/s11422-008-9090-4.

Turcotte, S. (2012). Computer-supported collaborative inquiry on buoyancy: A discourse analysis supporting the 'pieces' position on conceptual change. *Journal of Science Education and Technology, 21*(6), 808–825. doi:10.1007/s10956-012-9368-x.

van Aalst, J. (2013a). Assessment in collaborative learning. In C. E. Hmelo-Silver, C. A. Chinn, C. K. K. Chan, & A. M. O'Donnell (Eds.), *The international handbook of collaborative learning* (pp. 280–296). New York, NY: Routledge.

van Aalst, J. (2013b). Knowledge building and conceptual change: An epistemological resources perspective. In C. B. Lee & D. Jonassen (Eds.), *Fostering conceptual change with technology Asian perspectives* (pp. 173–198). Singapore: Cengage Learning.

van Aalst, J., & Chan, C. K. K. (2007). Student-directed assessment of knowledge building using electronic portfolios. *The Journal of the Learning Sciences, 16*(2), 175–220. doi:10.1080/10508400701193697.

van Aalst, J., & Truong, M. S. (2011). Promoting knowledge creation discourse in an Asian primary five classroom: Results from an inquiry into life cycles. *International Journal of Science Education, 33*(4), 487–515. doi:10.1080/09500691003649656.

Vosniadou, S. (2003). Exploring the relationships between conceptual change and intentional learning. In G. M. Sinatra & P. R. Pintrich (Eds.), *Intentional conceptual change* (pp. 377–406). Mahwah, NJ: Lawrence Erlbaum Associates.

Vosniadou, S. (2008). *International handbook of research on conceptual change*. New York, NY: Routledge.

Vosniadou, S. (2012). Reframing the classical approach to conceptual change: Preconceptions, misconceptions and synthetic models. In B. J. Fraser, K. Tobin, & C. J. McRobbie (Eds.), *Second International handbook of science education* (Vol. 24, pp. 119–130). Dordrecht, The Netherlands: Springer.

Vosniadou, S., & Kollias, V. (2003). Using collaborative, computer-supported, model building to promote conceptual change in science. In E. De Corte, L. Verschaffel, N. Entwistle, & J. van

Merrienboer (Eds.), *Powerful learning environments: Unravelling basic components and dimensions* (pp. 181–196). Amsterdam, The Netherlands: Pergamon.
Vosniadou, S., & Mason, L. (2012). Conceptual change induced by instruction: A complex interplay of multiple factors. In K. R. Harris, S. Graham, T. Urdan, S. Graham, J. M. Royer, & M. Zeidner (Eds.), *APA Educational psychology handbook* (Individual differences and cultural and contextual factors, Vol. 2, pp. 221–246). Washington, DC: American Psychological Association.
Vosniadou, S., Ioannides, C., Dimitrakopoulou, A., & Papademetriou, E. (2001). Designing learning environments to promote conceptual change in science. *Learning and Instruction, 11*(4–5), 381–419. doi: http://dx.doi.org/10.1016/S0959-4752(00)00038-4
Watkins, D. A., & Biggs, J. B. (2001). The paradox of the Chinese learner and beyond. In D. A. Watkins & J. B. Biggs (Eds.), *Teaching the Chinese learner: Psychological and pedagogical perspectives* (pp. 3–23). Hong Kong: CERC/ACER.
White, B. Y., & Frederiksen, J. R. (1998). Inquiry, modeling, and metacognition: Making science accessible to all students. *Cognition & Instruction, 16*(2), 3–118.
White, R., & Gunstone, R. (1992). *Probing understanding*. London, UK: Falmer Press.
Whitebread, D., & Cardenas, V. G. (2012). Self-regulated learning and conceptual development in young children: The development of biological understanding. In A. Zohar & Y. J. Dori (Eds.), *Metacognition in science education: Trends in current research* (pp. 101–131). Dordrecht, The Netherlands: Springer.
Zhang, J., Scardamalia, M., Lamon, M., Messina, R., & Reeve, R. (2007). Socio-cognitive dynamics of knowledge building in the work of 9- and 10-year-olds. *Educational Technology Research and Development, 55*(2), 117–145. doi: http://dx.doi.org/10.1007/s11423-006-9019-0
Zhang, J., Scardamalia, M., Reeve, R., & Messina, R. (2009). Designs for collective cognitive responsibility in knowledge-building communities. *The Journal of the Learning Sciences, 18*(1), 7–44. doi: http://dx.doi.org/10.1080/10508400802581676
Zhao, K., & Chan, C. K. K. (2014). Fostering collective and individual learning through knowledge building. *International Journal of Computer-Supported Collaborative Learning, 9*(1), 63–95. doi:10.1007/s11412-013-9188-x.
Zohar, A., & Barzilai, S. (2013). A review of research on metacognition in science education: Current and future directions. *Studies in Science Education, 49*(2), 121–169. doi:10.1080/03057267.2013.847261.

Chapter 21
Implementation of Standards-Based Curriculum by Chemistry Teachers: From Curriculum Materials to Teaching Practice

Bo Chen and Bing Wei

Introduction

Since the late 1980s, science curriculum development at the school level has come to a new era with the slogans such as "scientific literacy" used as the general purpose of science education. This kind of standards-based science curriculum initiatives, such as "Project 2061" (American Association for the Advancement of Science (AAAS), 1989), has promoted a new round of curriculum materials development after the curriculum reform that took place in the late 1950s and the early 1960s (Power & Anderson, 2002). According to Aikenhead (2006), these standards-based curriculum materials can be seen as having humanistic orientation, which means that they include inquiry as a part of science content, encourage a constructivist and student-centered approach to learning, and emphasize practical utility, human values, and a connectedness with societal events to achieve inclusiveness (Aikenhead, 2006; Powell & Anderson, 2002). Following the global tendency, the latest round of science curriculum reform was initiated in the beginning of the new millennium in China (Wei, 2010). As part of science curriculum reform, the official chemistry curricula were established with the national standards of chemistry curriculum at the stages of junior and senior secondary schools promulgated by the Ministry of Education (MoE) in 2001 and 2003, respectively. Subsequently, new chemistry textbooks, aligned with these curriculum standards, have been published and put into use. Researchers have found that these curriculum standards and chemistry textbooks in China exhibit new features that are akin to those of the standards-based science curriculum materials (Wei & Chen, in press; Wei & Thomas, 2006; Gao, 2007). Based on these studies, it could be concluded that the current chemistry textbooks in China have been transformed into the new type of curriculum materials.

B. Chen (✉) • B. Wei
Faculty of Education, University of Macau, Macau, China
e-mail: njcb0128@aliyun.com; bingwei@umac.mo

M.S. Khine (ed.), *Science Education in East Asia*,
DOI 10.1007/978-3-319-16390-1_21

Generally, there exists a basic assumption that standards-based approach to science education is the alignment among instruction, assessment, and the content standards in order to create equal opportunities for students to achieve expected learning outcomes (Herman & Webb, 2007). However, since the standards-based science curriculum requires the teachers to play a substantially different role in the classroom, to develop new knowledge, and to change their original beliefs, most of the teachers are reluctant to implement this kind of curriculum in their classrooms (Aikenhead, 2006). As such, the current situation is that although the standards-based curriculum materials are available, the implementation of these materials may not be adequate (e.g., Schneider, Krajcik, & Blumenfeld, 2005; Vos, Taconis, Jochems, & Pilot, 2011). Whereas the existing studies have been conducted in the western social context, little is known about the actual situation of the implementation of standards-based curriculum materials in eastern countries, especially in China, where the school system and the social context are different from the west. In this paper, we are interested to explore the degree to which the standards-based senior secondary school chemistry curriculum materials have been implemented in the classroom in China.

Literature Review

Generally speaking, curriculum materials refer to instructional resources such as textbooks, teacher's guides, lesson plans, worksheet, and other ancillary materials, which are adopted for use in schools for teachers to structure both their planned and enacted instruction (Forbes & Davis, 2010; Powell & Anderson, 2002). As we know, curriculum materials, particularly textbooks, have often determined the taught curriculum for many teachers, especially for young teachers (Ball & Feiman-Nemser, 1988; McNeil, 2003). Therefore, they have the potential to initiate and sustain reform in science education (Aikenhead, 2006; Powell & Anderson, 2002). However, curriculum materials themselves cannot generate changes in the classroom. It is dependent on teachers who can use them to enact changes in practice. This trait is called the "inert" character of curriculum materials (Powell & Anderson, 2002). In this sense, while recognizing that curriculum materials cannot provide guarantee of instructional change, Lloyd, Remillard, and Herbel-Eisenmann (2009) argued that teachers are central players in the process of transforming curriculum ideas into reality. Based on this assumption, a growing body of literature on teachers' use of curriculum materials appears in recent years, especially in the field of mathematics education (e.g., Remillard, Herbel-Eisenmann & Lloyd, 2009). According to Brown (2009), the relationship between teachers and curriculum materials involves bidirectional influences in which curriculum materials influence teachers through their affordances and constraints, while teachers mobilize curriculum materials through their perceptions and decisions. Given the interactive relationship, Powell and Anderson (2002) argued that "the interaction of the teacher with the materials determines what happens in the classroom" (p. 112). In this study, we focus on the interaction between senior secondary school chemistry teachers and the standards-based curriculum materials.

Based on Goodlad (1979)'s classification of curriculum representations and Van Hiele (1986)'s distinguishing on three "levels of thinking and acting," Vos, Taconis, Jochems, and Pilot (2010) developed a framework to examine the use of innovative context-based teaching materials by teachers in the classroom. This analytical framework consists of a nine-cell matrix, with "intended curriculum," "perceived curriculum," and "operational curriculum" as the rows and with "theoretical level," "descriptive level," and "grounded level" as the columns (Vos et al., 2010). At the rows, from the intended to operational curriculum, it represents a process in which what is intended by curriculum designers in curriculum materials is perceived by teachers and then manifested by them in their classrooms. At the columns, it distinguishes concrete teaching activities on ground level from teaching–learning strategies on descriptive level and aims and vision on theoretical level. This framework has been successfully employed to identify to what extent innovative context-based teaching materials are implemented adequately by teachers (Vos et al., 2011).

As mentioned earlier, this study is concerned with the degree to which the standards-based curriculum materials are implemented in the classroom, that is, how the intended curriculum is manifested in the classroom. Hence, we focus on two types of curriculum representations, i.e., "intended curriculum" and "operational curriculum." For each curriculum representation, according to the framework of Vos et al. (2010), we focus on three "levels of thinking and acting," which are "teaching objectives," "teaching strategies," and "teaching activities." Furthermore, consistent with Vos et al., the concepts of "curriculum emphases" and "instructional functions" are employed to code "teaching objectives" and "teaching strategies," respectively, in this study.

The concept of "curriculum emphases" was first proposed by Roberts (1982), which referred to the general purposes for teaching science curriculum. Originally, Roberts (1982) identified seven different "curriculum emphases." Since the features of "curriculum emphases" are dynamic and socially related (Roberts, 1988), elaborating on the work of Roberts in the context of analyzing chemistry curricula, scholars have categorized "curriculum emphases" afresh (e.g., Wei & Thomas, 2006; Van Driel, Bulte, & Verloop, 2005). In the present study, according to Van Driel et al. (2005) and Wei & Thomas (2006), four distinguished "curriculum emphases" are adopted, which are "fundamental chemistry" (FC); "chemistry, technology, and society" (CTS); "knowledge development in chemistry" (KDC); and "chemistry as inquiry" (CI). FC means that the focus is mainly on facilitating students' formal understanding of the canon of chemistry in order to prepare them for future formal chemical education. CTS implies that students should deepen the knowledge of chemistry and technology and the role of chemistry for present society, so that they can communicate and make decisions about issues involving chemical aspects. KDC is meant to present students with histories of chemistry and nature of science in order to help them understand what chemistry is and how knowledge in chemistry is developed in sociohistorical contexts. CI suggests that students should learn a variety of scientific methods and processes, so that their abilities of thinking and doing can be improved. The concept of "instructional functions" was introduced by Shuell (1996), which referred to teachers' adoptions of some kind of teaching methods and behavior in the process of teaching. The common "instructional functions" include "prior knowledge activation," "comparison," "integration," etc.

Research Method

If research aims to investigate how the intended curriculum is implemented, as suggested by Anderson and Helms (2001), then conducting research in school settings is necessary. This reason justifies why we conducted this study in four senior secondary schools in Nanjing, the capital of Jiangsu province in eastern China. According to Yin (2009), the case study is the preferred strategy when "how" or "why" questions are being posed, when the investigator has little control over events, and when the focus is on a contemporary phenomenon within some real-life context. Given that the present study coincides with the characteristics of case study and the use of curriculum materials of teachers is individualized, multiple case studies are employed in this study.

Context

The senior secondary school chemistry curriculum in China comprises required and selective course modules. Required course modules consist of *Chemistry 1* and *Chemistry 2* (*Chemistry 1* precedes *Chemistry 2*), which are required for all senior secondary school students. The six selective course modules are *Chemistry and Daily Lives*, *Chemistry and Technology*, *Particulate Structure and Properties of Substance*, *Chemical Reaction Mechanism*, *Basic Organic Chemistry*, and *Experimental Chemistry*, which are provided for students according to their needs and interests (MoE, 2003). In China, for a long time, textbooks have been thought of as constituent elements of the national curriculum development (Wu, Chen, & Lu, 1992). The circle of chemistry education in China specially recognizes that textbooks can be seen as the substantiation of the curriculum, and the ideas of the new curriculum should be delivered to practicing teachers through textbooks (Wang, 2010). In practice, chemistry teachers heavily rely on textbooks to determine their teaching content and sequences. Therefore, in the whole process of curriculum reform, no effort has been spared to compile materials and to publish new textbooks. Up to now, there have been three series of senior secondary school chemistry textbooks, which were written according to the national standards of the senior secondary school chemistry curriculum (MoE, 2003), have passed the official examination, and are currently used in schools.

In Nanjing, the series of chemistry textbooks published by the People's Education Press (PEP), which has been designed as the national education press to produce the syllabi and textbooks directly under the leadership of the MoE since the 1950s, is mandated in all senior secondary schools. This series comprises eight textbooks, each of them representing each of the eight curriculum modules. Units and sections constitute the main body of the textbooks. Each unit has three to six sections, which are the basic teaching units in class. Teaching a section usually takes one to three class sessions (40 min for each session). In most cases, some special columns, such as "experiments," "inquiry activities," "scientific perspectives," and "history of chemistry," are inserted in the texts. Main knowledge points are summed in the "summary of this unit." The units end with student exercises. Each textbook is accompanied by one teacher's guide, which is organized in the same sequence as the textbooks. For each unit, the general status and function of this unit, the teaching objectives of this unit, and the time allocation for each section of this unit are provided. For each section, the specific status and the function of the section and pedagogical suggestions are given. The chemistry textbooks and the accompanying teacher's guides constitute the curriculum materials in this study.

Participants

To more fully investigate the issue of chemistry teachers' use of the standards-based curriculum materials at senior secondary schools, we employed the sampling strategy of "maximum variation" (Marshall & Rossman, 2006) to select the participants. That is to say, teachers are varied in terms of school type,[1] teaching experience, and gender. In addition, teachers are also different in the content they taught, e.g., the type of chemistry curriculum (compulsory or elective), and the nature of the content of the unit (theoretical or descriptive chemistry). Finally, five teachers were invited as participants on a voluntary basis in this study. Table 21.1 shows the demographic information for the five teachers.

Data Collection

For each teacher, the research was focused on a whole unit to obtain a relatively complete picture of the related curriculum materials and the teacher's practice. The details of the curriculum materials we observed for each teacher are shown in Table 21.2.

[1] In China, senior secondary schools are classified into exemplary schools and ordinary schools. Exemplary schools have more resources and are able to recruit more competent students than ordinary schools.

Table 21.1 Demographic information for the five teachers

Teacher name	School name	School type	Educational background	Teaching experience	Teaching grade
Ms. Ai	Jiankang	Exemplary	BS/Chemistry Education	25 years	Year 1
Mr. Bi	Yingtian	Exemplary	BS/Chemistry Education	7 years	Year 1
Ms. Chang	Yingtian	Exemplary	BS/Chemistry Education	20 years	Year 2
Mr. Dong	Moling	Ordinary	BS, ME/Chemistry Education	14 years	Year 1
Mr. Fang	Tianjing	Ordinary	BS/Chemistry Education	15 years	Year 2

Note: Both teachers' and schools' names are anonymous

Table 21.2 Curriculum materials used by the five teachers

	Unit	Sections	Module
Ms. Ai	Metals and their compounds	1. Chemical properties of metals	Chemistry 1
		2. Several important metallic compounds	
		3. Metallic materials with wide usages	
Mr. Bi	Chemical substances and changes	1. Classifications of substances	Chemistry 1
		2. Ionic reactions	
		3. Oxidation and reduction reaction	
Ms. Chang	Basic organic substances in lives	1. Grease	Basic Organic Chemistry
		2. Saccharide	
		3. Protein and nucleic acid	
Mr. Dong	Chemistry reactions and energy	1. Chemical energy and thermal energy	Chemistry 2
		2. Chemical energy and electric energy	
		3. The rate and the limitation of chemical reaction	
Mr. Fang	Ionic equilibrium in aqueous solution	1. The ionization of weak electrolytes	Chemical Reaction Mechanism
		2. Water ionization and the acidity and alkalinity of solution	
		3. Hydrolysis of salts	
		4. The dissolution equilibrium of insoluble electrolytes	

For each unit, the teaching objectives, teaching strategies in the teacher's guide, and the special column, "inquiry activity," in the textbooks were selected as the components of the intended curriculum. To find the degree of discrepancies between the operational and intended curriculum, classroom observations and interviews were used as research methods. Each teacher was observed through the whole unit, which lasted approximately from 2 to 4 weeks. Specifically, classroom observations were focused to see what teaching strategies were used and how the special column, "inquiry activity," was implemented. All the observations were also videotaped by using an electronic camera. Moreover, the interviews were conducted after each section, which focused on what teaching objectives were set by the teacher for the section. The interviews were audiotaped and then transcribed into Chinese after the interviews. The transcripts were returned to each of the teachers for their confirmation and criticism. Overall, classroom observations were data sources of "teaching strategies" and "teaching activities" in teacher practices, and interviews were data sources of "teaching objectives" in the operational curriculum.

Data Analysis

For the purpose of this study, we first analyzed the curriculum materials to identify the "teaching objectives," "teaching strategies," and "teaching activities" in the intended curriculum. Then, we analyzed the teacher practices to identify the "teaching objectives," "teaching strategies," and "teaching activities" in the operational curriculum. Finally, we compared the operational curriculum with the intended curriculum to identify the degrees of the discrepancies between these two types of curriculum representations at the three levels.

For each case, the degrees of the implementation of the "teaching objectives" and "teaching strategies" were obtained by using a calculation scheme. Since teaching objectives and strategies are presented in teacher's guides as items, the degrees of the discrepancies between the operational and intended curriculum at these two levels were ranked in terms of the number of these items in which discrepancies were identified, with *none* indicating no item in which discrepancies were identified, *minor* indicating less than half of the number, *middle* indicating half of the number, *major* indicating more than half, and *complete* indicating all of the items in which discrepancies were identified.

For "teaching activities," a similar calculation was conducted to obtain the degree of the discrepancies. At first, each activity was disassembled as several procedures (experiments, activities), and then the degree of the discrepancies between the operational and the intended curriculum for each activity was ranked in terms of the number of these procedures in which discrepancies were identified, with *none* indicating no procedure in which discrepancies were identified, *minor* indicating less than half of the number, *middle* indicating half of the number, *major* indicating more than half, and *complete* indicating all of the procedures in which discrepancies were identified. It should be noted that students should conduct "inquiry activities." Thus, a procedure in an activity that was conducted by the teacher was not counted as a procedure. If there are two or more "inquiry activities" in a unit, the degree of the discrepancies at this level was calculated by using this scheme. First, a numerical value was given to the degree of discrepancies at each "inquiry activity," which are none (1), minor (2), middle (3), major (4), and complete (5). Then, the average score was calculated (if there was a fraction in the average, the principle of rounding off was used), and the calculated average indicated the degree of the discrepancies at this level.

To ensure reliability, the two authors analyzed the data together. Whenever disagreements occurred at any stage of analysis, we discussed our differences, eventually agreeing on one's ideas or a merging of both ideas.

Results

This section presents the five cases one by one. For each case, we will first describe the intended curriculum (curriculum materials) at the three levels of "teaching objectives," "teaching strategies," and "teaching activities," and then we will describe the teacher's operational curriculum (teaching practice) at the three levels. At the end of each case, we will compare the operational curriculum with the intended curriculum and sum up the degrees of the discrepancies between these two types of curriculum representations at the three levels. In the last part of this section, the degrees of discrepancies of the five cases at the three levels will be tabulated to abstract the pattern among the five cases.

Case 1: Ms. Ai

Intended Curriculum

In the teachers' guide, this unit has six teaching objectives (PEP, 2007a, p. 36). These objectives are:

1. *Understand the properties of metals (sodium, aluminum, iron, copper, etc.) and their compounds.*
2. *Tentatively understand the important role of metallic materials in the national economy and the close relationship between metallic materials and daily lives.*
3. *Through the experiments about the properties of metals and their compounds, students should cultivate experimental consciousness, manipulative skills, and abilities of analyzing and solving problems.*
4. *Through reading, searching information, discussions, and so on, students should cultivate the abilities of acquiring knowledge and information processing.*
5. *Through the learning of the properties of metals and their compounds and the applications of metallic materials in productions and daily lives, students' interest in learning chemistry should be improved.*
6. *Through a variety of activities, students should cultivate questioning spirit, innovative spirit, and cooperative spirit.*

According to the category of "curriculum emphases," the first one is about chemical knowledge, which belongs to FC. The second one is about the influence of chemistry on the society and daily lives, which is oriented at CTS. The third and fourth ones are about the cultivation of students' abilities, which can be classified as CI. The fifth and sixth ones are about emotion and attitude, which do not belong to any type of "curriculum emphases." Therefore, we will not be concerned about these two objectives in this study. Overall, the teaching objectives in the teacher's guide refer to three types of "curriculum emphases," namely, FC, CTS, and CI.

With regard to teaching strategies, in section 1, "chemical properties of metals," there are four teaching strategies suggested in the teacher's guides (PEP, 2007a, p. 39–40). The first one is "prior knowledge activation" – "leading students to recall the knowledge about chemical properties of metals and the sequence of metal reactivity which was taught in the junior secondary school before the new lesson." The second one is "experiment" – "to adequately make use of the experiments in the process of teaching." The third one is "inquiry" – "to actively carry out inquiry activities and to make students experience the processes of scientific inquiry." The fourth one is "integration" – "using the method of induction to help students have an overall understanding of the chemical properties of metals." In section 2, "several important metallic compounds," three teaching strategies are set in the teacher's guide (PEP, p. 44). The first one is "experiment" – "to conduct the experiments in the process of teaching." The second one is "picture" – "guiding students to observe the pictures in the textbook carefully and to find a wealth of information contained in the pictures." The third one is "relating" – "strengthening the relationship between the properties of metallic compounds and the sequence of metal reactivity." In section 3, "metallic materials with wide usages," there is only one teaching strategy set in the teacher's guide. The strategy is "social survey" – "to organize students to carry out social surveys so as to enhance their understanding of metallic materials" (PEP, p. 50).

In the textbook, four "inquiry activities" are set in this unit. These activities are (1) "the reaction of aluminum foil and oxygen" (section 1), (2) "the reaction of iron and water vapor" (section 1), (3) "the properties of sodium carbonate and sodium bicarbonate" (section 2), and (4) "the transformation of Fe^{3+} and Fe^{2+}" (section 2) (Song, 2007a). The first one contains two contrasting experiments, namely, heating the aluminum foil without grinding and heating the aluminum foil with grinding. The second one consists of three procedures, which are designing, experimenting, and communicating. The third one is composed of three experiments on the solubility, the acidity or alkalinity of the aqueous solution, and the thermal stability. The fourth one is divided into two steps, which are the reduction of Fe^{3+} into Fe^{2+} and the oxidation of Fe^{2+} into Fe^{3+}.

Operational Curriculum

At the level of "teaching objectives," in the interview after section 1, "chemical properties of metals," Ms. Ai summarized her objectives of this section into two aspects. Her first objective was about letting students know the physical and chemical properties of metals. Her second objective was about cultivating students' abilities in experimenting, observing, thinking, and acquiring knowledge. These two objectives could be coded as FC and CI, respectively. After the second section, "several important metallic compounds," Ms. Ai indicated that the teaching of the properties of several important metallic compounds was the focus of this section. Furthermore, she pointed that cultivating students' abilities in observing, inferring, thinking, and analyzing problems through practical work was another focus of this section. These two objectives belonged to FC and CI, respectively. After section 3, "metallic materials with wide usages," Ms. Ai noted that her objective of this section was about letting students understand the applications of metallic materials in productions and daily lives. This objective was oriented at CTS. Overall, Ms. Ai's objectives of this unit involved three types of "curriculum emphases," namely, FC, CI, and CTS.

At the level of "teaching strategies," in section 1, "chemical properties of metals," Ms. Ai made use of three strategies suggested in the teacher's guide, namely, "prior knowledge activation," "experiment," and "integration." However, the strategy of "inquiry" presented in the teacher's guide was not adopted by her. In section 2, "several important metallic compounds," Ms. Ai adopted the strategy of "experiment" set in the teacher's guide and added a strategy of "graph" – "drawing the graphs of the reactions of two types of binary mixtures (sodium carbonate and sodium hydroxide and sodium carbonate and sodium bicarbonate) and hydrochloric acid." However, she disregarded the strategies of "picture" and "relating" suggested in the teacher's guide. In section 3, "metallic materials with wide usages," Ms. Ai did not adopt the strategy of "social survey" presented in the teacher's guide. Furthermore, she also did not add any other strategy for this section.

For the "inquiry activity," "the reaction of aluminum foil and oxygen" in section 1, Ms. Ai conducted the lesson in this way. She arranged the students in groups, but they were only required to do the experiment on heating the aluminum foil without grinding. Compared with the intended curriculum, the contrasted experiment, that is, heating the aluminum foil with grinding, was not done. For the "inquiry activity," "the reaction of iron and water vapor" in section 1, Ms. Ai arranged it as a teacher's demonstration rather than as student experiments suggested in the textbook. The two "inquiry activities" in section 2 were carried out in this way. The activity on "the properties of sodium carbonate and sodium bicarbonate" was arranged as student experiments in groups. Students completed not only the three experiments on the solubility, the acidity or alkalinity of the aqueous solution, and the thermal stability arranged in the textbook but also an additional experiment, that is, the same mass of solid sodium carbonate and solid sodium bicarbonate reacting with hydrochloric acid separately. The other activity, "the transformation of Fe^{3+} and Fe^{2+}," was also arranged as student experiments as required in the textbook.

In sum, for Ms. Ai, the operational curriculum was consistent with the intended curriculum in terms of "teaching objectives," and thus, the degree of discrepancy was "none." However, at the level of "teaching strategies," the discrepancies existed between the operational and intended curriculum in all three sections, and the degree of discrepancy was "major." At the level of "teaching activities," except for "the transformation of Fe^{3+} and Fe^{2}," the discrepancies were between the operational and intended curriculum for other three activities, and the degree of discrepancy was "middle" at this level.

Case 2: Mr. Bi

Intended Curriculum

In the teacher's guide, there are six teaching objectives set for this unit (PEP, 2007a, p.21). These objectives are:

1. *Recognize the important role that the method of classification plays in chemical research and learning.*
2. *Know that the colloid is a common dispersion system and Tyndall effect.*
3. *Understand the concept of the electrolyte and know that the ionization can occur in acid, alkali, or salt solution.*
4. *Recognize the ionic reaction and its conditions through the experiments.*
5. *Realize that the nature of the oxidation and reduction reaction is electron transfer.*
6. *Recognize the value of the oxidation and reduction reaction in daily lives.*

According to the category of "curriculum emphases," the first one is about the learning of scientific method, which can be coded as CI. The second to fifth ones are about chemistry knowledge and concepts, which belong to FC. The sixth one is about the application of chemistry in daily lives, which is oriented at CTS. Overall, the teaching objectives in the teacher's guide involve three types of "curriculum emphases," namely, CI, FC, and CTS.

At the level of "teaching strategies," in the first section, "classification of substances," only one teaching strategy is suggested in the teacher's guide. The strategy is "inquiry" – "to engage students in the inquiry activity to compare the solution with the colloid with the purpose of being aware of the properties of colloid" (PEP, 2007a, p. 24). In the second section, "ionic reactions," two teaching strategies are set in the teacher's guide (PEP, p. 26). The first one is "experiment" – "doing the demonstration experiments in the process of teaching." The second one is "memorizing" – "leading students to memorize the insoluble substances (barium sulfate, silver halide, etc.)." In the third section, "oxidation and reduction reaction," two teaching strategies are suggested in the teacher's guide (PEP, p. 28). The first one is "analyzing" – "to explain the nature of the oxidation and reduction reaction from outward appearance to inner essence." The second one is "discussion" – "to design the topic of the discussion and to engage students in the discussion."

There is only one "inquiry activity" in this unit, which is set in section 1, "classifications of substances." The purpose of this activity is to let students know the properties of the colloid, and this activity contains three procedures, namely, "preparing the colloid," "observing Tyndall effect," and "filtrating the colloid and precipitation" (Song, 2007a).

Operational Curriculum

At the level of "teaching objectives," after the first section, "classification of substances," Mr. Bi reflected that the teaching objective of this section was about informing students on the features of colloids and Tyndall effect. This objective could be classified as FC. However, he did not mention the teaching objective about the method of classification, which is presented in the teacher's guide. In the interview after the second section, "ionic reactions," Mr. Bi summarized his objectives of this section into three aspects. The first one was about knowing the ionization of the electrolyte. The second one was about recognizing the ionic reaction. The third one was about writing the ionic equation. These three objectives all belonged to FC. After the third section, "oxidation and reduction reaction," Mr. Bi pointed that the teaching of the nature of the oxidation and reduction reaction was the focus of this section. Furthermore, he indicated that letting students understand the applications of the oxidation and reduction reaction in productions and daily lives was another focus of this section. These two objectives were oriented at FC and CTS, respectively. Overall, Mr. Bi's objectives of this unit referred to two types of "curriculum emphases," namely, FC and CTS.

With regard to teaching strategies, in section 1, "classification of substance," Mr. Bi added two strategies, namely, "experiment" and "former, discussion." In the Mr. Bi demonstrated the experiments on the properties of the colloid. Then, he gave students several minutes to discuss the phenomena observed in these experiments. However, he did not use the "inquiry" strategy suggested in the teacher's guide in his class. In section 2, "ionic reactions," Mr. Bi fully adopted the two strategies set in the teacher's guide, namely, "experiment" and "memorizing." In section 3, "oxidation and reduction reaction," Mr. Bi made use of the strategy of "analyzing" presented in the teacher's guide and added two strategies, namely, "experiment," "demonstrating the experiment of galvanic cell," and "exercise," "requiring students to complete the exercises related to oxidation and reduction reaction." However, the strategy of "discussion" suggested in the teacher's guide was not used by him.

For the "inquiry activity," Mr. Bi did not require students to conduct the experiment in groups. Instead, he conducted the experiment as a teaching demonstration.

In sum, the comparison between Mr. Bi's operational curriculum and the intended curriculum indicated discrepancies at all of the three levels. At the level of "teaching objectives," Mr. Bi added an objective about writing the ionic equation (FC) but disregarded an objective about the method of classification (CI), and the degree of discrepancy is "minor." At the level of "teaching strategies," the discrepancies were between the operational and intended curriculum in section 1 and section 3, and the degree of discrepancy was "major." At the level of "teaching activities," students' experiments were replaced by teacher's demonstrations; the degree of discrepancy was "complete."

Case 3: Ms. Chang

Intended Curriculum

In this unit, there are four teaching objectives presented in the teacher's guide (PEP, 2007b, p. 84). These objectives are:

1. *Know the composition, structure, and properties of grease, and tentatively understand the principle of the decontamination of the soap and the application of the synthetic detergent.*
2. *Recognize the composition and properties of saccharide. Understand the applications of saccharide in the food processing and the development of the biological energy. Through the investigative experiments of monosaccharide, disaccharide, and polysaccharide, enable students to further experience the processes of investigating chemical substances, understand the meaning of scientific inquiry, learn the basic methods of scientific inquiry, and enhance the abilities of doing scientific inquiry.*
3. *Know the composition, structure, and main chemical properties of amino acids, and understand the relationship between the protein and human health.*
4. *Recognize the composition, structure, and properties of the protein and the nucleic acid.*

According to the category of "curriculum emphases," the first one is about chemistry knowledge and the application of chemistry in daily lives, which can be classified as FC and CTS. The second one is about chemistry knowledge, the application of chemistry in productions and daily lives, and the learning of scientific method and the cultivation of students' abilities, which can be coded as FC, CTS, and CI. The third one is about chemistry knowledge and the relationship between chemistry and human, which belong to FC and CTS. The fourth one is about chemistry knowledge, which is oriented at FC. Overall, the teaching objectives in the teacher's guide refer to three types of "curriculum emphases," namely, FC, CTS, and CI.

With regard to teaching strategies, in section 1, "grease," there are two strategies set in the teacher's guide (PEP, 2007b, p. 86–87). The first one is "comparison" – "leading students to compare the concept of 'ester' with 'grease' and the concept of 'oil' with 'fat.'" The second one is "experiment" – "to organize students to conduct the experiment of making soap." In section 2, "saccharide," there are three strategies suggested in the teacher's guide (PEP, p. 89–91). The first one is "integration" – "leading students to generalize the structure of properties of monosaccharide, disaccharide, and polysaccharide in the form of the table." The second one is "experiment" – "to make use of the contrasting experiments (the contrasting experiments of the properties of glucose and fructose and the contrasting experiments of the properties of sucrose and maltose) in the process of teaching." The third one is "inquiry" – "to conduct the experiment of investigating the conditions of hydrolyzing starch through chemical methods and to let students experience the processes of scientific inquiry." In the third section, "protein and nucleic acid," there is only one strategy set in the teacher's guide, which is "experiment" – "to adequately conduct the experiments (the combustion, the salting-out, and the denaturation of the protein) in the process of teaching the properties of the protein" (PEP, p. 93–94).

This unit contains three "inquiry activities," which are (1) "the reducibility of fructose" (section 2); (2) "the reducibility of sucrose, maltose, and sucrose hydrolysates" (section 2); and (3) "the conditions for the chemical hydrolysis of starch" (section 2) (Song, 2007b). The first one includes two procedures, namely, designing the experiment and conducting the experiment. The second one contains three experiments on the reducibility of the sucrose solution, the maltose solution, and the sucrose hydrolysate solution. The third one also includes two procedures, which are designing the experiment and conducting the experiment.

Operational Curriculum

At the level of "teaching objectives," after section 1, "grease," Ms. Chang indicated that this section had two aspects of the objectives. One was about the concept, structure, and properties of grease. The other one was about the principle of the decontamination of the soap and the production of the synthetic detergent. These two objectives belonged to FC and CTS, respectively. After section 2, "saccharide," Ms. Chang summarized her objectives of this section into two aspects. The first one was about the structures and properties of monosaccharide, disaccharide, and polysaccharide. The second one was about the applications of saccharide in productions and daily lives. These two objectives could be classified as FC and CTS, respectively. However, Ms. Chang did not mention the teaching objective relevant to the learning of scientific method and the cultivation of students' abilities. In the interview after section 3, "protein and nucleic acid," Ms. Chang noted that informing students on the structures and properties of the amino acid, the protein, and the nucleic acid was the focus of this section. In addition, she pointed that letting students understand the relationship between the protein and lives was another focus of this section. These two objectives could be coded as FC and CTS, respectively. Overall, Ms. Chang's objectives of this unit involved two types of "curriculum emphases," namely, FC and CTS.

At the level of "teaching strategies," in the first section, "grease," Ms. Chang fully adopted the two strategies suggested in the teacher's guide, namely, "comparison" and "experiment." In section 2, "saccharide," Ms. Chang made use of two strategies of "integration" and "experiment" presented in the teacher's guide. However, the strategy of "inquiry" set in the teacher's guide was not adopted by her. In the third section, "protein and nucleic acid," Ms. Chang used the "experiment" strategy suggested in the teacher's guide.

In Ms. Chang's class, all three "inquiry activities" were arranged as student experiments in groups. The activity about "the reducibility of sucrose, maltose, and sucrose hydrolysates" was conducted according to the arrangement in the textbook with the three experiments carried out by students, respectively. For the activities on "the reducibility of fructose" and "the conditions of chemical hydrolysis of starch," Ms. Chang provided the procedures of the experiments instead of requiring students to design the experiments, and the tasks of students were to conduct the experiments according to these procedures.

In sum, for Ms. Chang, there were discrepancies between the operational and intended curriculum at all three levels. At the level of "teaching objectives," Ms. Chang did not identify the learning of scientific method and the cultivation of students' abilities (CI) as a teaching objective, and the degree of discrepancy was "minor." At the level of "teaching strategies," she disregarded the strategy of "inquiry," and the degree of discrepancy was "minor." At the level of "teaching activities," she adapted the investigative experiments into the confirmative students' experiments for two activities, namely, "the reducibility of fructose" and "the conditions of chemical hydrolysis of starch," and the degree of discrepancy was "minor."

Case 4: Mr. Dong

Intended Curriculum

In this unit, five objectives are arranged in the teacher's guide (PEP, 2007c, p. 15). These objectives are:

1. *Know that breakdown and formation of chemical bonds is the main reason for energy changes in chemical reactions. Understand that the comparison of the amounts of energy of reactants and products determines whether the reaction will absorb energy or release energy.*
2. *Understand the transformation of chemical energy and thermal energy and its applications in productions and daily lives.*
3. *Recognize the transformation of chemical energy and electric energy and its significance for the modernization.*
4. *Tentatively understand the rate of chemical reaction and its influencing factors. Know that the chemical reaction has a certain limitation. Recognize the significance of controlling the reaction conditions for productions, lives, and scientific research.*
5. *Realize the method and value of experiment in investigating the principles of chemical reactions.*

According to the category of "curriculum emphases," the first one is about chemistry knowledge and principles, which belongs to FC. The second to fourth ones are all about chemistry knowledge and the applications of chemistry in productions and daily lives, which can be coded as FC and CTS. The fifth one is about the learning of scientific method, which is oriented at CI. Overall, the teaching objectives in the teacher's guide involve three types of "curriculum emphases," namely, FC, CTS, and CI.

At the level of "teaching strategies," in the first section, "chemical energy and thermal energy," three strategies are suggested in the teacher's guide (PEP, 2007c, p. 19). The first one is "model" – "using the simulated courseware to visualize the abstract knowledge that breakdown and formation of chemical bonds is the main reason for energy changes in chemical reactions." The second one is "analogy" – "to use changes of water energy as an analogy to energy changes in chemical reactions." The third one is "experiment" – "to adequately make use of the experiments arranged in the textbook." In section 2, "chemical energy and electric energy," three strategies are presented in the teacher's guide (PEP, p. 25–27). The first one is "experiment" – "to make students realize the transformation of chemical energy and electric energy through experiments." The second one is "inquiry" – "to engage students in the investigative experiment of galvanic cell and let students master the manipulative and observational skills." The third one is "discussion" – "to organize students to discuss about the constitutions of galvanic cell in groups." In section 3, "the rate and the limitation of chemical reaction," also three strategies are set in the teacher's guide (PEP, p. 32–34). The first one is "context" – "designing the context to lead to the concept of the rate of chemical reaction." The second one is "experiment" – "to reveal the factors that affect the rate of chemical reaction through the experiments arranged in the textbook." The third one is "discussion" – "to organize students to discuss the changes of forward reaction rate and reverse reaction rate in the process of transiting to the state of equilibrium for the reversible reaction."

Only one "inquiry activity" is set for this unit, which is arranged in the second section, "chemical energy and electric energy." This activity mainly involves designing a galvanic cell and includes three procedures, namely, designing the experimental setting, conducting the experiment, and drawing the conclusion (Song, 2007c).

Operational Curriculum

With regard to teaching objectives, in the interview after section 1, "chemical energy and thermal energy," Mr. Dong summarized his objectives of this section into two aspects. The first one was about letting students know the transformation of chemical energy and thermal energy and recognize the reasons for energy changes in chemical reactions from the perspectives of macro and micro. The second one was about making students understand the applications of the transformation of chemical energy and thermal energy in daily lives. These two objectives could be classified as FC and CTS, respectively. After the second section, "chemical energy and electric energy," Mr. Dong pointed that the principle and constitution of galvanic cell was the focus of this section. Moreover, he indicated that the applications and development of galvanic cell was another focus of this section. These two objectives were oriented at FC and CTS, respectively. After the third section, "the rate and the limitation of chemical reaction," Mr. Dong stated that his teaching objectives had two aspects, which are the concepts of the chemical reaction rate and chemical equilibrium and the applications of these concepts in daily lives and industries. These two aspects of objectives belonged to FC and CTS, respectively. However, he did not mention the objective of the method of experiment in the interview. Overall, Mr. Dong's objectives of this unit referred to two types of "curriculum emphases," namely, FC and CTS.

At the level of "teaching strategies," in section 1, "chemical energy and thermal energy," Mr. Dong made use of two strategies presented in the teacher's guide, namely, "analogy" and "experiment." However, the "model" strategy suggested in the teacher's guide was not adopted by him. In the second section, "chemical energy and electric energy," Mr. Dong adopted the two strategies of "experiment" and "discussion" set in the teacher's guide and added a strategy of "exercise" – "making students complete the exercises related to galvanic cell." However, he did not use the "inquiry" strategy suggested in the teacher's guide. In the third section, "the rate and the limitation of chemical reaction," Mr. Dong made use of the two strategies of "context" and "discussion" set in the teacher's guide but did not adopt the strategy of "experiment" presented in the teacher's guide.

In the actual process of teaching, Mr. Dong ignored the "inquiry activity" arranged in the textbook.

In sum, comparing his operational curriculum with the intended curriculum, Mr. Dong demonstrated discrepancies at all three levels. At the level of "teaching objectives," he failed to identify the method of experiment (CI) as his teaching objective, and the degree of discrepancy was "minor." At the level of "teaching strategies," there existed the discrepancies between the operational and intended curriculum in all three sections, and the degree of discrepancy is "minor." At the level of "teaching activities," Mr. Dong disregarded the "inquiry activity" in the textbook; the degree of discrepancy was "complete."

Case 5: Mr. Fang

Intended Curriculum

In the teacher's guide, five teaching objectives are set for this unit (PEP, 2007d, p. 54). These objectives are:

1. *Understand the concepts of strong electrolytes and weak electrolytes and the ionization equilibrium of weak electrolytes.*

2. *Know that water is a kind of very weak electrolytes and the ion product of water is constant under a certain temperature.*
3. *Tentatively master the method of measuring pH of solution. Know that the control of pH of solution plays an important role in productions of industry and agriculture.*
4. *Recognize the principle of hydrolysis of salts and the factors that affect the degree of hydrolysis of salts. Understand the applications of hydrolysis of salts in productions and daily lives.*
5. *Know the concept of dissolution equilibrium and the nature of the transformation of precipitation.*

According to the category of "curriculum emphases," the first, second, and fifth ones are about chemistry knowledge and concepts, which are all oriented at FC. The third one is about the learning of scientific method and the applications of chemistry in productions, which is classified as CI and CTS. The fourth one is about chemical principle and theory and the applications of chemistry in productions and lives, which belongs to FC and CTS. Overall, the teaching objectives in the teacher's guide refer to three types of "curriculum emphases," namely, FC, CI, and CTS.

At the level of "teaching strategies," in the first section, "the ionization of weak electrolytes," there are two strategies suggested in the teacher's guide (PEP, 2007d, p. 56–57). The first one is "experiment" – "to make use of experiments to help students understand the concepts of strong electrolytes and weak electrolytes." The second is "graph" – "using the graph to help students recognize the process of the ionization equilibrium of weak electrolytes." In the second section, "water ionization and the acidity or alkalinity of solution," there are three strategies presented in the teacher's guide (PEP, p. 62–64). The first one is "discussion" – "to organize students to discuss what the impact of adding hydrochloric acid or sodium hydroxide solution into distilled water on the ionization equilibrium of water is." The second one is "exercise" – "making students complete the exercises related to the calculation of pH value of solution." The third one is "experiment" – "to organize students to conduct the experiment of acid–base neutralization titration." In the third section, "hydrolysis of salts," there are two strategies set in the teacher's guide (PEP, p. 67–68). The first one is "experiment" – "leading students to recognize the relationship between salt composition and acidity or alkalinity of salt solution through experiments." The second one is "inquiry" – "to organize students to investigate the factors that affect the shift of the hydrolysis equilibrium." In the fourth section, "the dissolution equilibrium of insoluble electrolytes," there is only one strategy suggested in the teacher's guide, which is "experiment" – "to make students understand the dissolution and transformation of precipitation through experiments" (PEP, p. 74).

This unit contains two "inquiry activities," which are "relationship between salt composition and acidity or alkalinity of salt solution" (section 3) and the "inquiring factors that affect the degree of hydrolysis of salts" (section 3) (Song, 2007d). The first one consists of three procedures, which are "determining acidity or alkalinity of salt solution," "distinguishing the types of salts," and "generalizing the relationship between the acidity or alkalinity of salt solution and types of salts." The second one is composed of four procedures, which are "analyzing the hydrolytic process of $FeCl_3$," "designing the experiment," "conducting the experiment," and "concluding and communicating."

Operational Curriculum

At the level of "teaching objectives," after the first section, "the ionization of weak electrolytes," Mr. Fang indicated that his objectives of this section had three aspects. The first one was about the concepts of strong electrolytes and weak electrolytes. The second one was about the ionization equilibrium of weak electrolytes. The third one was about the factors that affect the shift of the ionization equilibrium. These three objectives were all oriented at FC. In the interview after section 2, "water ionization and the acidity or alkalinity of solution," Mr. Fang summarized his objectives into three aspects, which are the ionization equilibrium and ion-product constant of water, the method of measuring pH of solution, and the applications of pH in productions of agriculture and human health. These three objectives could be coded as FC, CI, and CTS, respectively. After the third section, "hydrolysis of salts," Mr. Fang reflected that there were four objectives in this section. The first one was about the concept and principle of hydrolysis of salts. The second one was about the factors that affect the degree of hydrolysis of salts. The third one was about learning to determine the relationship between the concentrations of various ions in the solution. The fourth one was about the applications of hydrolysis of salts in productions and lives. The first three belonged to FC and the last one belonged CTS. In the interview after section 4, "the dissolution equilibrium of insoluble electrolytes," Mr. Fang noted that the concept of dissolution equilibrium of insoluble electrolytes was the focus of this section. Furthermore, he pointed that the nature of the formation, dissolution, and transformation of precipitation was another focus of this section. These two objectives were both oriented at FC. Overall, Mr. Fang's objectives of this unit involved three types of "curriculum emphases," namely, FC, CI, and CTS.

With regard to teaching strategies, in section 1, "the ionization of weak electrolytes," Mr. Fang made use of two strategies of "experiment" and "graph" set in the teacher's guide and added a strategy of "prior knowledge activation" – "leading students to review the concepts of electrolytes and nonelectrolytes before teaching the new lesson." In section 2, "water ionization and the acidity or alkalinity of solution," Mr. Fang fully adopted the three strategies suggested in the teacher's guide, which are "discussion," "exercise," and "experiment." In section 3, "hydrolysis of salts," Mr. Fang adopted the strategy of "experiment" presented in the teacher's guide and added a strategy of "exercise" – "make students complete the exercises related to writing hydrolyzed equations." However, the "inquiry" strategy set in the teacher's guide was not adopted by him. In section 4, "the dissolution equilibrium of insoluble electrolytes," Mr. Fang used the strategy of "experiment" suggested in the teacher's guide.

Table 21.3 The degrees of discrepancies of the five cases at the three levels

	Ms. Ai	Mr. Bi	Ms. Chang	Mr. Dong	Mr. Fang
Teaching objectives	None	Minor	Minor	Minor	Minor
Teaching strategies	Major	Major	Minor	Minor	Minor
Teaching activities	Middle	Complete	Minor	Minor	Major

For the "inquiry activity" of "relationship between salt composition and acidity or alkalinity of salt solution," the first procedure (determining acidity or alkalinity of salt solution) was adapted as a teaching demonstration. Students accomplished the other two procedures in groups. For another "inquiry activity," "inquiring factors that affect the degree of hydrolysis of salts," Mr. Fang disregarded it in the actual process of teaching.

In sum, the comparison between Mr. Fang's operational curriculum and the intended curriculum indicated discrepancies at all three levels. At the level of "teaching objectives," he added two objectives of FC, namely, "the factors that affect the shift of the ionization equilibrium" and "learning to determine the relationship between the concentrations of various ions in the solution," and the degree of discrepancy is "minor." At the level of "teaching strategies," the discrepancies were between the operational and intended curriculum in section 1 and section 3, and the degree of discrepancy was "minor." At the level of "teaching activities," Mr. Fang adapted an activity and ignored another activity, and the degree of discrepancy is "major."

Summary

Based on the foregoing analysis, the degrees of discrepancies of the five teachers' operational curricula compared with the intended curricula at the three levels are summarized in Table 21.3.

As shown in Table 21.3, at the level of "teaching objectives," the operational and intended curricula were consistent for Ms. Ai. However, the other four teachers (Mr. Bi, Ms. Chang, Mr. Dong, and Mr. Fang) exhibited "minor" discrepancies. At the "teaching strategies" level, three teachers (Ms. Chang, Mr. Dong, and Mr. Fang) exhibited "minor" discrepancies and the other two (Ms. Ai and Mr. Bi) exhibited "major" discrepancies. At the "teaching activities" level, Ms. Chang exhibited a "minor" discrepancy, and Ms. Ai exhibited a "middle" discrepancy. The degrees of discrepancies for the other three (Mr. Fang, Mr. Bi, and Mr. Dong) were either "major" or "complete." In general, distinct discrepancies were present between the operational and intended curricula among the five cases. Furthermore, for the three levels, "teaching objectives" exhibited a smaller degree of discrepancies, whereas "teaching activities" presented a larger degree of discrepancies, with "teaching strategies" being between the two.

Discussion and Implications

In this paper, we compared teachers' operational curricula with the intended curricula and identified the degrees of the discrepancies between these two types of curriculum representations at the three levels of "teaching objectives," "teaching strategies," and "teaching activities." As mentioned at the beginning of this paper, chemistry textbooks and teacher's guides developed for the secondary school level in China can be seen as the standards-based curriculum materials. Hence, this study provided empirical evidence for the actual use of this kind of curriculum materials in the classroom. The results revealed that the discrepancies between the operational and intended curricula were obvious. It embodied that the implementation of the new senior secondary school chemistry curriculum has not been going well in China, indicating the problematic interaction between teachers and the curriculum materials, which was akin to that in western countries (e.g., Schneider et al., 2005; Vos et al., 2011). It shows that the inconsistence between teaching practice and the standards-based curriculum materials is a global problem.

Furthermore, in this study, we found that "teaching objectives" presented a smaller degree of discrepancies among the three levels. We argue that this fact was due to the character of "teaching objectives." "Teaching objectives" is a more abstract level, which does not involve specific teaching behaviors or activities. In addition, the same teaching objective can be achieved by different teaching behaviors and activities. From this perspective, comparing with the other two levels, "teaching objectives" is not easier to exhibit discrepancies in the process of curriculum implementation. On the contrary, "teaching activities" presented a larger degree of discrepancies, and the discrepancies in this level mainly resulted from omissions rather than additions in the operational curricula. It indicates the situation of the implementation of curriculum materials with regard to scientific inquiry in the classroom is not ideal. This finding is consistent with those of previous research on the implementation of scientific inquiry, like Roehrig, Kruse, and Kern's (2007) and Zhang et al.'s (2003). It is worth mentioning that, different from the findings of this study, the degrees of discrepancies between curriculum representations at the three levels do not have distinct difference in Vos et al.'s (2011) study. We think it is related to the difference of grades of the degree of discrepancies set in the two studies. Vos et al. (2011) only distinguished the degree of discrepancies into three grades, which were "none," "partial," and "complete," whereas there were up to five grades of the degree of discrepancies set in this study, namely, "none," "minor," "middle," "major," and "complete." We argue that the more grades are set, the more findings will be obtained.

We confess that it is not our original intention to evaluate whether teachers' adaptations of curriculum materials impel or impede the implementation of the standards-based science curriculum. However, while taking a closer look at teachers' adaptations, it can be seen that teachers often disregard CI-emphasis, "inquiry" strategy, and some activity procedures, whereas they often add FC-emphasis and "exercise" strategy. Obviously, teachers' adaptations of curriculum materials have a definite direction, which is weakening the standards-based curriculum but strengthening the tradition-oriented curriculum. In other words, teachers' adaptations usually hinder the implementation of the standards-based science curriculum.

In this study, we adopted four "curriculum emphases" to code "teaching objectives," which were FC, CTS, KDC, and CI. It can be seen from the results that FC, CTS, and CI are all reflected in both curriculum materials and teachers' practice, but KDC is missing. As mentioned earlier, KDC emphasizes the learning of history of chemistry and nature of science. We argue that the absence of this emphasis is connected with the setting of the goals of the senior secondary school chemistry curriculum in China. In the national standard of senior secondary school chemistry curriculum, the curriculum goals are defined by three dimensions, namely, "knowledge and skills," "processes and methods," and "emotions, attitudes, and values" (MoE, 2003). Comparing these three dimensions with the four "curriculum emphases," it can be seen that "knowledge and skills" and "processes and methods" are roughly equal to the emphases of FC and CI, respectively, and CTS-emphasis is encompassed in the dimension of "emotions, attitudes, and values." However, KDC-emphasis is not involved in any dimensions of goals. Given that the curriculum goals seriously neglect KDC, it is not surprising to have found that this emphasis disappears in either curriculum materials or teaching practice.

As we know, history of science and nature of science are two interrelated important themes in contemporary science education reform (AAAS, 1989; NRC, 1996). According to McComas, Almazroa, and Clough (1998), the value of nature of science for teaching and learning included five aspects: (1) to enhance the learning of science content, (2) to enhance understanding of science, (3) to enhance interest in science, (4) to enhance decision making, and (5) to enhance instructional delivery. Driver, Leach, Millar, and Scott (1996) also systematically elaborated the necessity of incorporating history of science and nature of science into science education from five arguments, which were the utilitarian view, the democratic view, the culture view, the moral view, and the view of science learning. In this sense, in the future curriculum revision, curriculum materials written, and teachers' practice, KDC-emphasis should be taken seriously. It will facilitate students to understand the nature of science and improve their scientific literacy.

In this paper, we only disclosed that there existed discrepancies between the operational and intended curricula. However, the reasons for discrepancies between these two types of curriculum representations have not been investigated. According to Brown (2009) and Remillard (2005), the teacher's use of curriculum materials is usually influenced by many factors, such as the quality of the designs of curriculum materials, individual teacher characteristics, and features of the social context. That is to say, the interaction between the teacher and curriculum materials is a complex process which is influenced by multifaceted factors. Therefore, it can be inferred that there should be a variety of factors that lead to the discrepancies between teachers' operational curricula and the intended curricula. What are these factors? How do these factors lead to discrepancies? And what is the relationship between the various factors? We suggest that these questions merit to be explored in further studies. It will help us understand the interaction of the teacher with curriculum materials more deeply and give us some inspirations for enhancing the consistency between teaching practice and curriculum materials.

References

Aikenhead, G. S. (2006). *Science education for everyday life*. New York, NY: Teachers College Press.

American Association for the Advancement of Science (AAAS). (1989). *Science for all American: A project 2061 report on goals in science, mathematics, and technology*. Washington, DC: Author.

Anderson, R. D., & Helms, J. V. (2001). The ideal of standards and the reality of schools: Needed research. *Journal of Research in Science Teaching, 38*(1), 3–16.

Ball, D. L., & Feiman-Nemser, S. (1988). Using textbooks and teachers' guides: A dilemma for beginning teachers and teacher educators. *Curriculum Inquiry, 18*, 401–423.

Brown, M. W. (2009). The teacher-tool relationship: Theorizing the design and use of curriculum materials. In J. T. Remillard, B. A. Herbel-Eisenmann, & G. M. Lloyd (Eds.), *Mathematics teachers at work: Connecting curriculum materials and classroom instruction* (pp. 17–36). New York, NY: Routledge.

Driver, R., Leach, J., Millar, R., & Scott, P. (1996). *Young peoples' images of science*. Bristol, UK: Open University Press.

Forbes, C., & Davis, E. A. (2010). Curriculum design for inquiry: Preservice elementary teachers' mobilization and adaptation of science curriculum materials. *Journal of Research in Science Teaching, 47*, 820–839.

Gao, L. (2007). Reform and challenge: An review of new science curriculum reform in China. In Y. Sio & M. H. Cheung (Eds.), *The research of science education in Chinese society* (pp. 17–28). Hong Kong: Seedland Publishing Limited (in Chinese).

Goodlad, J. I. (1979). The scope of curriculum field. In J. I. Goodlad et al. (Eds.), *Curriculum inquiry: The study of curriculum practice* (pp. 17–41). New York, NY: McGraw-Hill.

Herman, J. L., & Webb, N. M. (2007). Alignment methodologies. *Applied Measurement in Education, 20*(1), 1–5.

Lloyd, G. M., Remillard, J. T., & Herbel-Eisenmann, B. A. (2009). Teachers' use of curriculum materials: An emerging field. In J. T. Remillard, B. A. Herbel-Eisenmann, & G. M. Lloyd (Eds.), *Mathematics teachers at work: Connecting curriculum materials and classroom instruction* (pp. 3–14). New York, NY: Routledge.

Marshall, C., & Rossman, G. B. (2006). *Designing qualitative research* (4th ed.). Thousand Oaks, CA: Sage Publications.

McComas, W. F., Almazroa, H., & Clough, M. P. (1998). The nature of science in science education: An introduction. *Science & Education, 7*, 511–532.

McNeil, J. D. (2003). *Curriculum: The teacher's initiative* (3rd ed.). Upper Saddle River, NJ: Pearson Education.

Ministry of Education (MoE). (2001). *The chemistry curriculum standard of compulsory education of full-time system*. Beijing, China: Beijing Normal University (in Chinese).

Ministry of Education (MoE). (2003). *National standards of general senior secondary school chemistry curriculum*. Beijing, China: People's Education Press (in Chinese).

National Research Council (NRC). (1996). *National science education standards*. Washington, DC: Author.

People Education Press (PEP). (2007a). *Chemistry 1: Teachers' guide* (3rd ed.). Beijing, China: Author (in Chinese).

People Education Press (PEP). (2007b). *Basic organic chemistry: Teachers' guide* (3rd ed.). Beijing, China: Author (in Chinese).
People Education Press (PEP). (2007c). *Chemistry 2: Teachers' guide* (3rd ed.). Beijing, China: Author (in Chinese).
People Education Press (PEP). (2007d). *Chemical reaction mechanism* (3rd ed.). Beijing, China: Author (in Chinese).
Powell, J. C., & Anderson, R. D. (2002). Changing teachers' practice: Curriculum materials and science education reform in the USA. *Studies in Science Education, 37*, 107–136.
Remillard, J. T. (2005). Examining key concepts in research on teachers' use of mathematics curricula. *Review of Educational Research, 75*, 211–246.
Remillard, J. T., Herbel-Eisenmann, B. A., & Lloyd, G. M. (Eds.). (2009). *Mathematics teachers at work: Connecting curriculum materials and classroom instruction*. New York, NY: Routledge.
Roberts, D. A. (1982). Developing the concept of 'curriculum emphasis' in science education. *Science Education, 66*(2), 243–260.
Roberts, D. A. (1988). What counts as science education? In P. Fensham (Ed.), *Development and dilemma in science education* (pp. 27–54). Philadelphia, PA: The Falmer.
Roehrig, G. H., Kruse, R. A., & Kern, A. (2007). Teacher and school characteristics and their influence on curriculum implementation. *Journal of Research in Science Teaching, 44*, 883–907.
Schneider, M. R., Krajcik, J., & Blumenfeld, P. (2005). Enacting reform-based science materials: The range of teacher enactments in reform Classrooms. *Journal of Research in Science Teaching, 42*, 283–312.
Shuell, T. J. (1996). Teaching and learning in a classroom context. In D. C. Berliner & R. C. Calfee (Eds.), *Handbook of educational psychology* (pp. 726–764). New York, NY: Simon and Schuster Macmillan.
Song, X. Q. (Ed.). (2007a). *Chemistry 1* (3rd ed.). Beijing, China: People Education Press (in Chinese).
Song, X. Q. (Ed.). (2007b). *Basic organic chemistry* (3rd ed.). Beijing, China: People Education Press (in Chinese).
Song, X. Q. (Ed.). (2007c). *Chemistry 2* (3rd ed.). Beijing, China: People Education Press (in Chinese).
Song, X. Q. (Ed.). (2007d). *Chemical reaction mechanism* (3rd ed.). Beijing, China: People Education Press (in Chinese).
Van Driel, J. H., Bulte, A. M. W., & Verloop, N. (2005). The conceptions of chemistry teachers about teaching and learning in the context of a curriculum innovation. *International Journal of Science Education, 27*(3), 303–322.
Van Hiele, P. M. (1986). *Structure and insight: A theory mathematics education*. New York, NY: Academic Press.
Vos, M. A. J., Taconis, R., Jochems, W. M .G., & Pilot, A. (2010). Teachers implementation context-based teaching materials: A framework for case-analysis in chemistry. *Chemistry Education Research and Practice, 11*, 193–206.
Vos, M. A. J., Taconis, R., Jochems, W. M. G., & Pilot, A. (2011). Classroom implementation of context-based chemistry education by teachers: The relation between experiences of teachers and the design of materials. *International Journal of Science Education, 33*, 1407–1432.
Wang, L. (2010). Progress and reflection of secondary chemistry curriculum reform in the past ten years (part A). *Chinese Journal of Chemical Education, 4*, 15–21 (in Chinese).
Wei, B. (2010). The changes in science curricula in China after 1976: A reflective review. In Y. J. Lee (Ed.), *Handbook of research in science education research in Asia* (pp. 89–102). Boston, MA: Sense Publishers.
Wei, B., & Chen, B. (in press). Examining the senior secondary school chemistry curriculum in China: In the view of scientific literacy. In L. Liang, X. F. Liu, & G. W. Fulmer (Eds.), Science education in China: Policy, research and practice. Berlin: Springer.

Wei, B., & Thomas, G. (2006). An examination of the change of the junior secondary school chemistry curriculum in the P. R. China: In the view of scientific literacy. *Research in Science Education, 36*, 403–418.

Wu, L., Chen, H., & Lu, D.(1992). Theories and practices in making teaching materials of primary and secondary schools in new China. In: The Chinese Association of International Education Exchange (Ed.), *Curriculum development and social progress: Selected papers in the international symposium on curriculum development and social progress* (pp. 15–22). Beijing, China: People's Education Press (in Chinese).

Yin, R. K. (2009). *Case study research: Design and methods* (4th ed.). Thousand Oaks, CA: Sage Publications.

Zhang, B., Krajcik, J. S., Sutherland, L. M., Wang, L., Wu, J., & Qian, Y. (2003). Opportunities and challenges of China's inquiry-based education reform in middle and high schools: Perspectives of science teachers and teacher educators. *International Journal of Science and Mathematics Education, 1*, 477–503.

Chapter 22
Supporting Socio-scientific Argumentation in the Classroom Through Automatic Group Formation Based on Students' Real-time Responses

Yu-Ta Chien and Chun-Yen Chang

Fostering citizens, who are capable for making thoughtful decisions on socio-scientific issues, which are multifaceted social issues with conceptual links to science, has become an important goal of science education in western countries (Driver, Newton, & Osborne, 2000; Kolstø, 2000; Kortland, 1996; Sadler, 2004). It is also currently one of the objectives of education reform in Taiwan (Ministry of Education, 2001, 2004). With the premise that a thoughtful decision on a socio-scientific issue should be made on the grounds of complete and coherent arguments, several studies in science education have investigated students' discourse patterns regarding socio-scientific issue to reveal the defects of students' reasoning processes. For instance, regarding waste management and recycling issues, Kortland (1996) examined the argumentation patterns of middle-school students. It was found that, although students in general had no problems in proposing arguments to support their decisions on the issues, most of them did not adequately take opposing positions into account. As a consequence, students' arguments lacked necessary rebuttals to fight against counter positions and to justify their decisions. Jiménez-Aleixandre, Rodríguez, and Duschl (2000) explored ninth grades' small-group argumentation patterns while students were resolving a socio-scientific issue involving genetic and environmental variability of farm-raised chickens. Echoing Kortland's study (1996), the group on which Jiménez-Aleixandre et al. (2000) focused did not demonstrate any rebuttal to respond to counterarguments. These findings signal a need to facilitate students' ability to attend to opposing positions and, accordingly, to defend contentious positions. This chapter, therefore, describes innovative instructional strategies, which are developed under the support of a national-level education reform project in Taiwan, to address the need.

Y.-T. Chien • C.-Y. Chang (✉)
Graduate Institute of Science Education, National Taiwan Normal University, Taipei, Taiwan
e-mail: changcy@ntnu.edu.tw

M.S. Khine (ed.), *Science Education in East Asia*,
DOI 10.1007/978-3-319-16390-1_22

Enhancing the Quality of Students' Argumentation

Possible reasons underlying the defects of students' argumentation regarding socio-scientific issues are identified from literature (Kortland, 1996; Sadler, 2004). The naive arguments presented by students are hypothetically due to two main factors: (1) a lack of knowledge about the socio-scientific issue of concern and (2) inexperience in the formulation of qualitative arguments. Two instructional orientations may help students overcome these difficulties. The first one is to activate and supply students' scientific knowledge regarding the socio-scientific issues of concern. This orientation can be effectively fulfilled by contingent teaching. Before engaging students in reasoning a socio-scientific issue, the teacher poses a scientific question related to that issue to activate students' knowledge which they may use to construct arguments. Depending on students' responses, the teacher then decides whether to provide a mini-lecture to supply and extend their knowledge bases. By executing the aforementioned steps iteratively, students should be better prepared for participating in socio-scientific issue argumentation. The second orientation is to provide students with ample opportunities to practice justifying claims, attending to counter positions, and dissecting contradictory opinions. This orientation is commonly fulfilled by having students engage in small-group discussion about socio-scientific issues. Students may become aware of that which constitutes well-reasoned arguments if they devote efforts to convincing and defending their peers. Students' argumentation skills may thus be improved after the small-group discussion session.

However, teachers will run into two main obstacles in a real classroom setting if they try to implement the aforementioned instructional strategies. When conducting contingent teaching, a teacher needs a real-time summary of students' responses to each in-class question to decide how the class should proceed. It is very time consuming to collect and synthesize every student's response during the lecture. The instructional time will become so compressed that the teacher may ignore how well or bad students understand the concept being taught and just follow an ossified teaching plan. In terms of orchestrating small-group argumentation activities, the teacher needs to assign students into groups first. The most commonly used method is to ask students to form dyads with their neighbors, which is termed as *convenient grouping* in this chapter. However, this method may not be adequate for directing students to attend to counter positions and dissect contradictory opinions. If the students with the same position are grouped into a dyad, as the convenient grouping method often does, they have to imagine what arguments may be proposed by an opponent. The *imaginary* opponent is constructed by students themselves and thus may not be effective to jump out of the box to attack students' blind side. Therefore, arranging students with conflicting positions into a dyad, which is termed as *heterogeneous grouping* in this chapter, should be a better way to orchestrate a small-group discussion. The heterogeneous grouping method makes sure that every student in a group has at least one *real* opponent to dialectically examine and refute their arguments. Students thus have an authentic opportunity to find the blind side

Fig. 22.1 Using CCR to pose and respond to in-class questions

of their own positions and strength it. However, identifying every student's position over the socio-scientific issue of concern and assigning them into heterogeneous groups during the lecture are very intractable and time-consuming tasks for teachers. Therefore, it is not surprising that most teachers would rather ask students to form dyads with neighbors, which is a convenient but not effective way, to proceed with small-group discussions.

We have developed a web-based instant response system, called Cloud Class Room (CCR), to assist teachers in conducting contingent teaching and heterogeneous grouping. CCR is written in jQuery and PHP scripts. It works on every Internet-capable device, such as PCs, laptops, PDAs, smartphones, or tablets, without further software or plug-in installation. As shown in the left panel of Fig. 22.1, once a teacher gets an Internet-capable device to log in to CCR, she/he can pose a true-false/multiple-choice/open-ended question to students. At the same time, as shown in the right panel of Fig. 22.1, students can respond to the question by clicking the options, or typing words, on their own handheld devices; CCR then instantly aggregates the answers from the entire class with a histogram as shown in Fig. 22.2. The teacher thus gets well informed about whether he/she should articulate the current learning unit again or proceed to the next stage. Another main functionality of CCR is automatic group formation. As shown in bottom of Fig. 22.2, once students respond to the question, CCR enables the teacher to

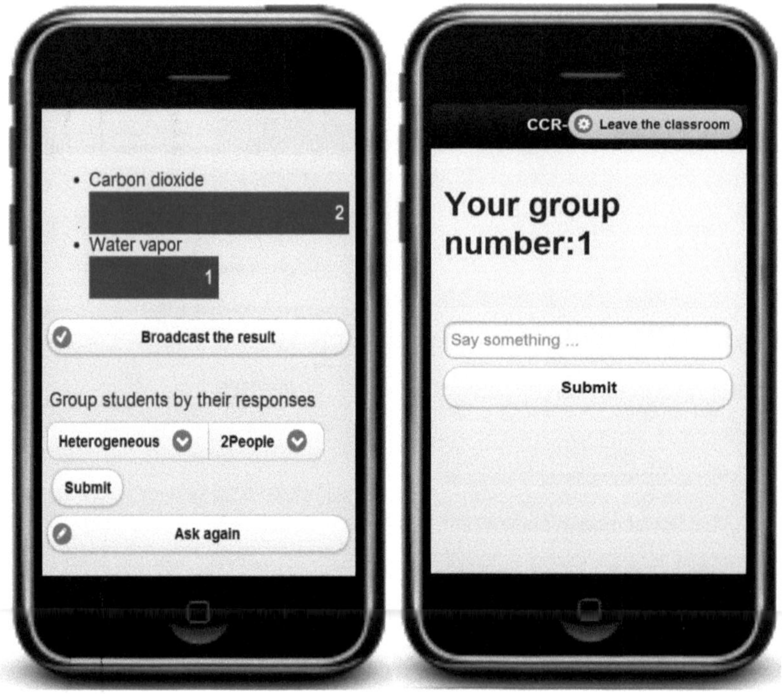

Fig. 22.2 Using CCR to aggregate students' responses and form heterogeneous groups

orchestrate small groups based on students' real-time responses. For instance, if the teacher poses a yes-no question and chose to form heterogeneous dyads, CCR will randomly assign students with conflicting answers (i.e., one of them responded *yes*, and the other responded *no*) into dyads. Students then follow the instructions provided by CCR, as shown in the right panel of Fig. 22.2, to find their group members and initiate peer discussions. Leveraging the aforementioned features of CCR, teachers certainly can save a lot of time collecting students' responses and arranging small groups. The classroom activities aiming to enhance the quality of students' argumentation may thus be implemented in a much smoother manner.

A Pilot Study to Examine the Effectiveness of CCR-Integrated Instructional Strategies

A pilot study was conducted in 2013 to explore the possible impacts of CCR-integrated instructional strategies on the quality of students' argumentation regarding a specific socio-scientific issue. The main focus of the study was to investigate whether different grouping methods, in conjunction with contingent teaching, produce different learning outcomes of socio-scientific argumentation. The research questions guiding the study were as follows: In comparison with the convenient grouping method, does the heterogeneous grouping method facilitate students to generate (1) more supportive arguments, (2) more counterarguments, and (3) more rebuttals, regarding a specific socio-scientific issue?

Method

Participants

A total of 65 students, from a college in northern Taiwan, participated in this pilot study. These students were recruited from a course of general education, called *Introduction to Natural Science*. None of the students majored in science. Every student was provided with an Apple iPod touch (Cupertino, CA, USA), connected with CCR, to interact with the teacher.

Procedure

Prior to the intervention, all the students took an argumentation pre-questionnaire to estimate their ability to generate arguments regarding whether the citizens of Taiwan should support space exploration programs. This issue was the core of the learning and evaluative materials. Students then received an interactive lecture, with the incorporation of CCR, to activate and supply their background knowledge of the issue. The lecture was about 100-min long, followed by a 20-min dyadic peer discussion section. How the lecture was conducted will be detailed in the next section. We manipulated how students formed small groups in the discussion section. By the use of CCR, each student was asked to answer a true-false question individually: *Do you support the Taiwan government to initiate space exploration programs*? Approximately half of the students ($n = 34$) formed dyads by using the convenient grouping method. Utilizing student's responses on CCR, the rest of students ($n = 31$) formed dyads by using the heterogeneous grouping method. The system recorded students' answers to the aforementioned question and randomly selected two students with conflicting answers (i.e., one of them responded *yes*, and the other responded *no*) to form a dyad. According to the system records, 15 students held a positive attitude to space exploration programs, and 16 students were negative. By the means of heterogeneous grouping, we got 15 dyads composed of different positions regarding the socio-scientific issue. The other one, who held a negative attitude, was randomly assigned to one of the 15 dyads. By the end of the discussion section, as a summative activity, students were invited to share their views with the whole class. An argumentation post-questionnaire, which was exactly the same as the pre-questionnaire, was then administrated to all students to evaluate the impact of different grouping methods on their learning outcomes. The time for students to complete a single questionnaire was limited to 50 min.

Interactive Lecture

An interactive lecture was conducted to help students prepare for the discussion on space exploration programs. The lecture was composed of two units. Within unit one, the teacher explained the concepts of biological diversity and migration. A short-answer question, *what factors are required for complex life*, was posed to students. Students were asked to type in their using CCR. The teacher then showed the list of students' answers and went through each answer. Additional scientific explanations were supplied by the teacher if any factor required for complex life did not appear in the list. The teacher also elaborated why some of the answers provided by students were not valid in science. In unit two, the teacher gave students an introduction on our own galaxy. A multiple-choice question, *how many Earth-size planets might exist in the galaxy*, was then posed. Students were asked to vote for answers individually, followed by a display of voting results. The teacher subsequently indicated the most possible answer and explained the possible relations between the planet size and the probability of life arising. An article about NASA's Kepler spacecraft, written by Cowen (2013), was detailed to make students have a preliminary understanding of what are space exploration programs for. As resources for students to construct arguments in the section of peer discussion, two opposed positions regarding whether NASA's Kepler would be able to find and characterize Earthlike planets orbiting small stars were introduced to them. Throughout the interactive lecture, CCR played a role in real time estimating how well students grasped the core concepts. By this means, the teacher was thus enabled to implement contingent teaching with ease. On the other hand, students were better prepared for participating in peer discussions because their background knowledge related to space exploration programs was activated and supplied through contingent teaching.

Instruments

The argumentation pre- and post-questionnaires, adapted from Wu and Tsai (2007), were used to evaluate students' capacity for constructing socio-scientific arguments. The evaluation focused on three basic argumentation components, including supportive arguments, counterarguments, and rebuttal arguments. The definitions of, and the questionnaire items corresponding to, these components are shown in Table 22.1. The pre- and post-questionnaires were open-ended and exactly the same as each other.

Data Analysis

Students' responses to the questionnaires were fractured into the components defined in Table 22.1. Two researchers independently counted students' scores according to the coding rubric shown in the left panel of Table 22.1. Each valid statement was worth one point. The inter-rater reliability was good, estimated by $\kappa = .82$. Any doubt in scoring results was resolved by discussion. Quantitative

Table 22.1 Components and items used to evaluate students' argumentation ability

Component	Corresponding item	Valid response		Invalid response	
		Definition	Exemplar	Definition	Exemplar
Supportive argument	Do you support the Taiwan government to initiate space exploration programs? If you want to convince your friend with your position, what arguments will you propose to convince him/her?	A position stated with explicit reasons	Yes, space exploration programs can help us find other environments which are suitable for people to live in	No response	(blank)
			No, there are many local crises on our Earth, such as climate change. These crises have a higher priority than space exploration	A position stated without any reason	I do not know
				A position stated with a self-contradicting reason	Yes (no), it is (not) worthy
					Yes (no), it should (not) be so
					Yes, space exploration is a waste of money
					No, it can bring a lot of benefits
Counterargument	If someone holds an opposite position with you on this issue, how may he/she challenge your arguments?	A statement casting doubt on the premises of the supportive argument, showing that the supportive argument does not follow from its premises in a valid manner or demonstrating that the truth of a conclusion is incompatible with that of the supportive argument	[The supportive argument previously stated is about leveraging space exploration programs to expand living environments and resources.] Today's Earth is shaped by an improbable combination of astrophysical and geological events and circumstances. It may be the unique one for humans to habitat. Seeking another planet for humans to live in might be just in vain	No response	(blank)
				An opposition stated without any reason	I do not know
				A opposition stated with a self-contradicting reason	But it is (not) worthy
					But it should (not) be so
					We should not support space exploration programs because their costs are not very high
					We should support space exploration programs because they give us no direct benefit

(continued)

Table 22.1 (continued)

Component	Corresponding item	Valid response		Invalid response	
		Definition	Exemplar	Definition	Exemplar
Rebuttal	Following the previous question, can you write down your opposing ideas to justify your position?	A statement refuting the counterargument	[The counterargument concerns the existence of Earthlike planets that are suitable for humans to live in.] There are billions of sunlike stars that have a system of planets in the observable universe. Even if we assume that only one out of a billion of these stars has planets supporting life, there would be some life-supporting planetary systems in the observable universe	No response	(Blank)
				A statement simply repeating the supportive arguments previously stated	I do not know
				An opposition stated without any reason	[A supportive argument has been stated that space exploration programs may expand living environments and resources.] Space exploration programs certainly can help us find other environments which are suitable for people to live in
				A new argument not directly refuting the counterarguments previously stated	[The counterargument concerns that space exploration is a waste of money.] It is worth it
					[The counterargument previously stated is about leveraging space exploration programs to expand living environments and resources.] Space exploration gives us no direct benefit

Table 22.2 Descriptive statistics of students' arguments about funding space exploration programs

		Pre			Post		
Type	Group	*n*	*M*	SD	*n*	*M*	SD
Supportive argument	Convenient grouping	34	1.71	0.63	34	1.82	0.63
	Heterogeneous grouping	31	1.68	0.65	31	1.94	0.68
Counterargument	Convenient grouping	34	0.79	0.88	34	1.41	0.82
	Heterogeneous grouping	31	0.84	0.86	31	1.58	0.89
Rebuttal	Convenient grouping	34	0.53	0.79	34	0.74	0.75
	Heterogeneous grouping	31	0.52	0.77	31	1.26	0.81

differences, if any, in the amounts of students' arguments between groups were used to evaluate the possible impact of different grouping methods on their learning outcomes. For instance, if the heterogeneous grouping method, in comparison with the convenient grouping method, was more effective to enhance students' ability to justify different positions regarding space exploration programs, students of heterogeneous groups should be able to write down more rebuttals to refute counterarguments than those of convenient groups. Students' responses to the pre-questionnaire were also examined to investigate whether there were any markedly differences in ability between groups prior to the intervention. Since the sample size was rather small, statistical hypothesis testing was not informative for interpreting the data. Students' scores of each group were counted and visualized as histograms for each argumentation component. Compared with statistical hypothesis testing, this approach enabled us to examine the difference in score distributions between groups with a higher resolution. The inferences drawn from the data were thus more reliable. Moreover, differences between groups were quantized as standardized effect sizes, which were calculated primarily based on Cohen's *d*; the mean of one group was subtracted from the mean of the other group, divided by the pooled standard deviation (Cohen, 1988). However, it was noted that the population variance would be underestimated when the pooled standard deviation was calculated. This implied that the effect size would be overestimated, especially when the sample size was small. Hedges's *g* (Hedges & Olkin, 1985), which used Bessel's correction to adjust the population variance, was thus applied instead. It provided a superior estimate of the standardized mean difference with small samples. Each effect size was reported with a 95 % confidence interval. The interpretation of effect sizes was adopted from Tallmadge (1977): The effect sizes greater than the interval between 0.25 and 0.33 were considered as "educationally significant."

Results

The scoring results of students' responses to the pre- and post- questionnaires are shown in Table 22.2. It was found that, prior to the intervention, students in general had no problems in proposing arguments to directly support their position over the

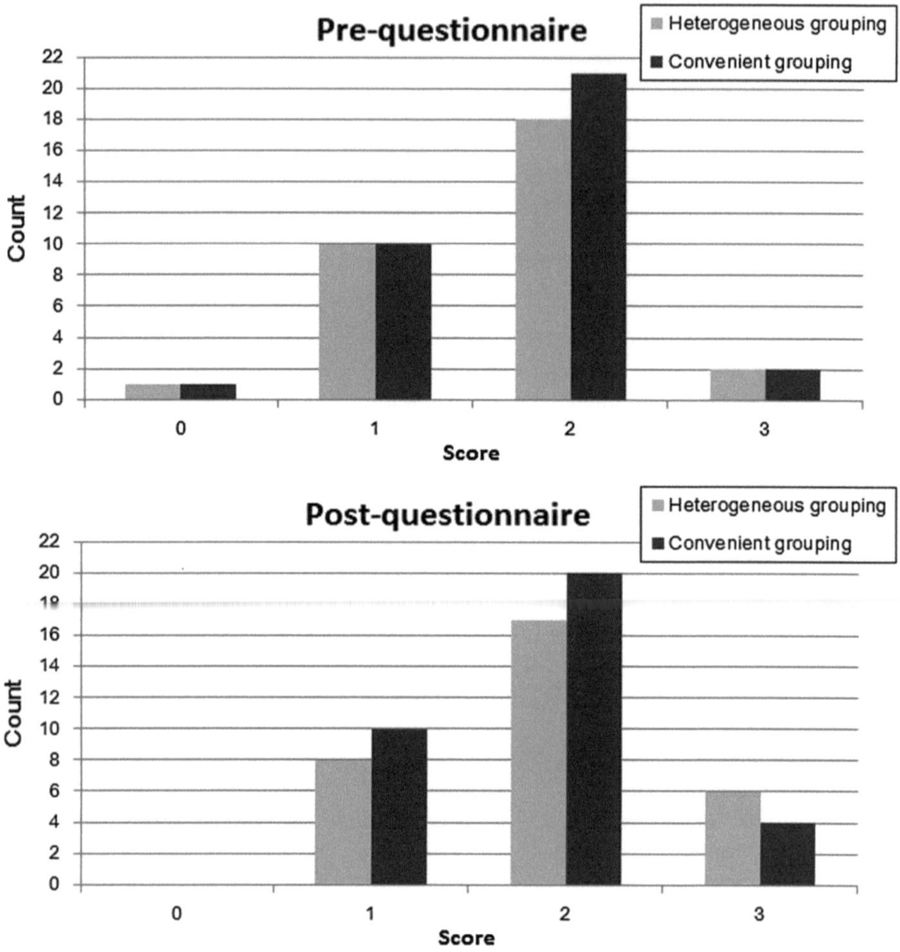

Fig. 22.3 Distributions of the amounts of supportive arguments generated by students

given issue. However, students' argumentation had a serious lack of two basic but important components: counterarguments and rebuttals. The quantitative differences in students' arguments are detailed in the following subsections in both pre-post and between-group manners.

Supportive Arguments

As shown in the upper panel of Fig. 22.3, prior to the intervention, the distribution of the amount of supportive arguments, which were generated by the students in convenient groups, was much alike to that of the students in heterogeneous groups. There was no significant difference in the ability to generate supportive arguments between groups, as evidenced by $g = -0.05$, 95 % CI [−0.53, 0.44]. After receiving the intervention, students in both convenient and heterogeneous groups generated slightly more supportive arguments than before ($M_{pre} = 1.71$, $M_{post} = 1.82$, for convenient groups; $M_{pre} = 1.68$, $M_{post} = 1.94$, for heterogeneous groups). However, no significant difference between groups was found after the intervention. Though the mean score of heterogeneous groups was slightly higher than the mean score of convenient groups, the magnitude of this difference was rather small, $g = 0.18$, 95 % CI [−0.30, 0.66]. The comparison between the distributions of argument amounts after the intervention, as shown in the lower panel of Fig. 22.3, also indicated that the difference between groups was insignificant. These results suggested that, in a practical sense, the heterogeneous grouping method was not more effective than the convenient grouping method to facilitate the amount of students' supportive arguments.

Counterarguments

Prior to the intervention, there was no significant difference in the ability to generate counterarguments between groups, as evidenced by $g = 0.06$, 95 % CI [−0.42, 0.54]. The distribution of the amount of counterarguments which were generated by the students in convenient groups, as shown in the upper panel of Fig. 22.4, was also much alike to that of the students in heterogeneous groups. After receiving the intervention, students in both convenient and heterogeneous groups generated significantly more counterarguments than before ($M_{pre} = 0.79$, $M_{post} = 1.41$, for convenient groups; $M_{pre} = 0.84$, $M_{post} = 1.58$, for heterogeneous groups). As shown in Fig. 22.4, the centers of the distributions of convenient and heterogeneous groups both moved from around 0.5 to around 1.5. Though the mean score of heterogeneous groups was slightly higher than the mean score of convenient groups, the magnitude of this difference was rather small, $g = 0.20$, 95 % CI [−0.29, 0.68]. The distribution of counterargument amounts of heterogeneous groups, as shown in the lower panel of Fig. 22.4, was still pretty much alike to that of convenient groups. These results suggested that, in a practical sense, the heterogeneous grouping method was not more effective than the convenient grouping method to facilitate the amount of students' counterarguments.

Rebuttals

Prior to the intervention, there was no significant difference in the ability to generate rebuttals between groups, as evidenced by $g = -0.01$, 95 % CI [−0.49, 0.47]. The distribution of the amount of rebuttals which were generated by the students in convenient groups, as shown in the upper panel of Fig. 22.5, was almost the same as that of the students in heterogeneous groups. After receiving the intervention, students in both convenient and heterogeneous groups generated significantly more rebuttals than before ($M_{pre} = 0.53$, $M_{post} = 0.74$, for convenient

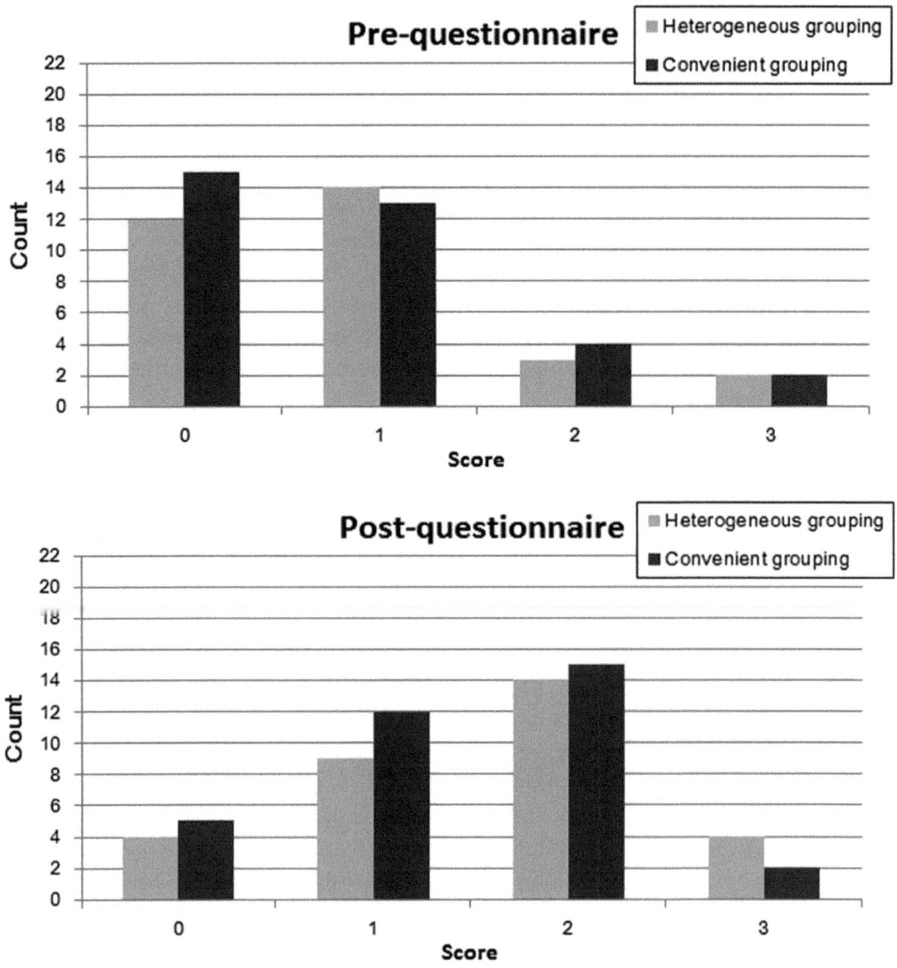

Fig. 22.4 Distributions of the amounts of counterarguments generated by students

groups; $M_{pre} = 0.52$, $M_{post} = 1.26$, for heterogeneous groups). As shown in Fig. 22.5, the centers of the distributions of convenient and heterogeneous groups both moved away from 0. Moreover, the mean score of heterogeneous groups was higher than the mean score of convenient groups, with a medium magnitude, $g = 0.66$, 95 % CI [0.17, 1.15]. As shown in the lower panel of Fig. 22.5, the center of the distribution of heterogeneous groups was located at the interval between 1 and 2. On the contrary, the distribution of convenient groups was centered at the interval between 0 and 1. These results suggested that, in a practical sense, the heterogeneous grouping method was more effective than the convenient grouping method to facilitate the amount of students' rebuttals.

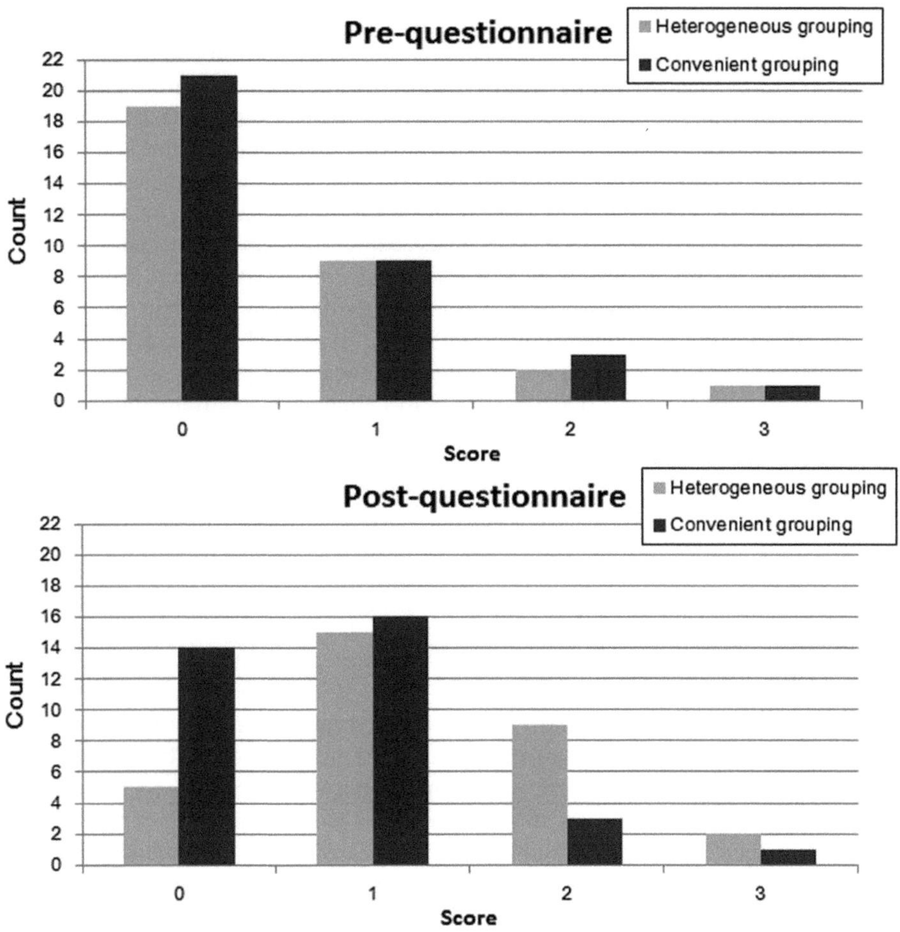

Fig. 22.5 Distributions of the amounts of rebuttals generated by students

Discussion

The results align with previous studies that, prior to our interventions, students' argumentation in general had a glaring lack of attention to opposing positions in the form of counterarguments and necessary rebuttals. The instructional strategies proposed in this chapter, either using convenient or heterogeneous grouping methods in conjunction with contingent teaching, were effective to facilitate students to generate counterarguments and rebuttals. Moreover, in comparison with the convenient grouping method, the heterogeneous grouping method was more effective to bring out students' rebuttals to refute counterarguments.

The superiority of heterogeneous grouping over convenient grouping may be resulted from the difference in the quality of opponents shaped by the grouping methods; the real opponents, who were assigned by the heterogeneous grouping method, were more effective than the imaginary opponents, who were constructed by students themselves in convenient groups, to attack students' supportive arguments. A closer inspection into students' responses to the post-questionnaire supports this explanation: The students in heterogeneous groups had a tendency to construct counterarguments that directly positioned doubt on the premises, validity, or coherence of supportive arguments, but the students in convenient groups did not. For instance, a student in a convenient group supported grant funds for space exploration programs. He stated that "space exploration programs can help us to find other environments which are suitable for people to live in" as a supportive argument. His counterargument was that "space exploration programs cost a lot of money." Although this counterargument indeed expressed an opposing position regarding the same issue, it did not directly interrogate the premise of the supportive argument, which is that "outside of the Earth, there is an environment(s) suitable for people to live in." He then just restated his position, "it (referring to the funding) is worthy," as a response to the counterargument. No rebuttal was proposed to justify his supportive argument. On the contrary, a student in a heterogeneous group stated a similar supportive argument, but he came up with a counterargument that "the emergence of complex life on Earth required a complex combination of circumstances. In other words, the Earth may be the only one suitable for human life. Seeking habitable environments outside of the Earth is a waste of time and money," to question the premise of the supportive argument. He then further articulated the rationale behind the premise, "even if we assume that the emergence of complex life on Earth was a host of fortuitous circumstances, there still would be some Earthlike planets habitable for humans in the entire universe, because billions of stars with planets orbiting them have been observed," to refute the counterargument. The students in heterogeneous groups tended to reflect on the premises of their supportive arguments, as delineated in the aforementioned case, and constructed rebuttals to strengthen them accordingly, which is a pattern absent in the post-questionnaire of the students in convenient groups.

In sum, the pilot study demonstrated that, in conjunction with contingent teaching, the heterogeneous grouping method was more effective than the convenient grouping method to facilitate students in constructing rebuttals to defend their positions over a socio-scientific issue. The depth of students' argumentation was deepened and its quality was thus enhanced. This effect was unlikely due to the difference in students' argumentation ability between groups prior to our intervention. As evidenced by the data presented in the upper panels of Figs. 22.3, 22.4, and 22.5, in terms of the amounts of supportive arguments, counterarguments, and rebuttal arguments, no remarkable difference between groups was found in students' responses to the pre-questionnaire. It can be concluded that, if the goal of instruction is to enhance the quality of students' argumentation about socio-scientific issues, the heterogeneous grouping method is a more ideal way than the convenient grouping method to orchestrate small-group discussions. A computerized system such as CCR introduced in this chapter, which is capable of instantly collecting, summarizing, and grouping students' responses to in-class questions, can help teachers to implement this strategy with ease.

Limitations and Future Directions

It should be noted that the inferences presented in the previous section are drawn from students' responses to the pre- and post-questionnaires. The pilot study reported in this chapter did not record the dynamics of students' discussions in small groups. In other words, how students interacted with peers in the process of discussion is unknown. Future studies, especially that go deeper into the processes and outcomes of peer discussion in the context of computer-supported socio-scientific argumentation, are warranted to verify our findings. Research of this line may better identify students' difficulties in computer-supported socio-scientific argumentation. Whether and how students' characteristics mediate the process and outcomes of peer discussion may be further understood as well. More specific suggestions to improve the practice of socio-scientific issue-based instruction can thus be made. With the use of CCR, the aforementioned tasks are currently undertaken in Taiwan.

References

Cohen, J. (1988). *Statistical power analysis for the behavioral sciences* (2nd ed.). Hillsdale, NJ: Lawrence Erlbaum Associates.

Cowen, R. (2013). The wheels come off Kepler. *Nature, 497*(7450), 417–418.

Driver, R., Newton, P., & Osborne, J. (2000). Establishing the norms of scientific argumentation in classrooms. *Science Education, 84*(3), 287–312.

Hedges, L. V., & Olkin, I. (1985). *Statistical methods for meta-analysis*. Orlando, FL: Academic.

Jiménez-Aleixandre, M. P., Rodríguez, A. B., & Duschl, R. A. (2000). "Doing the lesson" or "doing science": Argument in high school genetics. *Science Education, 84*(6), 757–792.

Kolstø, S. D. (2000). Consensus projects: Teaching science for citizenship. *International Journal of Science Education, 22*(6), 645–664.

Kortland, K. (1996). An STS case study about students' decision making on the waste issue. *Science Education, 80*(6), 673–689.

Ministry of Education. (2001). *The 1–9 grades science and life technology curriculum standards*. Taipei, Taiwan: Ministry of Education.

Ministry of Education. (2004). *The 10–12 grades science and life technology curriculum standards*. Taipei, Taiwan: Ministry of Education.

Sadler, T. D. (2004). Informal reasoning regarding socioscientific issues: A critical review of research. *Journal of Research in Science Teaching, 41*(5), 513–536.

Tallmadge, G. K. (1977). *The joint dissemination review panel idea book*. Washington, DC: U.S. Department of Health, Education and Welfare, National Institute of Education/U.S. Office of Education.

Wu, Y. T., & Tsai, C. C. (2007). High school students' informal reasoning on a socio-scientific issue: Qualitative and quantitative analyses. *International Journal of Science Education, 29*(9), 1163–1187.

Part VIII
International Benchmarking Tests

Chapter 23
Identifying Crucial and Malleable Factors of Successful Science Learning from the 2012 PISA

Chong Ho Yu, Shuang Frances Wu, and Casey Mangan

Introduction

Throughout the last several decades, results of international assessments, including the Trends in International Math and Science Study (TIMSS) and Program for International Student Assessment (PISA), consistently indicate that Asian students rank among the top in math and science performance. Take the 2012 PISA science test as an example. Among the top ten performers, six of them are Asian nations/regions: China, Shanghai (580), Hong Kong (555), Singapore (551), Japan (547), South Korea (538), and Vietnam (528) (Organization for Economic Cooperation and Development, 2013c). This chapter aims to identify malleable factors contributing to their success that are relevant and applicable to the West, particularly the United States. This study focuses on PISA only as it tests students on inferences and applications, whereas TIMSS is designed to follow school curricula. In other words, PISA emphasizes scenario-based problem-solving that goes beyond recalling textbook information. Therefore, PISA is considered fairer and more useful in comparison to TIMSS.

Numerous studies have been conducted with the goal of learning from the best practices of Asian education. There is no doubt these studies have merits and insightful findings, but they also have certain limitations. First, most studies concentrate on specific areas, as shown in the literature review section. Although each of these studies shed light on the issue and propose specific solutions, making reforms in one or a few areas might not be effective in the macro-environment

C.H. Yu (✉)
Faculty of Psychology, Azusa Pacific University, 901 East Alosta Ave., Azusa, CA 97102, USA
e-mail: chonghoyu@gmail.com; cyu@apu.edu

S.F. Wu • C. Mangan
Azusa Pacific University, Azusa, CA, USA
e-mail: swu@apu.edu

M.S. Khine (ed.), *Science Education in East Asia*,
DOI 10.1007/978-3-319-16390-1_23

where all factors are interdependent with one another. The improvement resulting from a change in minor factors might be undone by the status quo of major factors. In response to the 2012 PISA test result, the President of American Federation of Teachers Randi Weingarten (2013) declared that the United States failed to improve the performance of US students despite having mandated high-stake exams, publicizing proficiency rates, promoting charter schools, and other transformative policies created by *No Child Left Behind* and *Race to the Top*. Second, although several theories exist to explain why Asian students excel in science, some of those theories do not address malleable factors, and as a result, they do not necessarily lead to actionable items and practical applications. For example, some authors attribute Asian success to parent's social economic status (SES) and the prevailing cultural value on education (e.g., Ho, 2009, 2010a, 2010b; Tan, 2012). These explanations have theoretical significance, but it is difficult for policy makers to design a plan to improve parental SES. By the same token, it is also impractical to transform a cultural system in a short amount of time (e.g., infusing Confucian cultural elements into the United States).

Hence, these authors adopted a new approach as follows: the 2012 PISA data of science test scores, student variables, home-environment variables, and school-environment variables of Hong Kong, Singapore, Japan, South Korea, and Vietnam (top performers in 2012 PISA science test) were analyzed by the bootstrap forest approach. Shanghai, China, is excluded from the analysis because this sample from a single city is subject to a severe sampling bias. In the past, it was impossible to include all of the preceding variables in a single analysis because in conventional parametric procedures, estimating excessive parameters would result in lacking degrees of freedom. On the other hand, bootstrap forest, as a nonparametric data mining method, is immune to this shortcoming. When 399 variables are taken into account simultaneously, the most crucial predictors of success in science learning can be identified and ranked, resulting in a more parsimonious model. Next, each of these predictors was evaluated with the criterion of malleability. Factors that are unlikely to yield practical implications were put aside. Then the proposed reform efforts can be devoted to just a few manageable action items based on the findings. It was found that attitudinal variables of the students, home-environment variables, and technological factors outside of school, instead of teacher and school variables, were the determining factors of PISA science test performance.

Literature Review

A wide variety of factors influence the science achievement of primary and secondary school students. The evaluation of these factors has been established through the use of student assessments at the state/provincial, national, and international levels. The impact of different influences on student achievement has been explored empirically over the past four decades. Well-established factors affecting student achievement include socioeconomic status, the home environment, student affective variables, the school environment, curriculum, and teachers. Academic and government institutions have taken a keen interest in some of these factors because of their malleability. This literature review focuses on existing research on malleable factors on the science achievement of primary and secondary school students. This review is organized around exploration of variables at different levels and units of analysis: individual/affective, home/social environment, teachers and the classroom setting, and the broader school environment. While these topics are certainly different enough to deserve their own categories, it is important to note that they are by no means mutually exclusive or exhaustive. It is likewise necessary to note that variables affecting student science performance do so differentially among different cultures and nations.

There are also several methodological considerations that must be made regarding the research cited below. Across cultures, science examinations are more difficult to standardize than mathematics examinations (Wößmann, 2003). Likewise, the implications and weight of the results of international standardized examinations vary from country to country. Countries that have traditionally scored in the middle of the pack use the results as motivation to continue to improve their results. Some East Asian countries that have traditionally been in the top tier of international performers (e.g., China and Singapore) are seeking to catch up to the innovation and entrepreneurship of Western nations. Thus, despite their internationally enviable educational systems, these East Asian educational power houses are continually pursuing education reform (Zhao, 2012).

Individual-Level Variables

While there is certainly an entire world of factors that affect student performance, we need to look no further than the students themselves when searching for malleable factors related to science achievement. Studies about the effects of ability, motivation, attitude, and self-concept compose the majority of the current research on individual-level variables on student science achievement. Unlike the variables that affect student achievement above the individual unit of analysis (e.g., changes in socioeconomic status), affective and individual-level variables are much easier to influence and change. These variables can be changed for greater science achievement by individual students, families, and educators. Affective and individual-level variables have emerged as salient predictors of both math and science achievement in primary and secondary school students (Singh, Granville & Dika, 2002). In light of our emphasis on malleable factors, the potential for student achievement to develop over time is assumed. Yet it must be acknowledged that one of the most consistent determinants of future student science achievement is past achievement. Based on the TIMSS results of East Asian countries, Nabeshima (2003) concludes that the student's own ability, which is reflected partly in their past science achievement and partly in their academic behavior and cognitive abilities, carry a large effect on future achievement. The principle of the study of history, then, is relevant to predicting future science achievement from the individual to the national levels. While trends that shift achievement results over time certainly exist, high achievers tend to remain high achievers, average achievers tend to remain average achievers, and so on.

Early on, researchers discovered that the relationship between students and their academic achievement is not straightforward. Steinkamp and Maehr (1983) found that science achievement is more closely linked with affective than with cognitive variables but that the relationship is nonetheless complex. In their study, affect was operationalized as emotions, values, and feelings related to science. This operationalization is similar in studies that followed. In addition to affect and ability, the concept of self-efficacy has been researched in light of student science achievement. Britner and Pajares's study (2006) tested the extent to which Bandura's (1997) hypothesized sources of self-efficacy explained the science self-efficacy of middle school students. Science self-efficacy was operationalized as a combination of science grade/mark self-efficacy, science self-concept, science anxiety, self-efficacy for self-regulated learning, and science achievement. The results indicated that science self-efficacy was the most consistent predictor of students' science success. Student science achievement could be improved by parents and significant others encouraging students that they have the ability to master new and/or difficult concepts in science, teachers controlling student anxiety and fear related to science, and schools providing the opportunity for students to perceive similarities between themselves and those successful in science and STEM careers.

In her report on Singapore's continued topping of the list on international educational assessments, Davie (2013) notes that in high-performing countries, students report that their success is the result of hard work rather than intelligence. This suggests that schools and parents alike have the opportunity to influence the more malleable and more impactful factor of work ethic in their students. The 2012 PISA results also indicate that poor students have the opportunity to demonstrate their resilience and perform just as well as students with a higher socioeconomic status.

Home and Social Environment

After exploring student-level variables, it is likewise important to examine the impact of the social context on the science achievement of individuals, classrooms, and schools. We now turn our focus to the impact of the collection of individuals into social environments. The bulk of the focus for this section will be on parental involvement, parent education level, socioeconomic status (SES), and social support. Through decades of empirical studies, a positive relationship has become well established between SES and home possessions (Yang, 2003; Nabeshima, 2003), parental involvement (McNeal, 1999; Yalein, Aslan, & Usta, 2012), parent education level (Hampden-Thompson & Johnston, 2006), social support (Rice, Barth, Guadango, Smith, & McCallum, 2013), and student science achievement. While it is certainly true that some variables in the social environment are virtually non-malleable (e.g., parent income), there are still many variables related to social behavior (e.g., parental involvement) that are malleable. For example, Singaporean students who attended preschool performed better on the PISA than their peers who did not attend preschool (Davie, 2013).

Based on survey, data from education experts from Finland and Korea (who ranked first and third, respectively, on the 2003 PISA), Kim, Lavonen and Ogawa (2009) drew several conclusions about why these two nations had such high science achievement. The top two reasons experts gave for high achievement among Korean students were parental support of education (maternal support in particular) and Korean educational culture, which was defined as both enthusiastic and competitive. Parental involvement among Korean parents goes beyond support, with the sixth most important factor being parental influence over educational expenditures. The value that parents and communities from high-achieving nations place on education should not be overlooked when considering their high academic achievement (Aun, Riley, Atputhasamy, & Subramaniam, 2006).

Ho (2010a) goes a step further in showing that different types of parental involvement are more strongly related with science achievement than others. Examining the relationship between parental involvement and student science performance in Hong Kong using data from the 2006 PISA exam, Ho found that certain types of parental involvement are more strongly related to high science achievement. Specifically, the results indicated that students whose parents invested in cultural resources and organized science learning enrichment activities for their child at a young age had better cognitive and affective outcomes in science. Cultural resources were operationalized as the presence of classic literature, poetry books, and art work in the home. Science learning enrichment activities were noted to include watching programs about science, reading books about science, and watching, reading, or listening to science fiction films and/or books.

Teacher Variables

Educational factors affecting student science achievement will be explored through teacher variables and school-environment variables, respectively. But first, it must be established that neither teacher variables nor school-environment variables are mutually exclusive. The degree to which the two may be more exclusive varies by context. In her study on Japan, Korea, and Hong Kong, Ho (2006a) found that teacher participation in the implementation of curriculum is more influential than school autonomy over the same manner on student achievement. Additionally, varying degrees of centralization were found among three high-achieving countries, Japan, Korea, and Hong Kong. The results of Ho, therefore, indicate that high student science achievement can result from both school-driven and teacher-driven curriculums.

Like the social environment factors, both levels of educational variables vary in their malleability. Teacher quality, perhaps more so than any of the variables explored in this review, is the most difficult to operationalize. Different studies and researchers have given different weight to specific aspects of teaching. For example, certain researchers have concluded that educational background is less important than experience. In Kim et al.'s study (2009), both Finnish and Korean experts ranked science teacher education as a highly important factor for student science achievement. Despite their highly similar results, in Finland, science teachers are required to obtain a master's degree, while in Korea, they are not.

But even then, how much experience would be needed to make a teacher great? Akiba, LeTendre and Scribner (2007) concluded from the 2003 TIMSS study on 46 different countries that higher teacher quality results in higher student achievement cross-nationally. Therefore, investing in teacher quality, however one might choose to define it, is an important way to improve national achievement across academic subjects. This is an important distinction to make early on, as the conception of teacher quality will vary by school, region, and country.

School Environment

After examining different student and teacher-level variables on student science achievement, it becomes necessary to explore how the school context affects teachers and students alike. These factors vary, just like the previous three sets of factors, from school to school and country to country. A diverging trend that exemplifies how two high-achieving countries have success is that Korean experts cite their highly centralized curriculum and Finnish experts cite their devolution in curriculum as reasons for their similar levels of success (Kim et al., 2009). Generally speaking, variables at this level of analysis have a smaller effect size on student science achievement than others, and the effect is generally an indirect one. Despite their generally small effect sizes, it is still important to note that certain variables at this level do impact student affect, student science achievement, and the gathering of data regarding student science achievement. These variables are also generally less malleable than factors closer to the individual level. Perhaps, the most malleable and influential factor at the school and educational policy levels of analysis is the constant desire for self-improvement, which includes a consistent need for progress instead of simply dwelling on past successes and likewise looking outward to see how the best practices of other nations could be adopted for the Singaporean context (Low & Lee, 2012).

Another malleable school-level factor that affects both student affect and student science achievement is the frequency of high-stakes examinations. Results from the 2000 and 2003 PISA exams indicate that the use of high-stakes exams in Hong Kong does not contribute to higher academic achievement. Instead of improving academic achievement, PISA data indicates that using high-stakes exams has had numerous negative effects on Hong Kong students' achievement, the school system, and students' motivation and has burdened students and teachers alike. In fact, Hong Kong schools who use assessment results to guide their institutional level decision making perform no differently than schools that do not. It has also contributed to above average levels of anxiety in Hong Kong. Instead of high-stakes exams, Ho recommends the use of low-stakes exams to guide teaching, learning, and parental involvement towards the goal of increased student and school achievement by improving both affective and cognitive outcomes, not just increased examination scores (Ho, 2006b). What national and regional standardized exams intend to assess will not only encourage educators to prepare their students differently but will also affect how students perform on exams like the PISA. While the effect size is not quite as large as that of individual-level factors, it is worth noting that the Singaporean Education Ministry's strategy to expose students to applications of science learning to real-life situations was concluded to have been successful in light of the PISA 2012 results (Lee, 2013). In the next section, the authors evaluated the merits of the preceding factors altogether.

Methodology

Data Sources, Population, and Sample

In this study, the data are sourced from the OECD's PISA (2013a) website. The observations from Asian countries and regions that are among the top ten performers, namely, Hong Kong, Singapore, Japan, South Korea, Vietnam, and Japan, are extracted from the PISA online data bank. Shanghai, China, was excluded due to the limitation that all the students in this sample are from only one city of mainland China, and needless to say it is not a representative sample of mainland China. Although Hong Kong, a former British colony, resumed under the sovereignty of China in 1997, its culture is vastly different from that of mainland China. Hong Kong, as a special administrative region, has a high degree of autonomy in governance. Hence, in this study, Hong Kong is treated as an independent population. Additionally, the observations from the United States are also downloaded for comparison purposes.

OECD (2013b) employed multistage sampling to carpet the entire student population in the participating countries. The sampling process consisted of two stages. In the first stage, schools were selected, and in the second stage, students were sampled within the schools. It is a typical nested data set that violates the assumption of independence for parametric tests. The countermeasure will be discussed in the data analysis section.

It was required that at least 4,500 students and 150 schools were selected in each participating nation (National Center for Education Statistics, 2013). In this study, the sample size of Hong Kong is 4,670; Japan, 6,351; Korea, 5,033; Singapore, 5,546; United States, 4978; and Vietnam, 4,959. With such huge sample sizes, conventional hypothesis testing would face the problem of having too much statistical power, resulting in rejecting the null in spite of trivial effects. The remedy will also be discussed in the data analysis section.

Outcome Variable

To avoid overburdening the students with too many questions, PISA test developers chose to give different students different booklets or forms, which are subsets of all the questions. The missing values are imputed by imputation algorithms, and as a consequence, there is uncertainty in the estimation of student ability. Thus, instead of reporting one single test score, PISA test developers use five plausible values (PVs) from a probability distribution. Therefore, the actual ability of each student may be reflected by one of the five PVs (OCED, 2013b).

There are debates about how PVs should be analyzed. Some researchers suggest using the average value of all PVs, while others recommend using each of the PVs to compute the statistics of interest and then averaging the results over the five computations (von Davier, Gonzalez, & Mislevy, 2009). Although the latter approach seems to be thorough, it is not practical to merge five random forests, the data analysis approach chosen for this study, because the forest itself is already a product of merged results. There is no apparent benefit of doublemerging. Wu (2004) suggests a simpler yet more powerful approach: random selection from the PVs. According to Wu, each PV is equally good for estimating the population characteristics, as each set forms a sampling distribution of the estimated student ability. Following this line of reasoning, even if one PV per student is used, it can still yield an unbiased estimate of the population parameters, in contrast to using the mean of plausible values for each student, which may lead to biased estimates. In other words, using one randomly drawn PV from the probability distribution is better than using the mean of the distribution, in terms of getting an unbiased estimate. Consider this metaphor: If the police were searching for a missing person and there were five reports of sighting the person, it is definitely not a good strategy to calculate the centroid of the five locations and then deploy the search team there. Rather, the chance of spotting the missing person is higher if one of the five locations is randomly searched. Following the advice of Wu, this study uses a randomly selected PV of science test performance as the dependent or outcome variable.

Independent Variables

In addition to the test of scientific reasoning, PISA developers include several surveys to collect data that are relevant to students' learning and test performance (OECD, 2013a, 2013b). This cluster of surveys includes the student background information survey, the educational career questionnaire, the information and communication familiarity questionnaire, the parent questionnaire, and the school survey. The student survey is composed of items regarding demographics, family background, home environment, student attitude towards problem-solving, and experience at school. The educational career questionnaire is concerned with the student's career aspiration and preparation for future careers. The information and communication familiarity questionnaire consists of items regarding the availability of technology to students and their usage of technology. The parent survey entails questions pertaining to parental background, cost of educational service, attitudes towards child's school, parents' involvement with school, school choice, parental support for learning in the home, child's past academic performance, and child's career interests. The school questionnaire is completed by the principal or designee. It contains items related to school environment, school resources, school activities, teacher qualification, and teacher pedagogy.

Out of the above, the authors of this chapter exclude certain variables for a number of reasons. First, non-malleable factors, such as household income and parent professional qualification/career, are not included in this study. Even if income and professional qualification/career are found to be crucial factors to student test performance, it is unlikely to suggest a course of action based on this type of findings. To be more specific, it would be impossible for any policy makers to recommend parents finding a better job or making more money in order to improve their children's science learning. Second, in the student survey, some questions about attitude towards problem-solving are purely hypothetical. These hypothetical questions are also excluded from this study because when facing the challenge in real life, people usually do not do what they answer they would do in the hypothetical situation, a phenomenon known as the attitudinal fallacy (Jerolmack & Khan, 2014). In the end, 399 variables, including measures of student attitudes, work habits, career aspiration, access to technology, family background, and school background, were entered into the random forest model. It is important to point out that in these samples only Hong Kong and Korea participants answered all 399 items. All other samples took parts of the surveys only. Specifically, Japan participants were exposed to 331 items; Singapore, 357 items, Vietnam, 270 items; and United States, 269 items.

Data Analysis

In this study, the crucial predictors of success in science learning were identified by the bootstrap forest approach, which is a form of resampling (Yu, 2003, 2007). The bootstrap forest is based on the concept of the random forest (Breiman, 2000, 2001, 2004). Both of them are resampling techniques, in which the sample is treated as a virtual population, and is systematically reused by repeated sampling. The idea of bootstrapping originated from Efron (1979, 1981, 1982) and was further developed by Efron and Tibshirani (1993). "Bootstrap" means that one available sample gives rise to many others by resampling (a concept reminiscent of pulling yourself up by your own bootstrap). In bootstrapping, the original data set is duplicated numerous times, and then random samples are repeatedly drawn from this proxy or virtual population. The same statistical procedure is computed in each draw, and afterwards an empirical sampling distribution is created by putting all these statistics together. To put it simply, the researcher replicates the study to determine the bias and the confidence interval of the estimated parameters.

The random forest approach, which also utilizes repeated sampling from the virtual population, is a data-driven machine-learning algorithm (Breiman, 2001). Unlike the original simple bootstrapping, the random forest algorithm is able to fine-tune the model based on repeated analyses. Typically random forest starts with a "weak learner," which is a predictor with a low bias and a high variance, but the learner can get better and better. Hence, random forest is not subject to the potential problem that the number of parameters to be estimated might exceed the degrees of freedom, which is common in many parametric procedures. For example, in a study conducted by Breiman (2004), which involves 81 cases only and 4,682 variables, the error rate is as small as 1.23 % after 1,000 resampling. Random forest is so named because it assembles many decision trees to form a forest during the resampling process. The decision tree is a classification procedure, in which most important predictors are selected for their ability of partitioning the most homogeneous subgroup, as measured by the Gini Index (Breiman, Friedman, Olshen, & Stone, 1984). The partition process continues until subdivision of the data can no longer reduce the Gini Index.

Bootstrap forest is built on the idea of bootstrapping. While random forest randomly selects both predictors and subjects, bootstrap forest pick random subjects only. Like random forest, the bootstrap forest is an ensemble of many classification trees resulting from repeated sampling from the same data set. Afterwards, the results are combined to reach a converged conclusion. The output of the bootstrap forest is a table of column contributions, which rank order the importance of each predictor by its sum of squares and the number of split. The number of split is simply a vote count: How often does each variable appear in all decision trees?

The random forest and its variant bootstrap forest are considered superior to many other predictive modeling methods, such as regression and support vector machine for a number of reasons. First, the procedure is custom-made to handle large-scaled data sets, and thus, it is immune to the problem of over-powering and inflating type I error (Yu, 2003). Second, various studies confirm that the forest approach yields a higher predictive accuracy. Third, the forest-based modeling can rank the importance of the predictors but other procedures cannot. Fourth, the forest-based approach is flexible and versatile. Finally, this procedure is immune against missing data and noise (Cutler et al., 2007; Verikas, Gelzinis, & Bacauskiene, 2011). Taking all of the above into account, the authors determined that the bootstrap forest, which is accessible in JMP Pro (SAS Institute, 2013), is the most suitable method for this study.

There is, however, a limitation in the forest-based approach. This approach assumes independent observations, but in most countries, OECD collected the PISA data through multistage sampling, resulting in nested data. Without multilevel modeling, there is a risk of overfitting the data. To avoid overfitting, in each partition tree, about 20 % of the data were held back for validation. Twenty percent is an approximate figure, because in the random selection process, sometimes it is over 20 % and sometimes it is below 20 %. To be specific, in each run, the sample was divided into the training set and the validation set. As the name implies, in the training set the observations were utilized to train the algorithm to propose a model. Afterwards, cross-validation was performed with the rest of the observations. Additionally, there are two other remedies for the overfitting problem: use small trees and increase the number of trees (Fadaia, 2013). Both approaches are intuitive. By increasing the size of the forest (the number of trees), it creates more opportunities for machine learning, and thus an overfitted result that capitalized on chance can be avoided. Furthermore, when the size of each tree is minimized, it means variables that have trivial effects on the dependent variable cannot be considered. To implement the preceding proactive measures, the maximum number of tree size is set to 99. Moreover, rather than adopting the default number of trees in JMP PRO, which is 100, in this study the number of trees was set above the default. On one hand, going beyond the default could avoid overfitting. On the other hand, the early stopping rule was also applied to achieve optimality. When early stopping is turned on, the resampling process stops growing additional trees if adding more trees to the already saturated model does not improve the validation statistic. The authors found that in the PISA sample, most forest models were saturated before reaching 150 trees or less, and therefore the maximum number of tree was set to 200.

In bootstrap forest, there is a seemingly insurmountable shortcoming. Because the algorithm randomly divides the data set into the training and the validation sets and also randomly selects observations for growing each decision tree, the exact result cannot be replicated even if another bootstrap forest is run with the same data set. Researchers who are trained with the point-estimation approach, assuming that there is only one invariant, true parameter in the population, may find it unacceptable. Nonetheless, it is important to note that the very essence of resampling is to simulate replications. Even if researchers carry out actual replications, there would still be discrepancies among various results.

Results

Table 23.1 shows a summary of all six bootstrap forests using the observations of six nations/regions, respectively. The second column shows the maximum number of trees in the forest yielded by the algorithm. The program ceased running when the model was saturated and constructing additional trees could not improve the variance explained or predictive accuracy. The fourth column indicates the sample size used by the algorithm to "learn" about the data in each decision tree, whereas the fifth column shows the sample size for validation. The two rightmost columns depict the variance explained by the training set and the validation set. In the training set, the model goodness in terms of $R^2$was high due to overfitting. When the algorithm attempted to fit the proposed model into the remaining data set, inevitably there was discrepancy. Hence, the R^2 was substantively scaled back in order to avoid overfitting. As shown in Table 23.1, the original models yield R^2

Table 23.1 Summary of bootstrap forest

Nation/region	# of trees	Total *n*	Training set *n*	Validation set *n*	Training set R^2	Validation set R^2
Hong Kong	117	4,670	3,729	941	0.801	0.422
Japan	146	6,351	5,024	1,327	0.783	0.477
S. Korea	100	5,033	4,049	984	0.814	0.494
Singapore	94	5,546	4,474	1,072	0.853	0.565
United States	77	4,978	3,986	992	0.806	0.429
Vietnam	99	4,959	3,926	1,033	0.782	0.478

as high as .782–.853. After validation, they were humbled down to between .422 and .565.

Table 23.2 shows the bootstrap forest result of Hong Kong students. The importance of the predictors is ranked in order of their frequency in the forest. Although the bootstrap forest stops at 117 trees, the number of splits for a variable can exceed 117 because the same variable could recur in the tree several times. The two leftmost columns illustrate the relationship between the PISA science test score and the predictor. The demarcation point is suggested by the algorithm, not the authors. For example, the mean score of students who have 0–10 books at home is 511.227, while that of students who have more than 10 books is 566.798. It is because the decision algorithm based on the Gini Index determines that the cut point at 10 books can decisively partition the high-performing and low-performing students. By the same token, students who spend less than 2 h each week on homework have a mean score of 496.554, whereas students who spend 2 h or more on homework have an average of 562.116. It is noteworthy that with regard to time spent to access Internet out of school during weekdays, students who spend no time or too much time on the Internet tend to obtain a lower score than other students who do not have excessive use of the Internet. Another noteworthy result is that three out of the top ten predictors are related to openness for problem-solving.

Table 23.3 is a summary of the bootstrap forest result using the South Korean data set. South Korea is a technologically advanced nation, and thus it is not surprising to see that four out of the ten predictors are related to student usage of technology. Some of those outcomes are expected, such as the positive effects of browsing the Internet for schoolwork and reading news on the Internet outside of school and the negative effect of chatting online outside of school. However, the effects of playing games are astonishing. By common sense, playing games is considered detrimental to learning, but students who spend no time in playing, one-player games tend to have a lower PISA science test score (507.653) than those who spend time playing games (546.201). It is also noteworthy that two of the top ten variables are related to parents' high expectation of achievement and awareness of competition. Based on personal communications with Koreans, the authors found that Korean parental demands on their children for academic performance are extremely high.

Table 23.2 Important predictors of PISA science test scores yielded by bootstrap forest for Hong Kong students

PISA label	Variable	# of splits	Conditions	Mean score
ST28Q01	Number of books at home	392	0–10 books	511.227
			>10 books	566.798
ST57Q01	Out-of-school study time for homework or other study set by the teachers	336	<2 h	496.554
			≥2 h	562.116
ST91Q06	Perceived control: "I perform poorly at school whether or not I study for my exams."	318	Strong agree/agree	497.600
			Strongly disagree/disagree	561.471
ST94Q09	Openness for problem-solving: "I seek explanations for things"	314	Not at all like me/not much like me/somewhat like me	544.295
			Mostly like me/very much like me	578.628
ST94Q06	Openness for problem-solving: "I am quick to understand things"	285	Not at all like me/not much like me/somewhat like me	544.467
			Mostly like me/very much like me	578.278
IC06Q01	Time spent on accessing the Internet out of school during weekday	272	<30 min/more than 6 h	525.381
			Between 31 min and 6 h	564.029
ST70Q03	Number of class periods in science per week	268	<8	548.523
			≥8	593.658
ST94Q14	Openness for problem-solving: "I like to solve complex problems"	262	Not at all like me/not much like me	548.417
			Somewhat like me/mostly like me/very much like me	581.532
IC08Q03	Time spent on using e-mail out of school	258	Never or hardly ever	525.789
			Often	563.696
IC07Q01	Time spent on using the Internet outside of school on a typical weekend day	257	<30 min	509.211
			>30 min	561.815

Table 23.4 summarizes the findings of the bootstrap forest using the Japanese sample. Like the Hong Kong sample, the number of books at home was the most crucial factor contributing to better test performance in science among the Japanese students. Additionally, the type of books matters, too. Those students who have classic literature at home tend to outperform their counterparts that have no classic literature (576.154 vs. 516.979). Similar to South Korea, Japan is also a technologically advanced country, and this reflects in the finding that five out of the ten important variables are related to student usage of technology. As shown in Table 23.4, certain technology-related variables and test performance have a

Table 23.3 Important predictors of PISA science test scores yielded by bootstrap forest for South Korean students

PISA label	Variable	# of splits	Conditions	Mean score
ST94Q14	Openness for problem-solving: "I like to solve complex problems"	392	Not at all like me/not much like me	531.056
			Somewhat like me/mostly like me/very much like me	577.056
ST28Q01	Number of books at home	336	≤200 books	521.999
			>200 books	565.847
IC09Q01	Browsing the Internet for schoolwork (e.g., for preparing an essay or presentation)	318	Never or hardly ever	500.682
			Often	552.418
IC08Q08	Playing one-player games	314	Never or hardly ever	507.653
			Often	546.201
ST91Q06	Perceived control: "I perform poorly at school whether or not I study for my exams"	285	Not at all like me/not much like me	515.088
			Somewhat like me/mostly like me/very much like me	555.687
ST94Q06	Openness for problem-solving: "I am quick to understand things"	272	Strongly disagree/ disagree	488.731
			Strongly agree/agree	545.113
IC08Q07	Reading news on the Internet (e.g., current affairs) outside of school	268	Never or hardly ever	508.396
			Often	547.059
PA09Q02	High expectation: "Standards of achievement are high in my child's school"	262	Strongly disagree/ disagree	511.104
			Strongly agree/agree	553.179
PA11Q01	Awareness of competition: "There are two or more other schools in this area that compete with the school my child is currently attending"	258	No others/one other	529.520
			Two or more	549.309
IC08Q04	Chatting online outside of school	257	Every day/almost every day/once or twice a week	524.055
			Once or twice a month/never or hardly ever	549.314

curvilinear relationship. For example, if a student spends no time on using the Internet outside of school during a typical weekend day or the weekdays, it is likely that his or her test score is low. If the student spends between 1 min and 4 h per day on the Internet, test performance increases. But the relationship is reversed when the student spends 6 h or more. This pattern can also be found in the ownership of conventional electronics. If a student whose home has no TV set, his or her test scores may be relatively low (538.773). If the home has one or two TV sets, the test

Table 23.4 Important predictors of PISA science test scores yielded by bootstrap forest for Japanese students

PISA label	Variable	# of splits	Conditions	Mean score
ST28Q01	Number of books at home	831	≤25 books	500.968
			>26 books	560.318
IC07Q01	Time spent on using the Internet outside of school on a typical weekend day	645	No time/> 6 h	512.264
			Between 1 min and 4 h	557.882
IC08Q08	Obtaining practical information from the Internet (e.g., locations, dates of events) outside of school	635	Never or hardly ever	514.140
			Often	560.726
ST94Q14	Openness for problem-solving: "I like to solve complex problems"	634	Not at all like me/not much like me	536.784
			Somewhat like me/mostly like me/very much like me	573.202
ST26Q07	Have classic literature at home (e.g., Shakespeare)	619	No	516.979
			Yes	576.154
IC06Q01	Time spent accessing the Internet out of school during weekday	612	No time/> 6 h	509.562
			Between 1 min and 6 h	557.198
ST27Q02	Number of televisions at home	598	None/three or more	538.773
			One or two	554.809
ST11Q04	Who usually lives with the student at home? Sister(s) (including stepsisters)	588	Yes	546.776
			No	555.779
IC08Q07	Reading news on the Internet (e.g., current affairs) outside of school	582	Never or hardly ever	523.719
			Often	560.234
IC08Q09	Download music outside of school	544	Every day	511.769
			Almost every day/once a while/ never or hardly ever	553.126

scores increase to 554.809. But when there are three or more TV sets in the household, test performance drops.

Table 23.5 lists all the important predictors of PISA science test scores for the Singaporean sample. The number one predictor is directly related to the school, namely, the number of class periods per week in science. If the school schedules nine or more science lessons per week, students would gain more than 100 points than those who have fewer lessons (613.039 vs. 510.699). Student attitudes can also influence their test performance. Those who hold a negative attitude towards school and the Internet tend to have lower performance. While these attitudinal variables are within the control of the students, there are some variables outside of their control. Specifically, student test performance is largely determined by the age of first use of computer and first use of the Internet. These two important variables are not found in other samples in this study.

Table 23.5 Important predictors of PISA science test scores yielded by bootstrap forest for Singaporean students

PISA label	Variable	# of splits	Conditions	Mean score
ST70Q03	Number of class periods per week in science	588	<9	510.699
			≥9	613.039
ST57Q01	Out-of-school study time for homework or other study set by the teachers	482	<3 h	464.589
			≥3 h	558.281
ST28Q01	Number of books at home	414	≤25 books	502.104
			>26 books	569.042
ST88Q01	Attitude towards school: "School has done little to prepare me for adult life when I leave school"	347	Strongly agree/agree	499.353
			Strongly disagree/ disagree	575.709
IC03Q01	First use of computer	346	Never/10 years old or older	510.425
			9 years. old or younger	559.855
IC22Q08	Attitude towards Internet: "Information obtained from the internet is generally too unreliable to be used for school assignments"	326	Strongly agree/agree	508.913
			Strongly disagree/ disagree	572.048
IC08Q11	Uploading student created content for sharing (e.g., music, poetry, videos, computer programs)	313	Every day/almost every day/once or twice a week	507.545
			Once or twice a month/never or hardly ever	562.288
IC05Q01	Time spent on the Internet at school	312	From 31 min to 6 h and more	509.484
			No time/1–30 min	563.579
ST27Q03	Number of computers at home	311	None/one/two	523.181
			Three or more	573.686
IC04Q01	First access to the Internet	282	Never/10 years old or older	522.479
			9 years old or younger	562.695

Table 23.6 shows the result of the Vietnamese sample, which is vastly different from that of the previous four Asian samples. Interestingly, in this sample eight out of the top ten predictors are related to the family or the home environment of the students. For example, the test performance of a student is tied to whether the student lives with brothers, sisters, and grandparents. If a Vietnamese student lives with his or her siblings, he or she would have higher test scores. But if he or she lives with grandparents, this would negatively affect his or her test performance. A plausible explanation is that taking care of grandparents diverts the student's attention from study. In addition, the number of books at home and types of

Table 23.6 Important predictors of PISA science test scores yielded by bootstrap forest for Vietnamese students

PISA label	Variable	# of splits	Conditions	Mean score
ST55Q03	Hours spent per week attending out-of-school lessons for science	356	None/<2 h per week	516.965
			≥2 h per week	554.761
ST28Q01	Number of books at home	330	0–100 books/more than 500 books	526.021
			101–500 books	548.321
ST11Q03	Who usually lives at home with the student? Brother(s) (including stepbrothers)	318	No	523.541
			Yes	531.698
ST11Q04	Who usually lives at home with the student? Sister(s) (including stepsisters)	306	No	525.619
			Yes	530.069
ST94Q05	Openness for problem-solving: "I can handle a lot of information"	286	Not at all like me/not much like me/very much like me	512.059
			Somewhat like me/mostly like me	539.402
ST26Q07	Have classic literature at home (e.g., Shakespeare)	283	No	523.265
			Yes	533.752
ST11Q05	Who usually lives at home with the student? Grandparent(s)	272	Yes	522.242
			No	531.928
ST26Q11	Have technical reference books at home	269	No	522.136
			Yes	537.756
ST19Q01	Student's father work	265	Not working, but looking for a job	479.171
			Working full time, part time, at home, or retired	529.749
ST26Q02	Have a room of your own	258	No	525.540
			Yes	530.364

books (e.g., classical literature, technical reference) are also influential to their proficiency in science. Whether the student's father is working or not is a crucial factor, too. It could be explained by the likelihood that a non-working father is unable to provide the family with financial security, resulting in a home environment that is less conducive to learning. Lastly, if a student has his or her own room, it seems to be beneficial to his or her learning. Probably this could enable the student to study in a quiet environment. It is also possible that the students having his or her own room is a sign of higher socioeconomic status, resulting in more resources and higher performance.

The bootstrap forest result for US students is shown in Table 23.7. Although some important variables in this sample also appear in the Asian samples, such as the number of books and computers at home, the US and the Asian results are vastly

Table 23.7 Important predictors of PISA science test scores yielded by bootstrap forest for US students

PISA label	Variable	# of splits	Conditions	Mean score
ST28Q01	Number of books at home	577	≤25 books	458.227
			>26 books	527.462
ST27Q03	Number of computers at home	339	None/one or two	481.926
			Three or more	527.158
ST91Q06	Perceived control: "I perform poorly at school whether or not I study for my exams"	331	Strongly agree/agree	446.075
			Strongly disagree/disagree	506.169
ST57Q01	Out-of-school study time for homework or other study set by the teachers	316	<6 h	475.600
			≥6 h	533.841
ST71Q01	Number of all class periods a week	297	<10	437.470
			≥10	533.321
ST11Q04	Who usually lives at home with the student? Sister(s) (including stepsisters)	297	Yes	500.025
			No	510.965
ST11Q06	Who usually lives at home with the student? Others (e.g., cousin)	282	Yes	464.231
			No	514.746
ST11Q03	Who usually lives at home with the student? Brother(s) (including stepbrothers)	278	Yes	498.703
			No	514.095
ST26Q15	Have a guest room at home	275	No	500.508
			Yes	501.388

different. First, none of the attitudinal variables (e.g., openness to problem-solving, perceived control over school performance) are identified in the US sample. Second, at first glance, the US result resembles that of the Vietnamese result because in both cases three significant variables are about whether the student lives with certain family members. However, their directions of influence are opposite to each other. To be more specific, in the Vietnamese sample living with siblings is associated with higher test scores, but in the US sample, living with siblings or cousins is adversely correlated with test performance.

Discussion and Conclusion

One of the major objectives of this study is to identify malleable factors in the data sets and suggest actionable items to constituents. Three categories of malleable factors, i.e., attitudes, home environment, and technology, have been identified in the results.

Attitudinal Factors

The bootstrap forest results of the Asian countries/regions share several common threads. The variables pertaining to openness for problem-solving occur in four Asian countries/regions as crucial predictors of PISA science test performance (Hong Kong, South Korea, Japan, and Vietnam). These variables are especially salient for Hong Kong because three out of ten significant predictors are concerned with openness for problem-solving while other samples have one or two only. This finding is compatible with prior research, which substantiated that inquiry-based science (teaching students to ask and answer questions on their own) could enhance student higher reasoning and problem-solving skills (Gillies, Nichols, Burgh, & Haynes, 2012). Therefore, our research finding implies the need to create a culture of openness for problem-solving and to promote inquiry-based science curricula. Another attitudinal variable highlighted by the bootstrap forest is perceived control over science performance (for Hong Kong, South Korea, and the United States). Specifically, if students disagree with the concept that they would not succeed no matter if they study or not, they are more likely to earn higher scores. In light of this finding, it is important to empower students with the belief that they have control over their science performance and that if they invest in studying, they will achieve.

Factors on Home Environment

The strong presence of variables related to the home environment is worth noting. For example, the number of books at home is the number one, number two, or number three predictors of science performance in five Asian samples and the US sample. This finding corresponds with prior research on the US and Canadian students' science performance as measured by PISA (Yu, Kaprolet, Jannasch-Pennell, & DiGangi, 2012). The Japanese and Vietnamese samples further reveal the types of books matter (classic literature in Japan; classic literature and technical reference in Vietnam). One may argue that the number of books or the type of books does not constitute a causal explanation of test scores, because it may simply reflect the parents' educational levels. Nonetheless, it is a reasonable conjecture that parents who possess a higher number of books may value learning more. Therefore, it is advisable for parents to create a home environment that encourages student learning.

Additionally, the number of computers at home also plays an important role. OECD, the developer of PISA, had conducted an extensive study on the relationship between computer use and PISA test performance (Vincenzo, 2010). OECD found a positive and significant effect of the frequency of computer use on science test scores. However, in most countries, this effect is even stronger when the computer is used at home instead of at school. Based on this finding, OECD bluntly questions the effectiveness of educational policies aimed at promoting computer use at school as a tool for learning. Although having more computers at home does not necessarily imply that the students would use computers more often, it could still be a proxy measure of computer use at home.

Contradictory results on certain home-environment variables exist across different countries. Specifically, in Japan and the United States, students who live with siblings tend to have lower science scores, but in Vietnam, those who live with siblings tend to have higher science scores. A possible explanation is that Vietnam, as a developing country, is still very traditional in the sense that family members living together form a support system and that elder brothers and sisters tutor their younger siblings. In contrast, in Japan and the United States, two developed nations, test performance is possibly boosted when students enjoy more private space. It is important to emphasize that three out of the ten crucial factors in the United States sample are concerned with who lives with the student, i.e., brothers, sisters, or others such as cousin(s). Specifically, living with these family members or others is correlated with lower scores. At this point, it is premature to leap from this correlation to any causal inferences, and therefore this factor awaits further investigation.

Technological Factors

Student usage of technology out of school is found to be significant predictors of test performance in all Asian samples, except Vietnam. These variables include time to access the Internet outside of school during weekdays and the weekend, time to use e-mail outside of school, browsing the Internet for schoolwork, reading news on the Internet, chatting online, downloading music, and uploading content for sharing. As one may expect, some of the variables (e.g., browsing the Internet for schoolwork, reading news on the Internet) have positive effects on science learning, while chatting online has negative effects. Some of the variables (e.g., time to access the Internet outside of school during weekdays) and science test scores have a curvilinear relationship. Additionally, in the Singaporean sample, exposure to computers and the Internet at an early age is associated with high performance of science scores.

Today the ubiquitousness of technology is an irreversible trend, and it is possible that a decade from now technological factors might also become important predictors of Vietnamese student performance in PISA science test. Technology is a both a blessing and a curse, as indicated in the Asian experience. Nonetheless, it could be a malleable factor if teachers and parents could encourage children to use technology in ways that bring benefits to their science learning, discourage usage that is distracting to their academic studies, and guide them with optimal use of technologies that is curvilinearly correlated with science learning.

Concluding Thoughts

Although the significant variables recognized by the bootstrap forest could yield many practical implications for researchers and educators, the variables that are missing are equally revealing and intriguing. It is surprising to see that none of the variables on the school survey, such as school type (private or public), class size, student-computer ratio, teacher qualification and professional development, school activity, parent participation in school, school management, etc., had any effect on science performance in any chosen sample. Although some school-related variables were identified as crucial predictors, such as the number of class periods in science per week (in Hong Kong and Singapore) and the number of all class periods per week (in the United States), these are self-reported items in the student survey, not in the school questionnaire. In the South Korean sample, parents' level of agreement that the standards of achievement are high in their child's school and that competition exists from other schools is correlated with their child's test performance. However, this finding is parent-reported on their perceptions of the schools.

One may argue that the absence of school variables is not in alignment with prior research. For example, according to a comparison between Korea and the United States based upon the third TIMSS data set, family-school partnerships, early intervention in primary and middle schools, and after-school or weekend programs are found to be significant predictors of TIMSS test scores (Paik, 2004). However, unlike the bootstrap forest, regression analysis employed in Paik's study could not simultaneously evaluate all variables; rather, it had to keep home and school variables in isolation from others. Nonetheless, the authors of this chapter investigated only the PISA data set and thus whether the same result could be found in the TIMSS data set awaits further scrutiny.

Like other authors, Friedman (2013) suggested certain strategies to narrow the performance gap between American and Asian students. These recommended courses of action include a strong commitment to teacher training, peer learning and constant professional development, a deep involvement of parents in their children's learning, an insistence by the school's leadership on the highest standards, and a culture that prizes education and respects teachers. On one hand, the authors of this chapter have no intention to downplay the significance of the preceding recommendations. On the other hand, when 399 variables were evaluated simultaneously in the present large-scale study, another picture emerges from the data. As indicated in the results, no variables in the school questionnaire, such as teacher training, professional development, and school leadership, emerged in the bootstrap forest. Simply put, they paled in comparison to student-level variables, home-environment variables, and technological factors that are outside of school. The authors by no means attempt to dissuade government agencies from investing resources in schools or improving teacher qualifications. Neither do the authors assert that teachers and school administrators have no role in helping students achieve in their science learning. Nonetheless, given the higher ranks of importance of other variables, perhaps attention to these factors, may yield effective outcomes.

For the last several decades, US schools have been investing tremendous resources in instructional technologies, such as hypertext, multimedia, and mobile computing (e.g., iPad), but US student performance on both PISA and TIMSS stalled. Based on the salient findings of the present study, teachers, school administrators, and parents can work hand in hand to create a home and school culture that promotes openness for problem-solving and also guide students in the proper use of technology outside the school.

Acknowledgments Special thanks to Miss Danielle Reimer for her valuable input into this project.

References

Akiba, M., LeTendre, G. K., & Scribner, J. P. (2007). Teacher quality, opportunity gap, and national achievement in 46 countries. *Educational Researcher, 36*, 369–387.

Aun, T. K., Riley, J. P., Atputhasamy, L., & Subramaniam, R. (2006). School science achievement in Japan and Singapore: A tale of two cities. *Educational Research for Policy and Practice, 5* (1), 1–13.

Bandura, A. (1997). Self-efficacy: The exercise of control. New York, NY: Freeman.

Breiman, L. (2000) *Some infinity theory for predictor ensembles* (Technical Report 577), UC Berkeley. Retrieved from http://www.stat.berkeley.edu/~breiman

Breiman, L. (2001). Random forests. *Machine Learning, 45*, 5–32.

Breiman, L. (2004). *Consistency for a simple model of random forests* (Technical report 670), UC Berkeley. Retrieved from http://www.stat.berkeley.edu/~breiman

Breiman, L., Friedman, J. H., Olshen, R. A., & Stone, C. J. (1984). *Classification and regression trees*. Monterey, CA: Wadsworth International Group.

Britner, S., & Pajares, F. (2006). Sources of science self-efficacy beliefs of middle school students. *Journal of Research in Science Teaching, 43*, 485–499.

Cutler, D. C., Edwards, T. C., Beard, K. H., Cutler, A., Hess, K. T., Gibson, J., et al. (2007). Random forests for classification in ecology. *Ecology, 88*, 2783–2792.

Davie, S. (2013, December 5). Scaling education heights in PISA. *The Straits Times*, pp. 1–4.

Efron, B. (1979). Bootstrap methods: Another look at the jackknife. *The Annals of Statistics, 7*, 1–26.

Efron, B. (1981). Nonparametric estimates of standard error: The jackknife, the bootstrap and other methods. *Biometrika, 63*, 589–599.

Efron, B. (1982). *The jackknife, the bootstrap, and other resampling plans* (Society of industrial and applied mathematics CBMS-NSF monographs, 38). Philadelphia: Society for Industrial and Applied Mathematics.

Efron, B., & Tibshirani, R. J. (1993). *An introduction to the bootstrap*. New York: Chapman & Hall.

Fadaia, B. (2013). Random forest on grouped data. *Cross Validation*. Retrieved from http://stats.stackexchange.com/questions/19803/random-forest-on-grouped-data

Friedman, T. (2013, October 22). The Shanghai secret. *New York Times*. Retrieved from http://www.nytimes.com/2013/10/23/opinion/friedman-the-shanghai-secret.html?_r=0

Gillies, R., Nichols, K., Burgh, G., & Haynes, M. (2012). The effects of two strategic and meta-cognitive questioning approaches on children's explanatory behaviour, problem-solving, and learning during cooperative, inquiry-based science. *International Journal of Educational Research, 53*, 93–106.

Hampden-Thompson, G., & Johnston, J. S. (2006). Variation in the relationship between nonschool factors and student achievement on international assessments. *National Center for Education Sciences, 14*, 1–35.
Ho, E. S. (2006a). Educational decentralization in three Asian societies: Japan, Korea and Hong Kong. *Journal of Educational Administration, 44*, 590–603.
Ho, E. S. (2006b). High stakes testing and its impact on students and schools in Hong Kong: What we have learned from the PISA studies. *Journal of Educational Policy, 3*(1), 69–87.
Ho, E. S. (2009). Characteristics of East Asian learners: What we learned from PISA. *Educational Research Journal, 24*, 327–348.
Ho, E. S. (2010a). Family influences on science learning among Hong Kong adolescents: What we learned from PISA. *International Journal of Science and Mathematics Education, 8*, 409–428.
Ho, E. S. (2010b). Assessing the quality and equality of Hong Kong basic education results: From PISA 2000+ to PISA 2006. *Frontiers of Education in China, 5*, 238–257.
Jerolmack, C., & Khan, S. (2014). Talk is cheap: Ethnography and the attitudinal fallacy. Social *Sciences, Mathematical Methods, 9*. Advance online publication. doi:10.1177/0049124114523396.
Kim, M., Lavonen, J., & Ogawa, M. (2009). Experts' opinions on the high achievement of scientific Literacy in PISA 2003: A comparative study in Finland and Korea. *Eurasia Journal of Mathematics, Science & Technology Education, 5*, 379–393.
Lee, P. (2013, December 5). PISA score signals Singapore 'is on right path'. *The Straits Times*, pp. 1–2.
Low, E., & Lee, S. (2012). Bringing Singapore's teacher education beyond its shores. *Educational Research for Policy and Practice, 11*(1), 43–51.
McNeal, R. B. (1999). Parental involvement as social capital: Differential effectiveness on science achievement, truancy, and dropping out. *Social Forces, 78*(1), 117–144.
Nabeshima, K. (2003). *Raising the quality of secondary education in East Asia* (World bank policy research working paper 3140). http://dx.doi.org/10.1596/1813-9450-3140
National Center for Education Statistics. (2013). PISA methodology and technical notes – International requirements. Retrieved from http://nces.ed.gov/surveys/pisa/pisa2012/pisa2012highlights_9a.asp
Organization for Economic Co-operation and Development [OECD]. (2013a). *The PISA international database*. Retrieved from http://pisa2012.acer.edu.au/
Organization for Economic Co-operation and Development [OECD]. (2013b). *PISA 2012 assessment and analytical framework: Mathematics, reading, science, problem solving, and financial literacy*. Retrieved from http://www.oecd.org/pisa/pisaproducts/PISA%202012%20framework%20e-book_final.pdf
Organization for Economic Cooperation and Development [OECD]. (2013c). *Snapshot of student performance in mathematics, reading and science*. Retrieved from http://www.oecd.org/pisa/keyfindings/PISA-2012-results-snapshot-Volume-I-ENG.pdf
Paik, S. (2004). Korean and U.S. families, schools, and learning. *International Journal of Educational Research, 41*, 71–90.
Rice, L., Barth, J. M., Guadagno, R. E., Smith, G. P., & McCallum, D. M. (2013). The Role of social support in students' perceived abilities and attitudes toward math and science. *Journal of Youth and Adolescence, 42*, 28–40.
SAS Institute. (2013). *JMP Pro [Computer software]*. Cary, NC: Author.
Singh, K., Granville, M., & Dika, S. (2002). Mathematics and science achievement: Effects of motivation, interest, and academic engagement. *Journal of Educational Research, 95*, 323–332.
Steinkamp, M. W., & Maehr, M. L. (1983). Affect, ability, and science achievement: A quantitative synthesis of correlational research. *Review of Educational Research, 53*, 369–396.
Tan, C. (2012). "Our shared values" In Singapore: A Confucian perspective. *Educational Theory, 62*, 449–463.

Verikas A., Gelzinis, A., & Bacauskiene, M. (2011). Mining data with random forests: A survey and results of new tests. *Pattern Recognition, 44*, 330–349. doi:http://dx.doi.org/10.1016/j.patcog.2010.08.011

Vincenzo, S. (2010). Does computer use increase educational achievements? Student-level evidence from PISA. *OECD Journal: Economic Studies, 1*, 127–148.

von Davier, M., Gonzalez, E., & Mislevy, R. J. (2009). What are plausible values and why are they useful? In M. von Davier & D. Hastedt (Eds.), *IERI monograph series: Issues and methodologies in large-scale assessments* (Vol. 2, pp. 9–36). Princeton, NJ: Educational Testing Service.

Weingarten, R. (2013). AFT President Weingarten on PISA 2012 international results. *American Federation of Teachers*. Retrieved from http://www.aft.org/newspubs/press/2013/120313.cfm

Wößmann, L. (2003). Schooling resources, educational institutions and student performance: The international evidence. *Oxford Bulletin of Economics and Statistics, 65*(2), 117–170.

Wu, M. (2004). Plausible values. *Rasch Measurement Transactions, 18*(2), 976–978. Retrieved from http://www.rasch.org/rmt/rmt182c.htm

Yalein, M., Aslan, S., & Usta, E. (2012). Analysis of PISA 2009 exam according to some variables. *Mevlana International Journal of Education, 2*(1), 64–71.

Yang, Y. (2003). Dimensions of socio-economic status and their relationship to mathematics and science achievement at individual and collective levels. *Scandinavian Journal of Educational Research, 47*(1), 21–41.

Yu, C. H. (2003). Resampling methods: Concepts, applications, and justification. *Practical Assessment Research and Evaluation, 8*(19). Retrieved from http://pareonline.net/getvn.asp?v=8&n=19

Yu, C. H. (2007). Resampling: A conceptual and procedural introduction. In J. Osborne (Ed.), *Best practices in quantitative methods* (pp. 283–298). Thousand Oaks, CA: Sage Publications.

Yu, C. H., Kaprolet, C., Jannasch-Pennell, A., & DiGangi, S. (2012). A data mining approach to compare American and Canadian Grade 10 students in PISA 2006 Science test performance. *Journal of Data Science, 10*, 441–464. Retrieved from http://www.jds-online.com/file_download/362/JDS-1064.pdf

Zhao, Y. (2012, September 1). Flunking innovation and creativity: East Asia's highly touted test scores in math, science, and reading are masking important failures in developing innovators and entrepreneurs. *Phi Delta Kappan, 94*(1), 56.

Chapter 24
Dispositions Towards Science and Science Achievement in TIMSS 2011: A Comparison of Eighth Graders in Hong Kong, Chinese Taipei, Japan, Korea, and Singapore

Yoon Fah Lay, Shaljan Areepattamannil, Khar Thoe Ng, and Chwee Hoon Khoo

Introduction

The findings of the recent cycles of the Trends in International Mathematics and Science Study (TIMSS) suggest that East Asian education systems, such as Hong Kong, Chinese Taipei, Japan, Korea, and Singapore, tend to outperform other participating countries on the TIMSS mathematics and science assessments. Korea and Singapore were the top-performing countries on the TIMSS 2011 science assessment at the fourth grade, followed by Finland, Japan, Russian Federation, and Chinese Taipei. At the eighth grade, Singapore had the highest average science achievement. This was followed by the next highest-performing countries for science achievement that are Korea, Chinese Taipei, and Japan.

Factors that might have contributed to the outstanding performance of the aforementioned countries are multifaceted, and such factors have been widely researched recently, including cognitive, affective (i.e., interest, attitude, and motivation), and psychological/sociological aspects (see Ong & Gonzalez, 2012; Ong,

Y.F. Lay (✉)
Faculty of Psychology and Education, Universiti Malaysia Sabah, Kota Kinabalu, Sabah, Malaysia
e-mail: layyoonfah@yahoo.com.my

S. Areepattamannil
Curriculum, Instruction & Educational Technology Division, Emirates College for Advanced Education, Abu Dhabi, United Arab Emirates
e-mail: sareepattamannil@ecae.ac.ae

K.T. Ng
R&D Division, SEAMEO RECSAM, Gelugor, Penang, Malaysia
e-mail: cyberworldnkt@gmail.com

C.H. Khoo
Teacher Education Institute-Kent Campus, Tuaran, Sabah, Malaysia
e-mail: khoo8921@yahoo.com

M.S. Khine (ed.), *Science Education in East Asia*,
DOI 10.1007/978-3-319-16390-1_24

Gonzalez, & Shanmugam, 2013). The purpose of the present study, however, is to examine the predictive effects of an affective factor, i.e., students' dispositions towards science (including students like learning science, students value science, and students confident in science) on science achievement among eighth grade students in Chinese Taipei, Hong Kong, Japan, Korea, and Singapore who participated in the TIMSS 2011 assessment. There is a dearth of research on the predictive effects of students' dispositions towards science on science achievement among early adolescents in these high-performing East Asian education systems. The findings of such a study may indeed help educators and policymakers to identify and nurture the strong learning prerequisites of early adolescents in these education systems.

Review of Literature

Self-Beliefs in Motivational Constructs and Constructivist Theoretical Framework

How science is taught is believed to have influence on students' interest in science as a subject and their prospective career choice. The types of teaching approaches and influential philosophical framework or learning theories that are suitable for the acquisition of knowledge, learning, and assessment among students from diverse sociocultural background and levels of achievement are areas of concerns for many educational researchers. Much effort was made to review the contemporary issues and challenges faced in motivating students to like science and outline relevant factors contributing to the cognitive, psychomotor, and affective domains of science learning. For example, the various aspects of "behaviorism, cognitivism, and constructivism" were compared in the areas of "learning process, framework, principles, strategies and theorists" (Berns & Erickson, 2001; Purdue University, 2007). Three contemporary theories of learning, i.e., "cognitive science, constructivism, and social constructivism," to review "cognitive and social dimensions of learning" were also analyzed in "learning about learning" curriculum (University of Washington, 2001).

Constructivist and motivation theories recognize that motivation is influenced by how interesting and relevant the learners perceive the activities and information. According to educational psychologists, student's motivation is influenced by a number of beliefs, values, interests, and attitudes that can be positive or negative in their effects. The construct about "self" was grounded on "self-determination" theory (stating that students may do activities for interest or enjoyment, i.e., intrinsic motivation) and/or "self-belief" theory (including self-efficacy which stated when self-confidence is high, students will be more motivated to persist in a task until it is completed). The construct "belief on own coping ability" (intrinsic motivation) will be based on expectancy-value theory explaining that beliefs about one's ability to succeed are expectancy beliefs and beliefs about the extent to which the task is useful and enjoyable or relates to one's self-image are value beliefs (Glynn, Taasoobshirazi, & Brickman, 2007; Palmer, 2007; Phillips, 2007; Weiner, 1979). It is believed that students are motivated to learn when they value either the outcome or process of learning and they expect that they will be successful. As explained from psychological theories, students believe that the task is of value (value beliefs) and they believe he/she has the ability and confidence to succeed in the learning task (expectancy and self-efficacy beliefs) (Lefton, 1991; Phillips, 2007).

Relationship Between Self-Efficacy Beliefs and Achievement in Science

Self-efficacy is a very specific form of self-concept theory that refers to the people's beliefs about their mastery or capabilities to perform a task successfully at designated levels with convictions about their own effectiveness that can determine the types of behavior they will engage in or the amount of risk they will undertake. It is their belief about whether they can successfully engage in and execute a specific behavior or their confidence in their ability to behave in such a way to produce a desirable outcome (Bandura, 1977, 1997; Lefton, 1991). Self-efficacy determines and flows from the feelings of self-beliefs and self-worth. In other words, the people with self-efficacy consider themselves to be capable and worthy. A strong sense of self-efficacy allows people to feel free to select the influence, construct their own desirable lives, and even effect changes in themselves and persevere in tough time. Self-efficacy or self-beliefs make a difference in how people feel, think, and act, such as in science-related learning or activities. If people feel themselves to be able to control the situations, this increases their perceived self-efficacy or self-beliefs to manage it. In fact, people with a high level of self-efficacy is more likely to attribute success to variables within themselves rather than to chance factors and are more likely to pursue tasks (Bandura & Wood, 1989 in Lefton, 1991), subsequently striving towards accomplishing the tasks with commendable achievement.

Literature revealed that early experience with science is expected to have influence on students' self-belief or self-efficacy towards their involvement in science, for example, they will "like or be interested in" and "value" science learning, like to be "engaged" more in science-related lessons and even be more "confident" to pursue science-related studies and careers. A study was conducted by Conner et al. (2013) to evaluate secondary students' performance for scientific literacy, reading, and thinking skills in the Programme for International Student Assessment (PISA) 2009. The findings revealed that with greater emphasis on early years in scientific and reading literacy that promote student-centered learning and thinking skills, New Zealand emerged as one of the top-tenth high-performing country of PISA. In another study by Liu, Hsieh, Cho, and Schallert (2006) to examine the effect of a computer-enhanced problem-based learning (PBL) environment on middle school students' learning, the analysis of mixed-mode data revealed that self-efficacy was shown to be a statistically significant predictor of science achievement scores.

There are also other traits or distinguishing characteristics or qualities that are interrelated or overlapped to some extent and were identified as important for motivation towards science, technology, environment, and society (STES) education as explained from motivation and constructivist theories. For example, highly positively motivated students would find the task inherently enjoyable (intrinsic motivation), have an established long-term interest in particular topics (personal interest) based on interest theory, have a desire to fully understand the content (a mastery goal), and believe that success will be related to effort (an attribution). According to Weiner (1979), the attribution theory for motivation stated that students who believe that success is related to effort will be more likely to use effort than those who believe that success is due to inherent ability. When individuals have positive senses of effort, ability, and efficacy, they are more likely to choose to do the challenging tasks and excel in academic performance. Ng (2012) conducted a study on the effect of PBL using scaffolded instruction (PBL-SI) on students' motivation and higher-order thinking (HOT) using mixed-research methods. The findings revealed that more and moderately successful students demonstrated evidence of enhanced fluid intelligence as reflected in their "Fluid Intelligence Test" (FIT) and science achievement as reflected in their "Higher-Order Thinking Test in Science" (HOTTIS) scores. There was also enhanced motivation as reflected in their "Science Motivation Scale" (SMS) (which included self-efficacy as one of the motivational constructs) and "Motivation Towards Science Learning Survey" (MTSLS). For example, these students had shown intrinsic motivation and were involved actively in preparing projects for mini-science fair, project proposal competitions, and e-learning activities.

Background, Rationale, and Research Questions

Background Information of Top-Performing East Asian Countries in TIMSS 2011

East Asian countries consistently outperformed their counterparts in each and every cycle of the Trends in International Mathematics and Science Study (TIMSS) since its inception in 1995. In TIMSS 2011, at the eighth grade, Singapore had the highest average science achievement. This was followed by the next highest-performing countries for science achievement that are Korea, Chinese Taipei, and Japan. A summary of grade 8 students' science performance of East Asian countries in TIMSS 1995 to TIMSS 2011 is provided in Table 24.1.

Table 24.1 TIMSS (grade 8) science scores from 1995 to 2011 for Chinese Taipei, Hong Kong SAR, Japan, Korea, and Singapore

			Science grade 8			
No. of participating countries	Chinese Taipei	Hong Kong SAR	Japan	Korea	Singapore	
TIMSS 1995	45	–	510	554	546	580
TIMSS 1999	38	569	530	550	549	568
TIMSS 2003	46	571	556	552	558	578
TIMSS 2007	59	561	530	554	553	567
TIMSS 2011	63	564	535	558	560	590

Rationale

The main reason why these five countries were chosen in this study is that these are the five top-performing countries in TIMSS 2011. These five countries were also chosen for the following reasons. Firstly, the educational systems of these countries share some similarities in terms of sociocultural background. Secondly, the purpose of this study was to identify areas in which the educational systems of these countries can be improved after analyzing eighth grade students' dispositions towards science in these respective countries. Hence, this study examined the predictive effects of eighth graders' dispositions towards science (i.e., students like learning science, students value science, and students confident in science) on science achievement in five high-performing education systems: Hong Kong, Chinese Taipei, Japan, Korea, and Singapore, who participated in the TIMSS 2011.

Research Questions

In light of the scarcity of empirical research especially on the TIMSS 2011, the research question underpinned in this study was: How well do eighth graders' dispositions towards science predict their achievement in science in Hong Kong, Chinese Taipei, Japan, Korea, and Singapore?

Method

Data

Data for the study were drawn from the TIMSS 2011 database (http://timssandpirls.bc.edu/timss2011/international-database.html). A total of 24,564 eighth graders from Hong Kong ($n=4{,}015$), Chinese Taipei ($n=5{,}042$), Japan ($n=4{,}414$),

Korea ($n = 5{,}166$), and Singapore ($n = 5{,}927$) took part in the TIMSS 2011 science assessment.

Measures

Science Achievement

The TIMSS 2011 science achievement scale was based on 302 items spanning across content (i.e., biology, chemistry, physics, earth science) and cognitive (i.e., knowing, applying, reasoning) domains in science. TIMSS uses an imputation methodology, usually referred to as plausible values, to report student performance. The plausible values, an approach developed by Mislevy and Sheehan (1987, 1989) and based on the imputation theory of Rubin (1987), are random elements from the set of scores (i.e., random draws from the marginal posterior of the latent distribution) that could be attributed to each student. For each student, the mean plausible value was used as a measure of science achievement. The IEA's International Database (IDB) Analyzer for TIMSS, a plug-in for SPSS, was used to combine the five plausible values as well as to produce their average values and correct standard errors.

Students Like Learning Science

The Students Like Learning Science Scale was developed to measure students' interest in and liking of learning science. The TIMSS 2011 Students Like Learning Science Scale was based on five items (i.e., "I enjoy learning science," "I wish I did not have to study science," "Science is boring," "I learn many interesting things in science," "I like science"). All items were rated on a 4-point Likert type scale, ranging from "1" (*disagree a lot*) to "4" (*agree a lot*). The Cronbach's alpha reliability coefficients for the scale were 0.90, 0.87, 0.89, 0.90, and 0.89 in Chinese Taipei, Hong Kong, Japan, Korea, and Singapore, respectively.

Students Value Science

The TIMSS 2011 Students Value Science Scale addresses students' attitudes about the importance of the subject and usefulness of the subject, sometimes called attainment value and utility value (Wigfield & Eccles, 2000). The TIMSS 2011 Students Value Science Scale was based on six items (i.e., "I think learning science will help me in my daily life," "I need science to learn other school subjects," "I need to do well in science to get into the university of my choice," "I need to do well in science to get the job I want," "I would like a job that involves using science," "It is important to do well in science"). All items were rated on a 4-point Likert type scale, ranging from "1" (*disagree a lot*) to "4" (*agree a lot*). The Cronbach's alpha reliability coefficients for the scale were 0.89, 0.89, 0.85, 0.88, and 0.87 in Chinese Taipei, Hong Kong, Japan, Korea, and Singapore, respectively.

Students Confident in Science

The Students Confident in Science Scale assesses students' self-confidence or self-concept in their ability to learn science. The TIMSS 2011 Students Confident in Science Scale was based on nine items (i.e., "I usually do well in science," "Science is more difficult for me than for many of my classmates," "Science is not one of my strengths," "I learn things quickly in science," "Science makes me confused and nervous," "I am good at working out difficult science problems," "My teacher thinks I can do well in science with difficult materials," "My teacher tells me I am good at science," "Science is harder for me than any other subject"). All items were rated on a 4-point Likert type scale, ranging from "1" (*disagree a lot*) to "4" (*agree a lot*). The Cronbach's alpha reliability coefficients for the scale were 0.92, 0.87, 0.88, 0.91, and 0.91 in Chinese Taipei, Hong Kong, Japan, Korea, and Singapore, respectively.

In addition to these measures, student demographic characteristics such as gender (1 = "female"; 0 = "male"), home language (1 = "language of test"; 0 = "another language"), and parents' highest level of education (1 = "some primary or lower secondary or no school" to 5 = "university or higher") were included in the study as control variables.

Results and Discussion

Descriptive statistics for the predictor variables (students like learning science, students value science, and students confident in science) are presented in Table 24.2.

Tables 24.3, 24.4, and 24.5 show the percent of students who "like learning science, value science, and are confident in science" with their average science achievement in Chinese Taipei, Hong Kong, Japan, Korea, and Singapore.

In TIMSS 2011, internationally, at the eighth grade, about one-third (35 %) of the students, on average, "like learning science," and about one-fifth (21 %) "do not like learning science." Accompanying the decrease in liking learning science is a widening achievement gap between students who like learning science (515, on average) and those who do not (450) (Martin, Mullis, Foy, & Stanco, 2012). Internationally, on average, eighth grade students in general or integrated science countries placed a high value on science. Forty-one percent were in the "value" category and another 33 % were in the "somewhat value" category, on average. However, about one-fourth (26 %) were in the "do not value" category. Across the eighth grade, students who said they valued science typically had higher

Table 24.2 Descriptive statistics for the study variables

Variables	Hong Kong SAR		Chinese Taipei		Japan		Korea, Republic of		Singapore	
	M	SD	*M*	SD	*M*	SD	*M*	SD	*M*	SD
Students like learning science										
I enjoy learning science	3.06	.837	2.60	.915	2.74	.882	2.53	.841	3.25	.782
I wish I did not have to study science	2.06	.919	2.39	.993	2.25	.892	2.49	.883	1.96	.933
Science is boring	2.08	.910	2.26	.963	2.19	.850	2.40	.851	1.99	.808
I learn many interesting things in science	3.20	.786	2.89	.903	2.63	.901	2.81	.845	3.41	.723
I like science	3.01	.853	2.55	.966	2.55	.960	2.47	.874	3.17	.829
Students value science										
I think learning science will help me in my daily life	3.28	.748	2.93	.888	2.61	.881	2.83	.812	3.39	.708
I need science to learn other school subjects	2.84	.881	2.30	.896	2.29	.812	2.56	.818	2.99	.849
I need to do well in science to get into the university of my choice	2.95	.879	2.49	1.009	2.69	.962	2.75	.881	3.31	.795
I need to do well in science to get the job I want	2.79	.915	2.27	.959	2.50	.957	2.61	.916	3.12	.891
I would like a job that involves using science	2.53	.977	2.03	.922	1.93	.879	2.17	.910	2.73	.995
It is important to do well in science	3.10	.820	2.87	.955	3.00	.875	2.88	.827	3.51	.678
Students confident in science										
I usually do well in science	2.78	.800	2.38	.883	2.10	.790	2.31	.776	2.89	.831
Science is more difficult for me than for many of my classmates	2.25	.870	2.53	.967	2.38	.850	2.28	.791	2.19	.871

(continued)

Table 24.2 (continued)

Variables	Hong Kong SAR		Chinese Taipei		Japan		Korea, Republic of		Singapore	
	M	SD	*M*	SD	*M*	SD	*M*	SD	*M*	SD
Science is not one of my strengths	2.50	.913	2.74	1.002	2.65	.937	2.63	.835	2.38	.971
I learn quickly in science	2.66	.824	2.25	.871	2.21	.770	2.33	.746	2.78	.838
Science makes me confused and nervous	2.17	.868	2.46	.955	2.05	.862	2.25	.809	2.23	.899
I am good at working out difficult science problems	2.29	.840	2.01	.836	1.85	.787	2.02	.731	2.47	.860
My teacher thinks I can do well in science programs/clas-ses/lessons with difficult materials	2.33	.819	1.78	.806	1.78	.694	2.07	.739	2.65	.845
My teacher tells me I am good at science	2.14	.809	1.91	.850	1.72	.699	2.03	.730	2.44	.884
Science is harder for me than any other subject	2.21	.924	2.61	1.030	2.34	.914	2.34	.880	2.13	.941

Note: 1 = disagree a lot, 4 = agree a lot

achievement than students who only valued it somewhat, and those students, in turn, had higher achievement than students who did not value science (Martin et al., 2012). Internationally, on average, only 20 % of the eighth grade students in general or integrated science countries expressed confidence in their science ability, with 49 % "somewhat confident" and 31 % "not confident" (Martin et al. 2012).

To address the purpose of the study, separate simultaneous multiple regression analyses were conducted for each education system to determine whether students' dispositions towards science were predictive of science achievement (see Table 24.6). Eighth graders in Chinese Taipei, Hong Kong, Korea, and Singapore who liked learning science very much scored statistically and significantly higher on the TIMSS science assessment than their peers who did not like learning science. Similarly, eighth graders in Chinese Taipei, Hong Kong, Japan, Korea, and Singapore who greatly valued learning science and were more confident in learning science also scored statistically and significantly higher on the TIMSS science assessment than their peers who did not value learning science and were less confident in learning science.

Table 24.3 Students like learning science

Country	Like learning science		Somewhat like learning science		Do not like learning science		Average scale score
	Percent of students	Average achievement	Percent of students	Average achievement	Percent of students	Average achievement	
Chinese Taipei	17 (0.8)	618 (3.4)	43 (0.7)	571 (2.7)	40 (1.1)	534 (2.6)	9.0 (0.05)
Hong Kong	28 (1.2)	561 (4.1)	51 (0.9)	534 (3.3)	21 (1.1)	506 (4.9)	9.8 (0.06)
Japan	15 (0.8)	595 (3.7)	47 (1.1)	566 (2.2)	38 (1.5)	531 (3.1)	9.0 (0.06)
Korea	11 (0.5)	623 (3.8)	43 (0.9)	576 (2.1)	46 (1.1)	531 (2.2)	8.7 (0.04)
Singapore	38 (0.8)	617 (5.2)	46 (0.7)	584 (4.2)	16 (0.5)	542 (5.4)	10.2 (0.03)
International	35 (0.2)	515 (0.8)	44 (0.2)	472 (0.8)	21 (0.2)	450 (1.1)	

( ) Standard errors appear in parentheses

Table 24.4 Students value science

Country	Value		Somewhat value		Do not value		Average scale score
	Percent of students	Average achievement	Percent of students	Average achievement	Percent of students	Average achievement	
Chinese Taipei	12 (0.7)	612 (4.2)	30 (0.7)	586 (2.8)	58 (1.1)	543 (2.2)	8.5 (0.05)
Hong Kong	26 (1.0)	559 (4.1)	43 (0.8)	535 (3.8)	32 (1.1)	518 (4.0)	9.5 (0.04)
Japan	10 (0.7)	595 (4.9)	34 (1.0)	574 (2.7)	56 (1.1)	540 (2.7)	8.5 (0.04)
Korea	14 (0.6)	607 (4.1)	40 (0.9)	574 (2.3)	46 (1.0)	535 (2.2)	8.8 (0.03)
Singapore	41 (0.8)	616 (4.6)	43 (0.7)	583 (4.3)	17 (0.6)	546 (5.9)	10.2 (0.03)
International	41 (0.2)	502 (0.8)	33 (0.2)	477 (0.8)	26 (0.2)	457 (1.1)	

( ) Standard errors appear in parentheses

Table 24.5 Students confident in science

Country	Confident		Somewhat confident		Not confident		Average scale score
	Percent of students	Average achievement	Percent of students	Average achievement	Percent of students	Average achievement	
Chinese Taipei	6 (0.4)	648 (4.9)	27 (0.9)	599 (3.1)	67 (1.0)	543 (2.3)	8.3 (005)
Hong Kong	8 (0.6)	579 (4.9)	47 (1.1)	544 (4.1)	45 (1.3)	520 (3.4)	9.2 (0.04)
Japan	3 (0.3)	631 (7.7)	28 (0.9)	591 (2.6)	69 (1.1)	540 (2.6)	8.4 (0.04)
Korea	4 (0.3)	652 (4.6)	33 (0.8)	603 (2.1)	63 (0.9)	532 (1.9)	8.7 (0.03)
Singapore	14 (0.5)	630 (5.9)	48 (0.7)	600 (4.8)	37 (0.8)	562 (4.2)	9.6 (0.03)
International	20 (0.2)	536 (1.0)	49 (0.2)	482 (0.8)	31 (0.2)	450 (0.9)	

( ) Standard errors appear in parentheses

Table 24.6 Eighth graders' dispositions towards science predicting their science achievement

	Chinese Taipei		Hong Kong		Japan		Korea		Singapore	
B	SE	B	SE	B	SE	B	SE	B	SE	
Gender	9.11*	2.46	3.89	3.73	4.09	2.92	5.36*	2.74	0.17	3.14
Home language	14.74*	2.48	−5.42	3.01	21.00*	5.31	−5.89	2.62	12.59*	1.92
Parental education	20.42*	1.48	10.57*	1.88	16.56*	1.78	15.71*	1.37	21.48*	1.85
Students like science	4.68*	1.14	6.13*	1.33	0.65	1.43	4.24*	1.09	7.74*	1.21
Students value science	6.44*	1.00	3.00*	1.04	8.13*	1.21	5.97*	1.00	5.50*	1.11
Students confident in science	7.72*	0.91	3.85*	1.13	13.90*	1.23	14.48*	1.05	3.39*	1.07
Adjusted R^2	0.29		0.13		0.25		0.30		0.22	

*significant at $p < .05$

In terms of student demographic characteristics, female students significantly outperformed their male counterparts in Chinese Taipei and Korea. Eighth graders in Chinese Taipei, Japan, and Singapore who speak the language of the test at home scored significantly higher in science achievement as compared to those who speak other languages. Students whose parents obtained higher level of education scored significantly higher in science achievement than their peers whose parents obtained lower level of education.

TIMSS routinely presents very powerful evidence showing that within countries, students with more positive attitudes towards science have substantially higher achievement, and the results from TIMSS 2011 are consistent with previous assessments (Martin et al., 2012). There is extensive research showing that students with more positive attitudes towards mathematics and science have higher average achievement in mathematics and science (Martin et al., 2012; Mullis, Martin, Foy, & Arora, 2012). For example, a recent meta-analysis of student attitudes towards school found that attitudes towards mathematics or science were related to mathematics and science achievement across 288 studies (Hattie, 2009).

Conclusion

This study found that eighth graders in Chinese Taipei, Hong Kong, Korea, and Singapore, who liked learning science very much, scored statistically and significantly higher on the TIMSS science assessment than did their peers who did not like learning science. Similarly, eighth graders who greatly valued learning science and were more confident in learning science also scored statistically and significantly higher on the TIMSS science assessment than did their peers who did not value learning science and were less confident in learning science in Chinese Taipei, Hong Kong, Japan, Korea, and Singapore.

References

Bandura, A. (1977). *Social learning theory*. Englewood Cliffs, NJ: Prentice-Hall.

Bandura, A., & Wood, R. (1989). Effect of perceived controllability and performance standards on self-regulation of complex decision makes. *Journal of Personality and Social Psychology, 56*(5), 805–814.

Bandura, A. (1997). *Self-efficacy: The exercise of control*. New York, NY: W. H. Freeman.

Berns, R. G., & Erickson, P. M. (2001). *Contextual teaching and learning*. The Highlight Zone: Research @ Work No. 5. Retrieved October 1, 2007 from http://www.nccte.org/publications/infosynthesis/highlightzone/highlight05/

Conner, L., Ng, K. T., Ahmad, N. J., Ab Bakar, H., Parahakaran, S. & Lay, Y. F. (2013, November 11–14). *Evaluating students' performance for scientific literacy, reading and thinking skills in PISA 2009: Lessons learnt from New Zealand and Malaysia*. Paper presented and accepted for publication in science book for the fifth international conference on science and mathematics education (CoSMEd) for CoSMEd 2013 at RECSAM, Penang, Malaysia.

Glynn, S. M., Taasoobshirazi, G., & Brickman, P. (2007). Nonscience majors learning science: A theoretical model of motivation. *Journal of Research in Science Teaching, 44*(8), 1088–1107.
Hattie, J. (2009). *Visible learning: A synthesis of over 800 meta-analysis relating to achievement.* New York, NY: Taylor & Francis.
Lefton, L. A. (1991). *Psychology* (4th ed.). Boston, MA: Allyn and Bacon.
Liu, M., Hsieh, P. P-H, Cho, Y. & Schallert, D. L. (2006). Middle school students' self-efficacy, attitudes and achievement in a computer-enhanced problem-based learning environment. *Journal of Interactive Learning Research,* 17(3), 225–242. Retrieved January 9, 2014 from URL: http://jabba.edb.utexas.edu/minliu/liuHsieh06.pdf
Martin, M. O., Mullis, I. V., Foy, P., & Stanco, G. M. (2012). *Times roll international result in science.* Amsterdam: International Association for the Evaluation of Educational Achievement.
Mislevy, R. J., & Sheehan, K. M. (1987). Marginal estimation procedures. In A. E. Beaton (Ed.), *The NAEP 1983/84 technical report* (NAEP Report 15-TR-20, pp. 121–211). Princeton, NJ: Educational Testing Service.
Mislevy, R. J., & Sheehan, K. M. (1989). Information matrices in latent-variable models. *Journal of Educational and Behavioral Statistics, 14*(4), 335–350.
Mullis, I. V., Martin, M. O., Foy, P., & Arora, A. (2012). *TIMSS 2011 international results in mathematics.* Amsterdam, The Netherlands: International Association for the Evaluation of Educational Achievement.
Ng, K. T. (2012). *The effect of PBL-SI on secondary students' motivation and higher order thinking.* Unpublished doctoral thesis, Open University Malaysia, Kuala Lumpur.
Ong, S. L., Gonzalez, E., & Shanmugam, S. K. S. (Eds.). (2013). *TIMSS 2011: What can we learn together in reaching greater heights?* Penang, Malaysia: SEAMEO RECSAM.
Ong, S., & Gonzalez, E. J. (Eds.). (2012). *TIMSS 2007: What can we learn?* Penang, Malaysia: SEAMEO RECSAM.
Palmer, D. (2007, Autumn). What is the best way to motivate students in science? *Teaching Science: The Journal of the Australian Science Teachers Association. 53*(1), 38–42. Retrieved from http://www.aabri.com/manuscripts/11834.pdf
Phillips, J. A. (2007). *HMEF 5043: Psychology of learning and instruction.* Kuala Lumpur, Malaysia: UNITEM Sdn. Bhd.
Purdue University. (2007). *Learning theories for instructional designers.* EdTech Project. Retrieved October 1, 2007 from URL: http://web.ics.purdue.edu/~haagard/edci591t_web_02/compare/comparison.htm
Rubin, D. B. (1987). *Multiple imputation for nonresponse in surveys.* New York, NY: Wiley.
University of Washington. (2001). *T EDUC 502.* Retrieved October 1, 2008 from URL: http://www.washington.edu/students/crscatt/teduc.html
Weiner, B. (1979). A theory of motivation for some classroom experiences. *Journal of Educational Psychology, 71*(1), 3–25.
Wigfield, A., & Eccles, J. S. (2000). Expectancy-value theory of achievement motivation. *Contemporary Educational Psychology, 25*, 68–81.

Chapter 25
Correlates of Science Achievement in Singapore: A Multilevel Exploration

Shaljan Areepattamannil, Ching Leen Chiam, Daphnee H.L. Lee, and Helen Hong

The last two decades have witnessed a mushrooming of standardized international student assessment tests, such as the Trends in International Mathematics and Science Study (TIMSS), the Program for International Student Assessment (PISA), the Progress in International Reading Literacy Study (PIRLS), the International Civic and Citizenship Education Study (ICCS), and the International Computer and Information Literacy Study (ICILS). A large number of countries across the globe participate in these studies, and the number of participating countries has grown markedly over the years. However, the performance of several education systems on these international student assessment tests varies significantly from each other (see Martin, Mullis, Foy, & Stanco, 2012; Mullis, Martin, Foy, & Arora, 2012; Mullis, Martin, Foy, & Drucker, 2012; Organization for Economic Cooperation and Development [OECD], 2010a). As a result, international benchmarking has become a "buzzword" in these education systems, and educational policy makers are increasingly recognizing international benchmarking as a critical tool for better understanding the relative strengths and weaknesses of their education systems (see National Governors Association, 2008; Tucker, 2011).

The results of the recent cycles of the TIMSS, PISA, and PIRLS indicate that the world's best performing education systems are mainly a few East Asian education systems (see Martin et al., 2012; Mullis et al., 2012; OECD, 2010a). Hence, policy makers in low- and moderate-performing education systems are proposing that their governments benchmark their education systems against the standards of these high-performing East Asian education systems, such as Singapore (see National

S. Areepattamannil (✉)
Curriculum, Instruction & Educational Technology Division, Emirates College for Advanced Education, Abu Dhabi, United Arab Emirates
e-mail: sareepattamannil@ecae.ac.ae

C.L. Chiam • D.H.L. Lee • H. Hong
National Institute of Education, Nanyang Technological University, Singapore

M.S. Khine (ed.), *Science Education in East Asia*,
DOI 10.1007/978-3-319-16390-1_25

Governors Association, 2008; Tucker, 2011). However, there is sparse research on the factors contributing to high performance of successful education systems in East Asia; and within the region, there are significant differences in student performance between education systems. Indeed, all East Asian education systems are not top-performing education systems (see Martin et al., 2012; Mullis et al., 2012; OECD, 2010a). However, Singapore is recognized as one of the world's strong performers and successful reformers in education (see OECD, 2011).

On the PISA 2009 reading, mathematics, and science literacy scales, the 15-year-olds in Singapore had higher average scores than their counterparts in majority of the PISA participating countries (Walker, 2011). Furthermore, Singapore had a higher percentage of students performing at or above each of the six PISA proficiency levels in reading, mathematics, and science than most of the OECD member and nonmember economies (Walker, 2011). The glaring differences in student achievement between Singapore and the rest of the world are also evident from the recent cycles of the Trends in International Mathematics and Science Study (TIMSS). In Singapore, the 8th graders' average TIMSS mathematics and science scores increased between 2007 and 2011 by 18 and 23 points, respectively (Martin, Mullis, & Foy, 2008; Martin et al., 2012; Mullis, Martin, & Foy, 2008; Mullis et al., 2012). Moreover, Singapore had a higher percentage of students performing at or above each of the four TIMSS international mathematics and science benchmarks than several TIMSS participating countries at grades 4 and 8 (Martin et al., 2008, 2012; Mullis et al., 2008, 2012). Given the sterling performance of students in Singapore on international standardized assessments, it is crucial to examine the factors, particularly student- and school-level factors, contributing to variability in student performance. As Tucker (2011) posits,

> Countries that base their education strategies on the careful study of successful strategies employed by the leading nations are not as likely to go down blind alleys wasting large amounts of resources on initiatives that fail to pay off as are countries that base their strategies on untested theories. (p. 173)

Relationships Between Student-Level Factors and Academic Achievement

The student-level factors that might be linked to student achievement may include, among others, gender (e.g., Penner & CadwalladerOlsker, 2012), language spoken at home (e.g., Dustmann, Frattini, & Lanzara, 2012), immigration status (Areepattamannil & Freeman, 2008), family structure (e.g., Hampden-Thompson, 2013), family SES, including parental education, parental occupational status, home possessions, and availability of educational resources at home (e.g., Sirin, 2005), students' perceptions of classroom climate (e.g., Chionh & Fraser, 2009), teacher-student relationships (e.g., Roorda, Koomen, Spilt, & Oort, 2011), and students' attitudes toward school (e.g., Wang & Holcombe, 2010).

Gender Differences in Academic Achievement

The findings of hitherto research on gender differences in science achievement are a mixed bag. For example, Becker (1989) reanalyzed two meta-analyses of 30 studies on gender differences in science achievement and found that the magnitude of gender difference in science achievement varied by subject. Significant gender differences favoring males were found in biology, general science, and physics; but no significant differences were found in geology and earth sciences. OECD's (2009) secondary analysis of the PISA 2006 data revealed no significant gender difference in overall science performance in middle adolescence; however, significant gender differences were found for different science competencies such as identifying scientific issues (favoring females) and explaining phenomena scientifically (favoring males). Nonetheless, Haworth, Dale, and Plomin (2010) found no significant gender difference in science performance in middle childhood and early adolescence.

Home Language, Immigration Status, and Academic Achievement

A substantial body of research has documented the roles of home language and immigration status in the academic performance of school children (e.g., Areepattamannil & Freeman, 2008; Areepattamannil & Kaur, 2012; Areepattamannil & Lee, 2013; Dustmann et al., 2012; Martin, Liem, Mok, & Xu, 2012). Martin et al. (2012) demonstrated that language spoken at home was linked to achievement in mathematics and science among 113,767 first- and second-generation immigrant students aged 15–16 years in 17 countries, who took part in the second cycle of the PISA in 2003. Areepattamannil and Kaur (2012) also found among 19,002 15-year-old students from 842 schools in Canada that students who spoke the language of the test at home scored significantly higher on the PISA 2006 science assessment than did their peers who spoke a foreign language at home. The latest cycles of the TIMSS mathematics and science assessments as well indicate that students who speak the language of the test more often at home tend to perform better on the TIMSS mathematics and science assessments than students from homes where the language of the test is not often spoken (see Martin et al., 2012; Mullis et al., 2012).

Family SES, Family Structure, and Student Achievement

Parental education, parental occupation, the number of home possessions, and the number of educational resources at home constitute important components of family SES (Sirin, 2005). A growing corpus of research has examined the association between family SES and student achievement (e.g., Hattie, 2009; Sirin, 2005; White, 1982). White's (1982) meta-analysis of 200 studies published before 1980 indicated that the association between SES and academic achievement varied significantly by the type of SES and academic achievement measures. The typical SES measure was only weakly correlated with academic achievement; however, with aggregated units of analysis, the study found strong correlation between SES and academic achievement. In his meta-analysis based on 75 independent samples from 58 journal articles published between 1990 and 2000, Sirin (2005) showed that family SES was moderately associated with academic achievement at the student level, while it was strongly associated with academic achievement at the school level. Recent studies also suggest that students from higher SES backgrounds tend to perform better than their peers from lower SES backgrounds (e.g., Areepattamannil, 2012; Areepattamannil, Freeman, & Klinger, 2011; Caro, McDonald, & Willms, 2009; Hattie, 2009; Perry & McConney, 2010; Sousa, Park, & Armor, 2012).

Moreover, a large body of research has documented significant differences between students from single-parent families and those from two-parent families across a variety of academic outcomes (e.g., Hampden-Thompson, 2009, 2013; Hampden-Thompson & Pong, 2005; Hattie, 2009; Heuveline, Yang, & Timberlake, 2010; Marks, 2006; Pong, Dronkers, & Hampden-Thompson, 2003). Hampden-Thompson (2013), for example, drawing on data from PISA 2000, compared the reading, mathematics, and science literacy of students from two-parent families with their counterparts who reside in mother-only households among a sample of 115,000 fifteen-year-old students from 18 countries. For reading, mathematics, and science literacy, students from two-parent families, on average, outperformed their counterparts from mother-only families. In 12 of the countries, the difference in reading scores between the students from the two family structures was statistically significant. The difference in mathematics scores between the two family structures was statistically significant in 15 countries, while the difference in science scores between the two family structures was statistically significant in 10 countries.

The study further revealed that a family's economic resources (i.e., parental education, parental occupation, and number of books in the home) accounted for a significant amount of the reading, mathematics, and science literacy gaps between the students from the two family structures. Hampden-Thompson (2009) also found similar results for significant reading literacy gaps between students who reside in two-parent families and those who reside in single-mother families in 18 PISA participating countries. Heuveline et al. (2010), on the other hand, used the TIMSS 1995 data to test whether children growing up in two-parent families do better in school than children from single-parent families in 14 countries. They found that children from two-parent households tended to perform significantly better in mathematics and science than their counterparts from single-parent households.

Students' Perceptions of Classroom Environment and Attitudes Toward School

Students' perceptions of classroom environment and their attitudes toward school are important correlates of students' academic performance (see Chionh & Fraser, 2009; Fredricks, Blumenfeld, & Paris, 2004; Roorda et al., 2011; Wang & Holcombe, 2010). Classroom environment may constitute both disciplinary climate in the classroom and teacher-student relations. Positive classroom climate and positive teacher-student relations are linked to higher student achievement and lower student dropout rates (e.g., Barile, Donohue, Anthony, Baker, Weaver, & Henrich, 2012; Chionh & Fraser, 2009; Hattie, 2009; Reyes, Brackett, Rivers, White, & Salovey, 2012; Roorda et al., 2011).

In their meta-analysis of 99 studies including students from preschool to high school, Roorda et al. (2011) investigated the associations between positive teacher-student relationships and students' school engagement and achievement. The meta-analysis indicated that the association of positive teacher-student relationships with students' school engagement was medium to large, whereas the association with students' school achievement was small to medium. Recently, Barile et al. (2012) found among 7,779 high school students from 431 public schools in the United States that positive teacher-student relationships were associated with lower student dropout rates.

Numerous studies have also investigated the relationships between students' attitudes toward school and their academic achievement (e.g., Chiu, Pong, Mori, & Chow, 2012; Fredricks et al., 2004; Wang & Holcombe, 2010). Wang and Holcombe (2010) examined the effects of students' attitudes toward school on school engagement and achievement among 1,046 adolescents in the United States. The results of the study showed that students with positive attitudes toward school tended to have higher academic engagement and achievement.

In sum, several student-level factors, such as student background characteristics, students' positive perceptions of classroom climate, positive teacher-student relationships, and students' positive attitudes toward school are generally significantly associated with students' academic performance. In addition to these student-level factors, school-level factors may also be linked to students' academic performance. Hence, it is important to examine the relationships of school-level factors to student achievement as well.

Relationships Between School-Level Factors and Student Achievement

The school-level factors that might be linked to student achievement may include, among others, school size (see Leithwood & Jantzi, 2009), school organizational type (e.g., Jeynes, 2012), school SES (e.g., Perry & McConney, 2010), the availability and the quality of the school's educational resources (see Glewwe, Hanushek, Humpage, & Ravina, 2011), and school climate (see Cohen, McCabe, Michelli, & Pickeral, 2009).

School Background Characteristics and Student Achievement

There is mounting evidence that one of the school background characteristics, school size, matters for an array of student learning outcomes (e.g., Hattie, 2009; Leithwood & Jantzi, 2009; Sun, Bradley, & Akers, 2012; Weiss, Carolan, & Baker-Smith, 2010). Nevertheless, the findings of hitherto research on the impact of school size on student learning outcomes are inconsistent. For example, Leithwood and Jantzi's (2009) review of 57 post-1990 studies pertaining to the impact of school size on elementary and secondary students' academic engagement and achievement indicated that elementary students in smaller schools had better academic achievement than did their peers in larger schools, while secondary students in smaller to midsized schools had better academic achievement than did their counterparts in larger schools. Moreover, the review indicated that both elementary and secondary students in smaller schools had higher levels of academic engagement. However, Weiss et al. (2010) found that students in midsized high schools had higher levels of academic engagement, but school size was not significantly related to students' academic achievement. In contrast, Sun et al.'s (2012) study revealed a positive association between school size and student achievement, suggesting that students in larger schools tend to perform better than their peers in smaller schools.

In addition to school size effect research, numerous studies have been conducted over the years on the effects of school organizational type on student outcomes as well (e.g., Jeynes, 2012; Lubienski & Lubienski, 2006). Lubienski and Lubienski (2006), for example, using data from the 2003 National Assessment of Educational Progress (NAEP), compared fourth- and eighth-graders' mathematics achievement in public, charter, and private schools in the United States and found that public schools did not significantly differ from charter or private schools in terms of their students' mathematics achievement. A recent meta-analysis of 90 studies on the effects of religious private schools, charter schools, and public schools on student outcomes (see Jeynes, 2012), however, indicated that students who attended religious private schools had better academic and behavioral outcomes than did their peers in public schools. Nonetheless, no significant differences were found between charter schools and public schools with respect to their students' academic performance.

Furthermore, there is growing consensus that in addition to family SES, school SES may also be associated with student achievement (e.g., Palardy, 2008, 2013; Perry & McConney, 2010; Willms, 2010). Perry and McConney (2010), using the PISA 2003 data, examined the relationship between school SES and student outcomes in Australia. They found that school SES was strongly associated with reading, mathematics, and science literacy of 15-year-olds in Australia. Students from socioeconomically advantaged schools performed significantly better on the PISA 2003 reading, mathematics, and science assessments than did their peers from socioeconomically disadvantaged schools. Recently, Palardy (2013), using data from the 2002 Education Longitudinal Study, examined the association between high school socioeconomic segregation and student attainment outcomes in the United States. The results of the study indicated that the socioeconomic composition of the school was associated with student attainment outcomes (i.e., high school graduation and college enrollment).

Other school background characteristics that are critical for student engagement and achievement include the availability of school educational resources and the quality of the school's educational resources (Glewwe et al., 2011). There is growing evidence that the availability and the quality of educational resources, such as instructional materials (e.g., Fuller, 1987), laboratory equipment (e.g., Beese & Liang, 2010; Shin, Slater, & Backhoff, 2012), library materials (e.g., Small, Shanahan, & Stasak, 2010; Small, Snyder, & Parker, 2009), computers for instruction (e.g., Johnson, Perry, & Shamir, 2010; Martin et al., 2012; Mullis et al., 2012; Shegog et al., 2012; Sosa, Berger, Saw, & Mary, 2011), Internet access (e.g., Ellis, Goodyear, Bliuc, & Ellis, 2011), computer software for instruction (e.g., Papastergiou, 2009; Pilli & Aksu, 2013; Tüzün, Yilmaz-Soylu, Karakuş, İnal, & Kizilkaya, 2009), and audiovisual resources (e.g., DaSilva & Kvasnak, 2012), are crucial for student engagement and achievement. Finally, yet another school resource indicator that might be linked to student engagement and achievement is the availability of trained teachers. Prior research has demonstrated that a shortage of trained teachers may adversely affect student engagement and achievement (e.g., Beese & Liang, 2010; Mo, Singh, & Chang, 2013; Shin et al., 2012).

School Climate and Student Achievement

A large number of studies have shown that positive school climate is associated with greater student achievement (see Cohen et al., 2009, for a review). However, a number of student- and teacher-related factors may detrimentally affect school climate, which, in turn, may result in lower student achievement. Student-related factors affecting positive school climate and student achievement may include school absenteeism (e.g., Gu, Lai, & Ye, 2011), disruptive classroom behavior (e.g., Freiberg, Huzinec, & Templeton, 2009; Gaskins, Herres, & Kobak, 2012; Gu et al., 2011; Skaalvik & Skaalvik, 2011; Wang & Dishion, 2012; Way, 2011), skipping classes (e.g., Roeser & Eccles, 1998), disrespect toward teachers (e.g., Espelage et al., 2013; Gregory, Cornell, & Fan, 2012), use of alcohol or illegal drugs (e.g., Finn, 2012; Radliff, Wheaton, Robinson, & Morris, 2012; Sznitman, Dunlop, Nalkur, Khurana, & Romer, 2012), and intimidating or bullying other students (e.g., Gendron, Williams, & Guerra, 2011; Gu et al., 2011; Klein, Cornell, & Konold, 2012; Konishi, Hymel, Zumbo, & Li, 2010; Martin et al., 2012; Mehta, Cornell, Fan, & Gregory, 2013; Mullis et al., 2012).

Teacher-related factors affecting positive school climate and student achievement may include teachers' low expectations of students (e.g., Demanet & Van Houtte, 2012; Rubie-Davies, 2007), poor teacher-student relationships (e.g., Jennings & Greenberg, 2009; Sakiz, Pape, & Hoy, 2012), teachers not meeting individual students' needs (e.g., Finn et al., 2009; De Wit, Karioja, & Rye, 2010; Stewart, 2003), teacher absenteeism (e.g., Miller, Murnane, & Willett, 2008), teachers' resistance to change (see McKenzie & Scheurich, 2008; Moore, Edwards, Halpin, & George, 2002), teachers being too strict with students (e.g., McNeely, Nonnemaker, & Blum, 2002; Van Petegem, Aelterman, Van Keer, & Rosseel, 2008), and students not being encouraged to achieve their full potential (e.g., Jang, Reeve, & Deci, 2010).

In short, a myriad of student- and school-level factors are implicated in variability in students' academic performance. Nevertheless, the majority of the findings to date are the offspring of Western, particularly North American, research. There is a dearth of research on the relationships of these student- and school-level factors to academic performance among adolescent students in Singapore. Such an exploration is important, given the significant differences in academic performance between students in Singapore and the rest of the world. Moreover, benchmarking to one of the world's best education systems may help educational policy makers in underperforming education systems to understand the relative strengths and weaknesses of their education systems and to identify best practices and potential ways forward in improving the quality of their education systems. The present study, therefore, aimed to examine the student- and school-level factors related to science achievement in Singapore. The following three research questions addressed the purpose of the study:

1. How much of the variation in science achievement is within and between schools in Singapore?
2. What student-level factors are significantly associated with science achievement among 15-year-olds in Singapore?
3. What school-level factors are significantly associated with science achievement among 15-year-olds in Singapore?

Method

Data

Data for the study were drawn from the PISA 2009 database. PISA, administered in participating countries and economies every 3 years, is an international comparative survey of 15-year-old students' knowledge and skills in reading, mathematics, and science. Over 475,000 fifteen-year-old students, representing 26 million 15-year-olds, from 64 countries and economies participated in PISA 2009 (OECD, 2012). In Singapore, 5,283 students (male = 2,626, female = 2,657;

mean age = 15.73 years, SD = 0.28) from 171 schools took part in PISA in 2009. The number of students per school in the Singaporean sample varied between 24 and 35 (mean school size = 30.89, *SD* = 2.25).

Measures

Student-Level Factors

The student-level factors included student achievement, student background characteristics, students' attitudes toward school, students' perceptions of disciplinary climate in the classroom, and teacher-student relations. The student achievement measures included the PISA 2009 science achievement scales (see OECD, 2012). The PISA 2009 science assessment comprised a total of 53 cognitive items (OECD, 2012). PISA employed the mixed coefficients multinomial logit model (Adams, Wilson, & Wang, 1997), a multidimensional, generalized form of the Rasch model, to scale the PISA assessment data (OECD, 2012). Dichotomously scored items were scaled with the Rasch's logistic model (Rasch, 1960), and items with multiple score categories were scaled with Masters' partial credit model (Masters, 1982).

The student demographic characteristics included gender (1 = female, 0 = male), language spoken at home (1 = language of assessment, 0 = another language), immigration status (1 = native, 0 = immigrant), and family structure (1 = single-parent family, 0 = two-parent family). In addition to these variables, the PISA 2009 indices of highest educational level of parents (0 = *none* to 6 = *theoretically oriented tertiary and postgraduate*), highest occupational status of parents (an index based on the international socioeconomic index of occupational status [ISEI]), home educational resources (a summary index of educational resources at home including a desk and a quiet place to study, a computer that students can use for schoolwork, educational software, books to help with students' school work, technical reference books, and a dictionary), and home possessions (a summary index of all household items and the number of books at home) were included as proxy measures of socioeconomic status (SES) in the study (see OECD, 2012).

The index of attitude toward school was derived from students' level of agreement with the following four statements: (1) school has done little to prepare me for adult life when I leave school, (2) school has been a waste of time, (3) school has helped give me confidence to make decisions, and (4) school has taught me things which could be useful in a job (OECD, 2012). The index of disciplinary climate was derived from students' level of agreement with the following five statements: (1) students don't listen to what the teacher says, (2) there is noise and disorder, (3) the teacher has to wait a long time for students to quiet down, (4) students cannot work well, and (5) students don't start working for a long time after the lesson begins (OECD, 2012). The index of teacher-student relations was derived from students' level of agreement with the following five statements: (1) I get along well with most of my teachers; (2) most of my teachers are interested in my well-being; (3) most of my teachers really listen to what I have to say; (4) if I need extra help, I will receive it from my teachers; and (5) most of my teachers treat me fairly (OECD, 2012). All of these items were rated on a 4-point Likert-type scale ranging from 1 = *strongly disagree* to 4 = *strongly agree*. Negatively phrased items were inverted for item response theory (IRT) scaling, and higher values on these indices indicate more positive perceptions (OECD, 2012).

School-Level Factors

The school-level factors included school background characteristics and school principals' perceptions of student- and teacher-related factors affecting school climate. The school background characteristics included school size (continuous variable), school type (1 = public, 0 = private), school mean SES, student-teacher ratio, teacher shortage, and the quality of the school's educational resources. The school mean SES was derived from the PISA index of economic, social, and cultural status (ESCS). The index of ESCS was derived from the following three indices: highest occupational status of parents, highest educational level of parents, and home possessions (OECD, 2012). In the present study, we used the school aggregate of student-level ESCS as a proxy measure of school SES. The student-teacher ratio was calculated by dividing the school size by the total number of teachers in the school (OECD, 2012).

The index of teacher shortage was derived from four items measuring the school principal's perceptions of potential factors hindering instruction at school: (1) a lack of qualified science teachers, (2) a lack of qualified mathematics teachers, (3) a lack of qualified test language teachers, and (4) a lack of qualified teachers of other subjects (OECD, 2012). All items were rated on a 4-point Likert-type scale ranging from 1 = *not at all* to 4 = *a lot*. These items were not inverted for IRT scaling, and positive values indicate fewer teachers at a school (OECD, 2012). The index of quality of the school's educational resources was derived from seven items measuring the school principals' perceptions of potential factors hindering instruction at school: (1) shortage or inadequacy of science laboratory equipment, (2) shortage or inadequacy of instructional materials (e.g., textbooks), (3) shortage or inadequacy of computers for instruction, (4) lack or inadequacy of Internet connectivity, (5) shortage or inadequacy of computer software for instruction, (6) shortage or inadequacy of library materials, and (7) shortage or inadequacy of audiovisual resources (OECD, 2012). All of these items were rated on a 4-point Likert-type scale ranging from 1 = *not at all* to 4 = *a lot*. All items were inverted for IRT scaling, and positive values indicate positive evaluations (OECD, 2012).

The index of student-related factors affecting school climate was derived from school principals' reports on the extent to which the learning of students were hindered by the following factors in their schools: (1) student absenteeism, (2) disruption of classes by students, (3) students skipping classes, (4) student lacking respect for teachers, (5) student use of alcohol or illegal drugs, and (6) students intimidating or bullying other students (OECD, 2012). The index of teacher-related factors affecting school climate was derived from school principals' reports on the extent to which the learning of students were hindered by the following factors in their schools: (1) teachers' low expectations of students, (2) poor student-teacher relations, (3) teachers not meeting individual students' needs, (4) teacher absenteeism, (5) staff resisting change, (6) teachers being too strict with students, and (7) students not being encouraged to achieve their full potential (OECD, 2012). All of these items were rated on a 4-point Likert-type scale ranging from 1 = *strongly disagree* to 4 = *strongly agree*. These indices were constructed using IRT scaling, and positive values indicate positive evaluations (OECD, 2012).

Data Analytic Strategy

Given the hierarchical nature of the PISA sample (i.e., students nested within schools), two-level hierarchical linear modeling (HLM; Raudenbush & Bryk, 2002) analyses were conducted using HLM 7 for Windows (Raudenbush, Bryk, Cheong, Congdon, & du Toit, 2011). We employed HLM 7 for statistical analyses because PISA uses plausible values for reporting student performance on cognitive tests (see OECD, 2012). Plausible values, an approach developed by Mislevy and Sheehan (1987, 1989) and based on the imputation theory of Rubin (1987), are random elements from the set of scores (i.e., random draws from the marginal posterior of the latent distribution) that could be attributed to each student (OECD, 2012). HLM 7 is capable of handling plausible values as outcome measures (see Raudenbush et al., 2011).

In large-scale studies, such as PISA, missing data are common and unavoidable. Although HLM 7 can handle missing data at level-1, it cannot handle missing data at level-2 (see Raudenbush et al., 2011). In the present study, therefore, the expectation-maximization (EM) algorithm was used to handle missing data at level 1 and level 2 (see Dempster, Laird, & Rubin, 1977; Enders, 2001; Little & Rubin, 1987). The EM algorithm is considered to be one the best available methods for dealing with missing data (McCoach, 2010a). IBM SPSS Statistics 21 was used to run the EM algorithm.

The HLM models were built sequentially, using a series of models (see McCoach, 2010b). First, fully unconditional models (i.e., null or intercept-only models) were fitted to partition the total variance in outcome measures into within-school variance (i.e., variance attributable to students) and between-school variance (i.e., variance attributable to schools). A fully unconditional model is analogous to a one-way random effects ANOVA model (Raudenbush & Bryk, 2002). Second, a model that included all student-level variables was estimated. Finally, a full model that included both student- and school-level variables was estimated. Although there is growing empirical evidence that students' perceptions of classroom disciplinary climate, student-teacher relationships, and students' SES may vary across schools, these measures were not allowed to randomly vary across schools in the study because the level-2 model included measures such as student behavior, teacher behavior, and SES of school, which may explain the between-school variability in the slopes (see McCoach, 2010a). Therefore, the random intercepts and fixed slopes model was used to answer the second and third research questions.

Because HLM 7 is capable of handling sampling weights at both level 1 and level 2, sampling weights were specified at level 1 (i.e., final student weight: W_FSTUWT; OECD, 2012) and level 2 (i.e., final school weight: W_FSCHWT; OECD, 2012) to obtain unbiased parameter estimates (see Stapleton & Thomas, 2008). HLM analyses that incorporate sampling weights employ a pseudo-maximum likelihood estimation approach. Hence, HLM 7 automatically chose maximum likelihood as the estimation technique for HLM models. "The goal of maximum likelihood estimation is to find the set of parameter values that makes the actual data most likely to have been observed" (McCoach & Black, 2012, p. 24). Dichotomous student- and school-level variables were kept in their original metric. All continuous student- and school-level variables/measures were grand-mean centered. Centering not only helps in the interpretation of the model parameters (Ferron et al., 2008) but also "removes high correlations between random intercepts and slopes, and high correlations between first- and second-level variables and cross-level interactions" (Kreft & de Leeuw, 1998, p. 114). Grand-mean centering is preferred if the purpose of the study is to account for variability in a level 1 dependent variable using independent variables at more than one level (Bickel, 2007). The HLM mixed model equations and a description of the student- and school-level variables used in the HLM models are given in Appendices A and B.

Results

Partitioning of Variance in Reading, Mathematics, and Science Achievement

The intercept-only models (see Table 25.1) indicated that average science achievement ($\hat{y}_{00} = 535.02$) varied across schools in Singapore ($\chi^2(170, N = 5{,}283) = 3{,}168.70$, $p < .001$). Moreover, schools accounted for 36 % of the variance in science achievement ($\rho = 0.36$), and students within schools accounted for 64 % of the variance in science achievement.

Student- and School-Level Factors Predicting Science Achievement

Language spoken at home, highest educational level of parents, highest parental occupational status, home educational resources, students' perceptions of

Table 25.1 Results of hierarchical linear modeling analyses predicting science achievement in Singapore

	Unconditional model		Level-1 model		Full level-2 model	
	B	SE	*B*	SE	*B*	SE
Intercept	535.02***	5.29	533.60***	6.72	538.92***	19.52
Level 1						
Gender (female)			−6.53**	2.18	−5.94**	2.17
Home language (test language)			23.28***	3.67	21.66***	3.63
Immigration status (native)			−3.95	4.68	−2.96	4.52
Family structure (single-parent family)			−10.78*	4.90	−10.52*	4.87
Highest parental education			3.31**	1.12	3.14**	1.13
Highest parental occupational status			0.66***	0.09	0.64***	0.09
Home educational resources			13.84***	2.02	14.19***	2.01
Home possessions			3.06	1.97	1.82	2.04
Disciplinary climate			13.62***	1.74	13.37***	1.75
Teacher-student relations			3.37*	1.40	3.23*	1.39
Attitude toward school			4.48**	1.70	4.66**	1.71
Level 2						
School size					0.05***	0.01
School type (public)					−4.50	20.15
School mean ESCS					63.10***	11.58
Student-teacher ratio					−0.43	1.24
Teacher shortage					−10.39**	3.71
Quality of school's educational resources					2.40	3.25
Teacher behavior					−7.46	4.41
Student behavior					11.66	6.11
Intercept variance (τ_{00})	3,897.25		2,470.67		868.42	
Level 1 variance (σ^2)	6,870.90		6,029.00		6,027.49	
Intraclass correlation (ρ)	0.36		0.29		0.13	
Variance in achievement between schools explained (%)	NA		37		78	
Variance in achievement within schools explained (%)	NA		12		12	

Note: *ESCS* Economic, social, and cultural status
$*p < .05$, $**p < .01$, $***p < .001$

disciplinary climate in the classroom, teacher-student relations, students' attitudes toward school, school size, and school mean ESCS were statistically significantly positively related to science achievement (see Table 25.1). Students who spoke the language of assessment at home, students whose parents had higher levels of education and occupation, students who held more positive perceptions of their classroom climate, students who held more positive attitudes toward school, students who attended larger schools, and students who attended higher SES schools scored statistically significantly higher on the PISA science assessment. Gender, family structure, and teacher shortage were statistically significantly negatively associated with science achievement in Singapore, suggesting that female students, students from single-parent families, and students from schools facing higher rates of teacher shortage scored statistically significantly lower on the PISA science assessment. The final model for science achievement explained 78 % of the variance between schools and 12 % of the variance within schools.

Discussion

In the era of international benchmarking, it is crucial to examine what factors are linked to student achievement in high-performing education systems. Therefore, the objective of the present study was to examine the relationships of student- and school-level factors to science achievement among 15-year-old students in a high-performing education system, Singapore. The findings of the study indicated that schools accounted for more of the variability in science achievement than did the students within the schools in Singapore. However, several student-level factors were statistically significantly associated with science achievement in Singapore.

Student-Level Factors

Consistent with the findings of prior research (e.g.,Penner & CadwalladerOlsker, 2012), the results of the present study revealed significant gender differences in science achievement. Science was found to be male domain in Singapore. The male advantage in spatial skills is one of the important reasons implicated in male students' superiority in science-related tasks (see Uttal et al., 2013).

Further, speaking the language of the test at home was positively linked to science achievement in Singapore, a finding consistent with prior findings (e.g., Dustmann et al., 2012; Martin et al., 2012). In Singapore, the language of the PISA test was English; but only 41 % of the students spoke English at home (OECD, 2010b). Given the large number of students who speak a language other than English at home in Singapore, interventions targeted at improving conversational English alone may not guarantee school success. Instead, instructional interventions aimed at developing and enhancing academic language proficiency may be required for students who are at risk of failing to develop sufficient proficiency in academic English for school success (Slama, 2012).

Family SES was yet another factor positively associated with science achievement in Singapore. In particular, one of the proxy measures of family SES, home educational resources, was consistently positively associated with science achievement. In other words, Singapore is leaving some students behind in science, and they tend to be from less privileged socioeconomic backgrounds. This finding is consistent with prior findings indicating that students from low socioeconomic backgrounds tend to perform lower than their peers from high socioeconomic backgrounds (e.g., Hattie, 2009; Sirin, 2005). Indeed, educational policies targeted at improving the economic circumstances of students from less economically privileged families may help foster not only equity in learning opportunities but also equity in learning outcomes (OECD, 2010c).

Furthermore, in line with the findings of prior studies (e.g., Chionh & Fraser, 2009), students' positive perceptions of classroom disciplinary climate were also linked to science achievement in Singapore. Moreover, one of the important characteristics of classroom emotional climate, positive teacher-student relationships was related to science achievement in Singapore. Indeed, students may exhibit more engagement and less problem behaviors in socially and emotionally healthy classroom environments that are characterized by a sense of bonding—connectedness and belongingness, enjoyment and enthusiasm, and mutual respect (Brackett, Reyes, Rivers, Elbertson, & Salovey, 2011; Reyes et al., 2012).

Finally, students' positive attitudes toward school were associated with science achievement in Singapore. Promoting students' positive attitude toward school may be important because students with positive attitudes toward school are more likely to pursue schooling beyond high school (OECD, 2003). However, several contextual factors, such as peer relationships, home influences, teacher interactions, and the school-wide context, may play crucial roles in fostering school engagement and students' positive attitude toward school (Griffiths, Sharkey, & Furlong, 2009). Educational policies aimed at bolstering positive attitudes toward school may need to take into account these contextual influences impacting student engagement at school.

School-Level Factors

Of the school-level factors, school ESCS was positively linked to science achievement in Singapore. This finding concurs with prior research (e.g., Palardy, 2008, 2013; Perry & McConney, 2010; Willms, 2010) that suggests that students who attend socioeconomically advantaged schools tend to do well on standardized measures of academic achievement compared with their peers who attend socioeconomically disadvantaged schools. Because students who attend high SES schools are more likely to graduate from high school and enroll in postsecondary education than do their counterparts who attend low SES schools (Palardy, 2013), educational policies aimed at addressing the negative consequences of school socioeconomic segregation may be critical for fostering equity in learning opportunities and outcomes. One step toward achieving this end might be reducing differences in the educational context at low and high SES schools (Palardy, 2013). However, sound educational policies targeted at minimizing or eliminating school socioeconomic segregation might be more appropriate and effective in achieving equity in learning opportunities and outcomes (Perry & McConney, 2010).

School size was yet another school-level factor positively associated with science achievement in Singapore. Congruent with the findings of Sun et al.'s (2012) study, students who attended larger schools in Singapore performed significantly better than their peers who attended smaller schools. Because of the existence of economies of scale, larger schools can afford to provide a wide range of educational resources to foster student engagement and achievement, which, in turn, may encourage higher-performing students to enroll in these schools. No other school-level factors were significantly associated with science achievement in Singapore.

In conclusion, the present study explored the student- and school-level factors linked to science achievement among 15-year-olds in Singapore. Although several student-level factors were associated with science achievement in Singapore, only two school-level factors—school ESCS and school size—were related to science achievement in Singapore. Nevertheless, the results of the study should be treated with caution for two primary reasons. First, the student-level factors, excluding academic achievement measures, and the school-level factors were self-report measures. Even though self-report measures are not inherently inferior to other competing or alternative measures, critics of self-report measures have always challenged the reliability and validity of such measures (see Fulmer & Frijters, 2009). Second, because of the cross-sectional nature of the PISA data, the significant relationships reported in the current study are correlational in nature and should not be interpreted as causal relationships.

Appendix A: HLM Mixed Model Equations

Unconditional models

$$\text{SCIE}_{ij} = \gamma_{00} + u_{0j} + r_{ij}$$

Level-1 models

$$\begin{aligned}\mathrm{SCIE}_{ij} &= \gamma_{00} + \gamma_{10}{*}ST04Q01_{ij} + \gamma_{20}{*}ST19Q01 + \gamma_{30}{*}\mathrm{FAMSTRUC}_{ij}\\ &\quad + \gamma_{40}{*}\mathrm{HISCED}_{ij} \quad + \gamma_{50}{*}\mathrm{HISEI}_{ij} + \gamma_{60}{*}\mathrm{IMMIG}_{ij} + \gamma_{70}{*}\mathrm{ATSCHL}_{ij}\\ &\quad + \gamma_{80}{*}\mathrm{DISCLIMA}_{ij} + \gamma_{90}{*}\mathrm{HEDRES}_{ij} + \gamma_{100}{*}\mathrm{HOMEPOS}_{ij}\\ &\quad + \gamma_{110}{*}\mathrm{STUDREL}_{ij} + u_{0j} + r_{ij}\end{aligned}$$

Full level-2 models

$$\begin{aligned}\mathrm{SCIE}_{ij} &= \gamma_{00} + \gamma_{01}{*}SCHSIZE_j + \gamma_{02}{*}SCHTYPE_j + \gamma_{03}{*}STRATIO_J\\ &\quad + \gamma_{04}{*}SCMATEDU_j + \gamma_{05}{*}STUDBEHA_j + \gamma_{60}{*}TCSHORT_j\\ &\quad + \gamma_{07}{*}TEACBEHA_j + \gamma_{08}{*}ESCS\_MEAN_{j_{ij}} + \gamma_{10}{*}ST04Q01_{ij}\\ &\quad + \gamma_{20}{*}ST19Q01_{ij} + \gamma_{30}{*}FAMSTRUC_{ij} + \gamma_{40}{*}HISCED_{ij} + \gamma_{50}{*}HISEI_{ij}\\ &\quad + \gamma_{60}{*}IMMIG_{ij} + \gamma_{70}{*}ATSCHL_{ij} + \gamma_{80}{*}DISCLIMA_{ij} + \gamma_{90}{*}HEDRES_{ij}\\ &\quad + \gamma_{100}{*}HOMEPOS_{ij} + \gamma_{110}{*}STUDREL_{ij} + u_{0j} + r_{ij}\end{aligned}$$

Appendix B: Description of Level 1 and Level 2 Variables Used in the HLM Models

Variable/measure	Description
Outcome	Combined science achievement scale
SCIE	
Level 1 (student)	
ST04Q01	Gender (female)
ST19Q01	Language spoken at home (language of assessment)
IMMIG	Immigration status (native)
FAMSTRUC	Family structure (single-parent family)
HISCED	Highest educational level of parents (PISA index)
HISEI	Highest parental occupational status (PISA index)
HEDRES	Home educational resources (PISA index)
HOMEPOS	Home possessions (PISA index)
DISCLIMA	Disciplinary climate (PISA index)
STUDREL	Teacher student relations (PISA index)
ATSCHL	Attitude toward school (PISA index)
Level 2 (school)	
SCHSIZE	School size
SCHTYPE	School type (public)
ESCS_MEAN	School mean economic, social, and cultural status
STRATIO	Student-teacher ratio (PISA index)
TCSHORT	Teacher shortage (PISA index)
SCMATEDU	Quality of the school's educational resources (PISA index)
TEACBEHA	Teacher behavior (PISA index)
STUDBEHA	Student behavior (PISA index)

References

Adams, R. J., Wilson, M., & Wang, W. C. (1997). The multidimensional random coefficients multinomial logit model. *Applied Psychological Measurement, 21*, 1–23.

Areepattamannil, S. (2012). Effects of inquiry-based science instruction on science achievement and interest in science: Evidence from Qatar. *The Journal of Educational Research, 105*, 134–146.

Areepattamannil, S., & Freeman, J. G. (2008). Academic achievement, academic self-concept, and academic motivation of immigrant adolescents in the greater Toronto area secondary schools. *Journal of Advanced Academics, 19*, 700–743.

Areepattamannil, S., Freeman, J. G., & Klinger, D. A. (2011). Influence of motivation, self-beliefs, and instructional practices on science achievement of adolescents in Canada. *Social Psychology of Education, 14*, 233–259.

Areepattamannil, S., & Kaur, B. (2012). Factors predicting science achievement of immigrant and non-immigrant students: A multilevel analysis. *International Journal of Science and Mathematics Education*. Advance online publication. doi: 10.1007/s10763-012-9369-5

Areepattamannil, S., & Lee, D. H. L. (2013). Linking immigrant parents' educational expectations and aspirations to their children's school performance. *The Journal of Genetic Psychology*. Advance online publication. doi: 10.1080/00221325.2013.799061

Barile, J. P., Donohue, D. K., Anthony, E. R., Baker, A. M., Weaver, S. R., & Henrich, C. C. (2012). Teacher–student relationship climate and school outcomes: Implications for educational policy initiatives. *Journal of Youth and Adolescence, 41*, 256–267.

Becker, B. J. (1989). Gender and science achievement: A reanalysis of studies from two meta-analyses. *Journal of Research in Science Teaching, 26*, 141–169.

Beese, J., & Liang, X. (2010). Do resources matter? PISA science achievement comparisons between students in the United States, Canada, and Finland. *Improving Schools, 13*, 266–279.

Bickel, R. (2007). *Multilevel analysis for applied research: It's just regression*. New York, NY: Guilford.

Brackett, M. A., Reyes, M. R., Rivers, S. E., Elbertson, N. A., & Salovey, P. (2011). Classroom emotional climate, teacher affiliation, and student conduct. *Journal of Classroom Interaction, 46*, 27–36.

Caro, D. H., McDonald, J. D., & Willms, J. D. (2009). Socioeconomic status and academic achievement trajectories from childhood to adolescence. *Canadian Journal of Education, 32*, 558–590.

Chionh, Y. H., & Fraser, B. J. (2009). Classroom environment, achievement, attitudes and self-esteem in geography and mathematics in Singapore. *International Research in Geographical and Environmental Education, 18*, 29–44.

Chiu, M. M., Pong, S. L., Mori, I., & Chow, B. W. Y. (2012). Immigrant students' emotional and cognitive engagement at school: A multilevel analysis of students in 41 countries. *Journal of Youth and Adolescence, 41*, 1409–1425.

Cohen, J., McCabe, L., Michelli, N. M., & Pickeral, T. (2009). School climate: Research, policy, practice, and teacher education. *The Teachers College Record, 111*, 180–213.

DaSilva, E. B., & Kvasnak, R. N. (2012). Multimedia technology and students' achievement in geography. *The Geography Teacher, 9*, 18–25.

De Wit, D. J., Karioja, K., & Rye, B. J. (2010). Student perceptions of diminished teacher and classmate support following the transition to high school: Are they related to declining attendance? *School Effectiveness and School Improvement, 21*, 451–472.

Demanet, J., & Van Houtte, M. (2012). Teachers' attitudes and students' opposition. School misconduct as a reaction to teachers' diminished effort and affect. *Teaching and Teacher Education, 28*, 860–869.

Dempster, A. P., Laird, N. M., & Rubin, D. B. (1977). Maximum likelihood from incomplete data via the EM Algorithm. *Journal of the Royal Statistical Society: Series B Methodological, 39*, 1–38.

Dustmann, C., Frattini, T., & Lanzara, G. (2012). Educational achievement of second-generation immigrants: An international comparison. *Economic Policy, 27*, 143–185.

Ellis, R. A., Goodyear, P., Bliuc, A. M., & Ellis, M. (2011). High school students' experiences of learning through research on the Internet. *Journal of Computer Assisted Learning, 27*, 503–515.

Enders, C. K. (2001). A primer on maximum likelihood algorithms available for use with missing data. *Structural Equation Modeling, 8*, 128–141.

Espelage, D., Anderman, E. M., Brown, V. E., Jones, A., Lane, K. L., McMahon, S. D., ... & Reynolds, C. R. (2013). Understanding and preventing violence directed against teachers: Recommendations for a national research, practice, and policy agenda. *American Psychologist, 68*, 75–87.

Ferron, J. M., Hogarty, K. Y., Dedrick, R. F., Hess, M. R., Niles, J. D., & Kromrey, J. D. (2008). Reporting results from multilevel analyses. In A. A. O'Connell & D. B. McCoach (Eds.), *Multilevel modeling of educational data* (pp. 391–426). Charlotte, NC: Information Age.

Finn, A. N., Schrodt, P., Witt, P. L., Jernberg, K. A., Elledge, N., & Larson, L. (2009). A meta-analytical review of teacher credibility and its associations with teacher behaviors and student outcomes. *Communication Education, 58*, 516–537.

Finn, K. V. (2012). Marijuana use at school and achievement-linked behaviors. *The High School Journal, 95*, 3–13.

Fredricks, J. A., Blumenfeld, P. C., & Paris, A. H. (2004). School engagement: Potential of the concept, state of the evidence. *Review of Educational Research, 74*, 59–109.

Freiberg, H. J., Huzinec, C. A., & Templeton, S. M. (2009). Classroom management—A pathway to student achievement: A study of fourteen inner-city elementary schools. *The Elementary School Journal, 110*, 63–80.

Fuller, B. (1987). What school factors raise achievement in the Third World? *Review of Educational Research, 57*, 255–292.

Fulmer, S. M., & Frijters, J. C. (2009). A review of self-report and alternative approaches in the measurement of student motivation. *Educational Psychology Review, 21*, 219–246.

Gaskins, C. S., Herres, J., & Kobak, R. (2012). Classroom order and student learning in late elementary school: A multilevel transactional model of achievement trajectories. *Journal of Applied Developmental Psychology, 33*, 227–235.

Gendron, B. P., Williams, K. R., & Guerra, N. G. (2011). An analysis of bullying among students within schools: Estimating the effects of individual normative beliefs, self-esteem, and school climate. *Journal of School Violence, 10*, 150–164.

Glewwe, P. W., Hanushek, E. A., Humpage, S. D., & Ravina, R. (2011). *School resources and educational outcomes in developing countries: A review of the literature from 1990 to 2010*. Retrieved from http://www.nber.org/papers/w17554.pdf

Gregory, A., Cornell, D., & Fan, X. (2012). Teacher safety and authoritative school climate in high schools. *American Journal of Education, 118*, 401–425.

Griffiths, A. J., Sharkey, J. D., & Furlong, M. J. (2009). Student engagement and positive school adaptation. In R. Gilman, E. S. Huebner, & M. J. Furlong (Eds.), *Handbook of positive psychology in the schools* (pp. 197–211). New York, NY: Routledge.

Gu, H., Lai, S. L., & Ye, R. (2011). A cross-cultural study of student problem behaviors in middle schools. *School Psychology International, 32*, 20–34.

Hampden-Thompson, G. (2009). Are two better than one? A comparative study of achievement gaps and family structure. *Compare, 39*, 513–529.

Hampden-Thompson, G. (2013). Family policy, family structure, and children's educational achievement. *Social Science Research, 42*, 804–817.

Hampden-Thompson, G., & Pong, S.-L. (2005). Does family policy environment mediate the effect of single-parenthood on children's academic achievement? A study of 14 European countries. *Journal of Comparative Family Studies, 36*, 227–248.

Hattie, J. (2009). *Visible learning: A synthesis of over 800 meta-analyses relating to achievement*. New York, NY: Routledge.

Haworth, C. M. A., Dale, P. S., & Plomin, R. (2010). Sex differences in school science performance from middle childhood to early adolescence. *International Journal of Educational Research, 49*, 92–101.
Heuveline, P., Yang, H., & Timberlake, J. M. (2010). It takes a village (perhaps a nation): Families, states, and educational achievement. *Journal of Marriage and Family, 72*, 1362–1376.
Jang, H., Reeve, J., & Deci, E. L. (2010). Engaging students in learning activities: It is not autonomy support or structure but autonomy support and structure. *Journal of Educational Psychology, 102*, 588–600.
Jennings, P. A., & Greenberg, M. T. (2009). The prosocial classroom: Teacher social and emotional competence in relation to student and classroom outcomes. *Review of Educational Research, 79*, 491–525.
Jeynes, W. H. (2012). A meta-analysis on the effects and contributions of public, public charter, and religious schools on student outcomes. *Peabody Journal of Education, 87*, 305–335.
Johnson, E. P., Perry, J., & Shamir, H. (2010). Variability in reading ability gains as a function of computer-assisted instruction method of presentation. *Computers & Education, 55*, 209–217.
Klein, J., Cornell, D., & Konold, T. (2012). Relationships between bullying, school climate, and student risk behaviors. *School Psychology Quarterly, 27*, 54–169.
Konishi, C., Hymel, S., Zumbo, B. D., & Li, Z. (2010). Do school bullying and student-teacher relationships matter for academic achievement? A multilevel analysis. *Canadian Journal of School Psychology, 25*, 19–39.
Kreft, I., & de Leeuw, J. (1998). *Introducing multilevel modeling*. London, UK: Sage.
Leithwood, K., & Jantzi, D. (2009). A review of empirical evidence about school size effects: A policy perspective. *Review of Educational Research, 79*, 464–490.
Little, R. J., & Rubin, D. B. (1987). *Statistical analysis with missing data*. New York, NY: Wiley.
Lubienski, S. T., & Lubienski, C. (2006). School sector and academic achievement: A multilevel analysis of NAEP mathematics data. *American Educational Research Journal, 43*, 651–698.
Marks, G. N. (2006). Family size, family type and student achievement: Cross-national differences and the role of socioeconomic background and schools. *Journal of Comparative Family Studies, 37*, 1–24.
Martin, M. O., Mullis, I. V. S., & Foy, P. (2008). *TIMSS 2007 international science report: Findings from IEA's Trends in International Mathematics and Science Study at the eighth and fourth grades*. Chestnut Hill, MA: Boston College.
Martin, A. J., Liem, G. A., Mok, M., & Xu, J. (2012). Problem solving and immigrant student mathematics and science achievement: Multination findings from the Programme for International Student Assessment (PISA). *Journal of Educational Psychology, 104*, 1054–1073.
Martin, M. O., Mullis, I. V. S., Foy, P., & Stanco, G. M. (2012). *TIMSS 2011 international results in science*. Chestnut Hill, MA: TIMSS & PIRLS International Study Center, Boston College.
Masters, G. N. (1982). A Rasch model for partial credit scoring. *Psychometrika, 47*, 149–174.
McCoach, D. B. (2010a). Hierarchical linear modeling. In G. R. Hancock & R. O. Mueller (Eds.), *The reviewer's guide to quantitative methods in the social sciences* (pp. 123–140). New York, NY: Routledge.
McCoach, D. B. (2010b). Dealing with dependence (Part II): A gentle introduction to hierarchical linear modeling. *Gifted Child Quarterly, 54*, 252–256.
McCoach, D. B., & Black, A. C. (2012). Introduction to estimation issues in multilevel modeling. *New Directions for Institutional Research, 2012*, 23–39.
McKenzie, K. B., & Scheurich, J. J. (2008). Teacher resistance to improvement of schools with diverse students. *International Journal of Leadership in Education, 11*, 117–133.
McNeely, C. A., Nonnemaker, J. M., & Blum, R. W. (2002). Promoting school connectedness: Evidence from the national longitudinal study of adolescent health. *Journal of School Health, 72*, 138–146.
Mehta, S. B., Cornell, D., Fan, X., & Gregory, A. (2013). Bullying climate and school engagement in ninth-grade students. *Journal of School Health, 83*, 45–52.

Miller, R. T., Murnane, R. J., & Willett, J. B. (2008). Do teacher absences impact student achievement? Longitudinal evidence from one urban school district. *Educational Evaluation and Policy Analysis, 30*, 181–200.

Mislevy, R. J., & Sheehan, K. M. (1987). Marginal estimation procedures. In A. E. Beaton (Ed.), *The NAEP 1983–1984 technical report*. Princeton, NJ: Educational Testing Service.

Mislevy, R. J., & Sheehan, K. M. (1989). Information matrices in latent-variable models. *Journal of Educational Statistics, 14*, 335–350.

Mo, Y., Singh, K., & Chang, M. (2013). Opportunity to learn and student engagement: A HLM study on eighth grade science achievement. *Educational Research for Policy and Practice, 12*, 3–19.

Moore, A., Edwards, G., Halpin, D., & George, R. (2002). Compliance, resistance and pragmatism: The (re) construction of schoolteacher identities in a period of intensive educational reform. *British Educational Research Journal, 28*, 551–565.

Mullis, I. V. S., Martin, M. O., & Foy, P. (2008). *TIMSS 2007 international mathematics Report: Findings from IEA's Trends in International Mathematics and Science Study at the fourth and eighth grades*. Chestnut Hill, MA: Boston College.

Mullis, I. V. S., Martin, M. O., Foy, P., & Arora, A. (2012). *TIMSS 2011 international results in mathematics*. Chestnut Hill, MA: TIMSS & PIRLS International Study Center, Boston College.

Mullis, I. V. S., Martin, M. O., Foy, P., & Drucker, K. T. (2012). *PIRLS 2011 international results in reading*. Chestnut Hill, MA: TIMSS & PIRLS International Study Center, Boston College.

National Governors Association. (2008). *Benchmarking for success: Ensuring U.S. students receive a world-class education*. Retrieved from http://www.corestandards.org/assets/0812BENCHMARKING.pdf

Organization for Economic Cooperation and Development. (2003). *Student engagement at school: A sense of belonging and participation*. Retrieved from http://www.oecd.org/education/school/programmeforinternationalstudentassessmentpisa/33689437.pdf

Organization for Economic Cooperation and Development. (2009). *Equally prepared for life? How 15-year-old boys and girls perform in school*. Retrieved from http://www.oecd.org/pisa/pisaproducts/pisa2006/42843625.pdf

Organization for Economic Cooperation and Development. (2010a). *PISA 2009 results: What students know and can do: Student performance in reading, mathematics, and science*. Retrieved from http://www.oecd.org/dataoecd/10/61/48852548.pdf

Organization for Economic Cooperation and Development. (2010b). *Database – PISA 2009*. Retrieved from http://pisa2009.acer.edu.au/

Organization for Economic Cooperation and Development. (2010c). *PISA 2009 results: Overcoming social background – Equity in learning opportunities and outcomes*. Retrieved from http://www.oecd.org/pisa/pisaproducts/48852584.pdf

Organization for Economic Cooperation and Development. (2011). *Strong performers and successful reformers in education: Lessons from PISA for the United States*. Retrieved from http://www.oecd.org/dataoecd/32/50/46623978.pdf

Organization for Economic Cooperation and Development. (2012). *PISA 2009 technical report*. Retrieved from http://www.oecd.org/pisa/pisaproducts/pisa2009/50036771.pdf

Palardy, G. J. (2008). Differential school effects among low, middle, and high social class composition schools: A multilevel, multiple group latent growth curve analysis. *School Effectiveness and School Improvement, 19*, 21–49.

Palardy, G. J. (2013). High school socioeconomic segregation and student attainment. *American Educational Research Journal*. Advance online publication. doi: 10.3102/0002831213481240

Papastergiou, M. (2009). Digital game-based learning in high school Computer Science education: Impact on educational effectiveness and student motivation. *Computers & Education, 52*, 1–12.

Penner, A., & CadwalladerOlsker, T. (2012). Gender differences in mathematics and science achievement across the distribution: What international variation can tell us about the role of biology and society? In H. Forgasz & F. Rivera (Eds.), *Towards equity in mathematics education: Gender, culture, and diversity* (pp. 441–468). Berlin, Germany: Springer.

Perry, L., & McConney, A. (2010). Does the SES of the school matter? An examination of socioeconomic status and student achievement using PISA 2003. *The Teachers College Record, 112*, 1137–1162.

Pilli, O., & Aksu, M. (2013). The effects of computer-assisted instruction on the achievement, attitudes and retention of fourth grade mathematics students in North Cyprus. *Computers & Education, 62*, 62–71.

Pong, S. L., Dronkers, J., & Hampden-Thompson, G. (2003). Family policies and children's school achievement in single- versus two-parent families. *Journal of Marriage and Family, 65*, 681–699.

Radliff, K. M., Wheaton, J. E., Robinson, K., & Morris, J. (2012). Illuminating the relationship between bullying and substance use among middle and high school youth. *Addictive Behaviors, 37*, 569–572.

Rasch, G. (1960). *Probabilistic models for some intelligence and attainment tests.* Copenhagen, Denmark: Danish Institute for Educational Research.

Raudenbush, S. W., & Bryk, A. S. (2002). *Hierarchical linear models: Applications and data analysis methods* (2nd ed.). Newbury Park, CA: Sage.

Raudenbush, S. W., Bryk, A. S., Cheong, Y. F., Congdon, R., & du Toit, M. (2011). *HLM 7 for Windows [Computer software]*. Lincolnwood, IL: Scientific Software International.

Reyes, M. R., Brackett, M. A., Rivers, S. E., White, M., & Salovey, P. (2012). Classroom emotional climate, student engagement, and academic achievement. *Journal of Educational Psychology, 104*, 700–712.

Roeser, R. W., & Eccles, J. S. (1998). Adolescents' perceptions of middle school: Relation to longitudinal changes in academic and psychological adjustment. *Journal of Research on Adolescence, 8*, 123–158.

Roorda, D. L., Koomen, H. M., Spilt, J. L., & Oort, F. J. (2011). The Influence of affective teacher–student relationships on students' school engagement and achievement: A meta-analytic approach. *Review of Educational Research, 81*, 493–529.

Rubie-Davies, C. M. (2007). Classroom interactions: Exploring the practices of high-and low-expectation teachers. *British Journal of Educational Psychology, 77*, 289–306.

Rubin, D. B. (1987). *Multiple imputations for nonresponse in surveys.* New York, NY: Wiley.

Sakiz, G., Pape, S. J., & Hoy, A. W. (2012). Does perceived teacher affective support matter for middle school students in mathematics classrooms? *Journal of School Psychology, 50*, 235–255.

Shegog, R., Lazarus, M. M., Murray, N. G., Diamond, P. M., Sessions, N., & Zsigmond, E. (2012). Virtual transgenics: Using a molecular biology simulation to impact student academic achievement and attitudes. *Research in Science Education, 42*, 875–890.

Shin, S. H., Slater, C. L., & Backhoff, E. (2012). Principal perceptions and student achievement in reading in Korea, Mexico, and the United States: Educational leadership, school autonomy, and use of test results. *Educational Administration Quarterly*. Advance online publication. doi: 10.1177/0013161X12458796

Sirin, S. R. (2005). Socioeconomic status and academic achievement: A meta-analytic review of research. *Review of Educational Research, 75*, 417–453.

Skaalvik, E. M., & Skaalvik, S. (2011). Teacher job satisfaction and motivation to leave the teaching profession: Relations with school context, feeling of belonging, and emotional exhaustion. *Teaching and Teacher Education, 27*, 1029–1038.

Slama, R. B. (2012). A longitudinal analysis of academic English proficiency outcomes for adolescent English language learners in the United States. *Journal of Educational Psychology, 104*, 265–285.

Small, R. V., Shanahan, K. A., & Stasak, M. (2010). *The impact of New York's school libraries on student achievement and motivation: Phase III.* Retrieved from www.ala.org/aasl/slr/volume13/smallshanahan-stasak

Small, R. V., Snyder, J., & Parker, K. (2009). *The impact of New York's school libraries on student achievement and motivation: Phase I*. Retrieved from http://www.eric.ed.gov/PDFS/EJ859484.pdf

Sosa, G. W., Berger, D. E., Saw, A. T., & Mary, J. C. (2011). Effectiveness of computer-assisted instruction in statistics: A meta-analysis. *Review of Educational Research, 81*, 97–128.

Sousa, S., Park, E. J., & Armor, D. J. (2012). Comparing effects of family and school factors on cross-national academic achievement using the 2009 and 2006 PISA surveys. *Journal of Comparative Policy Analysis: Research and Practice, 14*, 449–468.

Stapleton, L. M., & Thomas, S. L. (2008). The use of national datasets for teaching and research: Sources and issues. In A. A. O'Connell & D. B. McCoach (Eds.), *Multilevel modeling of educational data* (pp. 11–57). Charlotte, NC: Information Age.

Stewart, E. A. (2003). School social bonds, school climate, and school misbehavior: A multilevel analysis. *Justice Quarterly, 20*, 575–604.

Sun, L., Bradley, K. D., & Akers, K. (2012). A multilevel modeling approach to investigating factors impacting science achievement for secondary school students: PISA Hong Kong sample. *International Journal of Science Education, 34*, 2107–2125.

Sznitman, S. R., Dunlop, S. M., Nalkur, P., Khurana, A., & Romer, D. (2012). Student drug testing in the context of positive and negative school climates: Results from a national survey. *Journal of Youth and Adolescence, 41*, 146–155.

Tucker, M. S. (2011). *Surpassing Shanghai: An agenda for American education built on the world's leading systems*. Cambridge, MA: Harvard University Press.

Tüzün, H., Yilmaz-Soylu, M., Karakuş, T., İnal, Y., & Kizilkaya, G. (2009). The effects of computer games on primary school students' achievement and motivation in geography learning. *Computers & Education, 52*, 68–77.

Uttal, D. H., Meadow, N. G., Tipton, E., Hand, L. L., Alden, A. R., Warren, C., et al. (2013). The malleability of spatial skills: A meta-analysis of training studies. *Psychological Bulletin, 139*, 352–402.

Van Petegem, K., Aelterman, A., Van Keer, H., & Rosseel, Y. (2008). The influence of student characteristics and interpersonal teacher behavior in the classroom on student's wellbeing. *Social Indicators Research, 85*, 279–291.

Walker, M. (2011). *PISA 2009 plus results: Performance of 15-year-olds in reading, mathematics and science for 10 additional participants*. Melbourne, Australia: ACER Press.

Wang, M. T., & Dishion, T. J. (2012). The trajectories of adolescents' perceptions of school climate, deviant peer affiliation, and behavioral problems during the middle school years. *Journal of Research on Adolescence, 22*, 40–53.

Wang, M. T., & Holcombe, R. (2010). Adolescents' perceptions of school environment, engagement, and academic achievement in middle school. *American Educational Research Journal, 47*, 633–662.

Way, S. M. (2011). School discipline and disruptive classroom behavior: The moderating effects of student perceptions. *The Sociological Quarterly, 52*, 346–375.

Weiss, C. C., Carolan, B. V., & Baker-Smith, E. C. (2010). Big school, small school: (Re) testing assumptions about high school size, school engagement and mathematics achievement. *Journal of Youth and Adolescence, 39*, 163–176.

White, K. (1982). The relation between socioeconomic status and academic achievement. *Psychological Bulletin, 91*, 461–481.

Willms, J. D. (2010). School composition and contextual effects on student outcomes. *Teachers College Record, 112*, 1008–1037.

Index

M.S. Khine (ed.), *Science Education in East Asia*,
DOI 10.1007/978-3-319-16390-1

D

E

MIX
Papier aus verantwortungsvollen Quellen
Paper from responsible sources
FSC® C105338

If you have any concerns about our products,
you can contact us on
ProductSafety@springernature.com

In case Publisher is established outside the EU,
the EU authorized representative is:
Springer Nature Customer Service Center GmbH
Europaplatz 3, 69115 Heidelberg, Germany

Printed by Libri Plureos GmbH
in Hamburg, Germany